Trophic Ecology

Bottom-Up and Top-Down Interactions across Aquatic and Terrestrial Systems

As researchers try to predict the effects of human modification at all trophic levels and mediate the impact of rapid environmental change, it has become clear that it is no longer a matter of agreeing that both bottom-up and top-down forces play important roles in diverse ecosystems. Rather, the question is: how do these forces interact across aquatic and terrestrial systems?

Bringing together the contributions of international experts in the field, this book presents a unique synthesis of trophic relationships within and across ecosystems that is a valuable foundation for the development of cross-system, multidisciplinary research. It also provides new insights into population biology and community ecology and examines the interactive effects of bottom-up and top-down forces on biodiversity at each trophic level.

A one-stop resource for learning about bottom-up and top-down interactions, this book encourages discussion and collaboration among researchers to identify similarities and differences in trophic interactions across aquatic and terrestrial systems.

TORRANCE C. HANLEY is an aquatic community ecologist, whose research focuses on the role of diversity in trophic interactions in freshwater and marine systems. She is also interested in how inter- and intra-specific diversity of producer and consumer species impacts population and community dynamics, trophic interactions, and ecosystem function.

KIMBERLY J. LA PIERRE is a terrestrial community ecologist, whose research focuses on the effects of global change on trophic interactions and ecosystem function. She is also interested in drivers of plant invasions, including the role of herbivory and microbial mutualisms.

Ecological Reviews

Ecological Reviews publishes books at the cutting edge of modern ecology, providing a forum for volumes that discuss topics that are focal points of current activity and likely long-term importance to the progress of the field. The series is an invaluable source of ideas and inspiration for ecologists at all levels from graduate students to more-established researchers and professionals. The series has been developed jointly by the British Ecological Society and Cambridge University Press and encompasses the Society's Symposia as appropriate.

Biotic Interactions in the Tropics: Their Role in the Maintenance of Species Diversity
Edited by David F. R. P. Burslem, Michelle A. Pinard, and Sue E. Hartley

Biological Diversity and Function in Soils
Edited by Richard Bardgett, Michael Usher, and David Hopkins

Island Colonization: The Origin and Development of Island Communities
By Ian Thornton
Edited by Tim New

Scaling Biodiversity
Edited by David Storch, Pablo Margnet, and James Brown

Body Size: The Structure and Function of Aquatic Ecosystems
Edited by Alan G. Hildrew, David G. Raffaelli and Ronni Edmonds-Brown

Speciation and Patterns of Diversity
Edited by Roger Butlin, Jon Bridle, and Dolph Schluter

Ecology of Industrial Pollution
Edited by Lesley C. Batty and Kevin B. Hallberg

Ecosystem Ecology: A New Synthesis
Edited by David G. Raffaelli and Christopher L. J. Frid

Urban Ecology
Edited by Kevin J. Gaston

The Ecology of Plant Secondary Metabolites: From Genes to Global Processes
Edited by Glenn R. Iason, Marcel Dicke, and Susan E. Hartley

Birds and Habitat: Relationships in Changing Landscapes
Edited by Robert J. Fuller

Trait-Mediated Indirect Interactions: Ecological and Evolutionary Perspectives
Edited by Takayuki Ohgushi, Oswald Schmitz, and Robert D. Holt

Forests and Global Change
Edited by David A. Coomes, David F. R. P. Burslem, and William D. Simonson

Trophic Ecology

Bottom-Up and Top-Down Interactions across Aquatic and Terrestrial Systems

Edited by

TORRANCE C. HANLEY
Northeastern University, USA

KIMBERLY J. LA PIERRE
University of California, Berkeley, USA

CAMBRIDGE
UNIVERSITY PRESS

University Printing House, Cambridge CB2 8BS, United Kingdom

Cambridge University Press is part of the University of Cambridge.

It furthers the University's mission by disseminating knowledge in the pursuit of education, learning and research at the highest international levels of excellence.

www.cambridge.org
Information on this title: www.cambridge.org/9781107077324

© Cambridge University Press 2015

First published 2015

Printed in the United Kingdom by TJ International Ltd. Padstow Cornwall

A catalog record for this publication is available from the British Library

Library of Congress Cataloging in Publication data
Trophic ecology: bottom-up and top-down interactions across aquatic and terrestrial systems / edited by Torrance C. Hanley, Northeastern University, USA, Kimberly J. La Pierre, University of California, Berkeley, USA.
    pages   cm. – (Ecological reviews)
Includes bibliographical references and index.
ISBN 978-1-107-07732-4 (hardback : alk. paper)   1. Multitrophic interactions (Ecology)
2. Food chains (Ecology)   I. Hanley, Torrance C., 1979– editor.   II. La Pierre, Kimberly J., editor.
QH541.15.F66T765   2015
577'.16 – dc23   2014036116

ISBN 978-1-107-07732-4 Hardback
ISBN 978-1-107-43432-5 Paperback

Additional resources for this publication at www.cambridge.org/9781107077324

Contents

List of contributors   *page* ix
Preface   xiii

Part I   Theory

1   Theoretical perspectives on bottom-up and top-down interactions
across ecosystems   3
Shawn J. Leroux and Michel Loreau

Part II   Ecosystems

2   The spatio-temporal dynamics of trophic control in large
marine ecosystems   31
Kenneth T. Frank, Jonathan A. D. Fisher, and William C. Leggett

3   Top-down and bottom-up interactions in freshwater ecosystems:
emerging complexities   55
Jason M. Taylor, Michael J. Vanni, and Alexander S. Flecker

4   Top-down and bottom-up interactions determine tree and
herbaceous layer dynamics in savanna grasslands   86
A. Carla Staver and Sally E. Koerner

5   Bottom-up and top-down forces shaping wooded ecosystems:
lessons from a cross-biome comparison   107
Dries P. J. Kuijper, Mariska te Beest, Marcin Churski, and Joris P. G. M. Cromsigt

6   Dynamic systems of exchange link trophic dynamics in freshwater
and terrestrial food webs   134
John L. Sabo and David Hoekman

7   Bottom-up and top-down interactions in coastal interface systems 157
*Jan P. Bakker, Karina J. Nielsen, Juan Alberti, Francis Chan, Sally D. Hacker,
Oscar O. Iribarne, Dries P. J. Kuijper, Bruce A. Menge, Maarten Schrama, and
Brian R. Silliman*

Part III   Patterns and Processes

8   Influence of plant defenses and nutrients on trophic control
of ecosystems   203
Karin T. Burghardt and Oswald J. Schmitz

9   Interactive effects of plants, decomposers, herbivores, and
     predators on nutrient cycling   233
     Sarah E. Hobbie and Sébastien Villéger

10 The role of bottom-up and top-down interactions in determining
     microbial and fungal diversity and function   260
     Thomas W. Crowther and Hans-Peter Grossart

11 The question of scale in trophic ecology   288
     Lee A. Dyer, Tara J. Massad, and Matthew L. Forister

12 The role of species diversity in bottom-up and top-down interactions   318
     Jerome J. Weis

13 Plant and herbivore evolution within the trophic sandwich   340
     Luis Abdala-Roberts and Kailen A. Mooney

14 Bottom-up and top-down interactions across ecosystems in an era
     of global change   365
     Kimberly J. La Pierre and Torrance C. Hanley

Index   407

Contributors

LUIS ABDALA-ROBERTS
Department of Ecology and
Evolutionary Biology, University of
California-Irvine, 321 Steinhaus Hall,
Irvine, CA 92697, USA
labdala@uci.edu

JUAN ALBERTI
Instituto de Investigaciones Marinas y
Costeras (IIMyC), Universidad
Nacional de Mar del Plata (UNMDP) –
Consejo Nacional de Investigaciones
Científicas y Técnicas (CONICET),
Argentina
jalberti@mdp.edu.ar

JAN P. BAKKER
University of Groningen, Community
and Conservation Ecology Group, P.O.
Box 11103, 9700 CC Groningen, The
Netherlands
j.p.bakker@rug.nl

KARIN T. BURGHARDT
Department of Ecology and
Evolutionary Biology, Yale University,
Osborn Memorial Laboratories, 165
Prospect Street, New Haven, CT
06511, USA
karin.burghardt@yale.edu

FRANCIS CHAN
Oregon State University, Department
of Integrative Biology, 3029 Cordley
Hall, OSU, Corvallis, OR 97331,
USA
chanft@science.oregonstate.edu

MARCIN CHURSKI
Mammal Research Institute, Polish
Academy of Sciences, ul.
Waszkiewicza 1, 17–230 Białowieża,
Poland
mchurski@ibs.bialowieza.pl

JORIS P.G.M. CROMSIGT
Department of Wildlife, Fish, and
Environmental Studies, Swedish
University of Agricultural Sciences,
901 83 Umeå, Sweden
jcromsigt@hotmail.com
and
Centre for African Conservation
Ecology, Department of Zoology,
Nelson Mandela Metropolitan
University, PO Box 77000, NMMU,
South Africa

THOMAS W. CROWTHER
School of Forestry and Environmental
Studies, Yale University, 195 Prospect
Street, New Haven, CT 06511, USA
thomas.crowther@yale.edu

LEE A. DYER
Biology Department, University of
Nevada, 1664 N. Virginia Street, Reno,
NV 89557–0314, USA
ecodyer@gmail.com

JONATHAN A.D. FISHER
Centre for Fisheries Ecosystems
Research, Fisheries and Marine
Institute of Memorial University of
Newfoundland, P.O. Box 4920, St.
John's, Newfoundland A1C 5R3,
Canada
jonathan.fisher@mi.mun.ca

ALEXANDER S. FLECKER
Department of Ecology and
Evolutionary Biology, Cornell
University, Ithaca, NY, USA
asf3@cornell.edu

MATTHEW L. FORISTER
Biology Department, University of
Nevada, Mail Stop: 0314, 1664 N.
Virginia Street, Reno, NV 89557–0314,
USA
mforister@unr.edu

KENNETH T. FRANK
Department of Fisheries and Oceans,
Bedford Institute of Oceanography,
Dartmouth, Nova Scotia B4A 3V4,
Canada
Kenneth.Frank@dfo-mpo.gc.ca

HANS-PETER GROSSART
Leiibniz Institute of Freshwater
Ecology and Inland Fisheries, Dept.
Experimental Limnology, Alte
Fischerhuette 2, D-16775 Stechlin,
Germany
hgrossart@igb-berlin.de

and
Potsdam University, Institute for
Biochemistry and Biology, Am Neuen
Palais 10, D-14460 Potsdam, Germany

SALLY D. HACKER
Oregon State University, Department
of Integrative Biology, 3029 Cordley
Hall, OSU, Corvallis, OR 97331, USA
hackers@science.oregonstate.edu

TORRANCE C. HANLEY
Department of Ecology, Evolution,
and Marine Biology, Northeastern
University, Marine Science Center,
430 Nahant Road, Nahant, MA 1908,
USA
t.hanley@neu.edu

SARAH E. HOBBIE
Department of Ecology, Evolution,
and Behavior, University of
Minnesota, St. Paul, MN 55108,
USA
shobbie@umn.edu

DAVID HOEKMAN
National Ecological Observatory
Network, 1685 38th St., Ste. 100,
Boulder, CO 80301, USA
dhoekman@neoninc.org
and
Department of Entomology,
University of Wisconsin, Madison,
WI 53706, USA
and
Department of Biology, Southern
Nazarene University, Bethany,
OK 73008, USA
dhoekman@mail.snu.edu

OSCAR O. IRIBARNE
Instituto de Investigaciones Marinas y
Costeras (IIMyC), Universidad
Nacional de Mar del Plata (UNMDP) –
Consejo Nacional de Investigaciones
Científicas y Técnicas (CONICET),
Argentina
osiriba@mdp.edu.ar

SALLY E. KOERNER
Department of Biology, Colorado
State University, Fort Collins, CO
80521, USA
sally.koerner@colostate.edu

DRIES P.J. KUIJPER
Mammal Research Institute, Polish
Academy of Sciences, ul.
Waszkiewicza 1, 17–230 Białowieża,
Poland
dkuijper@ibs.bialowieza.pl

KIMBERLY J. LA PIERRE
Department of Integrative Biology,
University of California, 1005 Valley
Life Sciences Bldg #3140, Berkeley, CA
94709, USA
kimberly.lapierre@berkeley.edu

WILLIAM C. LEGGETT
Department of Biology, Queen's
University, Kingston, Ontario
K7L 3N6, Canada
wleggett@queensu.ca

SHAWN J. LEROUX
Department of Biology, Memorial
University of Newfoundland, 232
Elizabeth Ave., St., John's, NL, Canada,
A1B 3X9
sleroux@mun.ca

MICHEL LOREAU
Centre for Biodiversity Theory and
Modelling, Station d'Ecologie
Expérimentale du CNRS, 09200
Moulis, France
michel.loreau@ecoex-moulis.cnrs.fr

TARA J. MASSAD
Chemistry Department, University of
Sao Paulo, Brazil
tmassad77@gmail.com

BRUCE A. MENGE
Oregon State University, Department
of Integrative Biology, 3029 Cordley
Hall, Corvallis, OR 97331, USA
mengeb@oregonstate.edu

KAILEN A. MOONEY
Department of Ecology and
Evolutionary Biology, University of
California-Irvine, 321 Steinhaus Hall,
Irvine, CA 92697, USA
mooneyk@uci.edu

KARINA J. NIELSEN
Sonoma State University, Department
of Biology, 1801 East Cotati Ave.,
Rohnert Park, CA 94928, USA
karina.nielsen@sonoma.edu

JOHN L. SABO
Faculty of Ecology, Evolution, and
Environmental Science, School of Life
Sciences, Arizona State University,
427 East Tyler Mall, Tempe, AZ
85287–4501, USA
john.l.sabo@asu.edu
and
Global Institute of Sustainability,
Arizona State University, Tempe, AZ,
USA

OSWALD J. SCHMITZ
School of Forestry & Environmental
Studies, Yale University, 370 Prospect
Street, New Haven, CT 06511, USA
oswald.schmitz@yale.edu

MAARTEN SCHRAMA
University of Manchester, Faculty of
Life Sciences, Soil and Ecosystem
Ecology Group, Oxford Road,
Manchester M13 9PT, UK
maartenschrama@gmail.com

BRIAN R. SILLIMAN
Division of Marine Science and
Conservation, Nicholas School of the
Environment, Duke University,
Beaufort, NC 28516, USA
brian.silliman@duke.edu

CARLA STAVER
Department of Ecology and
Evolutionary Environmental Biology,
Yale University, New Haven, CT
06511, USA
carla.staver@yale.edu

JASON M. TAYLOR
United States Department of
Agriculture, Agricultural Research
Service, National Sedimentation
Laboratory, Water Quality and

Ecology Research Unit, Oxford, MS,
USA
jason.taylor@ars.usda.gov

MARISKA TE BEEST
Department of Ecology and
Environmental Science, Umeå
University, SE-901 87 Umeå, Sweden
mariskatebeest@hotmail.com

MICHAEL J. VANNI
Department of Zoology, Miami
University, Oxford, OH, USA
vannimj@miamioh.edu

SÉBASTIEN VILLÉGER
CNRS (Centre National de la
Recherche Scientifique), Laboratoire
Écologie des Systèmes Marins Côtiers,
Université Montpellier 2, 34095
Montpellier, France
sebastien.villeger@univ-montp2.fr

JEROME J. WEIS
Department of Ecology and
Evolutionary Biology, Yale University,
New Haven, CT, USA
and
Department of Entomology,
University of Minnesota, St. Paul,
MN 55108, USA
weis0550@umn.edu

Preface

The idea for this book started as a series of lunch table conversations revolving around our respective research projects. Torrie's research at the time focused on trophic dynamics in lakes, looking at the interactive effects of food quality (specifically, algal carbon:phosphorus ratio) and predation on *Daphnia* life history and stoichiometry to better predict the effects of human modification of bottom-up and top-down forces in aquatic ecosystems. Kim's research examined the effects of nutrient availability and herbivore presence on grassland community composition and production across the broad precipitation gradient of the North American Great Plains. In discussing our respective studies, it became evident that our conversations about the interaction of bottom-up and top-down forces across aquatic and terrestrial ecosystems provided different perspectives and important insights that broadened our conceptual base and benefitted our research. This realization prompted us to organize a session for the annual meeting of the Ecological Society of America in 2011 to bring together aquatic and terrestrial ecologists studying the interaction of bottom-up and top-down processes in diverse ecosystems. During the session, similarities and differences in the strength and nature of trophic interactions across ecosystems became evident, stimulating dialogue between aquatic and terrestrial scientists. The success of this session and the satisfying exchange of ideas that resulted inspired this book.

The goal of this book is to provide a cohesive summary of the interaction of bottom-up and top-down processes across aquatic and terrestrial systems, which may serve as a basis for future cross-system studies examining patterns in these important drivers of community and ecosystems processes. In this book, the definitions of "bottom-up" and "top-down" are purposely broad to include a diverse group of studies and perspectives: bottom-up forces include nutrient and resource availability, and top-down forces include herbivores, predators, and parasites. Given the extent of human-induced global change, this topic is particularly timely and important. As we try to predict the effects of human modification at all trophic levels and mediate the impact of rapid environmental change, a better understanding of the interaction of bottom-up and top-down forces is instrumental to scientists and policymakers alike.

The independent effects of bottom-up and top-down forces are well understood in a diverse array of ecosystems. But it is widely accepted that trophic ecology is no longer a question of "bottom-up" or "top-down"; thus, it is important to examine how these often conflicting selection pressures interact, both in the laboratory and the field, to better understand the interactive effects of these factors in a variety of systems. The first section of this book ("Theory") describes the state of theory related to trophic interactions and highlights a number of approaches that apply across a wide variety of aquatic and terrestrial ecosystems, emphasizing the potential for cross-system comparison.

The second section of this book ("Ecosystems") focuses on aquatic and terrestrial ecosystems separately to clarify the state of our current understanding of trophic interactions in each system type and to highlight areas for future research. Each chapter describes the dominant bottom-up and top-down forces within the system and then explores the interaction of these factors and resultant effects on community and ecosystem processes. Despite the fact that this section is structured by ecosystem type, it begins to integrate across systems: Chapter 2 considers the strength and interaction of bottom-up and top-down processes in marine environments that range in scale from the open ocean to boundary upwelling systems to inland seas; Chapter 3 discusses bottom-up and top-down interactions in diverse freshwater environments, including ponds, lakes, streams, and rivers; Chapter 4 looks at trophic interactions in grasslands and savannas – related but distinct terrestrial environments; Chapter 5 compares the strength of bottom-up and top-down control in tropical and temperate forests; and Chapters 6 and 7 examine trophic dynamics at the aquatic–terrestrial border, including a diverse array of systems, such as lake-shore and stream-bank boundaries, the rocky intertidal, and salt marshes – all of which share important commonalities and differences.

The third section of this book ("Patterns and Processes") addresses how considering the interaction of bottom-up and top-down forces informs our understanding of a variety of ecological and evolutionary patterns and processes. In this section, each chapter focuses on a specific ecological or evolutionary process, comparing our current understanding of the role of trophic interactions in shaping these processes across ecosystems, and considering how these processes in turn shape trophic interactions. To facilitate communication across this extensive field, the chapters in this section encompass a broad range of aquatic and terrestrial ecosystems and cover a variety of observational, experimental, and theoretical approaches. First, the direct and indirect interactions of bottom-up and top-down forces are examined in detail: Chapter 8 highlights the importance of resource availability in mediating plant defenses in response to selection pressure from diverse consumers; and Chapter 9 discusses the role of herbivores and predators in determining nutrient cycling, and thus amounts and ratios of critical elements, such as carbon, nitrogen, and phosphorus. Next,

Figure 1 This image represents the challenges of considering the interaction of bottom-up and top-down processes across aquatic and terrestrial ecosystems; it illustrates the myriad components of aquatic and terrestrial food webs – including primary producers, consumers, and predators – the diversity of species within and across trophic levels and ecosystems, and the interconnectedness of these species, particularly at ecosystem boundaries. In addition, the image highlights the role of spatial scale (e.g., relative distance from the ecosystem boundary) and temporal scale (e.g., presence of diapausing eggs in the lake sediment) in trophic interactions, and most notably, it also represents the important similarities and differences that may emerge with cross-system comparisons of bottom-up and top-down processes. (Credit: Tanya L. Rogers, Northeastern University.)

Chapter 10 describes an oft-overlooked but critically important component of the food web – namely, microbial and fungal processes – including consideration of how bottom-up and top-down forces may affect their role in trophic dynamics and how they in turn may affect trophic interactions in aquatic and terrestrial systems. The aim of the next chapters is to highlight less commonly considered factors that likely play a key role in trophic interactions and may greatly improve our understanding of trophic dynamics: Chapter 11 discusses how consideration of spatial and temporal scale informs our understanding of the strength and interaction of bottom-up and top-down forces across ecosystems; Chapter 12 highlights how species diversity, both within and across trophic levels, may influence the relative importance of bottom-up and top-down forces and ultimately affect the nature of their interaction; and Chapter 13 addresses the importance of evolution occurring at ecological time scales and how this may alter trophic interactions and impact community and ecosystem processes. Lastly, Chapter 14 tackles the all too timely topic of human-induced, global environmental change, including altered nutrient availability, climate change, loss of top predators, changes in biodiversity, and species invasions. Given the extent of anthropogenic modification of aquatic and terrestrial ecosystems, it is necessary to consider the interaction of multiple stressors and how they impact the interaction of bottom-up and top-down forces. Looking across systems to better understand these important drivers of community and ecosystem processes is a critical first step to predicting and mediating the effects of anthropogenic activities.

In sum, the goal of this book is to prompt more lunch table discussions, joint lab meetings, and collaborations looking at the interaction of bottom-up and top-down forces across aquatic and terrestrial ecosystems. While a lot of progress has been made since the old "bottom-up" or "top-down" debate, there is still much to be learned about trophic interactions in a variety of systems. Just as thinking about these processes in a different system has greatly benefitted our research and understanding of bottom-up and top-down interactions, we hope that this book will stimulate cross-system studies that continue to explore and identify key commonalities and differences in aquatic and terrestrial systems.

Torrance Hanley
Kimberly La Pierre

PART I

Theory

Theoretical perspectives on bottom-up and top-down interactions across ecosystems

SHAWN J. LEROUX

Memorial University of Newfoundland, St. John's, Newfoundland, Canada

and

MICHEL LOREAU

Station d'Ecologie Expérimentale du CNRS, Moulis, France

Introduction

The study of the determinants of biomass pyramids (i.e., the patterns of biomass of organisms at different trophic levels of an ecosystem) within and across ecosystems is an enduring endeavor in the ecological sciences (Gripenberg and Roslin, 2007; Gruner et al., 2008). This classic ecological problem still fascinates ecologists worldwide and the lively debate on this question is an attestation of the complexity of ecological systems. The ecological literature reveals two main perspectives for predicting biomass pyramids; one perspective emphasizes the role of resources such as inorganic nitrogen (N) and phosphorus (P) or primary producers in determining the biomass of higher trophic levels, and the other perspective emphasizes the role of consumers such as herbivores and predators in determining the biomass of lower trophic levels (Oksanen and Oksanen, 2000; Gruner et al., 2008).

The resource-based hypothesis states that organisms are resource-limited, and therefore resources determine the shape of biomass pyramids (Elton, 1927; Lindeman, 1942; White, 1978; McQueen et al., 1986). Consistent with Elton's (1927) perspective, Lindeman (1942) and others (e.g., White, 1978; McQueen et al., 1986) argued that inorganic nutrients and solar radiation limit plant growth and subsequently the potential transfer of energy and nutrients from lower trophic levels to higher trophic levels in ecosystems. This bottom-up perspective has been expanded to consider the role of plant defense in limiting herbivory (Strong, 1992; Polis and Strong, 1996; also, see Chapter 8 and Chapter 13).

In contrast, the consumer-based hypothesis (i.e., Hairston Smith Slobodkin (HSS) Hypothesis) states that organisms are consumer-regulated, and therefore higher-level consumers determine biomass pyramids (Hairston et al., 1960).

Trophic Ecology: Bottom-Up and Top-Down Interactions across Aquatic and Terrestrial Systems, eds T. C. Hanley and K. J. La Pierre. Published by Cambridge University Press. © Cambridge University Press 2015.

Oksanen et al. (1981) further developed the consumer-regulated framework by developing the exploitation ecosystem hypothesis (EEH), which suggests that top-down control of ecosystems will vary along environmental gradients. Top-down perspectives gained additional support through Carpenter et al.'s (1985) empirical evidence of trophic cascades, whereby top predators have indirect positive effects on non-adjacent trophic levels. White (1978) referred to the debate on resource- versus consumer-based limitation as populations being "limited from below" or "controlled from above." McQueen et al. (1986) first introduced the terms "bottom-up" and "top-down" to describe White's (1978) use of resource-versus consumer-based limitation. For many years, ecologists have focused on demonstrating the primacy of their favorite hypothesis (White, 1978; McQueen et al., 1986). However, recent empirical results from a wide range of ecosystems, many of which are reviewed in this volume, provide unequivocal evidence that both resources and consumers interact to shape natural populations, communities, and ecosystems (e.g., Hunter and Price, 1992; Brett and Goldman, 1996; Hassell et al., 1998; Polis, 1999; Fath, 2004; Borer et al., 2006; Gruner et al., 2008; Polishchuk et al., 2013; Whalen et al., 2013). Ecological theory (e.g., Hairston et al., 1960; Oksanen et al., 1981) has been at the forefront of integrating our empirical knowledge of the interdependence of resource and consumer impacts on food webs and ecosystems (Table 1.1).

Contemporary ecological theory is now investigating the interrelationship and variability of bottom-up and top-down interactions in ecosystems in space and time. Building on Carpenter et al.'s (1985) foundational work and a plethora of empirical studies demonstrating the role of consumer-mediated recycling on ecosystem functioning, Leroux and Loreau (2009; 2010) and Schmitz et al. (2010) outline the many consumptive and non-consumptive mechanisms by which consumers can indirectly influence primary production and nutrient cycling. The key role consumers play in storing, recycling, and redistributing nutrients in ecosystems (Loreau, 1995; reviewed in Vanni, 2002; Schmitz et al., 2010; also, see Chapter 9) provides a mechanistic link between bottom-up and top-down forces in ecosystems. Specifically, organic nutrients recycled by organisms are mineralized by microorganisms and made available to plants, thus completing the energy cycle (Lindeman, 1942). Organismal material cycling has the potential to synthesize bottom-up and top-down processes, but it must overcome the current confusion surrounding these processes, which is evidenced by the fact that some authors refer to organism-mediated nutrient cycling as a bottom-up process (e.g., Northcote, 1988), while others call it a top-down process (e.g., Glaholt and Vanni, 2005).

Additional progress in bottom-up and top-down theory has occurred with the consideration of these processes along distinct energy pathways (e.g., brown versus green webs, Moore et al., 2004; Hulot and Loreau, 2006; Rooney et al., 2006). Indeed, a parsimonious explanation for the stability and dynamics of complex food webs is emerging based on two key ecosystem attributes: the

Table 1.1 *Chronological summary of the development of classic theories of community regulation; the table provides the citation, a brief summary of the history, general predictions, and original model system for each contribution*

Contribution	History	Prediction	Systems
Lindeman, 1942	Trophic-dynamic ecology	Inorganic nutrients and solar radiation fuel primary productivity, which provides energy for higher trophic levels. Death of higher order organisms provides a source of energy to decomposers, which make organic substances available for producers, thus completing the energy cycle	Lakes
Hairston et al., 1960	HSS or Green World Hypothesis (GWH) – based on ideas formulated in Elton (1927) regarding the structure of food webs	Predators have strong top-down regulation of herbivores, therefore releasing plants from herbivory. Plants are abundant because of this. An increase in plants will be passed on to the predators in a three-level food chain	Terrestrial
Rosenzweig, 1971	Paradox of enrichment	Increasing the resources to a system can be destabilizing and is known as the paradox of enrichment. Rip and McCann (2011) have generalized this concept as the principle of energy flux	Theoretical
Menge and Sutherland, 1976	Menge–Sutherland Hypothesis (MSH) – based on observations that omnivory is abundant in natural food webs	Predators regulate plant abundance not indirectly through consumption of herbivores but directly via omnivory on plants	Rocky intertidal and terrestrial

(cont.)

Table 1.1 *(cont.)*

Contribution	History	Prediction	Systems
Oksanen et al., 1981	Exploitation Ecosystem Hypothesis (EEH) – based on Fretwell (1977) and HSS	Similar to HSS, but incorporates productivity gradient. Stepwise accrual of plants and herbivores along a productivity gradient. At relatively high productivity ($700 \, g \, m^{-2} \, y^{-1}$), predators are present and regulate herbivores to a relatively constant biomass (converges with predictions from HSS). At low productivity, predators are absent and herbivores regulate plant biomass	Terrestrial, low productivity systems like Tundra and Boreal
Carpenter et al., 1985	Trophic Cascade Hypothesis (TCH) – based on HSS and EEH	Nutrient supply does not explain all the variation in plants. Cascading trophic interactions similar to HSS explain the differences in plants in systems with similar nutrient levels. First demonstration of this for a four-level food web	Lakes
McQueen et al., 1986	Bottom-up:Top-down hypothesis (BU:TD) – extension of EEH	Combines reciprocal effects of predators and resources. Biomass of plants is regulated by resources, and herbivores are regulated by predators, but both effects attenuate along food chains. At high resource levels, an increase in predators will have no effect on plants	Lakes
Arditi and Ginzburg, 1989	Ratio-Dependent Hypothesis (RDH)	The ratio of consumer to resource determines structure and abundance of different trophic levels. All trophic levels increase with an increase in primary production	Terrestrial
Strong, 1992; Polis and Strong, 1996	Diversity-Defense Hypothesis (DDH) – opposite to EEH	Strong cascading interactions are rare. Plants are abundant because of a diversity of defenses against herbivory	Grasslands

presence of mobile and generalist consumers that can couple energy pathways (e.g., McCann et al., 2005; Rooney et al., 2006; Wollrab et al., 2012) and the length of component food chains (Wollrab et al., 2012). Meta-ecosystem (i.e., a set of ecosystems connected by spatial flows of energy, materials, and organisms across ecosystem boundaries, sensu Loreau et al., 2003) theory provides another promising avenue to investigate variability in the spatial dynamics of resource limitation and consumer regulation (Loreau et al., 2003; Gravel et al., 2010; Massol et al., 2011). For example, Leroux and Loreau (2012) show how top-down regulation in one ecosystem can have indirect effects on the structure and dynamics of adjacent ecosystems (also, see Chapter 6 and Chapter 7).

In this chapter, we provide an overview of theoretical models and approaches that address the relative importance, variability, and interdependence of bottom-up and top-down forces in ecosystems and illustrate this theory with empirical examples from both aquatic and terrestrial realms. We begin by defining bottom-up and top-down processes independently; then we show how they can be related through material cycling. We review current work toward understanding spatial and temporal variability in bottom-up and top-down interactions and end with some future directions for bottom-up and top-down theory to pave the way for ecological synthesis on this matter. Ecosystems are complex, encompassing great horizontal (i.e., diversity within a single trophic level, e.g., competitors) and vertical diversity (i.e., diversity of food web interactions; Duffy et al., 2007). The bottom-up versus top-down debate was originally centered around vertical diversity; therefore, we focus on vertical diversity in this chapter, although we explore more complex ecosystems in later sections.

Defining bottom-up and top-down effects in ecosystems

Here, we derive ecosystem models to illustrate the basic definitions of bottom-up and top-down effects in ecosystems. Throughout this chapter, we consider a trophic level to consist of a group of species with shared resources. Since most ecosystems are thought to be limited by N or P (Elser et al., 2007; LeBauer and Treseder, 2008), we derive nutrient-limited ecosystem models. The same approach could be used for energy as long as nutrient recycling is ignored and the energetic content and stoichiometric composition of the various trophic levels are roughly equal. To illustrate bottom-up effects, consider a minimal ecosystem model with inorganic nutrient (N) as a basal resource for primary producers (P). Both trophic levels follow nutrient mass-balance constraints such that, at equilibrium, nutrient inputs balance nutrient outputs. This model tracks the basic ecosystem processes of consumption and production in each trophic level as follows (Loreau, 2010):

$$\frac{dN}{dt} = \Phi_N - \theta_N - \Phi_P \tag{1}$$

$$\frac{dP}{dt} = \Phi_P - \theta_P \tag{2}$$

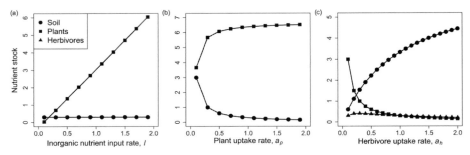

Figure 1.1 Illustration of bottom-up effects of nutrient increases (a), top-down effects of increased plant uptake rate (b), and the cascading effects of increased herbivore uptake rate (c). In a plant–soil ecosystem model, an increase in soil nutrients has a positive bottom-up effect on producer biomass (a) and an increase in producer nutrient uptake rate has a negative, top-down effect on soil nutrient stocks (b). In a herbivore–plant–soil ecosystem model, increasing herbivore nutrient uptake rate leads to negative, direct effects on plant biomass and positive indirect or cascading effects on soil nutrient stocks (c). α_P and α_H are the producer uptake rate and herbivore attack rate, respectively. e_H is the herbivore production efficiency, and m_N, m_P, and m_H are the mass-specific loss rates of soils, plants, and herbivores, respectively. Our results are not sensitive to particular model parameters, therefore, we selected arbitrary model parameter values of $\alpha_P = \alpha_H = 1$, $e_H = 0.1$, $m_N = m_P = m_H = 0.3$, $I = 2$ for illustration purposes.

where Φ_N and Φ_P are the production of the soil nutrient pool and primary producers, respectively, and θ_N and θ_P are the loss fluxes that include soil inorganic nutrients and plant senescence and mortality, respectively. The ecosystem is open at the basal level through a constant and independent input of inorganic nutrients; i.e., $\Phi_N = I$. We assume, as in classical theory of exploitation interactions (sensu Oksanen et al., 1981), that there is no interference among producer species so that the production of plants can be written as $\Phi_P = f_P(N)P$, where $f_P(N)$ is the functional response of plants. For simplicity, we use Lotka–Volterra functional responses for plants and herbivores throughout this chapter, as our main goal is to use simple models to illustrate bottom-up and top-down processes (see Loreau (2010) for generalized results to other functional responses). The Lotka–Volterra functional response for plants is $f_P(N)P = \alpha_P NP$, where α_P is the producer uptake rate. The loss flux is $\theta_N = m_N N$ from the soil nutrient pool and $\theta_P = m_P P$ from the primary producer pool, where m_N and m_P are the mass-specific loss rates of the soil and plants, respectively. At equilibrium, $N^* = \frac{m_P}{\alpha_P}$ and $P^* = \frac{I}{m_P} - \frac{m_N}{\alpha_P}$. We can investigate the bottom-up effect of increasing inorganic nutrients on the biomass of primary producers by taking the partial derivative of plant biomass with respect to the inorganic nutrient input rate, I (i.e., $\frac{\partial P^*}{\partial I}$). This partial derivative is positive ($\frac{\partial P^*}{\partial I} = \frac{1}{m_P}$), which demonstrates a positive, bottom-up effect of increasing basal resources on primary producer biomass (Fig. 1.1a). We obtain similar qualitative results for the bottom-up effect of increasing inorganic nutrients on primary production ($\frac{\partial \Phi_P}{\partial I} > 0$). Empirical evidence in support of this simple bottom-up effect of nutrients on plant biomass in aquatic and terrestrial ecosystems abounds (reviewed in Gruner et al., 2008). For example, Gratton and Denno (2003a) observed an increase in *Spartina alterniflora* production for 2–3 years after N fertilizer was added to their salt marsh study area, Rosemond et al. (1993) observed an increase in periphyton biomass and production after N and P additions to their woodland stream in eastern Tennessee, and in a meta-analysis of N and P fertilization experiments, Gruner et al. (2008) found an increase in producer biomass with fertilization in freshwater, marine, and terrestrial ecosystems.

The top-down effects of primary producers on soil inorganic nutrient stocks also can be elucidated through this simple ecosystem model. Top-down effects can occur via an increase in production (i.e., Φ_P) or a decrease in loss flux (i.e., θ_P). Consequently, the direction of top-down effects in this ecosystem can be determined by taking the partial derivative of soil inorganic nutrient stocks with respect to the producer uptake rate, α_P, or producer mass-specific loss rate, m_P. The top-down effect of increasing producer uptake rate has a negative effect on soil nutrient stocks ($\frac{\partial N^*}{\partial \alpha_P} = -\frac{m_P}{\alpha_P^2}$, Fig. 1.1b). Similarly, a decline in producer mass-specific loss rate leads to a negative top-down effect on soil nutrient stocks ($\frac{\partial N^*}{\partial m_P} > 0$). There is considerable empirical evidence in support of top-down effects of organisms on adjacent trophic levels in aquatic and terrestrial ecosystems. For example, Frank et al. (2007) presented evidence of top-down forcing (i.e., negative correlation between predator and prey abundance) in fish of the North Atlantic marine ecosystem and Creel et al. (2007) showed lower elk calf recruitment with the introduction of wolves to Yellowstone National Park.

Bottom-up and top-down interactions are meant to describe direct interactions among adjacent trophic levels. The trophic cascade is a concept for understanding indirect (i.e., non-adjacent) trophic interactions. By adding herbivores to the above ecosystem model, we can demonstrate the indirect, top-down effects of herbivores on soil nutrient stocks via a trophic cascade. Trophic cascades result in alternating abundance, biomass, or production across more than one trophic level in an ecosystem (Carpenter and Kitchell, 1993; Pace et al., 1999). To add herbivores (i.e., H) to this model, we must add an additional loss term to the primary producer trophic level (Eq. 2); herbivore production (Φ_H) scaled by the herbivore production efficiency (ε_H), which represents consumption by herbivores. Eq. 2 then becomes:

$$\frac{dP}{dt} = \Phi_P - \theta_P - \frac{\Phi_H}{\varepsilon_H} \tag{3}$$

The dynamical equation for the herbivore trophic level is as follows:

$$\frac{dH}{dt} = \Phi_H - \theta_H \tag{4}$$

and the dynamical equation for soil nutrients (Eq. 1) remains unchanged. Similar to primary producers, the production of herbivores can be written as $\Phi_H = f_H$ $(P)H = \alpha_H\, PH$, where α_H is the herbivore attack rate. The equilibrium stocks of this three-level ecosystem are $N^* = \frac{I}{m_N + \alpha_P P^*}$, $P^* = \frac{m_H}{\alpha_H}$, $H^* = \frac{\varepsilon_H(\alpha_P N^* - m_P)}{\alpha_H}$. Using partial derivatives as above we can show positive, indirect, bottom-up effects of increasing the inorganic soil nutrient input rate on herbivore stocks $(\frac{\partial H^*}{\partial I} = \frac{\alpha_P \varepsilon_H}{\alpha_H m_N + \alpha_P m_H})$, negative direct top-down effects of herbivore consumption on primary producer stocks $(\frac{\partial P^*}{\partial \alpha_H} = -\frac{m_H}{\alpha_H^2})$, and positive, indirect, top-down effects of increasing herbivore consumption on soil nutrient stocks via a trophic cascade $(\frac{\partial N^*}{\partial \alpha_H} = \frac{\alpha_P m_H I}{(\alpha_H m_N + \alpha_P m_H)^2}$, Fig. 1.1c).

Recent meta-analyses have demonstrated that top-down trophic cascades tend to be stronger in aquatic than terrestrial ecosystems (Schmitz et al., 2000; Shurin et al., 2002; Borer et al., 2005). The leading hypotheses to explain variation in the strength of trophic cascades across ecosystems include system differences in producer quality and defense (Borer et al., 2005; Hall et al., 2007; Cebrian et al., 2009), primary productivity (Borer et al., 2005; Shurin and Seabloom, 2005), ecosystem complexity (Strong, 1992; Hillebrand and Cardinale, 2004), behavioral avoidance of predation by herbivores (Persson, 1999; Schmitz et al., 2004), and rates of exogenous inputs (Leroux and Loreau, 2008). Overall, aquatic ecosystems tend to have producers with less structural material, receive higher quantities of external subsidies, and have less reticulated food webs, thus facilitating the propagation of indirect top-down interactions (Shurin et al., 2006).

While trophic cascades most often have been applied to explain indirect top-down effects in ecosystems, at its core, the concept is applicable to both bottom-up and top-down interactions. Broadly defined, trophic cascades simply refer to indirect effects of an ecosystem perturbation (i.e., change in soil nutrients or predation rate) throughout an ecosystem. Indeed, empirical studies have shown indirect effects originating from both bottom-up and top-down processes. For example, Gratton and Denno (2003a) demonstrate bottom-up cascading effects of increased nutrients on herbivorous planthoppers and carnivorous spiders in their mid-Atlantic salt marsh food web, whereas Myers et al. (2007) provide evidence of top-down cascading effects of a decline in great shark abundance on cownose ray and bay scallops in the Northwest Atlantic marine ecosystem.

Loreau (2010) has generalized the results we present here to show the functional consequences of bottom-up and top-down forces on biomass, production, and ecological efficiency of ecosystems with n trophic levels. Consistent with the non-nutrient based model of Oksanen et al. (1981), Loreau (2010) shows that an increase in soil nutrient levels will have positive bottom-up effects on trophic levels that lie at the top of ecosystems or at an even number of levels below it (see table 4.2. in Loreau, 2010). The number of trophic levels in an ecosystem and the position of a trophic level along the food chain will determine the relative effect of an increase in soil nutrient levels (i.e., bottom-up) versus the addition

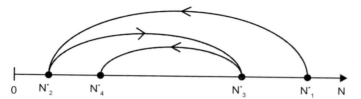

Figure 1.2 Effects of increasing food chain length, n, on the equilibrium amount of soil inorganic nutrient, N_n^*, in an ecosystem with a linear food chain. For example, N_2^* is the soil nutrient stock in a model with two trophic levels (i.e., plant–soil model), N_3^* is the soil nutrient stock for a model with three trophic levels, etc. Modified from Loreau (2010).

of a consumer trophic level (i.e., top-down). Specifically, in ecosystems with linear functional responses, the equilibrium amount of any target trophic level is highest when it lies at the top of the ecosystem, is lowest when it lies just below the top of the ecosystem (i.e., experiences direct top-down regulation), and alternates from low to high and high to low to eventually converge on an intermediate value as more trophic levels are added (Fig. 1.2). These general results suggest that top-down control exerted by top trophic levels cascades down the ecosystem, but becomes progressively weaker as the number of trophic levels above the target trophic level increases or, equivalently, as we move down the food web.

Material cycling for integrating bottom-up and top-down concepts in ecology

As defined with our very simple soil–plant ecosystem model above and classic theory on the topic, the concepts of bottom-up limitation and top-down regulation in ecosystems are very clear. Bottom-up effects are encountered when an increase (decrease) in the resource stock (e.g., via an increase in nutrient supply rate) leads to an increase (decrease) in the biomass of the next higher trophic level, and top-down effects occur when an increase (decrease) in the biomass of the higher trophic levels (e.g., via an increase in uptake rate or a decrease in loss rate) results in a decline (increase) in the biomass of the next lowest trophic level. In more complex ecosystems (i.e., ecosystems with more trophic levels), bottom-up and top-down processes can lead to cascading trophic interactions or indirect effects on non-adjacent trophic levels. Bottom-up, top-down, and cascading trophic interactions arise in models that track energy flux (e.g., Oksanen et al., 1981; Huxel and McCann, 1998; Rip and McCann, 2011) and material flux (e.g., DeAngelis, 1992; Loreau, 1995; Gravel et al., 2010; Leroux and Loreau, 2012) throughout ecosystems. The difference between these two classes of models is that material flux models include an explicit abiotic compartment at the bottom of the food web, whereas energy flux models only include biotic trophic levels. We favor the material cycling class of models because these models allow ecologists to explicitly incorporate important ecosystem feedbacks that arise from ecosystem nutrient cycling (Loreau, 2010; Leroux and Loreau, 2010). Ecological interactions must obey the basic laws of thermodynamics, in particular mass-balance constraints, and this puts fundamental constraints on ecosystem dynamics (Loreau and Holt, 2004). An issue that arises but is seldom discussed when trying to investigate the influence of bottom-up and top-down forces in ecosystems is that it is difficult to separate bottom-up and top-down processes in practice (Pace et al., 1999; Leroux and Loreau, 2010). Below we show how adopting a material cycling framework can allow us to understand the interrelationship between bottom-up and top-down forces in ecosystems.

The interdependence of bottom-up and top-down processes arises because plants and consumers can influence ecosystems via many different mechanisms (reviewed in Leroux and Loreau, 2010; Schmitz et al., 2010). Organismal nutrient recycling is one important feedback between plants and consumers and the soil nutrient pool. In a meta-analysis of the fate of primary production in different terrestrial ecosystems, Cebrian (1999) showed that up to 90% of terrestrial plant matter is not consumed by herbivores, but actually enters the dead organic matter pool as litterfall. It is therefore surprising (or concerning) that many food web and ecosystem models do not include this ecologically important feedback. As we show below, nutrient recycling can have very important impacts on ecosystem functioning (DeAngelis, 1992; Loreau, 2010; also, see Chapter 9). In real ecosystems, materials recycled from biotic compartments are first processed by detritivores and decomposers that mineralize nutrients and make them available for plant uptake (Loreau, 1995; reviewed in Moore et al., 2004). For simplicity, we do not include decomposition in our model, as our goal is simply to illustrate the connection between bottom-up and top-down processes via organismal nutrient recycling.

Let us start with the plant–soil model from above and add feedback loops from the plant trophic level back to the soil nutrient pool (Loreau, 2010). The plant feedback loop represents litterfall. To incorporate plant nutrient recycling, we add one parameter to the model: δ_P. This parameter describes the portion of plant nutrients lost from the ecosystem and $1 - \delta_P$ describes the portion of plant nutrients that is returned via senescence to the soil nutrient pool. Adding nutrient recycling to our simple plant–soil ecosystem model yields a modified Eq. 1 (see Eq. 5 below) but the same equation for the plant (Eq. 2) dynamics (Loreau, 2010).

$$\frac{dN}{dt} = \Phi_N - \theta_N - \Phi_P + (1 - \delta_P)\theta_P \tag{5}$$

At equilibrium $N^* = \frac{m_P}{\alpha_P}$ and $P^* = \frac{I - m_N N^*}{\delta_P m_P}$, and the effect of plant recycling on plant nutrient stocks (P^*), primary production (Φ_P), and plant nutrient recycling flux ($(1 - \delta_P)\theta_P$) can be seen by taking one minus the partial derivative of these ecosystem properties with respect to the fraction of plant nutrient that is lost from the ecosystem, δ_P. A positive effect of plant nutrient recycling on plant stocks (i.e., $1 - \frac{\partial P^*}{\partial \delta_P} > 0$), production (i.e., $1 - \frac{\partial \Phi_P}{\partial \delta_P} > 0$), and recycling flux (i.e., $1 - \frac{\partial \delta_P \theta_P}{\partial \delta_P} > 0$) emerges when the plant and soil trophic levels persist at equilibrium (i.e., the ecosystem is feasible, with feasibility condition $I > \frac{m_N m_P}{\alpha_P}$).

Organism-mediated nutrient recycling has important impacts on the dynamics of ecosystems, but is it a bottom-up or top-down process? In our simple plant–soil ecosystem model, plant nutrient recycling is not a clear bottom-up effect, as it has a positive influence on plant function via an indirect route through nutrients recycled from plants. However, plant nutrient recycling also is not a clear top-down effect in this model, as it does not represent a direct influence of plants on the soil nutrient pool as outlined for models without material recycling above. Recent work that has adopted a materials cycling framework (reviewed in Vanni, 2002; Leroux and Loreau, 2010; Schmitz et al., 2010) questions the usefulness of the bottom-up and top-down terminology. This work suggests that there is a need for caution when discussing bottom-up and top-down effects in ecology because organism-mediated nutrient recycling and other mechanisms (see Leroux and Loreau, 2010; Schmitz et al., 2010) do not fit within the classic definitions of bottom-up limitation or top-down regulation. We advise authors to be specific about the process of interest when using the terms bottom-up and top-down to avoid perpetuating the sometimes incorrect and confusing applications of these terms.

We can further investigate the indirect effects of organismal nutrient recycling by adding herbivores to the above model and a nutrient recycling feedback loop from the herbivore trophic level back to the soil nutrient pool, $1 - \delta_H$. The herbivore recycling feedback loop represents excretion and death. With the inclusion of this feedback loop, the dynamical equation for the soil nutrient pool is as follows:

$$\frac{dN}{dt} = \Phi_N - \theta_N - \Phi_P + (1 - \delta_P)\theta_P + (1 - \delta_H)\theta_H \tag{6}$$

The equilibrium stocks in this herbivore–plant–soil model are given by $N^* = \frac{\alpha_H H^* + \varepsilon_H m_P}{\varepsilon_H \alpha_P}$, $P^* = \frac{m_H}{\alpha_H}$, $H^* = \frac{\alpha_P(m_P P^* - (1 - \delta_P)m_P P^* - I) + m_N m_P}{\alpha_P m_H(\varepsilon_H(1 - \delta_H) - 1) - \alpha_H m_N}$. The addition of plant and herbivore nutrient recycling feedback loops has a positive effect on soil and herbivore stocks in ecosystems with Lotka–Volterra functional responses (Fig. 1.3). These feedback loops have no effect on the equilibrium plant stock, as the equilibrium plant stock is regulated by the herbivore mass-specific loss rate (m_H) and attack rate (α_H). Based on the above equilibrium, we can distinguish four distinct conditions; no organismal recycling ($\delta_P = \delta_H = 1$), plant-only nutrient recycling ($\delta_H = 1$), herbivore-only nutrient recycling ($\delta_P = 1$), and plant and herbivore nutrient recycling ($\delta_P = \delta_H \neq 1$). Soil and herbivore stocks are highest

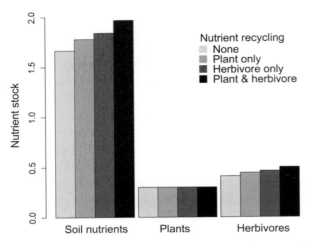

Figure 1.3 Effect of plant and herbivore nutrient recycling on the equilibrium amount of soil, plants, and herbivores in a herbivore–plant–soil ecosystem model. Results are for models with no organismal nutrient recycling $\delta_P = \delta_H = 1$, for models with plant-only nutrient recycling $\delta_P = 0.25$, $\delta_H = 1$, for models with herbivore-only nutrient recycling $\delta_P = 1$, $\delta_H = 0.25$, and for models with both plant and herbivore nutrient recycling $\delta_P = \delta_H = 0.25$. Model parameters are $\alpha_P = \alpha_H = 1$, $e_H = 0.3$, $m = m_P = m_H = 0.3$, $I = 1$.

in models with both plant and herbivore nutrient recycling, followed by models with only herbivore recycling, then models with only plant recycling, and lastly models with no organismal nutrient recycling (Fig. 1.3). Similar positive feedbacks of organismal nutrient cycling can be seen for herbivores on soil and plant resources via grazing optimization, i.e., the enhancement of primary production by herbivores (de Mazancourt et al., 1998). Grazing optimization requires that herbivores improve an ecosystem's nutrient conservation efficiency enough to compensate for loss of plant biomass to grazing.

The confusion surrounding the classification of organism-mediated nutrient recycling as a bottom-up or top-down process occurs because the classic definitions of bottom-up and top-down forces were applied to a limited number of mechanisms (i.e., nutrient enrichment, predator consumption) that did not consider the complex feedbacks involved in material cycling. Recent integrations of community and ecosystem ecology (e.g., Loreau, 2010; Schmitz, 2010) highlight a large number of mechanisms that can lead to indirect effects in ecosystems, particularly in the context of nutrient-limited ecosystems. The suite of mechanisms includes ones related to organismal consumption such as excretion and egestion (Vanni, 2002; McIntyre et al., 2008), translocation of nutrients (Moore et al., 2007; Abbas et al., 2012), and consumer-resource stoichiometric mismatches (Leroux et al., 2012; Cherif and Loreau, 2013). Other non-consumptive mechanisms like predation risk (Hawlena and Schmitz, 2010) and ecosystem engineering (Jones et al., 1994) can also lead to indirect effects in ecosystems. These consumptive and non-consumptive mechanisms can lead to complex indirect feedbacks on material cycling in ecosystems, which blur the lines between bottom-up and top-down processes. Adopting a material cycling framework encourages researchers to be clear about the processes of interest and allows us to explicitly track the feedbacks among processes, as well as the dynamical consequences they have in ecosystems.

For example, a material cycling framework has been instrumental in deciphering the keystone role of Pacific salmon (*Oncorhynchus* spp.) in freshwater and riparian ecosystems of the Pacific Northwest. Detailed empirical studies have demonstrated how salmon carcasses and gametes can provide nutrients for plants and consumers in freshwaters and riparian forests (e.g., Holtgrieve et al., 2009; Hocking and Reynolds, 2011; Field and Reynolds, 2011). Also, it has been demonstrated that salmon excretion and bioturbation can influence freshwater nutrient cycling and primary production (Moore et al., 2007; Verspoor et al., 2010). We encourage other empirical ecologists to adopt a material cycling framework as it can shed light on the key processes regulating ecosystems.

Bottom-up and top-down interactions across space and time

Early ecological theory investigated bottom-up and top-down processes in simple or relatively closed ecosystems at equilibrium (e.g., Hairston et al., 1960; Oksanen et al., 1981; Carpenter et al., 1985). Likewise, most empirical work on the importance of bottom-up versus top-down processes is done at very small spatial scales and short time frames (Gripenberg and Roslin, 2007). But with the emergence of landscape and spatial ecology as disciplines, there is mounting evidence for the role of space in shaping the dynamics of ecosystems. Indeed, there exists no natural ecosystem that is completely closed from outside influence, whether it be rainfall (e.g., Anderson et al., 2008), species dispersal (e.g., Cadotte, 2006), or nutrient flux (e.g., Vannote et al., 1980). What is more, most ecosystems are in a constant state of temporal flux driven by seasonality, life history dynamics, and environmental change, and the time scales of many ecological phenomena may be short (Hastings, 2004; 2010). Empirical evidence that the strength of bottom-up and top-down interactions can vary in space and time has helped to end the divisive debate on the dominance of either mechanism and instead shift our attention to explaining variation in the relative strength of each process in space and time (Gripenberg and Roslin, 2007; Leroux and Loreau, 2012; also, see Chapter 11). In this section, we review recent theoretical work on bottom-up and top-down interactions in spatially extended and temporally variable ecosystems.

The spatial flow of energy, materials, and organisms across ecosystem boundaries has the potential to influence the dynamics of donor and recipient ecosystems (Polis et al., 1997; Loreau et al., 2003). For example, consumers in streams

(Nakano et al., 1999; also, see Chapter 6), riparian forests (Murakami and Nakano, 2002), coastal marine zones (Rose and Polis, 1998; also, see Chapter 7), and pelagic ecosystems (Schindler and Scheuerell, 2002) can obtain a significant amount of their energy from food sources originating in neighboring ecosystems. Nowhere are the ecological implications of spatial subsidies more obvious than in the Pacific Northwest where large runs of Pacific salmon (*Oncorhynchus* spp.) provide a seasonal pulse of marine-derived nutrients to freshwater lakes and streams, and riparian forests. This nutrient pulse influences the distribution and abundance of many and diverse organisms, including plants (Hocking and Reynolds, 2011), fish (Flecker et al., 2010), birds (Field and Reynolds, 2011), and bears (Holtgrieve et al., 2009). Whereas Pacific salmon are mostly a nutrient subsidy (but see Moore et al., 2007) as it is their bodies and gametes that provide resources for plants and consumers in freshwaters and riparian forests, other consumers may functionally link distinct ecosystems by actively feeding on resources in multiple ecosystems. For example, bears in the Pacific Northwest may feed on stream and riparian forest resources thereby linking aquatic and terrestrial ecosystems (Helfield and Naiman, 2006). These mobile consumers couple neighboring ecosystems (McCann et al., 2005; Rooney et al., 2006) and have the potential to translocate nutrients among ecosystems (Leroux and Loreau, 2010; 2012). Other examples include roe deer (*Capreolus capreolus*), which translocate significant quantities of N and P between their primary foraging grounds (croplands) and their primary resting grounds (forests) (Abbas et al., 2012), and lake trout (*Salvelinus namaycush*), which feed in both pelagic and benthic habitats and effectively couple these distinct lake ecosystems (Vander Zanden and Vadeboncoeur, 2002). What is evident from empirical research on spatial subsidies and mobile consumers is that flows of energy, material, and organisms can drive bottom-up and top-down processes across ecosystem boundaries.

Theory on spatial subsidies has shown that ecosystems that receive spatial subsidies tend to have stronger top-down cascading trophic interactions (Leroux and Loreau, 2008). However, many subsidies are seasonal, occurring in pulses over short time frames (Baxter et al., 2005; Anderson et al., 2008; Yang et al., 2010). The seasonality inherent in many spatial subsidies (e.g., insect emergence, litterfall, anadromous salmon; Baxter et al., 2005) may stabilize ecosystem dynamics by complementing seasonal local resource deficiencies and by shifting consumer pressure between allochthonous and in situ resources (Takimoto et al., 2002; 2009; Leroux and Loreau, 2012). Similar to subsidies at lower trophic levels, mobile consumers can stabilize coupled ecosystems by generating asynchronous responses of resources to predation, thus spreading out the top-down impacts of predation (Rooney et al., 2006; McCann and Rooney, 2009). Two mechanisms that may explain the role of energy, material, and organism fluxes on cascading trophic interactions across ecosystems are increased energy flux up an ecosystem (Leroux and Loreau, 2008; Rip and McCann, 2011) and species competition mediated through a shared predator (i.e., apparent competition; *sensu* Holt, 1977).

If energy, material, or organism fluxes occur at lower trophic levels, these subsidies can flow up the ecosystem to support higher production of top consumers (i.e., top-heavy ecosystems), and thus larger indirect cascading effects of consumers in the recipient ecosystem (Leroux and Loreau, 2008). While some theoretical work that includes predator saturation has shown that top-heavy ecosystems can be mathematically unstable (Huxel and McCann, 1998; Rip and McCann, 2011), we argue that top-down trophic cascades are not necessarily a biologically destabilizing process but rather a fundamental component of many natural ecosystems. The availability of an allochthonous resource can lead to apparent competition among local and exogenous resources (Leroux and Loreau, 2010). This form of competition may be strongest if the subsidy is donor-controlled (i.e., consumer density does not affect the amount of resources consumed) and sustained, where a predator feeding on this subsidy may also exert strong top-down regulation of local resources. This indirect competition between the subsidy and local resource is an example of apparent competition. Subsidies, therefore, can result in trophic cascades across ecosystems, where resources in one ecosystem can influence the biomass, abundance, and production of neighboring ecosystems (Leroux and Loreau, 2012). Knight et al. (2005) provide convincing evidence of cross-ecosystem cascades where fish (*Centrachidae* spp.) prey on aquatic dragonfly larvae, which reduces terrestrial adult dragonfly (*Odonata* spp.) abundance. This reduction in terrestrial adult dragonfly abundance resulted in an increase in terrestrial dragonfly prey (*Hymenoptera* spp., *Diptera* spp., and *Lepidoptera* spp.) and an increase in terrestrial host plant (*Hypericum fasciculatum*) production.

The effects of allochthonous resources on top-down regulation across ecosystems can be illustrated by adding an allochthonous plant resource, A, available to the herbivores in our herbivore–plant–soil ecosystem model. An empirical example of such a case is presented in the roe deer example discussed above (Abbas et al., 2012). Also, we add a parameter, π, which describes the herbivore's preference for the local plant. We do not include organismal nutrient recycling in this model, as we wish to focus on the effects of allochthonous resources on top-down regulation. The revised dynamical equation for herbivores with access to local and allochthonous resources is:

$$\frac{dH}{dt} = \pi \Phi_H + (1 - \pi) \Phi_A - \theta_H \tag{7}$$

where $\Phi_A = f_A(A)H = \alpha_A AH$, and α_A is the herbivore uptake rate for allochthonous resources. Eq. 7 paired with Eq. 1 and 3 leads to the following equilibrium: $N^* = \frac{I}{m_N + \alpha_P P^*}$, $P^* = \frac{m_H - (1-\pi)\alpha_A A}{\pi \alpha_H}$, $H^* = \frac{\varepsilon_H(\alpha_P N^* - m_P)}{\alpha_H}$. An increase in the stock of allochthonous plants available to herbivores leads to an increase in the soil nutrient pools (i.e., $\frac{\partial N^*}{\partial A} > 0$, Fig. 1.4) and herbivore stocks (i.e., $\frac{\partial H^*}{\partial A} > 0$, Fig. 1.4),

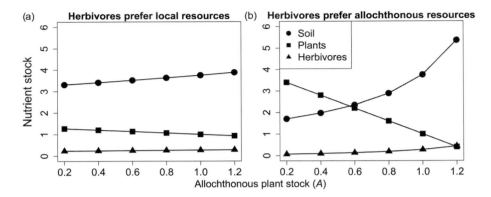

Figure 1.4 Effect of increasing allochthonous plant stock on herbivore, local plant, and soil nutrient stocks. The results are shown for (a) herbivores that prefer local resources ($\pi = 0.75$) and (b) herbivores that prefer allochthonous resources ($\pi = 0.25$). All other model parameters are $\alpha_P = \alpha_H = 0.4$, $e_H = e_A = 0.1$, $m = m_P = m_H = 0.4$, $I = 3$.

and a decline in plant stocks (i.e., $\frac{\partial P^*}{\partial A} < 0$, Fig. 1.4). These qualitative direct and indirect effects of allochthonous resources hold irrespective of the herbivore's preference for local or allochthonous resources (Fig. 1.4). The magnitude of the negative effect of increasing allochthonous resources on plant stocks, however, is larger when herbivores prefer allochthonous resources (Fig. 1.4).

The concept of meta-ecosystems (sensu Loreau et al., 2003) is a useful framework for understanding the dynamics of spatially extended ecosystems within a material cycling framework (Massol et al., 2011), and thus for understanding linkages between bottom-up and top-down interactions. Meta-ecosystem theory is particularly useful in the context of understanding bottom-up and top-down processes in space because source-sink dynamics emerge from meta-ecosystem models as a result of nutrient mass-balance constraints. For example, a flux of inorganic nutrients from an ecosystem with elevated nutrient biomass (i.e., source) to a local ecosystem with low-nutrient biomass (i.e., sink) has the potential to create conditions for switching the source-sink dynamics of this coupled ecosystem (Gravel et al., 2010; Loreau et al., 2013). What is more, if ecosystems are connected via material flows and consumers demonstrate a behavioral response to these resource flows, strong top-down forces in the recipient ecosystem may correspond to weaker top-down forces in its connected ecosystem (Leroux and Loreau, 2012). This occurs because most of the material flowing in this meta-ecosystem will be flowing in the recipient ecosystem. Resolving the cross-ecosystem effects of energy, material, and organism subsidies remains a challenge for both theoretical and empirical ecologists.

Classic theories of bottom-up, top-down, and trophic cascades make predictions for long-term or equilibrium dynamics (e.g., Rosenzweig, 1971; Oksanen et al., 1981). This is often done for convenience, as the mathematical tools for analyzing equilibrium conditions are better resolved than for analyzing non-equilibrium conditions. Most empirical studies, however, track short-term responses of ecosystems to perturbations (Gripenberg and Roslin, 2007; Marczak et al., 2007). There is a need to expand our knowledge of trophic dynamics to non-equilibrium conditions, particularly in light of rapid environmental change (Hastings, 2010; also, see Chapter 14). As discussed above, many spatial subsidies occur as pulses (Holt, 2008; Leroux and Loreau, 2012) due to seasonality (e.g., floods/droughts), life history traits (e.g., anadromy), or other abiotic factors (e.g., rainfall), therefore the effects of space and time on bottom-up and top-down forces in ecosystems are not independent. Resource pulses can be frequent, such as marine wrack from wave action linking oceans to islands (Polis and Hurd, 1995), or infrequent, such as the mass emergence of cicadas every 13–17 years linking belowground and aboveground temperate forest ecosystems (Yang, 2004). The temporally variable driving forces of ecosystems can lead to variation in the extent of bottom-up and top-down regulation over time (Gratton and Denno, 2003b; Denno et al., 2005; Kerimoglu et al., 2013; Whalen et al., 2013). For example, epiphytic algae production in a seagrass ecosystem in the York River, Virginia, USA showed temporal variation in response to mesograzer abundance. Specifically, epiphytic algae production increased in response to nutrient addition in the fall when mesograzer abundance was low, but algae were regulated by mesograzers in the summer when mesograzers were abundant (Whalen et al., 2013). Temporal variations in bottom-up and top-down regulation can even occur in the absence of variable external driving forces as a result of the spatial dynamics of nutrient flows within meta-ecosystems, which generate asynchronous fluctuations of trophic levels in connected patches (Marleau et al., 2010). The transient or short-term response of ecosystems to resource pulses may improve our understanding of the interrelationship between bottom-up and top-down processes. Theory is beginning to investigate temporal variation in the strength of bottom-up and top-down processes (Holt, 2008; Hastings, 2012; Leroux and Loreau, 2012), but more work needs to be done. Given that most empirical studies are conducted on short time frames, the development and application of mathematical methods for analyzing the short-term dynamics of ecosystems will be necessary for bridging the divide between theory and empirical research on bottom-up and top-down processes in ecology. The mathematical concepts of reactivity, maximum amplification, and timing of maximum amplification hold promise for analyzing and understanding the transient dynamics of ecosystems (Neubert and Caswell, 1997; Neubert et al., 2009).

Conclusions

The theory of resource versus consumer limitation of organisms has evolved from the consideration of very simple ecosystems to more complex and realistic cases that incorporate spatial and temporal variability of natural ecosystems. Guided by empirical research, which has presented a plethora of evidence showing that both bottom-up and top-down forces occur in most ecosystems, theory is now leading the way in our investigation of mechanisms to explain variation in trophic regulation. A better understanding of the relative roles and determinants of variation in bottom-up versus top-down regulation is critical as trophic-dynamic theory is the foundation for many real-world applications in natural resource and wildlife management. For example, the addition or removal of consumers is often used to control overabundant wildlife. Piscivorous fish are frequently added to lakes to indirectly control algal abundance (Findlay et al., 2005), and wolves have been added to Yellowstone National Park with indirect effects on aspen regeneration via predation on elk (Ripple and Beschta, 2005). Further understanding of bottom-up and top-down dynamics will improve our applications of these principles to natural resource management. Below we briefly outline five key areas for future directions for the development of theory on bottom-up, top-down, and cascading trophic interactions.

First, bottom-up, top-down, and in particular, cascading trophic interactions result in a diverse suite of indirect effects in ecosystems. There are a large number of consumptive and non-consumptive mechanisms (see Leroux and Loreau, 2010; Schmitz et al., 2010) that can lead to indirect effects in ecosystems, and theory can be useful in understanding the relative importance of these mechanisms in describing ecosystem dynamics. Such a mechanistic framework is necessary to improve our predictions of bottom-up, top-down, and cascading trophic dynamics in light of environmental change. For example, a meta-analysis of marine systems demonstrated that predator removal experiments show weak top-down effects of predators on marine phytoplankton (Shurin et al., 2002). Stibor et al. (2004), however, showed that top-down effects of jellyfish predation in their marine system are contingent upon the size of algae, with positive top-down effects occurring when large algae are initially abundant, negative effects when small algae are initially abundant, and no effect when algae are combined. This detailed study by Stibor et al. (2004) provides empirical evidence for the need for a better mechanistic understanding of the general patterns of top-down and bottom-up regulation. Consequently, understanding the structure of species interactions in this food web may help to interpret the resulting trophic dynamics, with implications for ecosystem functioning.

Second, ecological stoichiometry has the potential to shed light on the mechanisms for variation and interdependence in bottom-up and top-down processes in ecosystems. Stoichiometric models allow for explicitly tracking material cycling and for investigating the role of resource quality and consumer-resource quality mismatches on ecosystem functioning (Sterner and Elser, 2002). In addition, stoichiometric theory may help us identify bottlenecks for a switch between bottom-up and top-down processes in complex ecosystems. Ecological stoichiometry theory is still mostly applied to relatively simple ecosystems, but several studies are shedding light on the implications of stoichiometric constraints in more complex cases (e.g., Grover, 2003; Leroux et al., 2012; Cherif and Loreau, 2013). For example, Leroux et al. (2012) show that consumptive and non-consumptive effects of consumers (i.e., top-down effects) can have direct impacts on the stoichiometry of their prey and indirect impacts on soil nutrient composition. This stoichiometrically explicit approach to studying organism-mediated nutrient recycling shows the interdependence between bottom-up and top-down processes in ecosystems.

Third, as discussed in the previous section, the strength of bottom-up and top-down regulation can vary in space. Spatial subsidies and mobile consumers can influence regulation of coupled ecosystems and the meta-ecosystem framework may be useful for guiding progress on this matter (Loreau et al., 2003; Massol et al., 2011). In addition to spatial subsidies and mobile consumers, species with complex life histories may influence bottom-up and top-down regulation across ecosystems. Schreiber and Rudolf (2008) derive a stage-structured model with juveniles and adults occupying different habitats to show how a change in the abundance of one stage can lead to abrupt shifts in abundances across ecosystems. Similarly, McCoy et al. (2009) show how organisms with complex life histories can generate strong trophic linkages across ecosystem boundaries in what they term "predator shadows." Theory has an important role to play in deciphering the bottom-up and top-down dynamics of spatially connected ecosystems.

Fourth, there is ample empirical evidence of oscillating or temporally variable trophic control (e.g., Daskalov et al., 2007; Litzow and Ciannelli, 2007; Kerimoglu et al., 2013); therefore we must move beyond demonstrating patterns of trophic regulation to explaining variation in these patterns. As highlighted above, the meta-ecosystem framework (Loreau et al., 2003; Massol et al., 2011) and theory for temporally variable processes (Neubert and Caswell, 1997; Holt, 2008; Neubert et al., 2009; Hastings, 2012; Leroux and Loreau, 2012) will be useful for deciphering the mechanisms for variable trophic control and their implications for ecosystem functioning. In light of seasonality and rapid environmental change, the transient response of communities to perturbations (i.e., addition or removal of organisms, nutrients, etc.) may be just as important to the long-term average response (Hastings, 2004).

Finally, theory has an important role in generating and developing novel hypotheses, clarifying mechanisms, and providing testable predictions. But we must ensure continual feedback between theory and data to improve our understanding of bottom-up, top-down, and cascading trophic interactions in ecosystems. One way to facilitate this interaction is for theoreticians to use bottom-up and top-down metrics that can be easily measured in empirical settings. For example, response ratios are commonly used in experimental studies and increasingly they are being used in theory (e.g., Shurin and Seabloom, 2005; Leroux and Loreau, 2008; 2010). In addition, both theory and empirical researchers need to better communicate the processes they are studying when referring to bottom-up and top-down effects to facilitate the integration of theory and data.

We are at a critical juncture in the ecological study of biomass pyramids, where the body of empirical and theoretical research has enabled us to move beyond describing patterns of bottom-up limitation and top-down regulation within and across ecosystems to explaining and understanding spatial and temporal variations in the patterns of trophic control. Progress along these lines may be hampered by the inconsistent and often confusing application of the terms "bottom-up" and "top-down." We urge ecologists to be explicit about the processes they are studying as opposed to proliferating the general use of the bottom-up and top-down terms. We particularly believe that rapid progress in our understanding of trophic dynamics can occur by considering the multiple mechanisms of trophic interactions within a material cycling framework. The mathematical methods and tools now exist for studying spatially expansive ecosystems and temporal variability in trophic interactions. The integration of ecological theory with empirical studies will improve our predictions of trophic dynamics under global change.

Acknowledgments

We thank T. Hanley and K. La Pierre for the invitation to contribute to this book and for comments on a draft of this chapter. S. Leroux was supported by an NSERC Discovery grant, and M. Loreau by the TULIP Laboratory of Excellence (ANR-10-LABX-41).

References

Abbas, F., Merlet, J., Morellet, N., et al. (2012). Roe deer may markedly alter forest nitrogen and phosphorous budgets across Europe. *Oikos*, **121**, 1271–1278.

Anderson, W. B., Wait, D. A. and Stapp, P. (2008). Resources from another place and time: responses to pulses in a spatially subsidized system. *Ecology*, **89**, 660–670.

Arditi, R. and Ginzburg, L. R. (1989). Coupling in predator prey dynamics – ratio-dependence. *Journal of Theoretical Biology*, **139**, 311–326.

Baxter, C. V., Fausch, K. D. and Saunders, W. C. (2005). Tangled webs: reciprocal flows of invertebrate prey link streams and riparian zones. *Freshwater Biology*, **50**, 201–220.

Borer, E. T., Halpern, B. S. and Seabloom, E. W. (2006). Asymmetry in community regulation: effects of predators and productivity. *Ecology*, **87**, 2813–2820.

Borer, E. T., Seabloom, E. W., Shurin, J. B., et al. (2005). What determines the strength of a trophic cascade? *Ecology*, **86**, 528–537.

Brett, M. T. and Goldman, C. R. (1996). A meta-analysis of the freshwater trophic cascade. *Proceedings of the National Academy of Sciences of the USA*, **93**, 7723–7726.

Cadotte, M. W. (2006). Dispersal and species diversity: a meta-analysis. *American Naturalist*, **167**, 913–924.

Carpenter, S. R. and Kitchell, J. F. (1993). *The Trophic Cascade in Lake Ecosystems*. Cambridge: Cambridge University Press.

Carpenter, S. R., Kitchell, J. F. and Hodgson, J. R. (1985). Cascading trophic interactions and lake productivity. *Bioscience*, **35**, 634–639.

Cebrian, J. (1999). Patterns in the fate of production in plant communities. *American Naturalist*, **154**, 449–468.

Cebrian, J., Shurin, J. B., Borer, E. T., et al. (2009). Producer nutritional quality controls ecosystem trophic structure. *PLoS One*, **4**, e4929.

Cherif, M. and Loreau, M. (2013). Plant-herbivore-decomposer stoichiometric mismatches and nutrient cycling in ecosystems. *Proceedings of the Royal Society B: Biological Sciences*, **280**, 2012–2453.

Creel, S., Christianson, D., Liley, S. and Winnie Jr., J. A. (2007). Predation risk affects reproductive physiology and demography of elk. *Science*, **315**, 960.

Daskalov, G. M., Grishin, A. N., Rodionov, S. and Mihneva, V. (2007). Trophic casades triggered by overfishing reveal possible mechanisms of ecosystem regime shifts. *Proceedings of the National Academy of Sciences of the USA*, **104**, 10518–10523.

de Mazancourt, C., Loreau, M. and Abbadie, L. (1998). Grazing optimization and nutrient cycling: when do herbivores enhance plant production? *Ecology*, **79**, 2242–2252.

DeAngelis, D. L. (1992). *Dynamics of Nutrient Cycling and Food Webs*. New York: Chapman and Hall.

Denno, R. F., Lewis, D. and Gratton, C. (2005). Spatial variation in the relative strength of top-down and bottom-up forces: causes and consequences for phytophagous and insect populations. *Annales Zoologici Fennici*, **42**, 295–311.

Duffy, J. E., Carinale, B. J., France, K. E., et al. (2007). The functional role of biodiversity in ecosystems: incorporating trophic complexity. *Ecology Letters*, **10**, 522–538.

Elser, J. J., Bracken, M. E. S., Cleland, E. E., et al. (2007). Global analysis of nitrogen and phosphorous limitation of primary producers in freshwater, marine and terrestrial ecosystems. *Ecology Letters*, **10**, 1135–1142.

Elton, C. S. (1927). *Animal Ecology*. New York: Macmillan Co.

Fath, B. D. (2004). Distributed control in ecological networks. *Ecological Modelling*, **179**, 235–245.

Field, R. D. and Reynolds, J. D. (2011). Sea to sky: impacts of residual salmon-derived nutrients on estuarine breeding bird communities. *Proceedings of the Royal Society B: Biological Sciences*, **278**, 3081–3088.

Findlay, D. L., Vanni, M. J., Paterson, M., et al. (2005). Dynamics of a boreal lake ecosystem during a long-term manipulation of top predators. *Ecosystems*, **8**, 603–618.

Flecker, A. S., McIntyre, P. B., Moore, J. W., et al. (2010). Miigratory fishes as material and process subsidies in riverine ecosystems. *American Fisheries Society Symposium*, **73**, 559–592.

Frank, K. T., Petrie, B. and Shackell, N. L. (2007). The ups and downs of trophic control in continental shelf ecosystems. *Trends in Ecology and Evolution*, **22**, 236–242.

Fretwell, S. D. (1977). The regulation of plant communities by food chains exploiting them. *Perspectives in Biology and Medicine*, **20**, 169–185.

Glaholt, S. P. and Vanni, M. J. (2005). Ecological responses to simulated benthic-derived nutrient subsidies mediated by omnivorous fish. *Freshwater Biology*, **50**, 1864–1881.

Gratton, C. and Denno, R. F. (2003a). Inter-year carryover effects of a nutrient pulse on *Spartina* plants, herbivores, and natural enemies. *Ecology*, **84**, 2692–2707.

Gratton, C. and Denno, R. F. (2003b). Seasonal shift from bottom-up to top-down impact in phytophagous insect populations. *Oecologia*, **134**, 487–495.

Gravel, D., Guichard, F., Loreau, M. and Mouquet, N. (2010). Source and sink dynamics in meta-ecosystems. *Ecology*, **91**, 2172–2184.

Gripenberg, S. and Roslin, T. (2007). Up or down in space? Uniting the bottom-up versus top-down paradigm and spatial ecology. *Oikos*, **116**, 181–188.

Grover, J. P. (2003). The impact of variable stoichiometry on predator-prey interactions: a multinutrient approach. *American Naturalist*, **162**, 29–43.

Gruner, D. S., Smith, J. E., Seabloom, E. W., et al. (2008). A cross-system synthesis of consumer and nutrient resource control on producer biomass. *Ecology Letters*, **11**, 740–755.

Hairston, N. G., Smith, F. E. and Slobodkin, L. B. (1960). Community structure, population control, and competition. *American Naturalist*, **94**, 421–425.

Hall, S. R., Shurin, J. B., Diehl, S. and Nisbet, R. M. (2007). Food quality, nutrient limitation of secondary production, and the strength of trophic cascades. *Oikos*, **116**, 1128–1143.

Hassell, M. P., Crawley, M. J., Godfray, H. C. J. and Lawton, J. H. (1998). Top-down versus bottom-up and the Ruritanian bean bug. *Proceedings of the National Academy of Sciences of the USA*, **95**, 10661–10664.

Hastings, A. (2004). Transients: the key to long-term ecological understanding? *Trends in Ecology and Evolution*, **19**, 39–45.

Hastings, A. (2010). Timescales, dynamics, and ecological understanding. *Ecology*, **91**, 3471–3480.

Hastings, A. (2012). Temporally varying resources amplify the importance of resource input in ecological populations. *Biology Letters*, **8**, 1067–1069.

Hawlena, D. and Schmitz, O. J. (2010). Herbivore physiological response to predation risk and implications for ecosystem nutrient dynamics. *Proceedings of the National Academy of Sciences of the USA*, **107**, 15503–15507.

Helfield, J. M. and Naiman, R. J. (2006). Keystone interactions: salmon and bear in riparian forests of Alaska. *Ecosystems*, **9**, 167–180.

Hillebrand, H. and Cardinale, B. J. (2004). Consumer effects decline with prey diversity. *Ecology Letters*, **7**, 192–201.

Hocking, M. D. and Reynolds, J. D. (2011). Impacts of salmon on riparian plant diversity. *Science*, **331**, 1609–1612.

Holt, R. D. (1977). Predation, apparent competition, and the structure of prey communities. *Theoretical Population Biology*, **12**, 197–229.

Holt, R. D. (2008). Theoretical perspectives on resource pulses. *Ecology*, **89**, 671–681.

Holtgrieve, G. W., Schindler, D. E. and Jewett, P. K. (2009). Large predators and biogeochemical hotspots: brown bear (*Ursus arctos*) predation on salmon alters nitrogen cycling in riparian soils. *Ecological Research*, **24**, 1125–1135.

Hulot, F. D. and Loreau, M. (2006). Nutrient-limited food webs with up to three trophic levels: feasibility, stability, assembly rules, and effects of nutrient enrichment. *Theoretical Population Biology*, **69**, 48–66.

Hunter, M. D. and Price, P. W. (1992). Playing chutes and ladders – heterogeneity and the relative roles of bottom-up and top-down forces in natural communities. *Ecology*, **73**, 724–732.

Huxel, G. R. and McCann, K. (1998). Food web stability: the influence of trophic flows across habitats. *American Naturalist*, **152**, 460–469.

Jones, C. G., Lawton, J. H. and Shachak, M. (1994). Organisms as ecosystem engineers. *Oikos*, **69**, 373–386.

Kerimoglu, O., Straile, D. and Peeters, F. (2013). Seasonal, inter-annual and long-term variation in top-down versus bottom-up regulation of primary production. *Oikos*, **122**, 223–234.

Knight, T. M., McCoy, M. W., Chase, J. M., McCoy, K. A. and Holt R. D. (2005). Trophic cascades across systems. *Nature*, **437**, 880–883.

LeBauer, D. S. and Treseder, K. K. (2008). Nitrogen limitation of net primary productivity in terrestrial ecosystems is globally distributed. *Ecology*, **89**, 371–379.

Leroux, S. J. and Loreau, M. (2008). Subsidy hypothesis and strength of trophic cascades across ecosystems. *Ecology Letters*, **11**, 1147–1156.

Leroux, S. J. and Loreau, M. (2009). Disentangling multiple predator effects in biodiversity and ecosystem functioning research. *Journal of Animal Ecology*, **78**, 695–698.

Leroux, S. J. and Loreau, M. (2010). Consumer-mediated recycling and cascading trophic interactions. *Ecology*, **91**, 2162–2171.

Leroux, S. J. and Loreau, M. (2012). Dynamics of reciprocal pulsed subsidies in local and meta-ecosystems. *Ecosystems*, **15**, 48–59.

Leroux, S. J., Hawlena, D. and Schmitz, O. J. (2012). Predation risk, stoichiometric plasticity and ecosystem elemental cycling. *Proceedings of the Royal Society B: Biological Sciences*, **279**, 4183–4191.

Lindeman, R. L. (1942). The trophic-dynamic aspect of ecology. *Ecology*, **23**, 399–418.

Litzow, M. A. and Ciannelli, L. (2007). Oscillating trophic control induces community reorganization in a marine ecosystem. *Ecology Letters*, **10**, 1124–1134.

Loreau, M. (1995). Consumers as maximizers of matter and energy-flow in ecosystems. *American Naturalist*, **145**, 22–42.

Loreau, M. (2010). *From Populations to Ecosystems: Theoretical Foundations for a New Ecological Synthesis*. Princeton, NJ: Princeton University Press.

Loreau, M. and Holt, R. D. (2004). Spatial flows and the regulation of ecosystems. *The American Naturalist*, **163**, 606–615.

Loreau, M., Daufresne, T., Gonzalez, A., et al. (2013). Unifying sources and sinks in ecology and Earth sciences. *Biological Reviews*, **88**, 365–379.

Loreau, M., Mouquet, N. and Holt, R. D. (2003). Meta-ecosystems: a theoretical framework for a spatial ecosystem ecology. *Ecology Letters*, **6**, 673–679.

Marczak, L. B., Thompson, R. M. and Richardson, J. S. (2007). Meta-analysis: trophic level, habitat, and productivity shape the food web effects of resource subsidies. *Ecology*, **88**, 140–148.

Marleau, J. N., Guichard, F., Mallard, F. and Loreau, M. (2010). Nutrient flows between ecosystems can destabilize simple food chains. *Journal of Theoretical Biology*, **266**, 162–174.

Massol, F., Gravel, D., Mouquet, N., et al. (2011). Linking community and ecosystem dynamics through spatial ecology. *Ecology Letters*, **14**, 313–323.

McCann, K. S. and Rooney, N. (2009). The more food webs change, the more they stay the same. *Philosophical Transactions of the Royal Society of London B*, **364**, 1789–1801.

McCann, K. S., Rasmussen, J. B. and Umbanhowar, J. (2005). The dynamics of spatially coupled food webs. *Ecology Letters*, **8**, 513–523.

McCoy, M. W., Barfield, M. and Holt, R. D. (2009). Predator shadows: complex life histories as generators of spatially patterned indirect interactions across ecosystems. *Oikos*, **118**, 87–100.

McIntyre, P. B., Flecker, A. S., Vanni, M. J., et al. (2008). Fish distributions and nutrient cycling in streams: can fish create biogeochemical hotspots? *Ecology*, **89**, 2335–2346.

McQueen, D. J., Post, J. R. and Mills, E. L. (1986). Trophic relationships in fresh-water pelagic ecosystems. *Canadian Journal of Fisheries and Aquatic Sciences*, **43**, 1571–1581.

Menge, B. A. and Sutherland, J. P. (1976). Species-diversity gradients – synthesis of roles of predation, competition, and temporal heterogeneity. *American Naturalist*, **110**, 351–369.

Moore, J. C., Berlow, E. L., Coleman, D. C., et al. (2004). Detritus, trophic dynamics and biodiversity. *Ecology Letters*, **7**, 584–600.

Moore, J. W., Schindler, D. E., Carter, J. L., et al. (2007). Biotic control of stream fluxes: Spawning salmon drive nutrient and matter export. *Ecology*, **88**, 1278–1291.

Murakami, M. and Nakano, S. (2002). Indirect effects of aquatic insect emergence on a terrestrial insect population through predation by birds. *Ecology Letters*, **5**, 333–337.

Myers, R. A., Baum, J. K., Shepherd, T. D., Powers, S. P. and Peterson, C. H. (2007). Cascading effects of the loss of apex predatory sharks from a coastal ocean. *Science*, **315**, 1846–1850.

Nakano, S., Miyasaka, H. and Kuhara, N. (1999). Terrestrial-aquatic linkages: riparian arthropod inputs alter trophic cascades in a stream food web. *Ecology*, **80**, 2435–2441.

Neubert, M. G. and Caswell, H. (1997). Alternatives to resilience for measuring the responses of ecological systems to perturbations. *Ecology*, **78**, 653–665.

Neubert, M. G., Caswell, H. and Solow, A. R. (2009). Detecting reactivity. *Ecology*, **90**, 2683–2688.

Northcote, T. G. (1988). Fish in the structure and function of freshwater ecosystems: a "top-down" view. *Canadian Journal of Fisheries and Aquatic Sciences*, **45**, 361–379.

Oksanen, L., Fretwell, S. D., Arruda, J. and Niemela, P. (1981). Exploitation ecosystems in gradients of primary productivity. *American Naturalist*, **118**, 240–261.

Oksanen, L. and Oksanen, T. (2000). The logic and realism of the hypothesis of exploitation ecosystems. *American Naturalist*, **155**, 703–723.

Pace, M. L., Cole, J. J., Carpenter, S. R. and Kitchell, J. F. (1999). Trophic cascades revealed in diverse ecosystems. *Trends in Ecology and Evolution*, **14**, 483–488.

Persson, L. (1999). Trophic cascades: abiding heterogeneity and the trophic level concept at the end of the road. *Oikos*, **85**, 385–397.

Polis, G. A. (1999). Why are parts of the world green? Multiple factors control productivity and the distribution of biomass. *Oikos*, **86**, 3–15.

Polis, G. A., Anderson, W. B. and Holt, R. D. (1997). Toward an integration of landscape and food web ecology: The dynamics of spatially subsidized food webs. *Annual Review of Ecology and Systematics*, **28**, 289–316.

Polis, G. A. and Hurd, S. D. (1995). Extraordinarily high spider densities on islands: flow of energy from the marine to terrestrial food webs and the absence of predation. *Proceedings of the National Academy of Sciences of the USA*, **92**, 4382–4386.

Polis, G. A. and Strong, D. R. (1996). Food web complexity and community dynamics. *American Naturalist*, **147**, 813–846.

Polishchuk, L. V., Vijverberg, J., Voronov, D. A. and Mooij, W. M. (2013). How to measure top-down vs bottom-up effects: a new population metric and its calibration on Daphnia. *Oikos*, **122**, 1177–1186.

Rip, J. M. K. and McCann, K. S. (2011). Cross-ecosystem differences in stability and the principle of energy flux. *Ecology Letters*, **14**, 733–740.

Ripple, W. J. and Beschta, R. L. (2005). Linking wolves and plants: Aldo Leopold on trophic cascades. *Bioscience*, **55**, 613–621.

Rooney, N., McCann, K., Gellner, G. and Moore, J. C. (2006). Structural asymmetry and the stability of diverse food webs. *Nature*, **442**, 265–269.

Rose, M. D. and Polis, G. A. (1998). The distribution and abundance of coyotes: the effects of allochthonous food subsidies from the sea. *Ecology*, **79**, 998–1007.

Rosemond, A. D., Mulholland, P. J. and Elwood, J. W. (1993). Top-down and bottom-up control of stream periphyton: effects of nutrients and herbivores. *Ecology*, **74**, 1264–1280.

Rosenzweig, M. l. (1971). Paradox of enrichment – destabilization of exploitation ecosystems in ecological time. *Science*, **171**, 385–387.

Schindler, D. E. and Scheuerell, M. D. (2002). Habitat coupling in lake ecosystems. *Oikos*, **98**, 177–189.

Schmitz, O. J. (2010). *Resolving Ecosystem Complexity*. Princeton, NJ: Princeton University Press.

Schmitz, O. J., Hamback, P. A. and Beckerman, A. P. (2000). Trophic cascades in terrestrial systems: a review of the effects of carnivore removals on plants. *American Naturalist*, **155**, 141–153.

Schmitz, O. J., Hawlena, D. and Trussell, G. C. (2010). Predator control of ecosystem nutrient dynamics. *Ecology Letters*, **13**, 1199–1209.

Schmitz, O. J., Krivan, V. and Ovadia, O. (2004). Trophic cascades: the primacy of trait-mediated indirect interactions. *Ecology Letters*, **7**, 153–163.

Schreiber, S. and Rudolf, V. H. W. (2008). Crossing habitat boundaries: coupling dynamics of ecosystems through complex life cycles. *Ecology Letters*, **11**, 576–587.

Shurin, J. B., Borer, E. T., Seabloom, E. W., et al. (2002). A cross-ecosystem comparison of the strength of trophic cascades. *Ecology Letters*, **5**, 785–791.

Shurin, J. B., Gruner, D. S. and Hillebrand, H. (2006). All wet or dried up? Real differences between aquatic and terrestrial food webs. *Proceedings of the Royal Society B: Biological Sciences*, **273**, 1–9.

Shurin, J. B. and Seabloom, E. W. (2005). The strength of trophic cascades across ecosystems: predictions from allometry and energetics. *Journal of Animal Ecology*, **74**, 1029–1038.

Sterner, R. W. and Elser, J. J. (2002). *Ecological Stoichiometry: The Biology of Elements From Molecules to the Biosphere*. Princeton, NJ: Princeton University Press.

Stibor, H., Vadstein, O., Diehl, S., et al. (2004). Copepods act as a switch between alternative trophic cascades in marine pelagic food webs. *Ecology Letters*, **7**, 321–328.

Strong, D. R. (1992). Are trophic cascades all wet – differentiation and donor-control in speciose ecosystems. *Ecology*, **73**, 747–754.

Takimoto, G., Iwata, T. and Murakami, M. (2002). Seasonal subsidy stabilizes food web dynamics: balance in a heterogeneous landscape. *Ecological Research*, **17**, 433–439.

Takimoto, G., Iwata, T. and Murakami, M. (2009). Timescale hierarchy determines the indirect effects of fluctuating subsidy inputs on in situ resources. *American Naturalist*, **173**, 200–211.

Vander Zanden, M. J. and Vadeboncoeur, Y. (2002). Fishes as integrators of benthic and pelagic food webs in lakes. *Ecology*, **83**, 2152–2161.

Vanni, M. J. (2002). Nutrient cycling by animals in freshwater ecosystems. *Annual Review of Ecology and Systematics*, **33**, 341–370.

Vannote, R. L., Minshall, G. W., Cummins, K. W., Sedell, J. R. and Cushing, C. E. (1980). The river continuum concept. *Canadian Journal of Fisheries and Aquatic Sciences*, **37**, 130–137.

Verspoor, J. J., Braun, D. C. and Reynolds, J. D. (2010). Quantitative links between Pacific salmon and stream periphyton. *Ecosystems*, **13**, 1020–1034.

Whalen, M. A., Duffy, J. E. and Grace, J. B. (2013). Temporal shifts in top-down vs bottom-up control of epiphytic algae in a seagrass ecosystem. *Ecology*, **94**, 510–520.

White, T. C. R. (1978). Importance of a relative shortage of food in animal ecology. *Oecologia*, **33**, 71–86.

Wollrab, S., Diehl, S. and De Roos, A. M. (2012). Simple rules describe bottom-up and top-down control in food webs with alternative energy pathways. *Ecology Letters*, **15**, 935–946.

Yang, L. H. (2004). Periodical cicadas as resource pulses in North American forests. *Science*, **306**, 1565–1567.

Yang, L. H., Edwards, K., Byrnes, J. E., et al. (2010). A meta-analysis of resource pulse-consumer interactions. *Ecological Monographs*, **80**, 125–151.

PART II

Ecosystems

CHAPTER TWO

The spatio-temporal dynamics of trophic control in large marine ecosystems

KENNETH T. FRANK

Bedford Institute of Oceanography, Dartmouth, Nova Scotia, Canada

JONATHAN A. D. FISHER

Fisheries and Marine Institute of Memorial University of Newfoundland, St. John's, Newfoundland, Canada

and

WILLIAM C. LEGGETT

Queen's University, Kingston, Ontario, Canada

Introduction

The ways in which productivity, stability, population interactions, and community structure are regulated in ecosystems have been a central focus of ecology for over a century. At large spatial scales, major insights into these dynamics have been principally derived from analyses of changes induced from hunting, harvesting, and agricultural practices – so-called "natural experiments." In terrestrial ecosystems estimates of the fraction of land transformed or degraded by human activity fall within the range of 39 to 75% (Vitousek et al., 1997; Ellis et al., 2010). Equally profound is the reality that up to 75% of the global oceans and in particular the continental shelf, transitional slope water areas, and reef habitats have been strongly impacted by human activity (Halpern et al., 2008).

One of the most widely studied human impacts has been the over-exploitation of large-bodied species. Berger et al. (2001) estimated that the spatial distribution of large mammalian carnivores that once played a dominant role in terrestrial ecosystems has declined by 95–99%. In the global oceans large predatory fish biomass may be as low as 10% of pre-industrial levels (Myers and Worm, 2003). These changes have created a vertical compaction and blunting of the trophic pyramid (Duffy, 2003; Chapter 14, this volume). On a global scale, these losses are attributable to a positive association between body size and sensitivity to

Trophic Ecology: Bottom-Up and Top-Down Interactions across Aquatic and Terrestrial Systems, eds T. C. Hanley and K. J. La Pierre. Published by Cambridge University Press. © Cambridge University Press 2015.

population declines experienced by larger species which exhibit a greater susceptibility to decline or collapse as a consequence of their lower population densities, greater times to maturity, lower clutch sizes, and larger home ranges (Schipper et al., 2008). This reduction in the abundance of apex predators has led to abnormally high densities of their former prey in a wide range of ecosystems, which has, in turn, resulted in sometimes catastrophic changes in the ecosystems occupied. This has led some to conclude that large-bodied species are essential to the maintenance of ecosystem structure and stability (Hildrew et al., 2007; Estes et al., 2011).

Intense size-selective exploitation has also resulted in declines in the average body sizes of many terrestrial mammals and aquatic vertebrates (Darimont et al., 2009; Fisher et al., 2010a), which have major impacts on ecological processes (Strong and Frank, 2010). Because large-bodied individuals serve very different functional roles in ecosystems relative to their smaller-bodied counterparts, these intra-specific changes in body size can also produce trophic restructuring (Shackell et al., 2010; Estes et al., 2011).

Hairston, Smith, and Slobodkin (1960) were the first to formally establish the conceptual foundation for top-down control of ecosystem dynamics. They (HSS) concluded that in terrestrial ecosystems herbivore abundances were controlled by predation, thereby preventing over-grazing of the food supply. Their "green world" hypothesis refocused research on top-down interactions (predation and competition) as regulators of the structure and function of terrestrial ecosystems. This resulted in a voluminous body of new knowledge highlighting the importance of apex predators to the structure and stability of ecosystems ranging from deserts to tropical forests and, ultimately, to freshwater systems, the rocky intertidal, kelp forests, tropical reefs, and estuaries (e.g., Carr et al., 2003; Ritchie and Johnson, 2009; Estes et al., 2011).

An important manifestation of top-down control is the extension of trophic interactions beyond predators and prey to non-adjacent trophic levels – commonly referred to as a "trophic cascade." This theory was first evaluated experimentally in the rocky intertidal and in whole-lake ecosystems by Paine (1980) and Carpenter et al. (1985), respectively. Their studies demonstrated that changes in the intensity of predation by top predators can produce alternating patterns of abundance at consecutively lower trophic levels that ultimately influence primary production and nutrient cycling (Fig. 2.1). Top-down structuring of freshwater ecosystems was widely accepted by limnologists by the end of the 1980s. However, marine ecologists have generally been reluctant to embrace the concept (Frank et al., 2005; Baum and Worm, 2009) and the ubiquity of trophic cascades there remains a hotly debated subject; several reviews and meta-analyses have attempted to define which ecosystems are susceptible to such influences and what conditions lead to their formation (Strong, 1992; Shurin et al., 2002; Baum and Worm, 2009; Salomon et al., 2010).

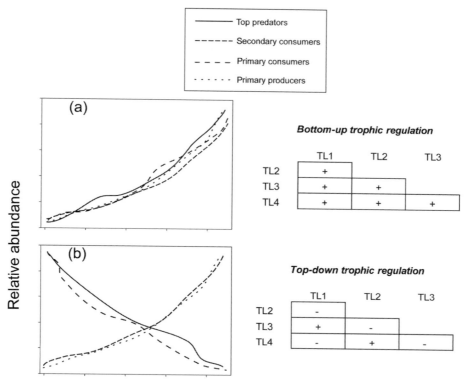

Figure 2.1 Hypothetical time series of data representative of four trophic levels (TL 1–4) illustrating (a) positive correlations among all trophic levels (indicative of bottom-up control) and (b) negative correlations between adjacent trophic levels (top-down control) and positive correlations between non-adjacent levels (indicative of a trophic cascade). Other configurations are possible depending on environmental context and the specificity of the perturbation.

The presumed primacy of bottom-up control

Despite the long-standing empirical evidence of top-down alterations to ecosystems, the belief that bottom-up processes (resource limitation; Fig. 2.1a) ultimately regulate marine ecosystem structure and function has deep and powerful roots. Hensen (1887) was among the first to view bottom-up processes in marine systems in the light of agricultural models that related basal resource inputs to the yields at higher trophic levels, a perspective that dominated marine research for a century (see Verity, 1998). Following Hensen's lead, Elton (1927) concluded that the "whole structure and activities of communities are dependent upon questions of food supply." White (1978) argued that "predators that prevent or cause substantial decreases in the abundance of their herbivorous prey are the exception rather than the rule" and that environmental factors are dominant in controlling herbivore population dynamics. Even more recently, Sinclair and

Krebs (2002) argued that bottom-up control must regulate all populations and that regulation through resources must be the "universal primary standard." This perspective has been reinforced by the rich tradition of marine research into the dynamics of physical and chemical conditions and their impact on lower trophic levels (i.e., phytoplankton, zooplankton), which has often ignored interactions with higher trophic levels (Franks, 2002). This bottom-up focus has garnered support from comparative cross-system analyses of multiple and diverse marine ecosystems that have revealed positive relationships between primary production and fish production (Nixon, 1988; Ware and Thompson, 2005; Chassot et al., 2007), between primary production and macrobenthic biomass (Kemp et al., 2005; Johnson et al., 2007), and between chlorophyll a and mesozooplankton (Finenko et al., 2003). Underlying this conclusion is the assumption that measures of biomass from different, often widely dispersed, ecosystems, and derived from singular snapshots or time-averaged data series, which are frequently non-contemporaneous, approximate the equilibrium characteristic (e.g., carrying capacity) of the systems studied, an assumption that has been challenged by Frank et al. (2006).

Several efforts to circumvent the above assumption have been made. Richardson and Schoeman (2004), who analyzed time series data from the Continuous Plankton Recorder program, reported positive correlations between annual estimates of phytoplankton and herbivorous zooplankton around the British Isles from 1958 to 2002, which they interpreted as evidence of bottom-up control (Fig. 2.1a). Time series analyses conducted in diverse ecosystems have also revealed several positive relationships between indices of primary or secondary production and the recruitment or biomass of adult fish (Oregon Continental Shelf: Peterson and Schwing, 2003; Peruvian Shelf: Ayón et al., 2004; Faroe Shelf: Eliasen et al., 2011; North Sea: Frederiksen et al., 2006). Drinkwater (2006) employed retrospective reconstructions of events in large marine ecosystems prior to the onset of large-scale industrial fisheries that led him to conclude that bottom-up forcing was the dominant control mechanism in the sub-arctic seas around Greenland, Iceland, and Norway during the 1920s and 1930s. Similarly, Bailey et al. (2006) analyzed long-term monitoring data for the unexploited abyssal northeast Pacific and concluded that bentho-pelagic fish abundance in that system was regulated from the base of the food chain.

Human impacts reveal top-down structure of marine food webs

Attempts to incorporate higher trophic level effects into large-scale models of marine ecosystems have been rendered difficult by challenges associated with adequate sampling and by the longer lifespans and complex behaviors of higher trophic level species relative to those at lower trophic levels (deYoung et al., 2004). These challenges are now being overcome by linking lower trophic level models (driven primarily by climate) to higher trophic level models (driven principally by exploitation) via the common currency of predation (Cury et al., 2008).

These "end-to-end" models are now yielding an improved understanding of the interrelationships between climate forcing, nutrient dynamics, production at the lower and upper food chain levels, and their compositional variability (Cury et al., 2008; Barange et al., 2010; Doney et al., 2012).

As a consequence, it is now becoming clear that while bottom-up forcing of community structure within large, relatively shallow (< 200 m) ecosystems may have dominated in the era of pre-industrial exploitation, industrial-scale exploitation has produced impacts that have resulted in global depletions and disproportionate losses of larger-bodied species and to the total collapse of many stocks (Pauly et al., 2002).

It is now estimated that the biomass of large predatory fish in the global oceans has declined to approximately 10–30% of pre-industrial levels (Christensen et al., 2003; Myers and Worm, 2003; Tremblay-Boyer et al., 2011). In the North Atlantic, 18 of the 21 populations of the once ecologically dominant cod declined by more than 90% (Myers and Worm, 2005). The magnitude of these biomass removals, their spatial extent, and the focused depletion of formerly dominant, large-bodied species makes fishing a strong potential driver of large-scale top-down changes in marine ecosystem structure and function.

Exploitation has also specifically targeted larger species and individuals within species (Hildrew et al., 2007; Fisher et al., 2010a). In most marine fish communities, body size and trophic level are strongly correlated (Romanuk et al., 2011). Thus, one might expect exploitation to shorten food chains, reduce connectivity, and weaken stability within the food web in ways that are analogous to the changes observed in terrestrial systems.

It is, however, becoming evident that excessive exploitation is not the only potential top-down driver of ecosystem structure and function. Bax (1998) has shown that even in exploited large marine ecosystems, predation removes between 2 and 35 times the total fish production removed by fishing. System energetic studies of the Georges Bank ecosystem yielded similar findings (Fogarty and Murawski, 1998). Given the scale of these influences, it is not unreasonable to expect major ecosystem responses to changes in the biomass of top predators resulting from natural as well as anthropogenic factors. The magnitude of the fishery-induced changes in the abundance and distribution of top predators that have occurred in recent decades, combined with evidence of the evolution of defensive adaptations by species occupying intermediate and lower trophic levels (e.g., vertical migrations and anatomical adaptations, Verity and Smetacek, 1996), have led several investigators to now argue that top-down interactions are not only central to the structuring and dynamics of large marine ecosystems, but also that trophic cascades (Fig. 2.1b) can and do exist in these large marine ecosystems as they do in terrestrial and freshwater systems. Unfortunately, this growing body of evidence of their existence has been largely ignored or discounted due, in part, to an overly strict adherence to the bottom-up model of ecosystem structuring.

Continuing doubts regarding top-down structuring and trophic cascades in marine systems

Notwithstanding the growing body of evidence in support of top-down control of large marine ecosystems (Frank et al., 2005; Baum and Worm, 2009; Strong and Frank, 2010), many researchers are unwilling to seriously consider the hypothesis that contemporary open ocean food chains function as predicted by the theory of trophic cascades (see Gislason et al., 2000; Pershing et al., 2010). In addition, some remain hesitant to embrace ecological theory *per se* as it relates to large marine ecosystems, in part because of its perceived limited explanatory power, and in part because, to date, the largest body of observational and experimental data that support this theory is drawn from terrestrial and freshwater systems (Gislason et al., 2000). This scepticism has been underpinned by the fact that many large marine ecosystems are highly complex and are characterized by high species diversity and mobile/opportunistic predators, qualities that have led many to conclude that these ecosystems are capable of absorbing any impacts they may experience. Consistent with this belief, Steele (1998) observed that he knew of no cases in which reductions in marine fish stocks had affected the abundance of their food supply or in which major shifts in species composition induced by fishing had resulted in ecosystem effects comparable to those observed in terrestrial systems. This, and the widespread belief that this diversity and spatial heterogeneity would inhibit the development of top-down control by predators and the existence of trophic cascades (Strong, 1992), has reinforced the widespread adherence of the primacy of bottom-up control. In support of this view, McCann et al. (2005) demonstrated that spatially confined consumers exert a stronger top-down control than do more mobile, wider ranging predators that forage on multiple, more widely dispersed prey populations. Heath et al. (2013) note that this confinement effect may contribute to enhanced density dependence – effects that are widely considered to be fundamental to the development of trophic cascades. This could explain why, to date, trophic cascades have been more frequently reported in lakes than in marine systems (Shurin et al., 2002).

In a similar vein, Micheli's (1999) meta-analysis of 47 marine experiments and time series of nutrients, plankton, and fish from 20 natural marine systems led her to conclude that consumer–resource interactions do not cascade down through marine pelagic food webs. It is also possible that the impact of trophic cascades on the overall dynamics of the food web in large marine ecosystems may be relatively minor and/or transient if the top predators affected are not dominant (Polis et al., 2000; Petrie et al., 2009).

Toward a greater acceptance of top-down forcing

Marine ecosystems beyond those immediately adjacent to the shore (see Chapter 7, this volume) represent the largest habitable volume on earth. Until recently, much of the research on marine ecosystems has focused on the development of inventories of existing and new species (e.g., Mora et al., 2011) and on devising solutions to the unique challenges of sampling these habitats effectively (Webb et al., 2010). There is now a discernible movement toward the development of a more comprehensive understanding of the unique trophody-namic roles of individual species and their contribution to the functioning of ecosystems. This, in turn, has led to an increased focus on both top-down (pre-dation/competition) and bottom-up (productivity influenced by environmental factors) drivers of ecosystem dynamics and their interactions. Recent events in the Northwest Atlantic, the Benguela Current, and elsewhere in the world have also dramatically demonstrated how fishery-induced alterations in the abundance and distribution of marine top predators can significantly affect the responses of ecosystems to both bottom-up and top-down forcing (Frank et al., 2005; Doney et al., 2012).

To date, most research into the dynamics of large marine ecosystems has focused primarily on subsets of interactions at either end of marine food webs: on bottom-up "N-P-Z" models of nutrients, phytoplankton, and zooplankton (Franks, 2002) and on top-down interactions commonly involving the removal of top predators by fishing and the responses of smaller-bodied predators or their prey (Baum and Worm, 2009). Greater attention is now being given to the simultaneous examination of data from several trophic levels (Frank et al., 2005; 2011) and the coupling of physical, biogeochemical, and higher trophic level models in an effort to achieve a more complete understanding of the responses of large marine ecosystems to the interaction of bottom-up and top-down forces at multiple spatial scales (Cury et al., 2008). As a consequence, the understand-ing of marine ecosystem trophodynamics has expanded rapidly in the past decade, aided by the integration and analyses of large, long-term datasets that incorporate trait data, diet information, and oceanographic observations and by small-scale experiments completed during large-scale management interven-tions involving the establishment of protected areas and fisheries closures – a form of whole-ecosystem manipulation.

Body size and trophodynamics

It is becoming increasingly evident that a size-based perspective is essential to the achievement of a more comprehensive understanding of the dynamics of marine ecosystems (Hildrew et al., 2007; Finkel et al., 2010). This requires a tran-sition from the historical species-based assessment of potential determinants of trophic level interactions to a more holistic analysis of the ontogeny of within- and between-species size ranges and their size-dependent roles as predators, prey, and competitors.

One fundamental and critical characteristic of size-structured ecosystems is their potential to experience predator–prey reversals. This occurs when the smaller-bodied prey of once dominant top predators themselves become the dominant controlling predators of the early life stages of their former top predators, a phenomenon linked to both the dramatic relative and absolute ontogenetic changes in body size that typify most marine organisms (see Hildrew et al., 2007) and to the over-exploitation of large-bodied species – conditions that allow these smaller-bodied species to persist and to dominate the ecosystem (Darimont et al., 2009; Shackell et al., 2010; Estes et al., 2011). Such a switching of roles (i.e., the hunter becoming the hunted) can have major implications for the recruitment (Swain and Sinclair, 2000; Fauchald, 2010) and time scale for recovery of depleted predator stocks (Frank et al., 2011). In this context, Walters and Kitchell (2001) hypothesized that the reproductive success/dominance of large predatory fish in unexploited systems may result from cultivation effects, in which large predators "cultivate" the ecosystem for their young by consuming or "cropping down" smaller species that otherwise operate as competitors/predators of their early life stages. The shifting dominance between predatory species, such as cod and their pelagic prey, as occurred in the eastern Scotian Shelf ecosystem and elsewhere in the Northwest Atlantic during the early 1990s (Frank et al., 2006) and the associated changes at lower levels of the food web (Frank et al., 2005), are consistent with the loss of this cultivation effect – a mechanism that appears to reinforce and sustain alternative community states through predation on or competition with the early life stages of once dominant large predatory fish (Frank et al., 2011; Minto and Worm, 2012; Fig. 2.2).

Recognizing the unique feature of marine ecosystems in which the early life stages of a diverse array of species are of a similar size and therefore compete directly for resources and are vulnerable to a range of common predators, Schipper et al. (2008), Fisher et al. (2010b), and Barton et al. (2013) mapped the geographical distribution of functional traits of some of the smallest and largest species groups, with the goal of understanding and, ultimately, predicting the impact of trait changes on the functioning of marine systems (Finkel et al., 2010). This more holistic approach has resulted in an increased acceptance of the possibility that trophic cascades can and do develop in both continental shelf and open ocean ecosystems (Frank et al., 2006; 2007) and in tropical (Micheli et al., 2005) and extra-tropical reef ecosystems (Shears et al., 2008). Moreover, a fuller understanding of the system characteristics and the heterogeneous responses to top-down and bottom-up forcing that predispose ecosystems to cascades or, alternatively, convey resistance and/or resilience in the face of altered states, is rapidly emerging (Frank et al., 2007; Petrie et al., 2009; Salomon et al., 2010; Heath et al., 2013). For example, Frank et al. (2007) demonstrated how warmer temperatures and higher species richness conveyed resilience and resistance to perturbation in ecosystems possessing these traits. Petrie et al. (2009), building on this insight, developed an empirical model of how these physical and biodiversity features operate to convey resistance and resilience and that facilitates the prediction of the susceptibility of large marine ecosystems to the development of altered ecosystem states. Additional ecosystem characteristics,

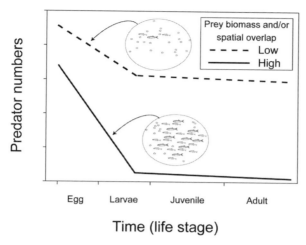

Figure 2.2 An illustration of the predator–prey reversal and its consequences for the recruitment dynamics of large-bodied predators. Under conditions of high predator biomass, prey species are maintained at low levels and the production and survival of the predator's eggs and larvae provides for high recruitment (dashed line). When predators are depleted through overfishing or other perturbations, their former prey become dominant, resulting in greater mortality induced by predation on, or competition with, the predator's eggs and larvae and, thereby, reduced recruitment levels (solid line).

including prey avoidance of predators (Solomon et al., 2010) and predator density dependence (Heath et al., 2013), have now been identified as being mediators of both bottom-up and top-down effects on marine ecosystem dynamics.

Structuring of ecosystems

Global patterns

The processes that regulate the expression of top-down structuring in marine systems and their variability are now becoming clearer. Caddy and Garibaldi (2000), utilizing data collected by the United Nation's Food and Agriculture Organization (FAO), highlighted the changing trophic composition of the world's harvest of marine fish by demonstrating that the ratio of landings of piscivores to zooplanktivores (equivalent to functional groups of predator and prey) exhibited a declining temporal trend from 1974 to 1997. These widespread changes are most notable in the North Atlantic (particularly area 21 in the Northwest Atlantic), Northeast Pacific (area 67), and eastern Indian Oceans (area 57; Fig. 2.3). In the same vein, Worm and Myers (2003), who conducted a meta-analysis of reciprocal changes in cod and shrimp abundance data across the North Atlantic, also concluded that release from predation following the collapse of cod contributed directly to dramatic increases in shrimp abundance.

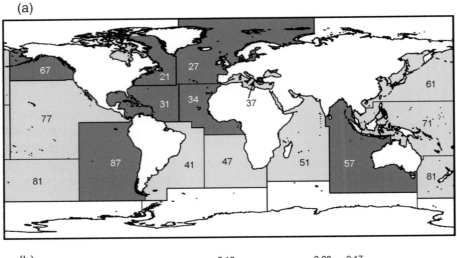

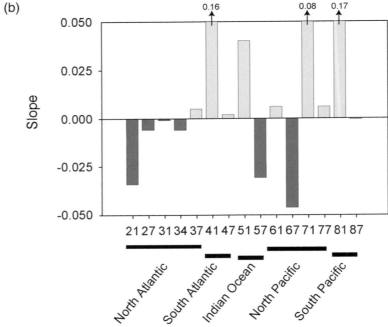

Figure 2.3 (a) Location and boundaries of 15 United Nations Food and Agriculture Organization fishing areas (excluding Arctic and Antarctic areas) spanning the global oceans. FAO member countries contribute taxonomically resolved fisheries landings data each year. Dark shaded areas are characterized by declining temporal trends in the ratio of piscivores to zooplanktivores, indicative of reciprocal trends between adjacent trophic levels and trophic unbalance; the remaining areas exhibited positive slopes, indicative of covariability between adjacent trophic levels (see (b) below for details). Piscivores were defined as those species whose primary prey items included finfish and pelagic cephalopods, while zooplanktivores were species that consumed mainly zooplankton and early live stages of fish. (b) Slope of the relationship between the ratio of piscivores to zooplanktivores versus time from each of the 15 FAO fishing areas during 1974 to 1997 (adapted from Caddy and Garibaldi, 2000).

Local patterns

Eastern Scotian Shelf

On the eastern Scotian Shelf (located within FAO area 21), analyses of fishery-independent scientific monitoring data revealed a rapid and dramatic increase in the biomass of forage fish species (namely capelin, sand lance, and herring) that followed the collapse of cod and other large-bodied piscivores (Frank et al., 2005). Bundy (2005) estimated that the biomass of forage fish increased from 2 to 20 mt per km^2 from the early 1980s to late 1990s. In turn, this appears to have fueled an exponentially growing and geographically expanding gray seal population (Fig. 2.4a). The biomass of northern shrimp (*Pandalus borealis*) and snow crab (*Chionoecetes opilio*), also former prey of cod and other large-bodied species, increased by 155% and 160%, respectively, relative to pre-collapse values (Fig. 2.4b) and resulted in the proliferation of highly lucrative commercial fisheries. Unexpectedly, cod and the other large-bodied predator species that simultaneously collapsed or experienced severe declines exhibited reduced physiological condition and poor growth, notwithstanding the increased abundance of forage fish. These changes led to the dysfunction of an entire trophic guild (Choi et al., 2004). Virtually identical changes in community structure, resulting from similar processes (i.e., predator depletion due to overfishing, prey outburst due to release from predation, and competitive release between large-bodied piscivores and seals), occurred in adjacent areas of the Northwest Atlantic (Savenkoff et al., 2007; Benoît and Swain, 2008).

The aforementioned changes in the structure of the eastern Scotian Shelf community following the collapse of the top predator guild and the resultant dominance of forage fish led to the development of a trophic cascade that involved four trophic levels and nutrients (Frank et al., 2005; Fig. 2.4). During the post-collapse period, the abundance of large-bodied herbivorous zooplankton declined while that of small-bodied zooplankton remained unchanged (Fig. 2.4c) – a result consistent with size-selective predation of zooplankton by forage fish and the early life stages of shrimp and crab, and reminiscent of the structuring effect of planktivorous fish on zooplankton community composition in lakes (Brooks and Dodson, 1965; Carpenter et al., 1985). Additionally, phytoplankton levels increased following the decline in large-bodied zooplankton and, in turn, nitrate concentration declined (Fig. 2.4d). Data limitations precluded assessment of the possibility that changes in the community of herbivorous zooplankton also regulated the species composition of phytoplankton. However, the amplitude of spring phytoplankton blooms on the eastern Scotian Shelf have been higher, and their duration longer, during the post-collapse years relative to earlier times (Head and Sameoto, 2007), a pattern consistent with a reduction in grazing pressure by herbivorous zooplankton on phytoplankton.

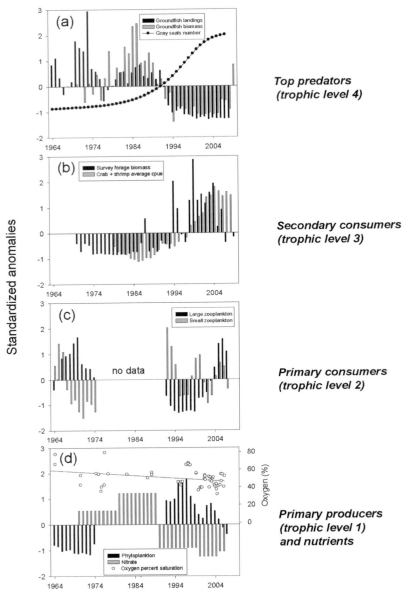

Figure 2.4 Temporal dynamics of the food chain on the eastern Scotian Shelf illustrating the development of a trophic cascade initiated by the collapse of large-bodied groundfish species, including cod (a), and dramatic increases in forage fish species (combined abundance of capelin, herring, and sandlance) and macroinvertebrates (combined catch rates of snow crab and shrimp (b)). (c) Large (> 2 mm) zooplankton were inversely correlated with forage fish biomass while there was no relationship with small (< 2 mm) zooplankton. (d) Indices of phytoplankton biomass varied inversely with large zooplankton, while nutrients decreased relative to the levels observed during the pre-collapse groundfish period. Also shown is the declining time trend of percent oxygen saturation. Note: missing data existed for the zooplankton and phytoplankton time series from 1977 to 1991; nutrient (nitrate) data were sparse so decadal averages are shown; and all time series (with the exception of oxygen saturation) are represented as standardized anomalies (units: s. d. units).

It is now well established that the size structure and abundance of zooplankton communities play a crucial role in determining the fate of primary production, the composition and sedimentation rate of sinking particles, and the flux of organic matter to the bottom of the ocean (Legendre and Michaud, 1998). On the Scotian Shelf, oxygen saturation has declined by 15% during the past 50 years (Fig. 2.4d). Current oxygen saturation in the central basin is now on the order of 40%, a stressful level given that maximum swimming speed, food ingestion, and growth of cod are reduced when dissolved oxygen is < 70% saturation (Claireaux et al., 2000). It is tempting to hypothesize that the paradox of the post-collapse-reduced physiological condition of large-bodied benthic fish predators in the face of an abundant and expanding food supply (Choi et al., 2004) resulted from an increased volume flux of ungrazed phytoplankton to the sediments and an accelerated oxygen demand associated with its decomposition, all originating from the depletion of top predators in the system.

The dramatic food chain restructuring of the eastern Scotian Shelf ecosystem was instrumental in overturning the prevailing, long-held view of the immunity of large marine ecosystems to such system level effects. The prophecy that "humans are well on their way to impacting oceans as if they were merely a very large lake" (Knowlton, 2004) had unfolded in a dramatic fashion. Principal components analyses of the empirical data series for the various trophic levels also revealed that this ecosystem had undergone a prolonged (∼ 20 year) transition to an alternate state (Choi et al., 2004; Fig. 2.5), one in which small pelagic species, once the prey of the collapsed large predator guild, assumed a more dominant position in the trophic hierarchy of the ecosystem.

Baltic Sea

The existence of trophic cascades and prolonged altered ecosystem states has now been documented in other ecosystems, particularly in inland seas, where species diversity is generally lower relative to open ocean systems (Frank et al., 2007). In the Baltic Sea, field data collected over a 33-year period revealed that the depletion of cod by fishing resulted in an explosive increase of its main prey, the zooplanktivorous clupeid *Sprattus sprattus*, a decline in the summer biomass of the once dominant zooplankton species (*Pseudocalanus acuspes*, *Acartia* spp., and *Temora longicornis*), and subsequent increases in phytoplankton biomass (Österblom et al., 2006; Casini et al., 2009). The sustained dominance of zooplanktivorous clupeids in this ecosystem and the corresponding failure of cod to recover are reflective of the strong regulation of the food resource for larval cod by sprat (Möllmann et al., 2008) and its role as an important predator of cod eggs and larvae (Köster and Möllmann, 2000) – a further example of predator–prey reversal and its role in hindering a return to the prior ecosystem state.

The depletion of Baltic Sea piscivores was also followed by declines in the fledgling body mass of the piscivorous seabird *Uria aalge*. This response is believed

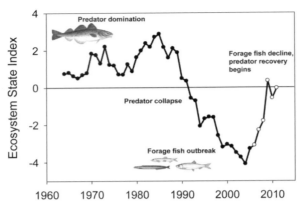

Figure 2.5 An index of the state of the eastern Scotian Shelf ecosystem based on the first mode of a principal components analysis of the biological time series data shown in Figure 2.4. Groundfish predators were dominant throughout the 1970s to mid 1980s, but their position weakened due to sustained overfishing, leading to the collapse of cod and other large-bodied species in the early 1990s. Following the collapse, forage fish abundance increased dramatically and dominated throughout the mid 1990s to early 2000s. Gradually, internal processes associated with density-dependent food limitation caused the forage fish biomass to plummet and then stabilize. In turn, this provided a window of opportunity for the recovery of their former predators. The start of the recovery phase is denoted by open circles.

to be linked to the density-dependent reduction in the energy content of individuals of the exploding sprat population (Österblom et al., 2006), a phenomenon now known to affect seabirds and marine mammals in a range of marine ecosystems (Österblom et al., 2008). Similar density-dependent reductions in body condition occurred in the forage fish complex on the eastern Scotian Shelf (Frank et al., 2011).

Österblom et al. (2006) hypothesized that top-down forces acting on plant biomass in pelagic habitats may be more important than commonly believed, a concept that contradicts the widely held view that eutrophication and changes in climate have been the primary drivers of recent increases in summer phytoplankton production and associated cyanobacterial blooms in the Baltic Sea (Finni et al., 2001). Consistent with this hypothesis, Eriksson et al. (2009) demonstrated that algal blooms in coastal areas of the Baltic Sea are linked to the composition of the fish community; piscivorous fish abundance exhibited a strong negative correlation with the large-scale distribution of bloom-forming macroalgae. Heck and Valentine (2007) reached similar conclusions following a review of the effects of overfishing of large-bodied consumers on plant biomass dynamics in coastal ecosystems, where such indirect consumer effects were as strong, or even stronger, than the reported effects of eutrophication. Thus, while eutrophication remains a major influence linking terrestrial and coastal marine food webs, the interaction between such bottom-up forces and their mediation by top-down processes is becoming increasingly recognized.

Black Sea

In the Black Sea, two major shifts in the pelagic food web occurred after 1950. The first occurred during the 1970s, when top predators, including bonito, tuna, swordfish, and bluefish, were depleted by overfishing. The second occurred in the 1990s, when the invasive ctenophore *Mnemiopsis leydii* reached extremely high levels and, given its generalist feeding mode, replaced fish as the dominant species controlling zooplankton abundances (Daskalov et al., 2007). Time series analyses of data from both periods revealed system-wide negative correlations between adjacent trophic levels (piscivores, planktivorous fish, jellyfish, zooplankton, and phytoplankton) and positive correlations between non-adjacent trophic levels (Daskalov et al., 2007), outcomes consistent with the existence of trophic cascades.

Trophic cascades are also hypothesized to favor the development of blooms of noxious phytoplankton species (Turner and Granéli, 2006), the mechanism being increased predation on herbivorous copepods by gelatinous predators, such as scyphomedusae and ctenophores, and small-bodied pelagic fish. This leads to decreased zooplankton grazing pressure and to an alteration in the species composition of phytoplankton, including a proliferation of harmful algal bloom species. Such mechanisms, identified in the Black Sea, directly link top-down alterations to lower trophic level dynamics. Separating the relative influences of trophic cascades and eutrophication on algal dynamics remains difficult, particularly so in systems that experience perturbations at both ends of the food web (Turner and Granéli, 2006).

Upwelling systems

Eastern boundary upwelling ecosystems are among the most productive in the world. They provide over 30% of the global landings of marine fish and are commonly believed to be strongly bottom-up regulated. They too have been dramatically impacted by top-down forces that result from over-exploitation of planktivorous predators (Fréon et al., 2009). The northern Benguela ecosystem off Namibia historically supported large stocks of sardines and anchovies. In the late 1970s, their annual catches declined from ~17 million tons to less than 1 million tons. In response, beginning in the early 1990s, the biomass of jellyfish (*Chyrsaora hysoscella* and *Aequorea forskalea*), supported by the increased availability of secondary production previously consumed by fish, increased dramatically (Lynman et al., 2006) and has since remained at very high levels – a change that some believe may be irreversible given that jellies are strong trophic competitors of sardines and anchovies (Lynman et al., 2006). Similar shifts in dominance from finfish to jellyfish have occurred in the Gulf of Mexico, the

Bering, Black, Caspian, Japan, and North Seas (Brodeur et al., 2008; Richardson et al., 2009).

An emerging synthesis: the importance of context

Not all open ocean ecosystems have responded to anthropogenic forcing as outlined above. In their review of top-down control in open ocean ecosystems, Baum and Worm (2009) characterized only 6 of 31 reported cascades as being true trophic cascades, while the remainder provided strong evidence only of meso-predator release (14 studies), prey declines (9 studies), and invertebrate release (2 studies) following pronounced reductions in predation by top predators.

Other North Atlantic ecosystems in which apex predator stocks have experienced fishing pressures equivalent to or greater than those known to have led to full cascades, most notably Georges Bank and the North Sea, have remained relatively resilient. These ecosystems are characterized by warmer water temperatures and relatively higher species diversity than those that have experienced trophic restructuring (Frank et al., 2007). Both population- and community-level mechanisms underlie these patterns. Myers et al.'s (1997) comparative analysis of demographic rates among 20 cod stocks across the North Atlantic revealed that population growth rates were higher and age at maturity lower in warmer water ecosystems (average growth rate = 40% per year, average age at maturity = 2–4 years) relative to colder water ($< 5\,°C$) systems (average growth rate = 18%, average age at maturity ≥ 6 years). The resulting higher reproductive potential of warmer water stocks facilitates greater resilience in the face of directed exploitation.

Warmer water ecosystems also typically exhibit greater species richness (Fisher et al., 2008). In these ecosystems, the depletion of one or a few dominant piscivore species has been shown to be readily compensated through corresponding increases in other functionally related species (Fogarty and Murawski, 1998; Reid et al., 2000; Duffy, 2002; Shackell and Frank, 2007; Gonzalez and Loreau, 2009; Laptikhovsky et al., 2013). Duffy (2002) hypothesized that this compensation is achieved by the release from competition of functionally equivalent species that results in increases in their density, population-level interaction strengths, and diet breadth.

Frank et al. (2007) examined this functional dependence in detail through a meta-analysis of more than 20 ecosystems. Their analyses confirmed that the resiliency of exploited large-bodied fish and fish assemblages that inhabited warmer, species-rich ecosystems was greater than that of those inhabiting cooler, species-poor systems, and that these warmer, species-rich ecosystems were more resistant to trophic restructuring (Fig. 2.6).

The linkage between structural resilience, local ocean conditions, and perturbation levels has several important implications: i) under warming (cooling)

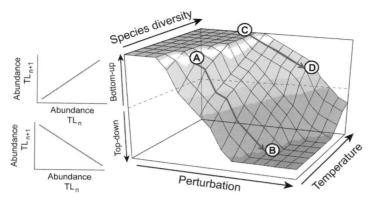

Figure 2.6 Conceptual diagram showing the tendency of trophic control to change rapidly from bottom-up to top-down with an increase in perturbation strength in colder water, less species-rich systems (A → B) in comparison to warmer water, species-rich systems (C → D). The type of trophic control is deduced from the sign of the correlation coefficient between time series of predators and prey, with positive (negative) correlations indicative of bottom-up (top-down) control (also see Fig. 2.1).

conditions, changes in trophic structure would be expected to become increasingly (decreasingly) resilient to a given level of perturbation (Fig. 2.6), (ii) warmer water, species-rich ecosystems can sustain much greater perturbation at higher trophic levels relative to those in colder water, less species-rich ecosystems, and (iii) during intervals typified by low or decreasing (high or increasing) levels of perturbation acting on large-bodied predators, a tendency toward positive (negative) covariability between adjacent trophic levels is to be expected (Fig. 2.6).

More than four decades of surveys of the Barents Sea and North Sea ecosystems provide a test of these expectations. In the Barents Sea, during the last decade which was characterized by the warmest temperature conditions on record, trophic control has shifted from top-down, which had persisted for the previous two decades, to bottom-up (Johannesen et al., 2012) – a result consistent with our first expectation. In the North Sea, temporal changes in the strength and sign of the relationship between herbivorous zooplankton and phytoplankton differed between two regimes (Llope et al., 2012): positive correlations typified the first regime (1958–1972) when fishing pressure was at a medium level, while during the second regime (1973–2004), correlations in many areas became negative as fishing pressure more than doubled (Daan et al., 2005) – a result consistent with our third expectation.

Management implications of trophic cascades

The development of trophic cascades and altered ecosystem states in large marine ecosystems can have serious biological and commercial consequences and long-lasting negative effects on the overall stability and productive capacity of the affected system. It is now clear that altered ecosystems do not respond well to conventional management approaches. For example, the longstanding scientific theory of stock and recruitment, which predicts recruitment rates should increase at low population levels (Hilborn and Walters, 1992), proved incompatible with the failed recovery of the top predators of the eastern Scotian Shelf and other Northwest Atlantic ecosystems. Culling of seals, the fishing down of small pelagics, stocking, and other approaches to restoring altered ecosystems have been proposed and their relative merits aggressively debated for decades (Yodzis, 2001; Bakun et al., 2009; Gerber et al., 2009). However, in the absence of an understanding of the fundamental dynamics of the undisturbed and altered ecosystem states, the potential for the expression of other equally undesirable states is real. For example, on the eastern Scotian Shelf, where seal population growth has closely tracked the increase in small pelagic fish biomass (Fig. 2.4), proposals are now being advanced to reduce seal abundance by culling to diminish their predation on cod and to facilitate the recovery of cod (Manning, 2012). However, culling also has the potential to reduce seal predation on the already large biomass of small pelagic fish, thereby strengthening their current dominant position in that food web (Swain and Sinclair, 2000) and further delaying the recovery of cod and other once dominant predators (Frank et al., 2011). In the future, if sustainable management of large marine ecosystems is to be achieved, it will be necessary to adjust harvesting strategies to account for the rapidly evolving knowledge of the complex bottom-up/top-down dynamics that regulate the structure and stability of these systems and their association with local environmental conditions (e.g., Petrie et al., 2009). In the meantime, until the requisite understanding of the linkages between top-down and bottom-up processes is achieved, a blend of traditional management measures and the establishment of a network of no-take zones (as implemented on Georges Bank and elsewhere) coupled with a lessened dependency on natural production systems (Duarte et al., 2009) may prove the best path forward. To date, and unfortunately, a rapid, global proliferation of fisheries focused on newly emerging (often compensating) species has been the rule rather than the exception (Anderson et al., 2008).

The good news is that the development and persistence of trophic cascades and altered ecosystem states in large marine ecosystems appears to be amenable to natural reversals, provided the major source of the perturbation has been reduced or eliminated. Based on the eastern Scotian Shelf experience, the time required for a return to the primary state appears to be functionally dependent on the time required for the irruptive/invasive species to reach and exceed their carrying capacity, to undergo a natural density-dependent decline, and then to stabilize. In that instance, this process required almost 15 years (Figs. 2.1 and 2.4; Frank et al., 2011; Swain and Mohn, 2012).

While the current trend is positive, complete recovery of the eastern Scotian Shelf ecosystem has yet to occur (Fig. 2.5), and interestingly and perhaps alarmingly, haddock now dominate the recovering biomass of the top predator complex rather than cod (Frank et al., 2011). Ultimately, a key goal of research in this area must be a clearer definition of the thresholds of the perturbation levels that any large marine ecosystem or group of ecosystems can withstand without breaching the critical point at which the development of an altered state occurs (Fig. 2.6).

Acknowledgments

This research was supported in part by the Natural Sciences and Engineering Research Council of Canada Discovery Grants program to KTF, JADF, and WCL and by the Department of Fisheries and Oceans, Canada. The Centre for Fisheries Ecosystem Research is supported by the Government of Newfoundland and Labrador and the Research and Development Corporation of Newfoundland and Labrador.

References

Anderson, S. C., Lotze, H. K. and Shackell, N. L. (2008). Evaluating the knowledge base for expanding low-trophic-level fisheries in Atlantic Canada. *Canadian Journal of Fisheries and Aquatic Sciences*, **65**, 2553–2571.

Ayón, P., Purca, S. and Guevara-Carrasco, R. (2004). Zooplankton volume trends off Peru between 1964 and 2001. *ICES Journal of Marine Science*, **61**, 478–484.

Bailey, D. M., Ruhl, H. A. and Snith, K. L. (2006). Long-term change in benthopelagic fish abundance in the abyssal northeast Pacific Ocean. *Ecology*, **87**(3), 549–555.

Bakun, A., Babcock, A. and Santora, C. (2009). Regulating a complex adaptive system via its wasp-waist: grappling with ecosystem-based management of the New England herring fishery. *ICES Journal of Marine Science*, **66**, 1768–1775.

Barange, M., Field, J. G., Harris, R. P., Hofmann, E. E. and Perry, R. I. (eds.) (2010). *Marine Ecosystems and Global Change*. Oxford: Oxford University Press.

Barton, A. D., Pershing, A. J., Litchman, E., et al. (2013). The biogeography of marine plankton traits. *Ecology Letters*, **16**, 522–534.

Baum, J. K. and Worm, B. (2009). Cascading top-down effects of changing oceanic predator abundances. *Journal of Animal Ecology*, **78**, 699–714.

Bax, N. J. (1998). The significance and prediction of predation in marine fisheries. *ICES Journal of Marine Science*, **55**, 997–1030.

Benoît, H. P. and Swain, D. P. (2008). Impacts of environmental change and direct and indirect harvesting effects on the dynamics of a marine fish community. *Canadian Journal of Fisheries and Aquatic Sciences*, **65**, 2088–2104.

Berger, J., Stacey, P. B., Bellis, L. and Johnson, M. P. (2001). A mammalian predator-prey imbalance: grizzly bear and wolf extinction affect avian neotropical migrants. *Ecological Applications*, **11**, 947–960.

Brodeur, R. D., Suchman, C. L., Reese, D. C., Miller, T. W. and Daly, E. A. (2008). Spatial overlap and trophic interactions between pelagic fish and large jellyfish in the northern California Current. *Marine Biology*, **154**, 649–659.

Brooks, J. L. and Dodson, S. I. (1965). Predation, body size, and composition of plankton. *Science*, **150**, 28–35.

Bundy, A. (2005). Structure and functioning of the eastern Scotian Shelf ecosystem before and after the collapse of groundfish stocks in the early 1990s. *Canadian Journal of Fisheries and Aquatic Sciences*, **62**(7), 1453–1473.

Caddy, J. F. and Garibaldi, L. (2000). Apparent changes in the trophic composition of world marine harvests: the perspective from the FAO capture database. *Ocean & Coastal Management*, **43**, 615–655.

Carpenter, S. R., Kitchell, J. F. and Hodgson, J. R. (1985). Cascading trophic interactions and lake productivity. *Bioscience*, **35**, 634–649.

Carr, M. H., Neigel. J. E., Estes, J. A., et al. (2003). Comparing marine and terrestrial ecosystems: implications for the design of coastal marine reserves. *Ecological Applications*, **13**, S90–S107.

Casini, M., Hjelm, J., Molinero, J.-C., et al. (2009). Trophic cascades promote threshold-like shifts in pelagic marine ecosystems. *Proceedings of the National Academy of Sciences of the USA*, **106**, 197–202.

Chassot, E., Mélin, F., Le Pape, O. and Gascuel, D. (2007). Bottom-up control regulates fisheries production at the scale of eco-regions in European seas. *Marine Ecology Progress Series*, **343**, 45–55.

Choi, J. S., Frank, K. T., Leggett, W. C. and Drinkwater, K. (2004). Transition to an alternate state in a continental shelf ecosystem. *Canadian Journal of Fisheries and Aquatic Sciences*, **61**, 505–510.

Christensen, V., Guenette, S., Heymans, J. J., et al. (2003). Hundred-year decline of North Atlantic predatory fishes. *Fish and Fisheries*, **4**, 1–24.

Claireaux, G., Webber, D. M., Lagardère, J.-P. and Kerr, S. R. (2000). Influence of water temperature and oxygenation on the aerobic metabolic scope of Atlantic cod (*Gadus morhua*). *Journal of Sea Research*, **44**, 257–265.

Cury, P. M., Shin, Y.-J., Planque, B., et al. (2008). Ecosystem oceanography for global change in fisheries. *Trends in Ecology and Evolution*, **23**(6), 338–346.

Daan, N., Gislason, H., Pope, J. G. and Rice, J. C. (2005). Changes in the North Sea fish community: evidence of indirect effects of fishing? *ICES Journal of Marine Science*, **62**, 177–188.

Darimont, C. T., Carlson, S. M., Kinnison, M. T., et al. (2009). Human predators outpace other agents of trait change in the wild. *Proceedings of the National Academy of Sciences of the USA*, **106**, 952–954.

Daskalov, G. M., Grishin, A. N., Rodionov, S. and Mihneva, V. (2007). Trophic cascades triggered by overfishing reveal possible mechanisms of ecosystem regime shifts. *Proceedings of the National Academy of Sciences of the USA*, **104**, 10518–10523.

deYoung, B., Heath, M., Werner, F., et al. (2004). Challenges of modelling ocean basin ecosystems. *Science*, **304**, 1463–1466.

Doney, S. C., Ruckelshaus, M., Duffy, J. E., et al. (2012). Climate change impacts on marine ecosystems. *Annual Reviews of Marine Science*, **4**, 11–37.

Drinkwater, K. F. (2006). The regime shift of the 1920s and 1930s in the North Atlantic. *Progress in Oceanography*, **68**, 134–151.

Duarte, C. M., Holmer, M., Olsen, Y., et al. (2009). Will the oceans help feed humanity? *BioScience*, **59**, 967–976.

Duffy, J. E. (2002). Biodiversity and ecosystem function: the consumer connection. *Oikos*, **99**, 201–219.

Duffy, J. E. (2003). Biodiversity loss, trophic skew and ecosystem functioning. *Ecology Letters*, **6**, 680–687.

Eliasen, K., Reinert, J., Gaard, E., et al. (2011). Sandeel as a link between primary production and higher trophic levels on the Faroe shelf. *Marine Ecology Progress Series*, **438**, 185–194.

Ellis, E. C., Goldewijk, K. K., Siebert, S., Lightman, D. and Ramankutty, N. (2010). Anthropogenic transformation of the biomes, 1700 to 2000. *Global Ecology and Biogeography*, **19**, 589–606.

Elton, C. (1927). *Animal Ecology*. London: Sidwick and Jackson.

Eriksson, B. K., Lunggren, L., Sandstrom, A., et al. (2009). Declines in predatory fish promote bloom-forming macroalgae. *Ecological Applications*, **19**, 1975–1988.

Estes, J. A., Terborgh, J., Brashares, J. S., et al. (2011). Trophic downgrading of planet Earth. *Science*, **333**, 301–306.

Fauchald, P. (2010). Predator-prey reversal: a possible mechanism for ecosystem hysteresis in the North Sea. *Ecology*, **91**, 2191–2197.

Finenko, Z. Z., Piontkovski, S. A., Williams, R. and Mishonov, A. V. (2003). Variability of phytoplankton and mesozooplankton biomass in the subtropical and tropical Atlantic Ocean. *Marine Ecology Progress Series*, **250**, 125–144.

Finkel, Z. V., Beardall, J., Flynn, K. J., et al. (2010). Phytoplankton in a changing world: cell size and elemental stoichiometry. *Journal of Plankton Research*, **32**, 119–137.

Finni, T., Kononen, K., Olsonen, R. and Wallström, K. (2001). The history of cyanobacterial blooms in the Baltic Sea. *Ambio* **30**, 172–178.

Fisher, J. A. D., Frank, K. T., Petrie, B., Leggett, W. C. and Shackell, N. L. (2008). Temporal dynamics within a contemporary latitudinal diversity gradient. *Ecology Letters*, **11**, 883–897.

Fisher, J. A. D., Frank, K. T. and Leggett, W. C. (2010a). Breaking Bergmann's rule: truncation of Northwest Atlantic marine fish body sizes. *Ecology*, **91**, 2499–2505.

Fisher, J. A. D., Frank, K. T. and Leggett, W. C. (2010b). Global variation in marine fish body size and its role in biodiversity-ecosystem functioning. *Marine Ecology Progress Series*, **405**, 1–13.

Fogarty, M. J. and Murawski, S. A. (1998). Large-scale disturbance and the structure of marine systems: fishery impacts on Georges Bank. *Ecological Applications*, **8**: S6–S22.

Frank, K. T., Petrie, B., Choi, J. S. and Leggett, W. C. (2005). Trophic cascades in a formerly cod-dominated ecosystem. *Science*, **308**, 1621–1623.

Frank, K. T., Petrie, B., Shackell, N. L. and Choi, J. S. (2006). Reconciling differences in trophic control in mid-latitude marine ecosystems. *Ecology Letters*, **9**, 1096–1105.

Frank, K. T., Petrie, B. and Shackell, N. L. (2007). The ups and downs of trophic control in continental shelf ecosystems. *Trends in Ecology and Evolution*, **22**(5), 236–242.

Frank, K. T., Petrie, B., Fisher, J. A. D. and Leggett, W. C. (2011). Transient dynamics of an altered large marine ecosystem. *Nature*, **477**, 86–89.

Franks, P. J. (2002). NPZ models of plankton dynamics: their construction, coupling to physics, and application. *Journal of Oceanography*, **58**, 379–387.

Frederiksen, M., Edwards, M., Richardson, A. J., Halliday, N. C. and Wanless, S. (2006). From plankton to top predators: bottom-up control of a marine food web across four trophic levels. *Journal of Animal Ecology*, **75**, 1259–1268.

Fréon, P., Barange, M. Arístegui, J. and McIntyre, A. D. (2009). Eastern boundary upwelling ecosystems: integrative and comparative approaches. *Progress in Oceanography*, **83**, 1–14.

Gerber, L. R., Morissette, L. and Pauly, D. (2009). Should whales be culled to increase fishery yield? *Science*, **323**, 880–881.

Gislason, H., Sinclair, M., Sainsbury, K. and O'Boyle, R. N. (2000). Symposium overview: incorporating ecosystem objectives within fisheries management. *ICES Journal of Marine Science*, **57**, 468–475.

Gonzalez, A. and Loreau, M. (2009). The causes and consequences of compensatory dynamics in ecological communities. *Annual Review of Ecology Evolution and Systematics*, **40**, 393–414.

Hairston, N. G., Smith, F. E. and Slobodkin, L. B. (1960). Community structure, population control, and competition. *American Naturalist*, **94**, 421–425.

Halpern, B. S., Walbridge, S., Selkoe, K. A., et al. (2008). A global map of human impact on marine ecosystems. *Science*, **319**, 948–952.

Head, E. J. H. and Sameoto, D. D. (2007). Inter-decadal variability in zooplankton and phytoplankton abundance on the Newfoundland and Scotian shelves. *Deep-Sea Research II*, **57**, 2686–2701.

Heath, M. R., Speirs, D. C. and Steele, J. H. (2013). Understanding patterns and processes in models of trophic cascades. *Ecology Letters* doi:10.1111/ele.12200

Heck, K. L. Jr. and Valentine, J. F. (2007). The primacy of top-down effects in shallow benthic ecosystems. *Estuaries and Coasts*, **30**, 371–381.

Hensen, V. (1887). *Ueber die Bestimmung des Plankton's oder des im Meere treibenden Materials an Pflanzen und Tieren*. Kommission zur wiss. Untersuchung der deutschen Meere, in Kiel, 1882–1886, Bericht 5, Vols. 12–16, pp. 1–107. Schmidt and Klaunig.

Hilborn, R. and Walters, C. J. (1992). *Quantitative Fisheries Stock Assessment: Choice, Dynamics and Uncertainty*. Norwell, MA: Kluwer Academic Publishers.

Hildrew, A., Raffaelli, D. and Edmonds-Brown, R. (2007). *Body Size: The Structure and Function of Aquatic Ecosystems*. Cambridge, UK: Cambridge University Press.

Johannesen, E., Ingvaldsen, R. B., Bogstad, B., et al. (2012). Changes in Barents Sea ecosystem state, 1970–2009: climate fluctuations, human impact, and trophic interactions. *ICES Journal of Marine Science*, **69**(5), 880–889.

Johnson, N. A., Campbell, J. W., Moore, T. S., et al. (2007). The relationship between the standing stock of the deep-sea macrobenthos and surface production in the western North Atlantic. *Deep-Sea Research I*, **54**, 1350–1360.

Kemp, W. M., Boynton, W. R., Adolf, J. E., et al. (2005). Eutrophication of Chesapeake Bay: historical trends and ecological interactions. *Maine Ecology Progress Series*, **303**, 1–29.

Knowlton, N. (2004). Multiple stable states and the conservation of marine ecosystems. *Progress in Oceanography*, **60**, 387–396.

Köster, F. W. and Möllmann, C. (2000). Trophodynamic control by clupeid predators on recruitment success in Baltic cod? *ICES Journal of Marine Science*, **57**, 310–323.

Laptikhovsky, V., Arkhipkin, A. and Brickle, P. (2013). From small bycatch to main commercial species: explosion of stocks of rock cod *Patagonotothen ramsayi* (Regan) in the southwest Atlantic. *Fisheries Research*, **147**, 399–403.

Legendre, L. and Michaud, J. (1998). Flux of biogenic carbon in oceans: size-dependent regulation by pelagic food webs. *Marine Ecology Progress Series*, **164**, 1–11.

Llope, M., Licandro, P., Chan, K.-S. and Stenseth, N. C. (2012). Spatial variability of the plankton trophic interaction in the North Sea: a new feature after the early 1970s. *Global Change Biology*, **18**, 106–117.

Lynman, C. P., Gibbons, M. J., Axelsen, B. E., et al. (2006). Jellyfish overtake fish in a heavily fished ecosystem. *Current Biology*, **16**, R492–493.

Manning, F. (2012). The sustainable management of grey seal populations: a path toward the recovery of cod and other groundfish stocks. Report of the Standing Senate Committee on Fisheries and Oceans. 42 pp.

McCann, K. S., Rasmussen, J. B. and Umbanhowar, J. (2005). The dynamics of spatially coupled food webs. *Ecology Letters*, **8**, 513–523.

Micheli, F. (1999). Eutrophication, fisheries, and consumer-resource dynamics in marine pelagic ecosystem. *Science*, **285**, 1396–1398.

Micheli, F., Benedetti-Cecchi, L., Gambaccini, S., et al. (2005). Cascading human impacts, marine protected areas, and the structure of Mediterranean reef assemblages. *Ecological Monographs*, **75**, 81–102.

Minto, C. and Worm, B. (2012). Interactions between small pelagic fish and young cod across the North Atlantic. *Ecology*, **93**, 2139–2154.

Möllmann, C., Müller-Karulis, B., Kornilovs, G. and St. John, M. (2008). Effects of climate and overfishing on zooplankton dynamics and ecosystem structure: regime shift, trophic cascade, and feedback loops in a simple ecosystem. *ICES Journal of Marine Scence*, **65**, 302–310.

Mora, C., Tittensor, D. P., Adl, S., Simpson, A. G. B. and Worm, B. (2011). How many species are there on Earth and in the ocean? *PLoS Biology*, **9**(8), e1001127.

Myers, R. A., Mertz, G. and Fowlow, P. S. (1997). Maximum population growth rates and recovery times for Atlantic cod, *Gadus morhua*. *Fishery Bulletin*, **95**, 762–772.

Myers, R. A. and Worm, B. (2003). Rapid worldwide depletion of predatory fish communities. *Nature*, **423**, 280–283.

Myers, R. A. and Worm, B. (2005). Extinction, survival, or recovery of large predatory fish. *Philosophical Transactions of the Royal Society, Series B*, **360**, 13–20.

Nixon, S. W. (1988). Physical energy inputs and the comparative ecology of lake and marine ecosystems. *Limnology and Oceanography*, **33**, 1005–1025.

Österblom, H., Casini, M., Olsson, O. and Bignert, A. (2006). Fish, seabirds and trophic cascades in the Baltic Sea. *Marine Ecology Progress Series*, **323**, 233–238.

Österblom, H., Olsson, O., Blenckner, T. and Furness, R. W. (2008). Junk-food in marine ecosystems. *Oikos*, **117**, 967–977.

Paine, R. T. (1980). Food webs: linkage, interaction strength and community infrastructure. *Journal of Animal Ecology*, **49**, 667–685.

Pauly, D., Christensen, V., Guenette, S., et al. (2002). Towards sustainability in world fisheries. *Nature*, **418**, 689–695.

Pershing, A. J., Head, E. H., Greene, C. H. and Jossi, J. W. (2010). Pattern and scale of variability among Northwest Atlantic shelf plankton communities. *Journal of Plankton Research*, **32**, 1661–1674.

Peterson, W. T. and Schwing, F. B. (2003). A new climate regime in northeast Pacific ecosystems. *Geophysical Resarch Letters*, **30**(17), 1896.

Petrie, B., Frank, K. T., Shackell, N. L. and Leggett, W. C. (2009). Structure and stability in exploited marine ecosystems: quantifying critical transitions. *Fisheries Oceanography*, **18**(2), 83–101.

Polis, G. A., Sears, A. L. W., Huxel, G. R., Strong, D. R. and Maron, J. (2000). When is a trophic cascade a trophic cascade? *Trends in Ecology and Evolution*, **15**(11), 473–475.

Reid, P. C., Battle, E. J. V., Batten, S. D. and Brander, K. M. (2000). Impacts of fisheries on plankton community structure. *ICES Journal of Marine Science*, **57**, 495–502.

Richardson, A. J. and Schoeman, D. S. (2004). Climate impact on plankton ecosystems in the Northeast Atlantic. *Science*, **305**, 1609–1612.

Richardson, A. J., Bakun, A., Hays, G. C. and Gibbons, M. J. (2009). The jellyfish joyride: causes, consequences and management responses to a more gelatinous future. *Trends in Ecology and Evolution*, **24**(6), 312–322.

Ritchie, E. G. and Johnson, C. N. (2009). Predator interactions, mesopredator release and biodiversity conservation. *Ecology Letters*, **12**, 982–998.

Romanuk, T. N., Hayward, A. and Hutchings, J. A. (2011). Trophic level scales positively with body size in fishes. *Global Ecology and Biogeography*, **20**, 231–240.

Salomon, A. K., Gaichas, S. K., Shears, N. T., et al. (2010). Key features and context-dependence of fishery-induced trophic cascades. *Conservation Biology*, **24**(2), 382–394.

Savenkoff, C., Swain, D. P., Hanson, J. M., et al. (2007). Effects of fishing and predation in a heavily exploited ecosystem: comparing periods before and after the collapse of groundfish in the southern Gulf of St. Lawrence (Canada). *Ecological Modeling*, **204**, 115–128.

Schipper, J., Chanson, J. S., Chiozza, F., et al. (2008). The status of the world's land and marine mammals: diversity, threat, and knowledge. *Science*, **322**, 225–230.

Shackell, N. L. and Frank, K. T. (2007). Compensation in exploited marine fish communities on the Scotian Shelf, Canada. *Marine Ecology Progress Series*, **336**, 235–247.

Shackell, N. L., Frank, K. T., Fisher, J. A. D., Petrie, B. and Leggett, W. C. (2010). Decline in top predator body size and changing climate alter trophic structure in an oceanic ecosystem. *Proceedings of the Royal Society, Series B*, **277**, 1353–1360.

Shears, N. T., Babcock, R. C. and Salomon, A. K. (2008). Context-dependent effects of fishing: variation in trophic cascades across environmental gradients. *Ecological Applications*, **18**, 1860–1873.

Shurin, J. B., Borer, E. T., Seabloom, E. W., et al. (2002). A cross-ecosystem comparison of the strength of trophic cascades. *Ecology Letters*, **5**, 785–791.

Sinclair, A. R. E. and Krebs, C. J. (2002). Complex numerical responses to top-down and bottom-up processes in vertebrate populations. *Philosophical Transactions of the Royal Society of London Series B*, **357**, 1221–1231.

Steele, J. H. (1998). Regime shifts in marine ecosystems. *Ecological Applications*, **8**, S33–S36.

Strong, D. R. (1992). Are trophic cascades all wet? Differentiation and donor-control in speciose ecosystems. *Ecology*, **73**, 747–754.

Strong, D. R. and Frank, K. T. (2010). Human involvement in food webs. *Annual Review of Environmental Resources*, **35**, 1–23.

Swain, D. P. and Mohn, R. K. (2012). Forage fish and the factors governing recovery of Atlantic cod (*Gadus morhua*) on the eastern Scotian Shelf. *Canadian Journal of Fisheries and Aquatic Sciences*, **69**, 997–1001.

Swain, D. P. and Sinclair, A. F. (2000). Pelagic fishes and the cod recruitment dilemma in the Northwest Atlantic. *Canadian Journal of Fisheries and Aquatic Sciences*, **57**, 1321–1325.

Tremblay-Boyer, L., Gascuel, D., Watson, R., Christensen, V. and Pauly, D. (2011). Modelling the effects of fishing on the biomass of the world's oceans from 1950 to 2006. *Marine Ecology Progress Series*, **442**, 169–185.

Turner, J. T. and Granéli, E. (2006). "Top-down" predation control on marine harmful algae. In *Ecology of Harmful Algae*. Ecological Studies, Vol. 189. Berlin Heidelberg: Springer-Verlag, pp. 355–366.

Verity, P. G. (1998). Why is relating plankton community structure to pelagic production so problematic? *South African Journal of Marine Science*, **19**, 333–338.

Verity, P. G. and Smetacek, V. (1996). Organism life cycles, predation, and the structure of marine pelagic ecosystems. *Marine Ecology Progress Series*, **130**, 277–293.

Vitousek, P. M., Mooney, H. A., Lubchenco, J. and Melillo, J. M. (1997). Human domination of earth's ecosystems. *Science*, **277**, 494–499.

Walters, C. and Kitchell, J. F. (2001). Cultivation/depensation effects on juvenile survival and recruitment: implications for the theory of fishing. *Canadian Journal of Fisheries and Aquatic Sciences*, **58**, 39–50.

Ware, D. M. and Thompson, R. E. (2005). Bottom-up ecosystem trophic dynamics determine fish production in the Northeast Pacific. *Science*, **308**, 1280–1284.

Webb, T. J., Berghe, E. V. and O'Dor, R. (2010). Biodiversity's big wet secret: the global distribution of marine biological records reveals chronic under-exploration of the deep pelagic ocean. *PLoS One*, **5**(8), e 10223.

White, T. C. R. (1978). The importance of a relative shortage of food in animal ecology. *Oecologia*, **33**, 71–86.

Worm, B. and Myers, R. A. (2003). Meta-analysis of cod-shrimp interactions reveals top-down control in oceanic food webs. *Ecology*, **84**, 162–173.

Yodzis, P. (2001). Must top predators be culled for the sake of fisheries? *Trends in Ecology and Evolution*, **16**, 78–84.

CHAPTER THREE

Top-down and bottom-up interactions in freshwater ecosystems: emerging complexities

JASON M. TAYLOR

United States Department of Agriculture, Oxford, MS, USA

MICHAEL J. VANNI

Miami University, Oxford, OH, USA

and

ALEXANDER S. FLECKER

University, Ithaca, NY, USA

Introduction

Lindeman (1942) made early distinctions between aquatic food webs and Elton's (1927) terrestrial biomass pyramids that firmly established the study of lakes and rivers as fertile ecosystems for examining the relative roles of resources and consumers in controlling energy flow and biomass. Building on these early observations, ecologists have established that energy transfers more efficiently through freshwater food webs than terrestrial food webs as a result of higher consumer-producer size ratios, higher producer growth rates and population turnover, and lower consumer-resource elemental imbalances, as compared to terrestrial systems (Shurin et al., 2006). Freshwater ecologists have confirmed the importance of nutrients in limiting primary production and the rapid transfer of energy to herbivores, thereby establishing the important role of bottom-up processes in regulating freshwater food webs (McQueen et al., 1986; Power, 1992). Freshwater ecologists have also recognized the role of top-down processes in freshwater ecosystems, and contributed substantially to demonstrating that higher trophic levels can influence primary producer biomass through trophic cascades (Carpenter et al., 1985; Power, 1992; Pace et al., 1999).

Clearly, both "top-down" (TD) and "bottom-up" (BU) regulation are pervasive in freshwater food webs (Shurin et al., 2006; Gruner et al., 2008), and these two processes do not act independently. For example, increasing nutrients can intensify

Trophic Ecology: Bottom-Up and Top-Down Interactions across Aquatic and Terrestrial Systems, eds T. C. Hanley and K. J. La Pierre. Published by Cambridge University Press. © Cambridge University Press 2015.

consumer control and the effects of trophic cascades on producer communities (Carpenter et al., 2001; Jeppesen et al., 2003), and increase overall contribution of animal-mediated nutrient recycling to ecosystem demand (Vanni et al., 2006; Wilson and Xenopoulos, 2011). Understanding mechanisms that facilitate interactions between resource and consumer control of food web structure is an important avenue of research. Moreover, the importance of BU and TD interactions also pervades applied aspects of ecology, including water quality management and biodiversity conservation. For example, BU and TD interactions are beginning to help conservationists predict consequences of changing species composition on ecosystem function (Eby et al., 2006; McIntyre et al., 2007; Vaughn, 2010). Additionally, manipulations of both BU and TD processes are actively viewed as management strategies for ameliorating the effects of cultural eutrophication of freshwaters (Carpenter et al., 1998; Søndergaard et al., 2008).

In this chapter, we first briefly review the evidence for BU and TD processes in freshwater systems. We then focus the second part of the chapter on relatively recent work on three integrative factors that are important in mediating BU and TD interactions in freshwater ecosystems (Fig. 3.1), but are largely absent from much of the foundational work examining TD and BU forces in freshwater ecosystems. First, it is now well-recognized that nutrient ratios, not just nutrient supply, regulate productivity in aquatic ecosystems, but the implications of nutrient ratios have not been fully incorporated into current views of BU regulation in food webs. This includes animal-mediated nutrient recycling, which integrates BU and TD processes by consumers re-mineralizing nutrients and making them available to primary producers; in this process, consumers can alter the ratios at which nutrient are supplied. Second, ecosystem subsidies of energy, nutrients, and organisms that move across the landscape can enhance or stabilize local food webs, but may also be modulated by consumers. Third, ecosystem engineers, that is organisms that physically modify their environment, are widespread and their activities in altering habitat and resource availability have strong consequences that can influence both TD and BU processes. Finally, we explore these three integrative factors within the context of applied examples related to biodiversity conservation and a water quality management case study.

Evidence for bottom-up and top-down control in aquatic ecosystems

Understanding the role of limiting nutrients in structuring freshwater ecosystems is a central theme in limnology. Early comparisons of lakes across differing nutrient regimes, as well as whole-lake manipulations, established phosphorus (P) as the ultimate limiting nutrient of primary production in many lakes (Dillon and Rigler, 1976; Schindler, 1977; 1978). However, bioassays and whole-lake experiments also demonstrate that primary producers can respond more strongly to increasing both nitrogen (N) and P compared to P alone (Elser et al.,

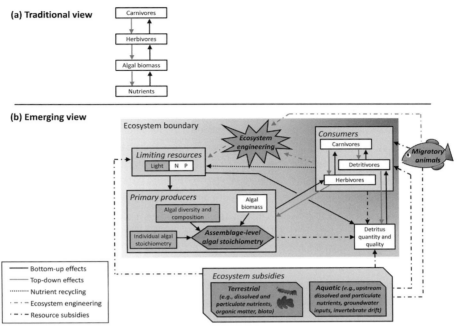

Figure 3.1 (a) Traditional view of bottom-up and top-down interactions in a linear food chain. (b) An expanded view of bottom-up and top-down interactions that incorporates ecological stoichiometry, animal-mediated nutrient cycling, resource subsidies, and ecosystem engineering. Bold outlined boxes and solid arrows represent traditional components associated with bottom-up and top-down processes. Dashed arrows and gray shapes represent newer concepts that expand our view of bottom-up and top-down interactions.

2007), and there is considerable debate over the single-limiting nutrient concept in aquatic ecosystems (Lewis and Wurtsbaugh, 2008; Conley et al., 2009). Additional data from streams further support co-limitation in many freshwater ecosystems (Francoeur, 2001; Dodds et al., 2002; Elser et al., 2007). BU effects induced by increased supply of limiting nutrients not only regulate primary productivity, but can also transfer through aquatic food webs and promote higher consumer biomass or growth in lakes (McQueen et al., 1986; Mills and Chalanchuk, 1987; Downing et al., 1990) and streams (Peterson et al., 1993; Forrester et al., 1999; Peckarsky et al., 2013). Nutrient enrichment can also influence consumer biomass through detrital pathways in some ecosystems. For example, in small forested streams, N and P can increase microbial respiration, fungal and bacterial biomass, and leaf and wood decomposition rates (Gulis and Suberkropp, 2003; Gulis et al., 2004), and result in increased allocthonous carbon (C) flow to primary and secondary consumers (Rosemond et al., 2002; Cross et al., 2006). Other limiting resources may include micronutrients such as silicon (Si) (Wetzel, 2001) and molybdenum (Mo) (Glass et al., 2012), light availability (Wootton and Power, 1993; Dickman et al., 2008), and in some freshwater ecosystems, allocthonous C subsidies exert BU control on consumers (Wallace et al., 1999; Johnson and Wallace, 2005).

While potential productivity is regulated by nutrients or other resources in aquatic ecosystems, actual productivity can be dependent on TD processes. Grazer effects in aquatic ecosystems are well documented (Hillebrand, 2002; Domis et al., 2013), and phytoplankton and periphyton experience the highest grazer effects compared to producers in other biomes (Cebrian, 1999). It is difficult, however, to consider direct TD effects of herbivory without considering the indirect effects of higher trophic levels on primary producer biomass through trophic cascades (Paine, 1980; Pace et al., 1999). Paine (1980) first introduced the term trophic cascade, and limnologists quickly embraced the concept (Carpenter et al., 1985), although it is well known that the concept was first raised for terrestrial ecosystems (Hairston et al., 1960). Since then, a large body of work has established that changes in predator biomass can cause substantial changes in lower trophic levels and ecosystem processes in lakes (Carpenter et al., 2010) and streams (Wootton and Power, 1993; Flecker and Townsend, 1994; Kurle and Cardinale, 2011). However, the strength of trophic cascades is dependent on a variety of factors that highlight the interactive nature of BU and TD controls, including prey edibility (Leibold, 1989) and vulnerability (Hambright et al., 1991), disturbance regimes (Power et al., 2008), water temperature (Kishi et al., 2005), depth (Jeppesen et al., 2003), and nutrient enrichment (e.g., Carpenter et al., 2001; Jeppesen et al., 2003; Peckarsky et al., 2013).

The relative importance of BU versus TD control has been of interest to freshwater ecologists for some time. Despite an early recognition that BU forces and trophic cascades in lakes are complementary, not contradictory (Carpenter et al., 1985), considerable interest and debate exists over the role of trophic cascades relative to resource availability in determining the structure of aquatic food webs (Brett and Goldman, 1996). McQueen et al. (1986) surmised that TD controls were prevalent at higher trophic levels but were overwhelmed by BU processes and quickly attenuated at lower trophic levels, and concluded that nutrients regulated primary production more than consumers. Meta-analyses of mesocosm experiments that manipulated both nutrients and planktivorous fish provide support for both TD and BU control, with primary producers more strongly regulated by resources and primary consumers by TD effects (Brett and Goldman, 1997; Gruner et al., 2008). However, trophic cascades depend on complex processes and interactions, making it difficult to predict the effect of system productivity on the strength of trophic cascades or grazer–algal interactions (*sensu* Carpenter et al., 2010; Peckarsky et al., 2013). Results from *in situ* experiments in streams demonstrate various outcomes including independent positive effects of fish-induced trophic cascades and nutrient enrichment on algal biomass (Forrester et al., 1999), dissipation of strong trophic cascades with nutrient enrichment (Riley et al., 2004), and enhancement of trophic cascades

with increasing resource supply (Peckarsky et al., 2013). Whole-lake manipulations have also demonstrated that trophic cascades can control primary
production across a wide gradient of nutrient enrichment (Pace et al., 1999;
Carpenter et al., 2001). Cross lake comparisons and more recent meta-analyses
have demonstrated that predators can have strong TD effects across entire food
chains, while BU processes primarily attenuate with primary producers (Borer
et al., 2006). These selected examples reveal that conclusions from the wide
breadth of literature seldom agree as to the relative importance of BU versus
TD control, nor the strength of the interaction between the two within aquatic
ecosystems. Part of this variability is no doubt due to actual variation in the
relative strength of BU and TD processes, but is also likely due to the wide range
of approaches and spatio-temporal scales used in different studies (Kurle and
Cardinale, 2011). Moreover, other factors, which we describe below, add further
complexity in efforts to parse out the roles of TD/BU processes.

The role of ecological stoichiometry in mediating bottom-up and top-down interactions

Ecological stoichiometry (ES) has emerged as a powerful conceptual framework
(Sterner and Elser, 2002), and posits that the ratios of elements supplied to, stored
by, and recycled by organisms can regulate food webs and ecosystem processes.
However, the importance of nutrient ratios has not been fully incorporated
into general BU frameworks. Rather, the conventional view of BU regulation is
that biomass and productivity of upper trophic levels are regulated by nutrient
supply rates, not ratios (Fig. 3.1a). In addition, TD forces can alter nutrient ratios,
resulting in potentially important feedbacks to algal production, community
composition, and stoichiometry (e.g., Hall et al., 2007), but these also have not
been integrated into general TD models. Here, we discuss the importance of
ecological stoichiometry in mediating BU and TD interactions.

Resource ratio effects on algal community composition and stoichiometry

Resource ratios can have many effects on algae (Fig. 3.1b). Nutrient ratios (e.g.,
Si:P, N:P) are known to determine algal species composition, because species
differ in their optimal cell C:N:P ratios (Sterner and Elser, 2002). For example, an
accepted paradigm in limnology is that N:P supply ratio determines the relative
abundance of cyanobacteria (Smith, 1983). Because some cyanobacteria taxa fix
N, this group often dominates when nutrients are supplied at low N:P (although
cyanobacterial dominance is also strongly affected by temperature, light, and
overall nutrient supply; Downing et al., 2001; Havens et al., 2003). Variation in
optimal N:P is partly due to interspecific variation in the subcellular distribution
of structures dedicated to resource acquisition (high N, low P) or high growth
rate (low N, high P) (Klausmeier et al., 2004). However, algal cells are quite
plastic with regard to the ratios of C, N, and P in their cells, in part because they

can store elements supplied in excess. Thus, for example, the N:P ratio in algal cells can vary as a function of the N:P ratio supplied, and the linkage between nutrient supply ratio and algal stoichiometry are evident both within a single species and at the level of entire algal assemblages (Fig. 3.1b).

The importance of resource ratios also applies to light. Thus, because C fixation increases with light intensity, algae tend to have higher cellular C:nutrient ratios when light supply is high. The light:nutrient ratio hypothesis (LNH, Sterner et al., 1997) posits that algal C:nutrient ratios increase with light and decrease with nutrients. The LNH is well supported in nature for phytoplankton (Sterner et al., 1997), but evidence for periphyton is more mixed (e.g., Hillebrand et al., 2004; Liess and Kahlert, 2007). However, considerable variation exists among studies in the magnitude of the response of phytoplankton C:N:P ratios to variations in light and nutrients (Dickman et al., 2006; Hall et al., 2007; Mette et al., 2011). One possible factor mediating this variation is the particular species pool present and the optimal cell C:N:P ratios of the constituent species. However, it is clear that variation in light and nutrient ratios can affect stoichiometry both within a species and at the assemblage level (Fig. 3.1b). Several recent studies reveal linkages between algal species composition, diversity, stoichiometry, and productivity. These linkages can result from, and modulate, BU interactions or TD forces, illustrating potentially complex feedbacks that necessitate a new view of TD and BU forces (Fig. 3.1b). For example, phytoplankton assemblages with high species richness can display a greater range of stoichiometric ratios (C:N, C:P, and N:P) than low-richness assemblages (Striebel et al., 2008; 2009; Mette et al., 2011), thus illustrating one way that biodiversity can impact BU/TD interactions (see Chapter 12). This richness effect may arise because diverse assemblages exhibit more complementarity among taxa, and hence greater efficiency, in light harvesting capacity and secondary pigments, leading to higher C:nutrient at the assemblage level (Striebel et al., 2009). High richness potentially leads to more variation in cellular N:P at the assemblage level. Additionally, algal taxonomic composition is likely linked to stoichiometry; e.g., two studies show that seston C:P is positively correlated with the relative biomass of chlorophytes and negatively correlated with relative diatom biomass (Hall et al., 2007; Mette et al., 2011). The mechanisms accounting for the links between algal species composition, diversity, and stoichiometry are not immediately evident and potentially complex, but experimental studies show that manipulation of algal species richness can directly affect assemblage-level C:P (Striebel et al., 2009).

Resource ratio effects on consumers

Variation in algal C:N:P ratios has profound implications for both BU and TD processes, namely energy flow to consumers and the regulation of lower trophic levels via trophic cascades and nutrient cycling (Fig. 3.1b). It is well known that plants (including algae) are C-rich but nutrient-poor compared to animals (i.e., plants have much higher C:nutrient ratios), and herbivore growth is therefore often limited by N or P (Cebrian and Lartigue, 2004). Effects of algal stoichiometry on herbivore growth are documented for zooplankton feeding on phytoplankton (Sterner and Elser, 2002) and benthic invertebrates feeding on periphyton (e.g., Frost and Elser, 2002) or detritus (Danger et al., 2013). However, some studies show no effect of algal C:nutrient ratios on the growth of benthic invertebrate herbivores (e.g., Hill et al., 2010; Liess and Lange, 2011). Information on resource ratio effects on fishes is scant outside of the aquaculture literature, but P limitation in wild fish populations is likely for herbivorous fishes with high body P content (Hood et al., 2005). Nutrient supply ratios can also affect herbivores via changes in algal community composition. For example, cyanobacteria are relatively poor food for herbivores, and can even be toxic; therefore low N:P supply can reduce energy flow to upper trophic levels by promoting dominance of cyanobacteria (Brett et al., 2000).

Variation in algal stoichiometry can also indirectly regulate carnivore growth and productivity. High algal C:N:P ratios may affect carnivore production simply by modulating herbivore production. However, recent studies also show that nutrient limitation can "travel up the food chain" (Boersma et al., 2008). Thus, herbivorous zooplankton feeding on low-quality (high C:P) phytoplankton are in turn low-quality food for carnivores, i.e., carnivore production per unit zooplankton production is lower when zooplankton consume phytoplankton with high C:P (compared to when zooplankton feed on phytoplankton with low C:P). These "carryover effects" have been demonstrated in lab and field experiments, and for several carnivore species, including fish (Malzahn et al., 2007; Dickman et al., 2008) and invertebrates (lobster larvae; Schoo et al., 2012). However, not all carnivores respond similarly. For example, jellyfish growth rates are relatively insensitive to herbivore quality, possibly because they have lower P demands (high body C:P) than other carnivores and consequently are less likely to be P-limited (Malzahn et al., 2010).

Feedbacks between algal diversity, composition, and stoichiometry via BU and TD forces

In intact food webs, myriad direct and indirect interactions occur simultaneously. The potential feedbacks among these interactions make it difficult to pinpoint mechanisms and also blur the distinction between BU and TD interactions (Fig. 3.1b). Resource supply rates and ratios influence algal stoichiometry at the assemblage level, and these BU forces may propagate to herbivores and ultimately carnivores via the mechanisms discussed above. However, simultaneous TD interactions will also affect algal stoichiometry. Herbivores can directly affect algal biomass, composition, and diversity via consumptive effects; in turn, algal biomass can regulate dissolved nutrient availability via nutrient uptake and light availability via shading. Benthic herbivores and omnivores also increase light by removing sediments from biofilms (see Case Study 3.1). Furthermore, herbivores and carnivores recycle nutrients, and may do so at ratios that differ from other nutrient sources; if so, they can alter algal communities via this mechanism. These interactions set up potential feedbacks. For example, repeated cycles of grazing and nutrient cycling can shape algal biomass and stoichiometry (Sterner and Elser, 2002; Tessier and Woodruff, 2002), thereby further modifying food quantity and quality for herbivores. This can change herbivore biomass and species composition, and affect the next cycle of grazing and nutrient cycling. At longer time scales, changes in herbivore productivity resulting from changes in algal quantity and quality can affect carnivore growth, and thus the strength of the TD control emanating from top predators. Only a few studies have experimentally manipulated light, nutrients, and consumers (either herbivore or carnivores) simultaneously (Hall et al., 2007; Liess and Kahlert, 2007; Mette et al., 2011). In all of these studies, consumers affected algal stoichiometry and biomass, but consumer effects also interacted with light, nutrients, or both. All studies attribute some of the consumer effects on algal biomass and stoichiometry to grazing, but none of these studies explicitly separated the role of animal-mediated nutrient cycling. The complexity of these interactions and the paucity of studies make it difficult to draw conclusions as to the overall importance of different interactions, and highlight the need for carefully designed experiments that will elucidate the importance of various mechanisms.

Case Study 3.1

Grazing minnows and BU–TD interactions in streams

Central stonerollers (*Campostoma anomalum*) are grazing minnows that can influence stream nutrient cycling through complex interactions between BU and TD forces in stream food webs. Stonerollers modify benthic habitats by consuming sediment and detritus while grazing, which increases the relative proportion of algae and increases sequestration of nutrients within benthic habitats (Taylor et al., 2012a). Increased autotrophy and nutrient content within the benthos likely increases BU flow to stonerollers and other benthic grazers. However, grazing invertebrates may also compete with stonerollers for resources, and TD control of invertebrates by insectivorous fish (*Etheostoma spectabile*) can enhance BU flow of resources to stonerollers (Hargrave et al., 2006). Stonerollers also modulate BU forces by interacting with nutrients via animal-mediated recycling pathways. Taylor et al. (2012b) demonstrated experimentally that periphyton in downstream habitats adjacent to caged stonerollers in N-enriched stream mesocosms had lower C:N ratios, higher algal biomass, and increased autotrophic and heterotrophic production. Within the periphyton matrix, algal-bacterial production was more strongly coupled downstream of stonerollers in N-enriched streams (Taylor et al., 2012b). Increased algal biomass and N content in response to fish-mediated nutrient cycling likely increased P limitation within downstream periphyton communities, leading to increased coupling of algal-bacterial production via microbial-mediated P-cycling.

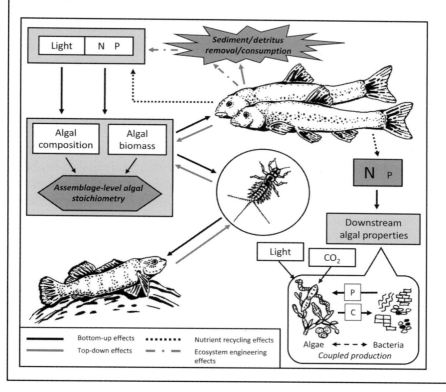

Animal-mediated nutrient cycling: an interaction of BU and TD forces

Animals can modulate nutrient cycling through several mechanisms (Vanni, 2002). They can directly regulate nutrient fluxes by consuming, storing, transporting, and releasing nutrients. Animals also have many indirect effects on nutrient cycling (Fig. 3.1b). Trophic cascades initiated by top predators can alter the rates (and ratios) by which lower trophic levels use and cycle nutrients, and modification of physical habitats (ecosystem engineering) by animals can also induce changes in nutrient cycling. Finally, animals can move nutrients between habitats and ecosystems, often at long distances. This "nutrient translocation" is a unique characteristic of animals in regards to nutrient cycling, as plants and microbes have limited mobility (see Case Studies 3.1 and 3.2). In this section, we discuss the direct regulation of nutrient cycling by animals, as well as indirect effects mediated through trophic cascades, within the context of BU/TD regulation.

Nutrient excretion by animals has been well studied in freshwater ecosystems, probably more than any other biome. Metabolic ecology and ecological stoichiometry have emerged as valuable frameworks to predict nutrient animal excretion rates and ratios (although these two frameworks have not been well merged; Allen and Gillooly, 2009). As predicted by metabolic ecology (Gillooly et al., 2001), excretion rates of individual animals are often closely correlated with animal body size and temperature, but the slopes and intercepts of these relationships vary greatly among species in nature (e.g., Hall et al., 2007). Thus, no universal size-based animal model of nutrient excretion is yet evident.

Ecological stoichiometry (ES) theory predicts that animal excretion rates are functions of the concentrations (or ratios) of elements in animals' diets and in their bodies (Sterner and Elser, 2002). Specifically, an individual's excretion rate should increase with the concentration of that element in the consumer's food, all else being equal. In terms of ratios, an animal consuming food with a low N:P ratio should excrete at a lower N:P than the same animal consuming food with a high N:P. However, ES also predicts that excretion N:P is a function of animal body nutrient ratios. Animals with a low body N:P will excrete at a high N:P, because they need to sequester large amounts of P compared to animals with high body N:P with the same diet. Thus, excretion N:P should be positively correlated with food N:P, but negatively correlated with consumer body N:P (Sterner and Elser, 2002). The relative importance of food versus animal body elemental composition in predicting excretion ratios varies among studies, and may be a function of the relative variation in the two drivers. Studies showing a strong (negative) correlation between body N:P and excretion N:P tend to be those in which there is relatively great variation among species in body N:P (e.g., Vanni et al., 2002; McManamay et al., 2011). In contrast, when interspecific variation in body N:P is not as great, body and food N:P are not correlated (e.g., Elser and Urabe, 1999; Torres and Vanni, 2007).

At the ecosystem scale, excretion by animals can be an important flux compared to other nutrient sources (or compared to nutrient demand) and can affect nutrient recipients such as algae (Fig. 3.1b). However, there is tremendous variation in the importance of animals across studies. For example, excretion by fish can sustain anywhere from < 5% to > 80% of nutrient demand, depending on the ecosystem (e.g., Grimm, 1988a; Vanni et al., 2006; McIntyre et al., 2008;

Small et al., 2011; Wilson and Xenopoulos, 2011; Capps and Flecker, 2013). Similar ranges exist for zooplankton in lakes (e.g., Schindler et al., 1993; Johnson et al., 2010) and benthic invertebrates in streams (Grimm, 1988b; Hall et al., 2003; Benstead et al., 2010; Moslemi et al., 2012). Little is known about the mechanisms accounting for this variation, and the importance of animal-mediated nutrient cycling can vary even along the same environmental gradient. For example, in both lakes (Vanni et al., 2006) and streams (Wilson and Xenopoulos, 2011), the proportion of nutrient demand sustained by fish excretion increased as watershed agriculture increased, due mainly to increases in fish biomass (see Case Study 3.2). In contrast, Spooner et al. (2013) found that the importance of nutrient cycling by unionid mussels decreased with increasing agriculture, because agricultural nutrients were far greater than the recycled nutrients contributed by mussels. In general, we are just beginning to learn about the mechanisms underlying variation in the importance of animal-mediated nutrient cycling. However, studies that have experimentally separated and quantified the effects animals have on algae via nutrient cycling versus direct consumption show that nutrient cycling by fish and invertebrates can affect the size and N:P ratio of nutrient pools and exports, the relative severity of N versus P limitation of algae, algal community composition, and microbial production (e.g., Knoll et al., 2009; Taylor et al., 2012b; Atkinson et al., 2013).

Case Study 3.2

Implications of BU–TD interactions for water quality management in lakes

Eutrophication of freshwater ecosystems remains a major environmental issue, despite decades of attention (Carpenter et al., 1998; Smith and Schindler, 2009). Although reducing nutrient inputs can reverse eutrophication, several factors can inhibit or delay response to nutrient reductions, including internal loading of P from sediments to water, which maintains a turbid stable state dominated by phytoplankton (Scheffer and Carpenter, 2003). Benthivorous fish may be especially important to maintaining turbid states because they translocate "new nutrients" (*sensu* Dugdale and Goering, 1967) from sediments, increasing nutrient availability in the pelagic zone. Gizzard shad (*Dorosoma cepedianum*) is a widespread fish in lakes and reservoirs of eastern USA. Gizzard shad consume terrestrial- and phytoplankton-derived sediment detritus and translocate excreted nutrients into the pelagic habitat. Because gizzard shad biomass is often very high in eutrophic reservoirs, this pathway can significantly enhance primary productivity in these ecosystems (Schaus et al., 1997; Shostell and Bukaveckas, 2004).

Fish-mediated nutrient translocation supports a larger proportion of reservoir primary production as agricultural land use increases within watersheds, likely because gizzard shad biomass is much higher in eutrophic reservoirs (Vanni et al., 2006). Removing benthivorous fish such as gizzard shad may reduce bioturbation and nutrient transfer from benthic to pelagic habitats (Horppila et al., 1998; Søndergaard et al., 2008), ideally shifting lakes closer to a clear water state characterized by macrophyte dominance and benthic production (Scheffer and Carpenter, 2003; Vadeboncoeur et al., 2003).

In an effort to mitigate eutrophication, gizzard shad are being removed from some subtropical lakes in Florida. Significant quantities of large individuals have been removed since 1993 using a commercial gill net fishery, to reduce both nutrient translocation (excretion) and nutrients sequestered in gizzard shad biomass, which can represent a substantial P pool. Removal of shad has resulted in P reductions equal to approximately 20% of external P loading reductions (excretion) and similar amounts of P removed by a treatment wetland (fish tissue) during a period from 1993 to 2005 (Schaus et al., 2010). However, an experimental manipulation in a nearby lake did not demonstrate significant changes in eutrophication indicators, likely due to low biomass reduction (< 40%; Catalano et al., 2010). Long-term, aggressive removal of gizzard shad biomass will be necessary considering that gizzard shad life history traits facilitate compensation through density-dependent, pre-recruitment survival (Catalano and Allen 2010; Catalano et al., 2010). Prey-switching in response to more abundant and highly nutritious *Daphnia* after biomass reductions (Schaus et al., 2002) can also restore biomass quickly. Despite mixed results regarding gizzard shad fish removal as a lake management strategy, recognizing their role in transferring nutrients from sediments is important for lake restoration, as this flux can potentially offset and delay improvements in water quality following reduction of external inputs (Schaus et al., 2010). Despite substantial lags, fish biomass tends to decline with shifts toward lower trophic states in response to decreased nutrient loading (Jeppesen et al., 2005). While reductions of external P loads provide the greater gains in long-term lake ecosystem recovery, reductions in both external loading and gizzard shad should act synergistically to improve water quality (Vanni et al., 2005; Schaus et al., 2010). The role of gizzard shad and benthic-feeding fish in general needs to be assessed in other systems as well. For example, gizzard shad and smallmouth buffalo make up a considerable portion of fish biomass in oxbow lakes within the floodplains of subtropical lowland rivers in North America (Winemiller et al., 2000). These small, shallow lakes are often embedded in intensive agricultural landscapes, and numerous best management strategies are employed to reduce agricultural inputs from watersheds that can have some water quality benefits (Locke et al., 2008). However, understanding how internal animal-mediated processes influence water quality goals is an important step in developing appropriate restoration and management strategies.

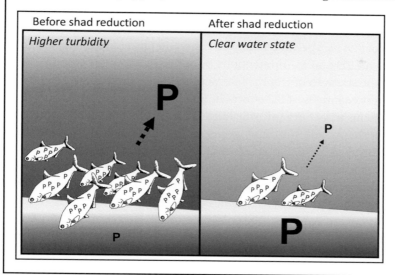

In addition to direct effects via nutrient excretion, animals have many indirect effects on nutrient cycling rates (Fig. 3.1b). For example, herbivores can regulate N-fixation rates by altering the relative biomass of N-fixing cyanobacteria by selective feeding (Arango et al., 2009), excreting nutrients at a particular N:P ratio (MacKay and Elser, 1998), or both (Hambright et al., 2007). Grazers can also influence nutrient dynamics by modifying the extent of nutrient limitation of primary producers. Experiments in a Venezuelan stream revealed that the response by autotrophs to nitrogen enrichment was significantly greater on substrates accessible to fishes compared to substrates where grazing fishes were excluded (Flecker et al., 2002). Moreover, trophic cascades induced by carnivores can also indirectly modulate nutrient cycling. For example, many aquatic predators selectively prey on the largest individuals they can consume and shift prey assemblages to smaller individuals; this can increase nutrient cycling by the prey assemblage because smaller individuals have higher mass-specific excretion rates (Bartell, 1981). However, predators often decrease the total biomass of prey assemblages, possibly offsetting the increase in mass-specific excretion rates (e.g., for fish feeding on zooplankton; Schindler et al., 1993; Johnson et al., 2010). Thus, top predators can induce changes in size structure and biomass in lower trophic levels, with concomitant alterations in nutrient cycling. Such indirect effects on nutrient cycling probably occur in most trophic cascades, but are usually not explicitly quantified.

Ecosystem subsidies and their influence on top-down and bottom-up processes

Over the last two decades, there has been increasing appreciation of the importance of ecosystem subsidies (i.e., landscape-level flows of energy, nutrients, and organisms) in driving local ecosystem dynamics and linking together disparate parts of land- or river-scapes in a food web context (Polis et al., 1997). Subsidies link directly with TD and BU interactions by acting as inputs of resources and by modifying the strength of trophic interactions (Fig. 3.1b; also, see Chapter 6). Some of the best examples of BU effects in food webs have been demonstrated in streams, systems where organic subsidies are key to their structure and function. Although theory and terminology regarding ecosystem subsidies have a more recent history, there is a long legacy in stream ecology that recognizes the importance of ecosystem subsidies from terrestrial zones to running waters. An early key concept of stream ecology was recognition of the fundamental role of riparian zones in fueling forested headwater streams (Hynes, 1975; Vannote et al., 1980; Meyer and Wallace, 2001). Wallace et al. (1999) provided a direct experimental demonstration that allochthonous inputs of terrestrial litter controlled multiple trophic levels via BU forces. They showed that the exclusion of leaf litter over a 3-year period from a small, forested headwater stream resulted in a strong BU effect on invertebrate secondary production, which propagated upward through the food web from detritivores to predators. More recently, lake studies have also come to appreciate the importance of allochthonous inputs as key subsidies of organic C to invertebrates and fish (Pace et al., 2004; Carpenter et al., 2005).

The River Continuum Concept (RCC) further elaborated the fundamental role of ecosystem subsidies in streams, not only in terms of terrestrial inputs to headwaters, but also via upstream to downstream linkages in energy sources (Vannote et al., 1980; also, see Chapter 6). Thus, the RCC envisaged that inefficiencies in organic matter processing resulted in flows downstream of key energy sources originating higher in the drainage network, i.e., upstream regions can subsidize downstream food webs. The Flood Pulse Concept, an alternative integrative framework of stream ecosystem structure and function, also is predicated on subsidies as key energetic linkages in riverine ecosystems; however, for large rivers it focuses on lateral exchanges with adjacent floodplains (Junk et al., 1989; Bayley, 1995) rather than on the longitudinal links from upstream of the RCC.

BU subsidies can also modify the strength of TD controls. Polis et al. (1997) posited that prey subsidies of top consumers should strengthen trophic cascades by increasing predator biomass based on energy from prey imported from elsewhere in the landscape. Models developed by Leroux and Loreau (2008) provide theoretical support for this subsidy hypothesis, by examining how allochthonous inputs can influence the strength of trophic cascades (see

Chapter 1). Consistent with the subsidy hypothesis prediction, Vander Zanden et al. (2011) have shown that in many lakes, fish get the majority of their energy from benthic sources. This should promote higher biomass of fish that are at least partially zooplanktivorous, resulting in stronger TD effects in the pelagic food web. On the other hand, the work of Nakano et al. (1999) in a Japanese stream provides a contrasting perspective on the influence of resource subsidies on TD controls. Nakano and colleagues found that native Dolly Varden (*Salvelinus malma*) relied heavily on inputs of terrestrial insects as their major food sources, and under natural conditions exerted little TD control of aquatic insects and algae. In contrast, if terrestrial subsidies were interrupted using experimental greenhouse covers, a much different set of trophic dynamics was observed. Dolly Varden shifted their diets to aquatic insects, which in turn resulted in a strong set of cascading effects with significant increases in epilithon biomass. These studies reinforce the idea that food web dynamics often cannot be dichotomized as a set of TD versus BU effects; instead, major controls can be complex and interactive.

Migratory species as material subsidies

Migratory fish provide additional examples of external subsidies in freshwater ecosystems that can increase resource availability and propagate upwards from the base of riverine and lake food webs (Fig. 3.1b; Flecker et al., 2010). Pacific salmon, which are diadromous (i.e., migrate between freshwater and oceanic environments) and spend most of their lives at sea where they gain > 99% of their mass before migrating to natal rivers and lakes, are the best studied case of migratory fish as material subsidies in freshwaters (e.g., Willson et al., 1998; Naiman et al., 2002; Schindler et al., 2003). Pacific salmon have the potential to act as especially effective material subsidies due to their semelparous life histories that concentrate the release of nutrients and energy in space and time. These marine-derived subsidies have received the most attention, where large biomass migrations are coupled with the low-nutrient status of oligotrophic freshwater systems in locations such as southeastern Alaska. Under conditions of high biomass, subsidies from salmon potentially can be transmitted across multiple trophic levels, including prey that sustain juvenile salmon, although there remains considerable uncertainty about the conditions in which broad BU effects of subsidies are transmitted to top trophic levels of rivers and lakes (Naiman et al., 2009).

Less well studied are the effects of potamadromous migratory species (i.e., migrations confined within freshwaters), although there is reason to believe that they can also be important vectors of materials. Fittkau (1970) posited that subsidies of migratory fish to large predators such as black caiman (*Melanosuchus niger*) were critically important to the productivity and nutrient dynamics of many floodplain lagoons and tributaries of the central Amazon. He reasoned that extirpation of black caiman from severe overharvest in many locations resulted in declines in productivity and Amazon riverine fisheries, as crocodilian predators were no longer present to intercept, assimilate, and remineralize nutrients originating elsewhere that became tied up in the bodies of migratory fishes. Likewise, Winemiller and Jepsen (2004) used stable isotopes to document the importance of migratory detritivores (*Semaprochilodus*) in subsidizing populations of piscivorous peacock cichlids (*Cichla temensis*) in a Venezuelan river. *Semaprochilodus* spawn in whitewater rivers, where young-of-the-year (YOY) gain body mass in productive floodplains before migrating into nutrient-poor blackwater rivers. YOY *Semaprochilodus* comprised nearly 50% of ingested biomass in *C. temensis* diets, presumably key for sustaining peacock cichlid populations. Winemiller and Jepsen speculated that *Semaprochilodus* subsidies strengthen TD effects; thus, detrital subsidies from whitewater floodplains could have local consequences in blackwater rivers that extend across multiple trophic levels. Clearly, large subsidies in streams and lakes, whether migratory fish or other sources, are likely to have measurable effects in food webs as external material inputs; nevertheless, there remains considerable need for elucidating different ways and sets of conditions in which they can modulate both TD and BU processes.

Ecosystem engineering

Although TD and BU forces are usually viewed within the context of trophic interactions, non-trophic interactions can also influence the strength of trophic dynamics and BU controls. Earlier we discussed how consumers in freshwater ecosystems can influence nutrient availability (i.e., their chemical environment) via processes such as excretion. Moreover, ecosystem engineering, whereby organisms alter their physical environment, is another class of non-trophic interaction that potentially can have significant consequences on TD and BU controls (Fig. 3.1b). By definition, ecosystem engineers are organisms that modulate resource availability to other organisms by modifying, creating, or maintaining habitats (Jones et al., 1994). By influencing resource availability, engineers can have strong effects that can be transmitted to other trophic levels. The classic ecosystem engineer is the beaver (*Castor* spp.), which transforms freshwater environments through its dam-building activities. In ecosystems where beaver are present, it is difficult to consider TD and BU forces in the absence of the environmental template associated with beaver activity. For example, Naiman et al. (1986) assembled nutrient budgets for stream reaches with versus without beaver, and surmised that beaver increased primary production by augmenting light availability through opening riparian canopies, in addition to modifying other C fluxes. At the same time, beaver dams can improve habitat quality for many stream fishes, thereby enhancing fish populations (Pollock et al., 2003). This greater diversity and abundance of fishes at the top of stream food webs provide the potential to strengthen TD controls in habitats containing these keystone ecosystem engineers.

Organisms that process organic-rich sediments, including fishes and tadpoles (e.g., Power, 1990; Matthews, 1998; Flecker et al., 1999; Flecker and Taylor, 2004), and invertebrates such as crayfish (Creed and Reed, 2004), freshwater shrimp (e.g., Pringle et al., 1993), and bivalves (e.g., Strayer et al., 1999; Sousa et al., 2009; Allen and Vaughn, 2011) provide further examples of ecosystem engineers in freshwaters. These organisms consume, re-package, or re-suspend large volumes of organic matter and can influence TD and BU processes via bioturbation and nutrient recycling in a suite of different freshwater ecosystems. For example, the central stoneroller is a widespread grazing minnow in temperate streams of eastern and midwestern North America, and a large body of work has demonstrated that stoneroller grazing influences periphyton composition and structure, overall organic matter dynamics, and sediment accumulation on grazed substrates in streams (Gelwick and Matthews, 1992; Matthews, 1998), all of which are potentially modulated through trophic cascades initiated by large-bodied piscivores (Power et al., 1985). Recent experimental work has shown that stonerollers are involved in complex BU and TD interactions through enhancement of autotrophic biomass via sediment removal, indirect effects of trophic cascades on competitors, and nutrient recycling (see Case Study 3.1).

Migratory species as ecosystem engineers

Migratory fish can exert strong effects extending far beyond simple vectors of material subsidies. In fact, some of the best documented impacts of migratory fish are consequences of their capacity as ecosystem engineers (a process subsidy, cf. Flecker et al., 2010) rather than as external sources of materials per se. For example, migratory flannelmouth characin (*Prochilodus mariae*) modify benthic environments by processing and re-suspending sediments, thereby reducing the availability of organic-rich sediments. One outcome of engineering by flannelmouth characin is to remove sediments that can otherwise bury cyanobacteria, potentially enhancing species that are important N-fixers (Flecker, 1996). At the same time, removal of organic-rich sediments by *P. mariae* modifies rates of insect secondary production, presumably via reductions in resource availability (Hall et al., 2011). Moreover, nest-digging activities of migratory fish such as spawning salmon can represent a significant source of physical disturbance, which can strongly influence standing stocks of multiple trophic levels (e.g., Moore, 2006; Moore and Schindler, 2008). Some nests are as large as $17\,\text{m}^2$ and in some cases bioturbation by salmon can move more sediments than floods (Gottesfeld et al., 2004; Hassan et al., 2008). This disturbance to the stream bed may not only result in higher export of organic matter, but also decreases in periphyton, reductions in benthic invertebrates, and temporary increases in prey availability for drift-feeding fishes (e.g., Peterson and Foote, 2000; Scheuerell et al., 2007;

Moore and Schindler, 2008). These studies serve to illustrate that ecosystem engineering, whether by resident or migratory species, can exert strong non-trophic effects, which can be manifested across multiple trophic levels. Non-trophic effects are largely neglected in studies of TD and BU processes in food webs. Nevertheless, their importance with respect to BU and TD processes is likely to be widespread, and hopefully an increasing focus of food web studies in the future.

BU and TD interactions provide context for ecosystem consequences of altering freshwater communities

Biodiversity loss

Freshwaters support a disproportionate amount of global biodiversity, representing less than 1% of the world's surface area but approximately 10% of all described species (Strayer and Dudgeon, 2010). Evidence is mounting in freshwater ecosystems for linkages between declines in freshwater biodiversity and changing ecosystem processes (McIntyre et al., 2007; Vaughn, 2010). Changes in species assemblages may influence ecosystem function through interactions between BU and TD controls (see Chapter 14) related to changes in resource ratios, animal-mediated nutrient recycling, ecosystem subsidies, or processes mediated by ecosystem engineers.

Declines in freshwater biodiversity likely have detrimental effects on ecosystem function by altering nutrient turnover in aquatic systems. Thus, animal-mediated nutrient recycling is an ideal quantitative framework for directly linking changes in ecosystem function to declines in biodiversity (McIntyre et al., 2007). Using a combination of experiments, field data, and modeling, research summarized by Vaughn (2010) demonstrates that contributions of mussel nutrient recycling to stream nutrient dynamics depends on expression of species traits regulated by species composition and abundance, as well as interactions between trait expression and environmental factors. Field studies by Vaughn and colleagues demonstrated that common mussel species are declining in abundance, leading to shifts in dominance and alteration of overall assemblage-level nutrient cycling (Vaughn, 2010). A similar approach has been employed with fish assemblages that demonstrated assemblage-level nutrient recycling was dominated by a few abundant and large-sized species, characteristics that account for disproportionate contributions to assemblage-level nutrient recycling, but are also targeted by fisherman (McIntyre et al., 2007). One species in particular, the flannelmouth characin (*Prochilodus mariae*) is currently overfished (Taylor et al., 2006), yet accounts for almost half of the N recycled by fish assemblages in a South American stream. Simulated extinction scenarios that included non-random extinction of *P. mariae* based on fishermen creel surveys predict highly divergent nutrient cycling patterns (McIntyre et al.,

2007). Changes in C-cycling are also expected with population declines of this migratory detritivore. Large-scale experimental removal demonstrates that loss of *P. mariae* decreases downstream transport of organic C, and increases primary production and respiration (Taylor et al., 2006). Consequently, loss of one functionally unique species alters interactions between BU and TD controls via nutrient recycling, benthic feeding (ecosystem engineering), and river network linkages (process subsidies).

There are few examples where aquatic ecologists have measured changes in ecosystem function related to "actual" loss of functionally important consumers. A rare case involves research documenting structural and functional changes to Neotropical stream ecosystems associated with rapid extinctions of amphibians due to chytrid fungus (Whiles et al., 2006). Healthy larval amphibian populations contribute significantly to stream N turnover through consumption and nutrient recycling (Whiles et al., 2013). Relaxation of TD control after amphibian extinctions has resulted in major increases in epilithon (benthic algae) and fine particulate organic matter (FPOM) accrual. Critical links in N movement through stream food webs have been severed, resulting in slower N turnover and greater uptake lengths (Whiles et al., 2013). An important factor to recognize in this and the previous *P. mariae* example is the lack of functional redundancy, which is expected to buffer ecosystem effects of species loss in these diverse tropical systems (Hooper et al., 2005).

Species introductions

Almost all freshwater ecosystems throughout the world are threatened by invasive aquatic species, and many invaders can have ecologically important effects on recipient ecosystems (Strayer, 2010). While marine and freshwater systems are suffering food web consequences of the "fishing down" of large-bodied fish within food webs (Allan et al., 2005; also, see Chapter 2 and Chapter 14), freshwater ecosystems worldwide have also experienced intensive stocking of exotic game fishes to enhance fisheries (Eby et al., 2006). Obvious predator effects on prey species are inherent with game fish introductions, but ecologists know relatively little about impacts on ecosystem function. Grazing invertebrates and fishes can also have significant effects on ecosystem function in invaded habitats (Hall et al., 2003; Johnson et al., 2009; Strayer, 2012; Capps and Flecker, 2013).

Salmonids represent a widely transplanted fish family, and a handful of studies suggest that these introduced predators may alter BU processes (e.g., primary productivity, nutrient cycling, export of subsidies to terrestrial systems) through changes in TD effects. Trout introductions to streams and alpine lakes can alter BU processes in lakes and riparian food webs by intercepting emerging insects, thus translocating benthic P to pelagic communities (Schindler et al., 2001) and directly reducing reciprocal subsidies of emerging aquatic insects from aquatic habitats to riparian predators (Finlay and Vredenburg, 2007; Epanchin et al., 2010). Trout introductions also indirectly influence aquatic and riparian food webs through displacement of native trout species (Benjamin et al., 2011) or by competing with and shifting feeding behavior of native salmonids from reliance on terrestrial prey subsidies to benthic prey (Baxter et al., 2004). Additionally, non-native salmonids may displace native fish predators, alter strength of trophic cascades, and ultimately influence N cycling in streams (Townsend, 2003). For example, trout can achieve higher biomass and initiate stronger trophic cascades than native *Galaxis* spp. in New Zealand streams. Major shifts in trophic structure and energy flow, higher annual net production across all trophic levels, and greater N uptake and retention have been observed in streams invaded by trout (Flecker and Townsend, 1994; Huryn, 1998; Simon et al., 2004).

Invasive herbivores may alter nutrient dynamics through consumption and nutrient remineralization simply by attaining high biomass relative to other species, or by modifying the flux of limiting nutrients in invaded ecosystems (Strayer, 2012; Johnson et al., 2009). In some cases, invasive herbivores may act as ecosystem engineers and induce ecosystem state changes by drastically altering the producer community. For example, invasion by the macrophytophagous golden apple snail (*Pomacea canaliculata*) can cause a collapse of the aquatic plant community, strongly modify the flux of nutrients by translocating plant-bound nutrients to the water column via nutrient excretion, and shift systems from a plant-dominated, clear water state to a turbid, algae-dominated state (Carlsson et al., 2004). In contrast, invasion of filter-feeding mussels can induce changes in phytoplankton, zooplankton, and organic and inorganic particles, and redirect production and nutrients out of the water column and into benthic habitats (Strayer et al., 1999; Sousa et al., 2013).

Invasive herbivores that achieve high secondary production can dominate nutrient fluxes in streams. For example, in an invaded stream with high primary production and tight N cycling, New Zealand mud snails (*Potamopyrgus antipodarum*) dominate consumption of gross primary production (GPP) and regeneration of ammonium (NH_4), and represent N standing stocks equal to plant biomass (Hall et al., 2003). The high proportion of N demand met by this single introduced species is similar to entire assemblage estimates in other ecosystems (McIntyre et al., 2008). Invasions by herbivores that achieve high biomass and have unique body stoichiometry can have further impacts on nutrient cycles in streams. In a Mesoamerican river, introduced armored catfish (Loricariidae) biomass is two orders of magnitude greater than the native fish assemblage and loricariid excretion dominates the nutrient flux in the system, resulting in as much as a 95% reduction in the estimated distance required for fish excretion to turn over ambient pools of NH_4 and soluble reactive P (Capps and Flecker, 2013). Moreover, armored catfish growth is P-limited (Hood et al., 2005), and the loricariid population may act as a net N remineralizer and a net P sink (Capps and Flecker, 2013). Invasive herbivore impacts on nutrient cycling in aquatic habitats may also act synergistically with other anthropogenic drivers. For example, in Trinidad, experimental removal of riparian vegetation adjacent to stream reaches reduced light limitation of in-stream production, increased densities of invasive snails, and increased mass-specific excretion rates relative to shaded sites (Moslemi et al., 2012).

Managing eutrophication

One of the largest obstacles to protecting freshwater and downstream coastal marine systems is eutrophication resulting from anthropogenic inputs of N and P (Carpenter et al., 1998; Smith et al., 2006). Here again, understanding BU and TD interactions can be applied to an environmental issue threatening freshwater ecosystems. While manipulating BU processes through nutrient load reductions is an obvious strategy for controlling eutrophication, TD manipulations can have positive benefits as well. Biomanipulation (Shapiro et al., 1975) involves manipulating TD control in lakes either by selectively removing planktivorous fish or by inducing trophic cascades by enhancing predator biomass, which theoretically increases the presence of large-bodied *Daphnia* and subsequent grazing pressure on phytoplankton, resulting in a clear water state characteristic of pristine oligotrophic conditions (Carpenter et al., 1985; Hansson et al., 1998). More recently, biomanipulation has been applied to controlling benthivorous fish to reduce bioturbidation and nutrient cycling between benthic and pelagic habitats, as well as to remove substantial P stored in fish biomass (Case Study 3.2). Long-term data demonstrate that high removal rates of planktivorous and benthivorous fish can decrease chlorophyll a, TP, TN, and suspended solids for extended periods of time (6–8 years) in shallow eutrophic lakes (Søndergaard et al., 2008), but require continuous manipulation to maintain decreased productivity (Jeppesen et al., 2007). However, manipulating BU and TD forces in eutrophic freshwater ecosystems offers a promising tool for not only managing eutrophication, but also understanding long-term recovery, as internal fish and invertebrate mediated nutrient cycling can delay ecosystem responses to nutrient reductions through complex BU and TD interactions.

Conclusions

The formal study of BU and TD interactions is perhaps more advanced in freshwater ecosystems than any other biome, and much has been learned over the past few decades in terms of the magnitudes of, and interactions between, BU and TD forces. Yet we are just beginning to scratch the surface in terms of understanding the relative importance of various BU and TD processes, and how this understanding can be applied to help maintain sustainable aquatic ecosystems. The three integrative areas we focus on here represent emerging research directions and illustrate the complex ways in which BU and TD interactions are manifested and how our view of BU/TD interactions has changed over the years

(Fig. 3.1a,b). Recognizing these complex interactions should aid in the design of future research programs and help in the quest to understand variation among ecosystems in how top consumers and basal resources affect food webs and ecosystem processes. Along the same lines, recognizing that animals can affect ecosystems via a wide array of BU and TD processes will improve our ability to understand invasive species effects, biodiversity loss, and eutrophication. We also hope that this recognition can serve as a framework for future research. For example, studies examining the effect of a top predator (native or invasive) on aquatic ecosystems should benefit from considering the interactions and feedbacks we have discussed here. In many regards, the study of freshwaters led the way in demonstrating quantitatively and experimentally that both top predators and nutrient additions can regulate community and ecosystem properties. Similarly, there are many opportunities for freshwater studies to continue to unravel the myriad mechanisms and feedbacks that account for the complex interactions among these effects.

Acknowledgments

We gratefully acknowledge valuable comments and suggestions from Torrance Hanley, Kimberly La Pierre, Jeffrey Back, and Richard Lizotte. MJV's efforts were supported in part by NSF grants DEB 0918993 and 1255159, and ASF was supported in part by NSF grants EF 0623632 and DEB 1045960.

References

Allan, J. D., Abell, R., Hogan, Z., et al. (2005). Overfishing of inland waters. *Bioscience*, **55**, 1041–1051.

Allen, A. P. and Gillooly, J. F. (2009). Towards an integration of ecological stoichiometry and the metabolic theory of ecology to better understand nutrient cycling. *Ecology Letters*, **12**, 369–384.

Allen, D. C. and Vaughn, C. C. (2011). Density-dependent biodiversity effects on physical habitat modification by freshwater bivalves. *Ecology*, **92**, 1013–1019.

Arango, C. P., Riley, L. A., Tank, J. L. and Hall Jr., R. O. (2009). Herbivory by an invasive snail increases nitrogen fixation in a nitrogen-limited stream. *Canadian Journal of Fisheries and Aquatic Sciences*, **66**, 1309–1317.

Atkinson, C. L., Vaughn, C. C., Forshay, K. J. and Cooper, J. T. (2013). Aggregated filter-feeding consumers alter nutrient limitation: consequences for ecosystem and community dynamics. *Ecology*, **94**, 1359–1369.

Bartell, S. M. (1981). Potential impact of size-selective planktivory on phosphorus release by zooplankton. *Hydrobiologia*, **80**, 139–145.

Baxter, C. V., Fausch, K. D., Murakami, M. and Chapman, P. L. (2004). Fish invasion restructures stream and forest food webs by interrupting reciprocal prey subsidies. *Ecology*, **85**, 2656–2663.

Bayley, P. B. (1995). Understanding large river floodplain ecosystems. *Bioscience*, **45**, 153–158.

Benjamin, J. R., Fausch, K. D. and Baxter, C. V. (2011). Species replacement by a nonnative salmonid alters ecosystem function by reducing prey subsidies that support riparian spiders. *Oecologia*, **167**, 503–512.

Benstead, J. P., Cross, W. F., March, J. G., et al. (2010). Biotic and abiotic controls on the ecosystem significance of consumer excretion in two contrasting tropical streams. *Freshwater Biology*, **55**, 2047–2061.

Boersma, M., Aberle, N., Hantzsche, F. M., et al. (2008). Nutritional limitation travels up the food chain. *International Review of Hydrobiology*, **93**, 479–488.

Borer, E. T., Halpern, B. S. and Seabloom, E. W. (2006). Asymmetry in community regulation: effects of predators and productivity. *Ecology*, **87**, 2813–2820.

Brett, M. T. and Goldman, C. R. (1996). A meta-analysis of the freshwater tropic cascade. *Proceedings of the National Academy of Sciences of the USA*, **93**, 7723–7726.

Brett, M. T. and Goldman, C. R. (1997). Consumer versus resource control in freshwater pelagic food webs. *Science*, **275**, 384–386.

Brett, M. T., Muller-Navarra, D. C. and Park, S.-K. (2000). Empirical analysis of the effect of phosphorus limitation on algal food quality for freshwater zooplankton. *Limnology and Oceanography*, **45**, 1564–1575.

Capps, K. A. and Flecker, A. S. (2013). Invasive aquarium fish transform ecosystem nutrient dynamics. *Proceedings of the Royal Society B*, **280**, 20131520.

Carlsson, N. O. L., Brönmark C. and Hansson, L.-A. (2004). Invading herbivory: the golden apple snail alters ecosystem functioning in Asian wetlands. *Ecology*, **85**, 1575–1580.

Carpenter, S. R., Kitchell, J. F. and Hodgson, J. R. (1985). Cascading trophic interactions and lake productivity. *Bioscience*, **35**, 634–639.

Carpenter, S. R., Caraco, N. F., Correl, D. L., et al. (1998). Nonpoint pollution of surface waters with phosphorus and nitrogen. *Ecological Applications*, **8**, 559–568.

Carpenter, S. R., Cole, J. J., Hodgson, J. R., et al. (2001). Trophic cascades, nutrients, and lake productivity: whole-lake experiments. *Ecological Monographs*, **71**, 163–186.

Carpenter, S. R., Cole, J. J., Pace, M. L., et al. (2005). Ecosystem subsidies: terrestrial support of aquatic food webs from 13C addition to contrasting lakes. *Ecology*, **86**, 2737–2750.

Carpenter, S. R., Cole, J. J., Kitchell, J. F. and Pace, M. L. (2010). Trophic cascades in lakes: lessons and prospects. In *Trophic Cascades: Predators, Prey, and Changing Dynamics of Nature*, ed. J. Terborgh and J. A. Estes. Washington, DC: Island Press, pp. 55–69.

Catalano, M. J. and Allen, M. S. (2010). A whole-lake density reduction to assess compensatory responses of gizzard shad *Dorosoma cepedianum*. *Canadian Journal of Fisheries and Aquatic Sciences*, **68**, 955–968.

Catalano, M. J., Allen, M. S., Schaus, M. H., Buck, D. G. and Beaver, J. R. (2010). Evaluating short-term effects of omnivorous fish removal on water quality and zooplankton at a subtropical lake. *Hydrobiologia*, **655**, 159–169.

Cebrian, J. (1999). Patterns in the fate of production in plant communities. *American Naturalist*, **154**, 449–468.

Cebrian, J. and Lartigue, J. (2004). Patterns of herbivory and decomposition in aquatic and terrestrial ecosystems. *Ecological Monographs*, **74**, 237–259.

Conley, D. J., Paerl, H. W., Howarth, R. W., et al. (2009). Controlling eutrophication: nitrogen and phosphorus. *Science*, **323**, 1014–1015.

Creed Jr., R. P. and Reed J. M. (2004). Ecosystem engineering by crayfish in a headwater stream community. *Journal of the North American Benthological Society*, **23**, 224–236.

Cross, W. F., Wallace, J. B., Rosemond, A. D. and Eggert, S. L. (2006). Whole-system nutrient enrichment increases secondary production in a detritus-based ecosystem. *Ecology*, **87**, 1556–1565.

Danger, M., Funck, J. A., Devin, S., Heberle, J. and Felten, V. (2013). Phosphorus content in detritus controls life-history traits of a detritivore. *Functional Ecology*, **27**, 807–815.

Dickman, E. M., Vanni, M. J. and Horgan, M. J. (2006). Interactive effects of light and nutrients on phytoplankton stoichiometry. *Oecologia*, **149**, 676–689.

Dickman, E. M., Newell, J. M., González, M. J. and Vanni, M. J. (2008). Light, nutrients, and food-chain length constrain planktonic energy transfer efficiency across multiple trophic levels. *Proceedings of the National Academy of Sciences of the USA*, **105**, 18408–18412.

Dillon, P. J. and Rigler, F. H. (1976). The phosphorus-chlorophyll relationship in lakes. *Limnology and Oceanography*, **19**, 767–773.

Dodds, W. K., Smith, V. H. and Lohman, K. (2002). Nitrogen and phosphorus relationships to benthic algal biomass in temperate streams. *Canadian Journal of Fisheries and Aquatic Sciences*, **59**, 865–874.

Domis, L. N. D., Elser, J. J., Gsell, A. S., et al. (2013). Plankton dynamics under different climatic conditions in space and time. *Freshwater Biology*, **58**, 463–482.

Downing, J. A., Plante, C. and Lalonde, S. (1990). Fish production correlated with primary productivity, not the morphoedaphic index. *Canadian Journal of Fisheries and Aquatic Sciences*, **47**, 1929–1936.

Downing, J. A., Watson, S. B. and McCauley, E. (2001). Predicting Cyanobacteria dominance in lakes. *Canadian Journal of Fisheries and Aquatic Sciences*, **58**, 1905–1908.

Dugdale, R. C. and Goering, J. J. (1967). Uptake of new and regenerated forms of nitrogen in primary productivity. *Limnology and Oceanography*, **12**, 196–206.

Eby, L. A., Roach, W. J., Crowder, L. B. and Stanford, J. A. (2006). Effects of stocking-up freshwater food webs. *Trends in Ecology and Evolution*, **21**, 576–584.

Elser, J. J. and Urabe, J. (1999). The stoichiometry of consumer-driven nutrient recycling: theory, observations, and consequences. *Ecology*, **80**, 735–751.

Elser, J. J., Bracken, M. E. S., Cleland, E. E., et al. (2007). Global analysis of nitrogen and phosphorus limitiation of primary producers in freshwater, marine, and terrestrial ecosystems. *Ecology Letters*, **10**, 1135–1142.

Elton, C. (1927). *Animal Ecology*. London, UK: Sidgwick and Jackson.

Epanchin, P., Knapp, R. and Lawler, S. (2010). Nonnative trout impact an alpine-nesting bird by altering aquatic insect subsidies. *Ecology*, **91**, 2406–2415.

Finlay, J. C. and Vredenburg, V. T. (2007). Introduced trout sever trophic connections in watersheds: consequences for a declining amphibian. *Ecology*, **88**, 2187–2197.

Fittkau, E. J. (1970). Role of caimans in the nutrient regime of mouthlakes of Amazon affluents (a hypothesis). *Biotropica*, **2**, 138–142.

Flecker, A. S. (1996). Ecosystem engineering by a dominant detritivore in a diverse tropical stream. *Ecology*, **77**, 1845–1854.

Flecker, A. S. and Taylor, B. W. (2004). Tropical fishes as biological bulldozers: density effects on spatial heterogeneity and species diversity. *Ecology*, **85**, 2267–2278.

Flecker, A. S. and Townsend, C. R. (1994). Community-wide consequences of trout introduction in New Zealand streams. *Ecological Applications*, **4**, 798–807.

Flecker, A. S., Feifarek, B. P. and Taylor, B. W. (1999). Ecosystem engineering by a tropical tadpole: density-dependent effects on habitat structure and larval growth rates. *Copeia*, **1999**, 495–500.

Flecker, A. S., Taylor, B. W., Bernhardt, E. S., et al. (2002). Interactions between herbivorous fishes and limiting nutrients in a tropical stream ecosystem. *Ecology*, **83**, 1831–1844.

Flecker, A. S., McIntyre, P. B., Moore, J. W., et al. (2010). Migratory fishes as material and process subsidies in riverine ecosystems. In *Community Ecology of Stream Fishes: Concepts, Approaches, and Techniques*, ed. K. B. Gido and D. A. Jackson. Bethesda, MD: American Fisheries Society, Symposium 73, pp. 559–592.

Forrester, G. E., Dudley, T. L. and Grimm, N. B. (1999). Trophic interactions in open systems: effects of predators and nutrients on stream food chains. *Limnology and Oceanography*, **44**, 1187–1197.

Francoeur, S. N. (2001). Meta-analysis of lotic nutrient amendment experiments:

detecting and quantifying subtle responses. *Journal of the North American Benthological Society*, **20**, 358–368.

Frost, P. C. and Elser, J. J. (2002). Growth responses of littoral mayflies to the phosphorus content of their food. *Ecology Letters*, **5**, 232–240.

Gelwick, F. P. and Matthews, W. J. (1992). Effects of an algivorous minnow on temperate stream ecosystem properties. *Ecology*, **73**, 1630–1645.

Gillooly, J. F., Brown, J. H., West, G. B., Savage, V. M. and Charnov, E. L. (2001). Effects of size and temperature on metabolic rate. *Science*, **293**, 2248–2251.

Glass, J. B., Axler, R. P., Chandra, S. and Goldman, C. R. (2012). Molybdenum limitation of microbial nitrogen assimilation in aquatic ecosystems and pure cultures. *Frontiers in Microbiology*, **3**, 331.

Gottesfeld, A. S., Hassan, M. A., Tunnicliffe, J. F. and Poirier, R. W. (2004). Sediment dispersion in salmon spawning streams: the influence of floods and salmon redd construction. *Journal of American Water Resources Association*, **40**, 1071–1086.

Grimm, N. B. (1988a). Feeding dynamics, nitrogen budgets, and ecosystem role of a desert stream omnivore, *Agosia chrysogaster* (Pisces, Cyprinidae). *Environmental Biology of Fishes*, **21**, 143–152.

Grimm, N. B. (1988b). Role of macroinvertebrates in nitrogen dynamics of a desert stream. *Ecology*, **69**, 1884–1893.

Gruner, D. S., Smith, J. E., Seabloom, E. W., et al. (2008). A cross-system synthesis of consumer and nutrient resource control on producer biomass. *Ecology Letters*, **11**, 740–755.

Gulis, V. and Suberkropp, K. (2003). Leaf litter decomposition and microbial activity in nutrient-enriched and unaltered reaches of a headwater stream. *Freshwater Biology*, **48**, 123–134.

Gulis, V., Rosemond, A. D., Suberkropp, K., Weyers, H. S. and Benstead, J. P. (2004). Effects of nutrient enrichment on the decomposition of wood and associated microbial activity in streams. *Freshwater Biology*, **49**, 1437–1447.

Hairston, N. G., Smith, F. E. and Slobodkin, L. B. (1960). Community structure, population control, and competition. *The American Naturalist*, **94**, 421–425.

Hall Jr., R. O., Tank, J. L. and Dybdahl, M. F. (2003). Exotic snails dominate nitrogen and carbon cycling in a highly productive stream. *Frontiers in Ecology and Environment*, **1**, 407–411.

Hall, Jr., R. O., Taylor, B. W. and Flecker, A. S. (2011). Detritivorous fish indirectly reduce insect secondary production in a tropical river. *Ecosphere*, **2**, 135. DOI: 10.1890/ES11-00042.1

Hall, S. R., Leibold, M. A., Lytle, D. A. and Smith, V. H. (2007). Grazers, producer stoichiometry, and the light: nutrient hypothesis revisited. *Ecology*, **88**, 1142–1152.

Hambright, K. D., Drenner, R. W., McComas, S. R. and Hairston, N. G. (1991). Gape-limited piscivores, planktivore size refuges, and the trophic cascade hypothesis. *Archiv Fur Hydrobiologie*, **121**, 389–404.

Hambright, K. D., Zohary, T. and Gude, H. (2007). Microzooplankton dominate carbon flow and nutrient cycling in a warm subtropical freshwater lake. *Limnology and Oceanography*, **52**, 1018–1025.

Hansson, L.-A., Annadotter, H., Bergman, E., et al. (1998). Biomanipulation as an application of food chain theory: constraints, synthesis and recommendations for temperate lakes. *Ecosytems*, **1**, 558–574.

Hargrave, C. W., Ramírez, R., Brooks, M., et al. (2006). Indirect food web interactions increase growth of an algivorous stream fish. *Freshwater Biology*, **51**, 1901–1910.

Hassan, M. A., Gottesfeld, A. S., Montgomery, D. R., et al. (2008). Salmon-driven bed load transport and bed morphology in mountain streams. *Geophysical Research Letters*, **35**, L04405.

Havens, K. E., James, R. T., East, T. L. and Smith, V. H. (2003). N : P ratios, light limitation, and cyanobacterial dominance in a subtropical lake impacted by non-point source nutrient pollution. *Environmental Pollution*, **122**, 379–390.

Hill, W. R., Smith, J. G. and A. J. Stewart. (2010). Light, nutrients, and herbivore growth in oligotrophic streams. *Ecology*, **91**, 518–2750.

Hillebrand, H. (2002). Top-down versus bottom-up control of autotrophic biomass: a meta-analysis on experiments with periphyton. *Journal of the North American Benthological Society*, **21**, 349–369.

Hillebrand, H., de Montpellier, G. and Liess, A. (2004). Effects of macrograzers and light on periphyton stoichiometry. *Oikos*, **106**(1), 93–104.

Hood, J. M., Vanni, M. J. and Flecker, A. S. (2005). Nutrient recycling by two phosphorus-rich grazing catfish: the potential for phosphorus-limitation of fish growth. *Oecologia*, **146**, 247–257.

Hooper, D. U., Chapin, F. S., Ewel, J. J., et al. (2005). Effects of biodiversity on ecosystem functioning: a consensus of current knowledge. *Ecological Monographs*, **75**, 3–35.

Horppila, J., Peltonen, H., Malinen, T., Luokkanen, E. and Kairesalo, T. (1998). Top-down or bottom-up effects by fish: issues of concern in biomanipulation of lakes. *Restoration Ecology*, **6**, 20–28.

Huryn, A. D. (1998). Ecosystem-level evidence for top-down and bottom-up control of production in a grassland stream system. *Oecologia*, **115**, 173–183.

Hynes, H. B. N. (1975). The stream and its valley. *Proceedings of the International Association of Theoretical and Applied Limnology*, **19**, 1–16.

Jeppesen, E., Jensen, J. P., Jensen, C., et al. (2003). The impact of nutrient state and lake depth on top-down control in the pelagic zone of lakes: a study of 466 lakes from the temperate zone to the Arctic. *Ecosystems*, **6**, 313–325.

Jeppesen, E., Jensen, J. P., Søndergaard, M. and Jauridsen, T. L. (2005). Response of fish and plankton to nutrient loading reduction in eight shallow Danish lakes with special

emphasis on seasonal dynamics. *Freshwater Biology*, **50**, 1616–1627.

Jeppesen, E., Meerhoff, M., Jacobsen, B. A., et al. (2007). Restoration of shallow lakes by nutrient control and biomanipulation: the successful strategy varies with lake size and climate. *Hydrobiologia*, **581**, 269–285.

Johnson, B. R. and Wallace, J. B. (2005). Bottom-up limitation of a stream salamander in a detritus-based food web. *Canadian Journal of Fisheries and Aquatic Sciences*, **62**, 301–311.

Johnson, C. R., Luecke, C., Whalen, S. C. and Evans, M. A. (2010). Direct and indirect effects of fish on pelagic nitrogen and phosphorus availability in oligotrophic Arctic Alaskan lakes. *Canadian Journal of Fisheries and Aquatic Sciences*, **67**, 1635–1648.

Johnson, P. T. J., Olden, J. D., Solomon, C. T. and Vander Zanden, M. J. (2009). Interactions among invaders: community and ecosystem effects of multiple invasive species in an experimental aquatic system. *Oecologia*, **159**, 161–170.

Jones, C. G., Lawton, J. H. and Shachak, M. (1994). Organisms as ecosystem engineers. *Oikos*, **69**, 373–386.

Junk, W., Bayley, P. and Sparks, R. (1989). The flood-pulse concept in river-floodplains systems. In *Proceedings of the International Large River Symposium*, ed. D. Dodge. Canada: Canadian Special Publication of Fisheries and Aquatic Sciences, pp. 110–127.

Kishi, D., Murakami, M., Nakano, S. and Maekawa, K. (2005). Water temperature determines strength of top-down control in a stream food web. *Freshwater Biology*, **50**, 1315–1322.

Klausmeier, C. A., Litchman, E., Daufresne, T. and Levin, S. A. (2004). Optimal nitrogen-to-phosphorus stoichiometry of phytoplankton. *Nature*, **429**, 171–174.

Knoll, L. B., McIntyre, P. B., Vanni, M. J. and Flecker, A. S. (2009). Feedbacks of consumer nutrient recycling on producer biomass and stoichiometry: separating direct and indirect effects. *Oikos*, **118**, 1732–1742.

Kurle, C. M. and Cardinale, B. J. (2011). Ecological factors associated with the strength of trophic cascades in streams. *Oikos*, **120**, 1897–1908.

Leibold, M. A. (1989). Resource edibility and the effects of predators and productivity on the outcome of trophic interactions. *The American Naturalist*, **134**, 922–949.

Leroux, S. J. and Loreau, M. (2008). Subsidy hypothesis and strength of trophic cascades across ecosystems. *Ecology Letters*, **11**, 1147–1156.

Lewis Jr., W. M. and Wurtsbaugh, W. A. (2008). Control of lacustrine phytoplankton by nutrients: erosion of the phosphorus paradigm. *International Review of Hydrobiologia*, **93**, 446–465.

Liess, A. and Kahlert, M. (2007). Gastropod grazers and nutrients, but not light, interact in determining periphytic algal diversity. *Oecologia*, **152**, 101–111.

Liess, A. and Lange, K. (2011). The snail *Potamopyrgus antipodarum* grows faster and is more active in the shade, independent of food quality. *Oecologia*, **167**(1), 85–96.

Lindeman, R. L. (1942). The trophic-dynamic aspect of ecology. *Ecology*, **23**, 399–417.

Locke, M. A., Knight, S. S., Smith Jr., S., et al. (2008). Environmental quality research in the Beasley Lake watershed, 1995–2007: succession from conventional to conservation practices. *Journal of Soil and Water Conservation*, **63**, 430–442.

MacKay, N. A. and Elser, J. J. (1998). Nutrient recycling by Daphnia reduces N-2 fixation by cyanobacteria. *Limnology and Oceanography*, **43**, 347–354.

Malzahn, A. M., Aberle, N., Clemmesen, C. and Boersma, M. (2007). Nutrient limitation of primary producers affects planktivorous fish condition. *Limnology and Oceanography*, **52**, 2062–2071.

Malzahn, A. M., Hantzsche, F., Schoo, K. L., Boersma, M. and Aberle, N. (2010). Differential effects of nutrient-limited primary production on primary, secondary or tertiary consumers. *Oecologia*, **162**, 35–48.

Matthews, W. J. (1998). *Patterns in Freshwater Fish Ecology*. New York: Chapman & Hall.

McIntyre, P. B., Jones, L. E., Flecker, A. S. and Vanni, M. J. (2007). Fish extinctions alter nutrient recycling in tropical freshwaters. *Proceedings of the National Academy of Sciences of the USA*, **104**, 4461–4466.

McIntyre, P. B., Flecker, A. S., Vanni, M. J., et al. (2008). Fish distributions and nutrient cycling in streams: can fish create biogeochemical hotspots? *Ecology*, **89**, 2335–2346.

McManamay, R. A., Webster, J. R., Valett, H. M. and Dolloff, C. A. (2011). Does diet influence consumer nutrient cycling? Macroinvertebrate and fish excretion in streams. *Journal of the North American Benthological Society*, **30**, 84–102.

McQueen, D. J., Post, J. R. and Mills, E. L. (1986). Trophic relationships in freshwater pelagic ecosystems. *Canadian Journal of Fisheries and Aquatic Sciences*, **43**, 1571–1581.

Mette, E. M., Vanni, M. J., Newell, J. M. and Gonzalez, M. J. (2011). Phytoplankton communities and stoichiometry are interactively affected by light, nutrients, and fish. *Limnology and Oceanography*, **56**, 1959–1975.

Meyer, J. L. and Wallace, J. B. (2001). Lost linkages and lotic ecology: rediscovering small streams. In *Ecology: Achievement and Challenge*, ed. M. Press, N. Huntly and S. Levin. Oxford, UK: Blackwell Science, pp. 295–317.

Mills, K. H. and Chalanchuk, S. M. (1987). Population-dynamics of lake whitefish (*Coregonus clupeaformis*) during and after the fertilization of Lake 226, The Experimental Lakes Area. *Canadian Journal of Fisheries and Aquatic Sciences*, **44** (supplement 1), 55–63.

Moore, J. W. (2006). Animal ecosystem engineers in streams. *BioScience*, **56**, 237–246.

Moore, J. W. and Schindler, D. E. (2008). Biotic disturbance and benthic community dynamics in salmon-bearing streams. *Journal of Animal Ecology*, **77**, 275–284.

Moslemi, J. M., Snider, S. B., MacNeill, K., Gilliam, J. F. and Flecker, A. S. (2012). Impacts of an invasive snail (*Tarebia granifera*) on nutrient cycling in tropical streams: the role of riparian deforestation in Trinidad, West Indies. *PLos One*, **7**, e38806.

Naiman, R. J., Melillo, J. M. and Hobbie, J. E. (1986). Ecosystem alteration of boreal forest streams by beaver (*Castor canadensis*). *Ecology*, **67**, 1254–1269.

Naiman, R. J., Bilby, R. E., Schindler, D. E. and Helfield, J. M. (2002). Pacific salmon, nutrients, and the dynamics of freshwater and riparian ecosystems. *Ecosystems*, **5**, 399–417.

Naiman, R. J., Helfield, J. M., Bartz, K. K., Drake, D. C. and Honea, J. M. (2009). Pacific salmon, marine-derived nutrients, and the characteristics of aquatic and riparian ecosystems. *American Fisheries Society Symposium*, **69**, 395–425.

Nakano, S., Miyasaka, H. and Kuhara, N. (1999). Terrestrial-aquatic linkages: riparian arthropod inputs alter trophic cascades in a stream food web. *Ecology*, **80**, 2435–2441.

Pace, M. L., Cole, J. J., Carpenter, S. R. and Kitchell, J. F. (1999). Trophic cascades revealed in diverse ecosystems. *Trends in Ecology and Evolution*, **14**, 483–488.

Pace, M. L., Cole, J. J., Carpenter, S. R., et al. (2004). Whole-lake carbon-13 additions reveal terrestrial support of aquatic food webs. *Nature*, **427**, 240–243.

Paine, R. T. (1980). Food webs: linkage, interaction strength and community infrastructure. *The Journal of Animal Ecology*, **49**, 666–685.

Peckarsky, B. L., McIntosh, A. R., Álvarez, M. and Moslemi, J. M. (2013). Nutrient limitation controls the strength of behavioral trophic cascades in high elevation streams. *Ecosphere*, **4**, 110.

Peterson, B., Fry, B., Deegan, L. and Hershey, A. (1993). The trophic significance of epilithic algal production in a fertilized tundra river ecosystem. *Limnology and Oceanography*, **38**, 872–878.

Peterson, D. P. and Foote, C. J. (2000). Disturbance of small-stream habitat by spawning

sockeye salmon in Alaska. *Transactions of the American Fisheries Society*, **129**, 924–934.

Polis, G. A., Anderson, W. B. and Holt, R. D. (1997). Toward an integration of landscape and food web ecology: the dynamics of spatially subsidized food webs. *Annual Review of Ecology and Systematics*, **28**, 289–316.

Pollock, M. M., Heim, M. and Naiman, R. J. (2003). Hydrologic and geomorphic effects of beaver dams and their influence on fishes. In *The Ecology and Management of Wood in World Rivers*, ed. S. V. Gregory, K. Boyer and A. Gurnell. Bethesda, MD: American Fisheries Society, pp. 213–234.

Power, M. E. (1990). Resource enhancement by indirect effects of grazers: armored catfish, algae, and sediment. *Ecology*, **71**, 897–904.

Power, M. E. (1992). Top-down and bottom-up forces in food webs: do plants have primacy? *Ecology*, **73**, 733–746.

Power, M. E., Matthews, W. J. and Stewart, A. J. (1985). Grazing minnows, piscivorous bass, and stream algae: dynamics of a strong interaction. *Ecology*, **66**, 1448–1456.

Power, M. E., Parker, M. S. and Dietrich, W. E. (2008). Seasonal reassembly of a river food web: floods, droughts, and impacts of fish. *Ecological Monographs*, **78**, 263–282.

Pringle, C. M., Blake, G. A., Covich, A. P., Buzby, K. M. and Finley, A. (1993). Effects of omnivorous shrimp in a montane tropical stream: sediment removal, disturbance of sessile invertebrates and enhancement of understory algal biomass. *Oecologia*, **93**, 1–11.

Riley, R. H., Townsend, C. R., Raffaelli, D. A. and Flecker, A. S. (2004). Sources and effects of subsidies along the stream-estuary continuum. In *Food Webs at the Landscape Level*, ed. G. A. Polis, M. E. Power and G. R. Huxel. Chicago, IL: The University of Chicago Press, pp. 241–267.

Rosemond, A. D., Pringle, C. M., Ramirez, A., Paul, M. J. and Meyer, J. L. (2002). Landscape variation in phosphorus concentration and effects on detritus-based tropical streams. *Limnology and Oceanography*, **47**, 278–289.

Schaus, M. H., Vanni, M. J., Wissing, T. E., Bremigan, M. T., Garvey, J. E. and Stein, R. A. (1997). Nitrogen and phosphorus excretion by detritivorous gizzard shad in a reservoir ecosystem. *Limnology and Oceanography*, **42**, 1386–1397.

Schaus, M. H., Vanni, M. J. and Wissing, T. E. (2002). Biomass-dependent diet shifts in omnivorous gizzard shad: implications for growth, food web, and ecosystem effects. *Transactions of the American Fisheries Society*, **131**, 40–54.

Schaus, M. H., Godwin, W., Battoe, L., et al. (2010). Impact of the removal of gizzard shad (*Dorosoma cepedianum*) on nutrient cycles in Lake Apopka, Florida. *Freshwater Biology*, **55**, 2401–2413.

Scheffer, M. and Carpenter, S. R. (2003). Catastrophic regime shifts in ecosystems: linking theory to observation. *Trends in Ecology and Evolution*, **18**, 648–656.

Scheuerell, M. D., Moore, J. W., Schindler, D. E. and Harvey, C. J. (2007). Varying effects of anadromous sockeye salmon on the trophic ecology of two species of resident salmonids in Southwest Alaska. *Freshwater Biology*, **52**, 1944–1956.

Schindler, D. E., Kitchell, J. F., He, X., Carpenter, S. R., Hodgson, J. R. and Cottingham, K. L. (1993). Food-web structure and phosphorus cycling in lakes. *Transactions of the American Fisheries Society*, **122**, 756–772.

Schindler, D. E., Knapp, R. A. and Leavitt, P. R. (2001). Alteration of nutrient cycles and algal production resulting from fish introductions into mountain lakes. *Ecosystems*, **4**, 308–321.

Schindler, D. E., Scheuerell, M. D., Moore, J. W., et al. (2003). Pacific salmon and the ecology of coastal ecosystems. *Frontiers in Ecology and the Environment*, **1**, 31–37.

Schindler, D. W. (1977). Evolution of phosphorus limitation in lakes: natural mechanisms compensate for deficiencies of nitrogen and carbon in eutrophied lakes. *Science*, **195**, 260–262.

Schindler, D. W. (1978). Factors regulating phytoplankton production and standing crop in the world's freshwaters. *Limnology and Oceanography*, **23**, 478–486.

Schoo, K. L., Aberle, N., Malzahn, A. M. and Boersma, M. (2012). Food quality affects secondary consumers even at low qualities: an experimental test with larval European lobster. *PLoS One*, **7**, e33550.

Shapiro, J., Lamarra, V. and Lynch, M. (1975). Biomanipulation: an ecosystem approach to lake restoration. In *Symposium on Water Quality Management through Biological Control, Gainesville, FL.* ed. P. L. Brezonik and J. L. Fox. Gainesville, FL: University of Florida, pp. 85–96.

Shostell, J. and Bukaveckas, P. A. (2004). Seasonal and interannual variation in N and P fluxes associated with tributary inputs, consumer recycling and algal growth. *Aquatic Ecology*, **38**, 359–373.

Shurin, J. B., Gruner, D. S. and Hillebrand, H. (2006). All wet or dried up? Real differences between aquatic and terrestrial food webs. *Proceedings of the Royal Society of London B*, **273**, 1–9.

Simon, K. S., Townsend, C. R., Biggs, B. J. F., Bowden, W. B. and Frew, R. D. (2004). Habitat-specific nitrogen dynamics in New Zealand streams containing native or invasive fish. *Ecosystems*, **7**, 777–792.

Small, G. E., Pringle, C. M., Pyron, M. and Duff, J. H. (2011). Role of the fish *Astyanax aeneus* (Characidae) as a keystone nutrient recycler in low-nutrient Neotropical streams. *Ecology*, **92**, 386–397.

Smith, V. H. 1983. Low nitrogen to phosphorus ratios favor dominance by blue-green-algae in lake phytoplankton. *Science*, **221**, 669–671.

Smith, V. H. and Schindler, D. W. (2009). Eutrophication science: where do we go from here? *Trends in Ecology and Evolution*, **24**, 201–207.

Smith, V. H., Joye, S. B. and Howarth, R. W. (2006). Eutrophication of freshwater and marine ecosystems. *Limnology and Oceanography*, **51**, 351–355.

Søndergaard, M., Liboriussen, L., Pedersen, A. R. and Jeppesen, E. (2008). Lake restoration by fish removal: short- and long-term effects in 36 Danish lakes. *Ecosystems*, **11**, 1291–1305.

Sousa, R., Gutiérrez, J. L. and Aldridge, D. C. (2009). Non-indigenous invasive bivalves as ecosystem engineers. *Biological Invasions*, **11**, 2367–2385.

Sousa, R., Novais, A., Costa, R. and Strayer, D. L. (2013). Invasive bivalves in fresh waters: impacts from individuals to ecosystems and possible control strategies. *Hydrobiologia*, in press.

Spooner, D. E., Frost, P. C., Hillebrand, H., et al. (2013). Nutrient loading associated with agriculture land use dampens the importance of consumer-mediated niche construction. *Ecology Letters*, **16**(9), 1115–1125.

Sterner, R. W. and Elser, J. J. (2002). *Ecological Stoichiometry: the Biology of Elements from Molecules to the Biosphere*. Princeton, NJ: Princeton University Press.

Sterner, R. W., Elser, J. J., Fee, E. J., Guildford, S. J. and Chrzanowski, T. H. (1997). The light:nutrient ratio in lakes: the balance of energy and materials affects ecosystem structure and process. *American Naturalist*, **150**, 663–684.

Strayer, D. L. (2010). Alien species in fresh waters: ecological effects, interactions with other stressors, and prospects for the future. *Freshwater Biology*, **55**, 152–174.

Strayer, D. L. (2012). Eight questions about invasions and ecosystem functioning. *Ecology Letters*, **15**, 1199–1210.

Strayer, D. L. and Dudgeon, D. (2010). Freshwater biodiversity conservation: recent progress and future challenges. *Journal of the North American Benthological Society*, **29**, 344–358.

Strayer, D. L., Caraco, N. F., Cole, J. J., Findlay, S. and Pace, M. L. (1999). Transformation of freshwater ecosystems by bivalves: a case study of zebra mussels in the Hudson River. *Bioscience*, **49**, 19–27.

Striebel, M., Sporl, G. and Stibor, H. (2008). Light-induced changes of plankton growth and stoichiometry: experiments with natural phytoplankton communities. *Limnology and Oceanography*, **53**, 513–522.

Striebel, M., Behl, S. and Stibor, H. (2009). The coupling of biodiversity and productivity in phytoplankton communities: consequences for biomass stoichiometry. *Ecology*, **90**, 2025–2031.

Taylor, B. W., Flecker, A. S. and Hall Jr., R. O. (2006). Loss of harvested fish species disrupts carbon flow in a diverse tropical river. *Science*, **313**, 833–836.

Taylor, J. M., Back, J. A. and King, R. S. (2012a). Grazing minnows increase benthic autotrophy and enhance the response of periphyton elemental composition to experimental phosphorus additions. *Freshwater Science*, **31**, 451–462.

Taylor, J. M., Back, J. A., Valenti, T. W. and King, R. S. (2012b). Fish-mediated nutrient cycling and benthic microbial processes: can consumers influence stream nutrient cycling at multiple spatial scales? *Freshwater Science*, **31**, 928–944.

Tessier, A. J. and Woodruff, P. (2002). Cryptic trophic cascade along a gradient of lake size. *Ecology*, **83**, 1263–1270.

Torres, L. E. and Vanni, M. J. (2007). Stoichiometry of nutrient excretion by fish: interspecific variation in a hypereutrophic lake. *Oikos*, **116**, 259–270.

Townsend, C. R. (2003). Individual, population, community, and ecosystem consequences of a fish invader in New Zealand streams. *Conservation Biology*, **17**, 38–47.

Vadeboncoeur, Y., Jeppesen, E., Vander Zanden, M. J., et al. (2003). From Greenland to green lakes: cultural eutrophication and the loss of benthic pathways in lakes. *Limnology and Oceanography*, **48**, 1408–1418.

Vander Zanden, M. J., Vadeboncoeur, Y. and Chandra, S. (2011). Fish reliance on littoral-benthic resources and the distribution of primary production in lakes. *Ecosystems*, **14**, 894–903.

Vanni, M. J. (2002). Nutrient cycling by animals in freshwater ecosystems. *Annual Review of Ecology and Systematics*, **33**, 341–370.

Vanni, M. J., Flecker, A. S., Hood, J. M. and Headworth, J. L. (2002). Stoichiometry of nutrient cycling by vertebrates in a tropical stream: linking species identity and ecosystem processes. *Ecology Letters*, **5**, 285–293.

Vanni, M. J., Arend, K. K., Bremigan, M. T., et al. (2005). Linking landscapes and food webs: effects of omnivorous fish and watersheds on reservoir ecosystems. *BioScience*, **55**, 155–167.

Vanni, M. J., Bowling, A. M., Dickman, E. M., et al. (2006). Nutrient cycling by fish supports relatively more primary production as lake productivity increases. *Ecology*, **87**, 1696–1709.

Vannote, R. L., Minshall, G. W., Cummins, K. W., Sedell, J. R. and Cushing, C. E. (1980). The river continuum concept. *Canadian Journal of Fisheries and Aquatic Sciences*, **37**, 130–137.

Vaughn, C. C. (2010). Biodiversity losses and ecosystem function in freshwaters: emerging conclusions and research directions. *Bioscience*, **60**, 25–35.

Wallace, J. B., Eggert, S. L., Meyer, J. L. and Webster, J. R. (1999). Effects of resource limitation on a detrital-based ecosystem. *Ecological Monographs*, **69**, 409–442.

Wetzel, R. G. (2001). *Limnology – Lake and River Ecosystems*. San Diego, CA: Elsevier/Academic Press.

Whiles, M. R., Lips, K. R., Pringle, C. M., et al. (2006). The effects of amphibian population declines on the structure and function of Neo-tropical stream ecosystems. *Frontiers in Ecology and the Environment*, **4**, 27–34.

Whiles, M. R., Hall Jr., R. O., Dodds, W. K., et al. (2013). Disease-driven amphibian declines alter ecosystem processes in a tropical stream. *Ecosystems*, **16**, 146–157.

Willson, M. F., Gende, S. M. and Marston, B. H. (1998). Fishes and the forest. *Bioscience*, **48**, 455–462.

Wilson, H. F. and Xenopoulos, M. A. (2011). Nutrient recycling by fish in streams along a gradient of agricultural land use. *Global Change Biology*, **17**, 130–139.

Winemiller, K. O. and Jepsen, D. B. (2004). Migratory neotropical fish subsidize food webs of oligotrophic blackwater rivers. In *Food Webs at the Landscape Level*, ed. G. A. Polis, M. E. Power and G. R. Huxel. Chicago, IL: University of Chicago Press, pp. 115–132.

Winemiller, K. O., Tarim, S., Shormann, D. and Cotner, J. B. (2000). Fish assemblage structure in relation to environmental variation among Brazos River oxbow lakes. *Transactions of the American Fisheries Society*, **129**, 451–468.

Wootton, J. T. and Power, M. E. (1993). Productivity, consumers, and the structure of a river food chain. *Proceedings of the National Academy of Sciences of the USA*, **90**, 1384–1387.

CHAPTER FOUR

Top-down and bottom-up interactions determine tree and herbaceous layer dynamics in savanna grasslands

A. CARLA STAVER

Yale University, New Haven, CT, USA

and

SALLY E. KOERNER

Colorado State University, Fort Collins, CO, USA

Introduction

The savanna biome is defined, broadly, as comprising ecosystems in which trees and grass coexist, where the grass layer is approximately continuous and the tree layer discontinuous (Scholes and Archer, 1997; Bond, 2008). This definition encompasses a diverse range of vegetation types, from savanna woodlands at the woody extreme to open savannas or even grasslands at the other. A growing body of evidence from small (Hennenberg et al., 2006) to large scales (Staver et al., 2011a; 2011b) suggests that savanna is an ecologically meaningful biome categorization. Where savannas transition to forests, savanna boundaries are abrupt and discontinuous, marked by thresholds in canopy cover associated with grass persistence (Hennenberg et al., 2006; Lloyd et al., 2008) and fire spread (Archibald et al., 2009; Staver et al., 2011b). The degree to which savannas differ qualitatively from grasslands at the biome level – whether at their arid, flooding, or elevational extremes – is less clear. Some argue that the transition from savanna to grassland is also abrupt and discontinuous (Hirota et al., 2011), while others suggest, based on similarities in ecological processes, that grasslands may represent the treeless extreme of savannas (Bond, 2008; Wakeling et al., 2012). Here, on the basis of these similarities in trophic ecology, we discuss interactions between top-down and bottom-up ecological processes in savannas and grasslands together, and hereafter we refer to the entire gradient as savanna grassland (*sensu* Scholes and Walker, 1993).

Savanna grasslands represent a fundamental challenge for ecology in two ways. First, the climatic and edaphic variables that have historically been used to

Trophic Ecology: Bottom-Up and Top-Down Interactions across Aquatic and Terrestrial Systems, eds T. C. Hanley and K. J. La Pierre. Published by Cambridge University Press. © Cambridge University Press 2015.

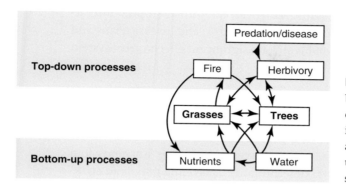

Figure 4.1 Interactions among bottom-up and top-down determinants of tree–grass ratios in savanna and grassland are abundant, resulting in substantial unpredictability in ecosystem structure.

map global biome distributions (Holdridge, 1947; Whittaker, 1975) do a uniquely poor job in describing the limits of savanna grasslands (Whittaker, 1975; Bond et al., 2005). Second, the long-term coexistence of two vastly different functional types – trees and grasses – presents a challenge not just for savanna ecology (Scholes and Archer, 1997), but for ecology more generally (Hutchinson, 1959; Tilman and Pacala, 1994).

These same challenges make savanna grasslands of particular interest in thinking about interactions between top-down and bottom-up ecological processes. Among the likely drivers of savanna grassland dynamics and distributions are processes related to resource availability – including both water and nutrients (Scholes and Archer, 1997; Knapp et al., 2001). However, top-down processes are also thought to be particularly important in these systems (Fig. 4.1). First, large herbivores are abundant (Olff et al., 2002) and may play an unparalleled role in the ecology of savanna grasslands (Knapp et al., 1999; Bond, 2008). Second, fire represents an alternative top-down ecological process in savanna grasslands (Bond, 2005; Bond and Keeley, 2005), as it does in other fire-prone systems (e.g., Mediterranean systems like fynbos and chaparral, as well as boreal and coniferous forests; see Chapter 5). Fires are frequent (Archibald et al., 2009) and probably fundamental both in expanding savanna grassland distributions (Bond et al., 2005; Staver et al., 2011a; 2011b) and in shaping their dynamics (Trollope and Tainton, 1986; Collins, 1992; Higgins et al., 2000; Hoffmann and Moreira, 2003).

Only recently have savanna grassland ecologists begun to explicitly consider how interactions among top-down and bottom-up processes, although they clearly occur, shape the distribution and structure of the biome as a whole (Sankaran et al., 2004; Koerner and Collins, 2014). The relative importance of water limitation versus fire in determining ecosystem structure and dynamics, for instance, clearly changes along gradients of water limitation (Bond et al., 2005; Staver et al., 2011a). However, we have, as yet, no comprehensive understanding of how interactions between resource availability, herbivory, and fire change along environmental gradients. Moreover, additional trophic interactions due to, for instance, predation (Loarie et al., 2013) and disease (Prins and van der Jeugd, 1993; Holdo et al., 2009b) can have significant effects on ecosystem structure, yet are rarely considered.

Bottom-up ecological processes in savanna grassland

Effects of soil resources on trees and grass in savanna grassland

Although not perfectly predictive, soil resources – including nutrients and water – are among the fundamental determinants of savanna grassland structure and dynamics, via effects on both the tree and grass layers. Perhaps the most striking pattern evident in savanna grassland is a strong response of tree cover to rainfall at continental scales (Sankaran et al., 2005; Bucini and Hanan, 2007; Good and Caylor, 2011; Hirota et al., 2011; Staver et al., 2011a; 2011b), which suggests that water availability directly limits tree cover in areas that receive up to 650 mm or 1000 mm mean annual precipitation (MAP) (Bond et al., 2005; Sankaran et al., 2005; Staver et al., 2011b), depending on intra-annual variability (Good and Caylor, 2011; Staver et al., 2011a). Above 1000 mm MAP, savanna grasslands exist in abundance, but fire – probably not water – limits tree cover and prevents the formation of a closed canopy (Bond et al., 2005; Staver et al., 2011a). Soil effects are harder to generalize (Staver et al., 2011a), and the relationship of nutrient availability to tree cover is largely unknown (Bond, 2010). Soil can clearly affect tree density and cover (Williams et al., 1996). Moreover, detailed experimental work, although scarce, suggests that nitrogen (N) may limit tree growth in African and North American savanna grasslands in particular (Reich et al., 1997; Cramer et al., 2007; 2010; Cech et al., 2008), while South American savanna grasslands, by contrast, are probably limited by phosphorus (P) availability (Bustamante et al., 2004; 2006; Nardoto et al., 2006; Viani et al., 2011). Nonetheless, our current understanding of the role and dynamics of nutrient limitation in savanna grasslands is preliminary.

The response of grass biomass to precipitation is equally striking and even more predictable. Over large spatial gradients, grass biomass increases linearly from c. 200 to 1300 mm MAP (Sala et al., 1988; Paruelo et al., 1998; Balfour and Howison, 2002; Govender et al., 2006; Smit et al., 2012), depending on interactions with soil nutrient availability. Savanna grassland can also be limited by both N and P, either independently or in co-limitation (Elser et al., 2007), although nutrient effects are often evident only in mesic systems (Ladwig et al., 2012). Nutrient additions, especially in grasslands, often result in increases in productivity. However, unlike in many terrestrial ecosystems, where increases in productivity can increase species richness (Pianka, 1966; Tilman et al., 1996; Hawkins et al., 2003), increasing productivity often decreases herbaceous species richness in savanna grassland (Huenneke et al., 1990; Kirkman et al., 2014). This suggests an evolved response by grasses to nutrient limitation – species drop out either as a direct result of nutrient availability or as a byproduct of higher biomass and associated light limitation.

Bottom-up mechanisms for tree–grass coexistence in savanna grassland

Long-term coexistence of species – or even in this case functional types – on the basis of resource competition is a classic question in ecology (Hutchinson, 1959; Tilman and Pacala, 1994). Perhaps the oldest explanation, and one that has received abundant attention in the savanna grassland literature, is that species or functional types coexist because each uses a different resource, thereby occupying different niches (Macarthur and Levins, 1967; Tilman and Pacala, 1994). In savanna grassland, the idea of niche separation has been predominantly considered in regard to competition for soil resources, including water and nutrients; the "Walter hypothesis" suggests that trees and grasses can coexist because grasses use water and nutrients near the soil surface, while trees use water from deeper soil layers (Walter, 1971).

In some contexts, trees and grass do appear to use water from different soil layers (February and Higgins, 2010; Kulmatiski and Beard, 2012). Root niche differentiation may even contribute to species coexistence within the herbaceous layer (Rossatto et al., 2012). However, evidence of vertical root niche separation is not sufficient evidence that trees and grasses do not compete; the vertical movement of water in soil is substantial, such that roots in shallow soil layers can have considerable effects on soil water in deeper layers (D'odorico et al., 2007; Caylor et al., 2009). In fact, experimental and observational evidence suggest that competition between trees and grasses can be significant (Cramer et al., 2007; 2010; February et al., 2013) and that grasses frequently outcompete trees in both nutrient and water uptake (Cramer et al., 2010; February et al., 2013). Moreover, trees appear to allocate increasing amounts of carbon to roots along broad-scale resource gradients, a pattern highly inconsistent with theories of resource competition and limitation (Bhattachan et al., 2012); this suggests that while these may be important driving factors, they are certainly not the only determinants of tree strategy in savanna grassland.

Others have suggested that instead of – or possibly in addition to – vertical root niche separation, trees and grasses occupy somewhat different temporal niches in savanna grassland systems (Scholes and Walker, 1993). Leaf-out phenology appears to be somewhat different in trees than it is in grasses (Archibald and Scholes, 2007; Higgins et al., 2011); trees may use stored nutrients and water to flush earlier and thus exploit early rains more than grasses (Scholes and Walker, 1993). However, detailed examination suggests that temporal separation in tree and grass resource use is slight and incomplete, at best providing a weak mechanism to reinforce tree–grass coexistence (Higgins et al., 2011).

The storage effect provides an alternative framework for considering functional coexistence in the face of significant environmental variability (Chesson, 2000a). Temporal variability, in which some periods favor populations of one type and other periods another, promotes the coexistence of long-lived organisms (or those with long-lived seeds) or, in spatial contexts, coexistence more generally (Horn and Macarthur, 1972; Chesson, 2000b). Savanna grasslands certainly occur in regions of the world with high levels of temporal variability (Scholes and Archer, 1997; Sankaran et al., 2004), and the storage effect provides a plausible mechanism to promote coexistence among grass species and of grass and forb species within grasslands (Facelli et al., 2005). However, the storage effect may also be relevant more broadly to functional type coexistence between trees and grasses (Higgins et al., 2000).

The effects of rainfall variability have been closely considered in savanna grassland, but variation in nutrient availability is neither well quantified nor its effects well understood. In arid savanna grasslands, seedling establishment occurs primarily in years with exceptionally high rainfall (O'Connor, 1995; Watson et al., 1997; Wiegand et al., 2004). Rainfall variability also directly affects the growth rates of savanna trees once they have established (Miller et al., 2001). In more mesic savanna grasslands, large-tree emergence rates can vary tremendously over long periods of time, potentially driven in part by variability in rainfall (Staver et al., 2011c). Droughts can also drive large mortality events, with major effects on ecosystem structure, in both arid (Gillson, 2006; Fensham et al., 2009) and mesic systems (Phillips et al., 2009). Thus, it is fairly clear that trees in many savanna grasslands seem to follow a storage effect population dynamic with respect to resource availability: they rely on periods of relatively high rainfall to persist, and populations persist in the long term because trees are long-lived.

Grass biomass responses to both rainfall and nutrients are also positive, although grass responses are much more predictable than those of trees. Grass productivity responds on a relatively short time scale (within a single growing season) to changes in water and nutrient availability (Knapp et al., 2001; February et al., 2013). However, variability in savanna grassland annual net primary productivity may also be subject to time lags (Sala et al., 2012; but see Nippert et al., 2005), probably due to legacy effects of rainfall on grass tiller density (Reichmann et al., 2013). Grasses, like trees, may also rely on wet years for seedling establishment (O'Connor, 1993) and for proliferation of belowground grass meristems (i.e., vegetative reproduction; Dalgleish and Hartnett, 2006), which maintains grass biomass over long time scales in perennial grasslands. Thus, grasses behave much as trees do: they appear to benefit from periods of

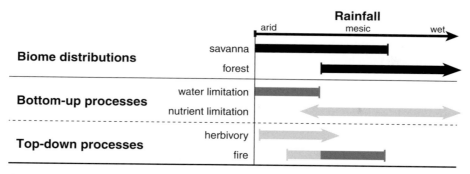

Figure 4.2 Bottom-up versus top-down limitations on vegetation growth and cover along a continental rainfall gradient. Black bars denote the distributions of savanna and forest; dark gray bars denote processes that define biome distributions; light gray bars denote processes that operate but probably do not limit biome distributions. Bars ending in arrows denote that there is uncertainty about the climatic range over which the process shapes vegetation structure.

relatively high rainfall and may even respond to environmental variability on relatively long time scales.

What this means for resource-mediated coexistence is unclear. There is strong evidence for coexistence mediated neither by niche separation (Riginos et al., 2009; February et al., 2013) nor by the storage effect. Grasses may do relatively better than trees under drier conditions, especially in particularly arid areas, or additional mechanisms may be necessary to maintain functional-level coexistence in savanna systems.

Top-down ecological processes in savanna grassland

The ecological role of fire in savanna grassland

Fire is an additional strong predictor of savanna grassland biome distributions globally (Bond et al., 2005; Staver et al., 2011a; 2011b), maintaining savanna grassland where climate is sufficient to support forest (Fig. 4.2; Bond and Keeley, 2005). Fire tends to promote the dominance of C_4 grasses at the expense of other plant functional groups, such as shrubs and trees (Briggs et al., 2005) and even forbs (Collins, 1992; Knapp et al., 1998). Paleoecological analyses suggest that fire probably played a major role in promoting the global dominance of C_4 grasses (Keeley and Rundel, 2005; Beerling and Osborne, 2006; Visser et al., 2012). C_4 grasses are morphologically well adapted to fire (Bond, 2008): they are highly flammable because they accumulate instead of decomposing litter and can resprout quickly after fire from belowground reserves (Ripley et al., 2010). Typically, more frequently burned grasslands have lower herbaceous plant community diversity and richness than those burned less frequently (Collins et al., 1995; Smith et al., 2012 – but see Kirkman et al., 2014), and fire can have major effects on grass productivity (Knapp and Seastedt, 1986; Johnson and Matchett, 2001), generally increasing grass biomass production (Buis et al., 2009) by removing moribund material and increasing light availability at the soil surface (Johnson and Matchett, 2001).

Fire also has huge effects on tree cover (Peterson and Reich, 2001; Briggs et al., 2005) and is a fundamental determinant of the adaptive strategies and population dynamics of trees (Schutz et al., 2009; Bhattachan et al., 2012; Werner and Prior, 2013). Within savanna grassland, fire limits tree cover not by killing savanna trees, but by limiting the recruitment of saplings into adult trees. Fires rarely kill adult savanna trees, which are protected from extensive fire damage both by their height (Trollope and Tainton, 1986; Hanan et al., 2008) and the accumulation of bark with size (Hoffmann et al., 2012). However, savanna tree saplings are susceptible to fire – aboveground shoots are often consumed or killed by fires – although they resprout readily from belowground root reserves (Higgins et al., 2000; Hoffmann et al., 2009; Schutz et al., 2009). Thus, fire strongly limits the recruitment of tree saplings into adults (Higgins et al., 2000; Hoffmann et al., 2009), with major potential effects on total tree cover in savanna grassland.

Fire provides an alternative plausible mechanism for the long-term stable coexistence of trees and grasses in savanna grasslands (Higgins et al., 2000; Bond et al., 2005), driven more explicitly by the storage effect. Grasses clearly benefit from repeated and frequent fires (Keeley and Rundel, 2005; Beerling and Osborne, 2006), and tree populations persist because trees can survive periods with frequent fires until a stochastic or periodic reduction in fire frequency (Higgins et al., 2000). This raises interesting – and largely unanswered – questions about the evolutionary dynamics of flammability that have given rise to the global prevalence of savanna grasslands and C_4 grasses more generally (Keeley and Rundel, 2005; Beerling and Osborne, 2006; Hoetzel et al., 2013).

The role of fire in expanding the global distribution of savanna grasslands (Staver et al., 2011a) makes understanding the factors that limit fire of fundamental importance. In this respect, fire interacts strongly with climate, because the weather can affect fire severity and the likelihood of spread (Govender et al., 2006), and because fire intensity increases strongly with fuel load – i.e., grass biomass (Govender et al., 2006; Archibald et al., 2009). However, because fire spread depends on fuel loads, grass exclusion by forest trees has the potential to stop the spread of fire as well. Empirical analyses suggest that the response of fire spread to tree cover – probably via the effects of trees on grass cover – behaves as a threshold: below c. 40% tree cover, fire spreads readily, but tree cover above c. 40% completely stops fire from spreading in landscapes (Archibald et al., 2009; Staver et al., 2011b). This is consistent with models that consider fire spread as an infection process, with a percolation-type response to landscape connectivity (Archibald et al., 2012). Integrating this spatial dynamic into the ecology and management of savanna grasslands remains a challenge.

The response of fire spread to the spatial arrangement of fuels in the landscape suggests another possible interpretation for the coexistence of trees and grasses in savanna grassland (Staver et al., 2011b; Staver and Levin, 2012). Fire limits tree cover, which promotes fire, resulting in a positive feedback, such that savanna may actually represent a stable alternative to closed-canopy forest over some ranges of rainfall (Fig. 4.2; Beckage et al., 2009; Staver et al., 2011b). Savanna grasslands clearly also dominate in areas where water, not fire, prevents the formation of a closed canopy (Bond et al., 2005; Sankaran et al., 2005; Staver et al., 2011b). Thus, evidence from both analyses of global distributions (Hirota et al., 2011; Staver et al., 2011a; 2011b) and of the existence of a fire-adapted flora (Gignoux et al., 1997; Hoffmann et al., 2003; Bond et al., 2008) suggests that savanna grasslands are not simply a transient vegetation type that occur when forests are disturbed, but rather a distinct and stably distributed flammable biome. Potential implications of this alternative stable state framework remain poorly explored. Feedbacks between fire and nutrients, which are potentially volatilized or exported as ash in fires (Reich et al., 2001; Hartshorn et al., 2009), and even climate (Sternberg, 2001) may potentially establish additional positive feedbacks that maintain savanna grassland as a stable and distinct biome.

The ecological role of herbivory in savanna grassland

The effects of herbivory in savanna grassland are much more difficult to gener-alize than those of fire but can be, at least locally, as significant. The difficulty lies in the fact that herbivores are diverse (Olff et al., 2002; Greve et al., 2012), possibly because of the abundance and variety of forage in savanna grasslands (Bond, 2005; Beerling and Osborne, 2006; Greve et al., 2012): grazers feed mostly on grasses, browsers on trees/shrubs, and mixed feeders on both. Their effects depend on their diet preferences (O'Kane et al., 2012), size (Laca et al., 2010; Sensenig et al., 2010; Hopcraft et al., 2011), and other aspects of their ecology, such as habitat preferences, herding behavior, and migration (Brashares et al., 2000; Holdo et al., 2009a; Hopcraft et al., 2010).

However, despite community-level variability among herbivores, some gen-eral patterns do emerge. Grazers can have significant effects on the grass layer in savanna grassland. High grazing pressure can trigger the formation of grazing lawns, which are characterized by low standing grass biomass, high productiv-ity, and sometimes a shift in grass community composition (McNaughton, 1984; Cromsigt and Olff, 2008; Waldram et al., 2008). Grazing lawns attract herbivores, possibly increasing carrying capacities, and persist as a potential alternative sta-ble state while regional-scale herbivore densities are high (McNaughton, 1984). These grazing lawns are associated with very high primary production (Frank and Groffman, 1998), due to increases in the availability of light, water, and/or nitrogen to the remaining plants (McNaughton, 1985; Frank and Groffman, 1998; Knapp et al., 1998), and are often associated with unique tree (Fornara and Toit, 2007; Staver and Levin, 2012) and faunal communities (Waldram et al., 2008; Krook et al., 2013). Grazers generally increase plant diversity in mesic savanna grassland, where herbivory prevents light competition (Belsky, 1992; Hartnett et al., 1996), but have negative or neutral effects on plant diversity in arid areas, where plants are unable to recover from grazing due to harsher environmental conditions (Yeo, 2005; Bakker et al., 2006; Olff and Ritchie, 1998).

Grazer effects on the herbaceous layer can result in significant interactions between herbivory and fire spread as well. Grazers consume biomass and can maintain very short-grass grazing lawns (Archibald et al., 2005), thus sometimes replacing fire as a dominant ecological top-down process (Archibald et al., 2005; Holdo et al., 2009b; Staver et al., 2012). Grazing pressure can release woody vegetation from fire and even from the effects of grass competition for resources (Walter, 1971; Scholes and Archer, 1997; Holdo et al., 2009b), such that intense grazing can sometimes be the cause of woody encroachment (Naito and Cairns, 2011). However, grazer removal results in the fairly rapid recovery of grass biomass (Waldram et al., 2008; Staver et al., 2014), potentially resulting in more frequent fires and renewed competition from grasses – an alternative type of trophic cascade (Holdo et al., 2009b).

In areas with extant diverse herbivore communities – mostly in Africa – grazing pressure can often go hand-in-hand with browsing pressure, either because browsers are attracted to areas with good visibility and a lower risk of predation (Riginos and Grace, 2008; Eby and Ritchie, 2013) or because grazing pressure and browsing pressure arise from one and the same functional group, the mixed feeders (O'Kane et al., 2012; Staver et al., 2014). An extensive collection of local herbivore exclusion studies in Africa have shown that the net effect of herbivore removal is often huge growth by trees that have been released from the effects of browsing (Augustine and McNaughton, 2004; Sharam et al., 2006; Fornara and Toit, 2007; Staver et al., 2009; Midgley et al., 2010; Moncrieff et al., 2011). Observational studies suggest that there is a "browse trap," analogous to the "fire trap": trees appear to be able to escape the effects of herbivores by growing tall (Bond and Loffell, 2001; Moncrieff et al., 2011; see also, Chapter 5 of this volume). Observational work (Prins and van der Jeugd, 1993; Holdo et al., 2009b; Staver et al., 2011c) and limited experimental work (Staver et al., 2014) suggest that a temporary reduction in browsing pressure, e.g., due to a herbivore population crash, can trigger the establishment of even-aged cohorts of trees in the landscape that can persist for decades and even centuries. In this way, the response of tree populations to variable herbivory, although relatively poorly understood, may again promote tree–grass coexistence via the storage effect.

But what drives variability in herbivore populations globally? This is not a trivial question, especially in the context of pervasive anthropogenic effects on herbivore populations; humans have driven herbivore extinctions, herbivore population crashes due to hunting and disease, and the collapse of historical migrations (Harris et al., 2009). However, it is clear that herbivore populations are subject to extreme variability, much of which is also driven by climatic variability, especially drought (Young, 1994).

Disease is also a major driver of that variability: in Africa and elsewhere, massive herbivore population crashes have resulted from outbreaks of rinderpest (Dobson, 1995), anthrax (Prins and van der Jeugd, 1993; Turner et al., 2013), and, recently, tuberculosis (Michel et al., 2006). Epidemics of rinderpest and tuberculosis have been the result of anthropogenic changes in disease ecology, but others – including anthrax – are endemic and appear to be subject to "natural" epidemic dynamics. These epidemics have been responsible, variously, for the establishment of regional-scale even-aged cohorts of trees after their release from browsing (Prins and van der Jeugd, 1993) and, elsewhere, for the suppression of tree establishment following grazer population crashes, which have resulted in grass release and, thus, more frequent fires (Holdo et al., 2009b).

Predation, meanwhile, affects not just herbivore population sizes (Hopcraft et al., 2010; Grange et al., 2012) but herbivore landscape use as well (Riginos and Grace, 2008; Anderson et al., 2010; Hopcraft et al., 2010), with demonstrable effects on grass (Waldram et al., 2008) and tree population dynamics (Ripple and Beschta, 2004). In African savanna grasslands, where herbivores and predators are hyper-abundant, predators can be responsible for 100% of mortality events of small herbivores and substantial mortality even of herbivores as large as giraffe (Sinclair et al., 2003). This can have radical effects on herbivore behavior, increasing herbivore vigilance and decreasing the time they spend foraging (Eby and Ritchie, 2013), and promoting herding behaviors, which have been tied to the evolution of grazing lawns (McNaughton, 1984). Predation can also induce herbivores to graze or browse in open patches with better visibility, setting up a positive feedback that maintains grazing lawns as areas that favor herbivores. However, this "landscape of fear" phenomenon also has potential interactions with fire, which creates relatively open landscapes with high-quality forage, at least temporarily (Archibald et al., 2005). The strength of this interaction has been shown to depend both on herbivore population dynamics and on the amount of recently burned area in the landscape (Archibald et al., 2005; Cromsigt and Olff, 2008).

Herbivore-driven landscapes have resulted in the evolution of novel adaptive strategies in plants that both promote a diverse assemblage of herbivores and allow them to persist in the face of chronic and continuous disturbance (see Chapter 13, this volume). Trees maximize their structural defenses, resulting in highly branched structures that minimize potential biomass removal by browsers (Archibald and Bond, 2003; Staver et al., 2012), while grasses have evolved similarly branched lateral growth forms that allow them to survive intensive grazing, but which make them less competitive in dense grass canopies (Coughenour, 1985; Briske, 1991). These plant strategies may have feedbacks on herbivore diversity as well; herbivore-adapted plants often promote, as well as benefit from, high herbivore diversity (McNaughton, 1984; Beerling and Osborne, 2006; Waldram et al., 2008; Greve et al., 2012), such that herbivore-adapted plants and herbivores have evolved a mutualistic relationship in savanna grasslands. The evolutionary and ecological consequences of herbivory in savanna grassland merit careful further consideration.

Top-down/bottom-up interactions in savanna grassland

Savanna grassland ecology has long focused on a discussion of which factors determine vegetation structure and promote tree–grass coexistence, whether bottom-up determinants (i.e., water and nutrient limitation) or top-down processes (e.g., fire and herbivory). Part of the debate has focused on the type of dynamic that characterizes functional coexistence – whether coexistence is stable or primarily a product of transient dynamics and whether niche differentiation or the storage effect is responsible. However, synthesizing the results discussed above reveals that interactions between bottom-up and top-down processes are abundant (Fig. 4.1), such that ecosystem-level outcomes are varied. Prediction, therefore, presents a significant challenge.

The field has increasingly begun to acknowledge possible changes in the importance of various drivers along ecological gradients as interaction strengths change (Sankaran et al., 2004; Bakker et al., 2006; Bond, 2008). The most obvious and well-studied of these is a change in the role of fire versus water limitation in determining ecosystem structure along climatic gradients (Fig. 4.2). Water limitation clearly plays a strong role at lower rainfall (Sankaran et al., 2005). Meanwhile, although it also plays an ecological role in arid savanna grasslands, fire is clearly both more frequent (Archibald et al., 2009; Smit et al., 2010) and more fundamental to ecosystem dynamics in mesic savanna grasslands, especially in terms of its effects on tree cover (Bond et al., 2005; Staver et al., 2011a; 2011b). Eventually, however, sufficient rainfall effectively excludes fire. Weather may directly limit fire (Govender et al., 2006), or tree cover itself may effectively exclude fire (Staver et al., 2011a; Hoffmann et al., 2012). Thus, it is only where fires can no longer limit tree cover that there exists a climatic limit on the distribution of savanna grasslands.

Similar interactions with soil nutrients are also possible, especially if fast tree growth rates can ultimately limit the potential for fire to impact tree cover. However, this is complicated by direct interactions between fire and nutrient cycles: nutrient limitation may play a more important role in mesic than in arid savanna grassland (Bustamante et al., 2004; 2006; Nardoto et al., 2006; Viani et al., 2011), due in part to leaching, but also due to increased fire frequency in mesic savanna grassland compared to arid savanna grassland (Reich et al., 2001; Coetsee et al., 2010).

Similar analyses of the changing role of herbivore effects on tree dynamics along resource gradients are currently lacking, although grass responses to grazing are relatively well synthesized. Some authors have argued that herbivores may play a stronger role in arid than in mesic savanna (Illius and O'Connor, 2000). Grazing tends to either not affect or decrease richness in more arid savanna grasslands, possibly because rates of herbaceous growth are relatively low (Balfour and Howison, 2002). Meanwhile, in more mesic areas, rates of grass biomass accumulation are high enough that grazers can only exclude fire under special conditions (Archibald et al., 2005; Waldram et al., 2008), and trees grow quickly enough to escape the effects of herbivory (Staver et al., 2009), at least during episodic reductions in herbivore pressure (Fig. 4.2). However, herbivore densities and diversity can even be high in mesic areas, and herbivory can potentially have ecosystem-level impacts. Grazing tends to increase herbaceous layer richness (Bakker et al., 2006; Young et al., 2013) and impacts functional diversity even where it does not have biome-level impacts on ecosystem structure (Buis et al., 2009), and browsing can aggravate the effects of fire on tree cover (Barnes, 2001; Staver et al., 2009).

Interactions between herbivory and nutrient availability may also be important; while fires are most frequent at relatively high rainfall, herbivores usually prefer nutrient-rich vegetation because of its nutritional value (Illius et al., 1999; Anderson et al., 2010; Hopcraft et al., 2010). Richness and productivity responses to nutrients parallel responses to rainfall: in nutrient-rich systems where grasses grow quickly, grazing increasing richness (Hillebrand et al., 2007; Koerner et al., in press) and changes community composition, but grazing decreases productivity and richness in low-nutrient areas (Proulx and Mazumder, 1998), where belowground competition dominates coexistence processes (Osem et al., 2002).

Conclusions

The degree of interaction between top-down and bottom-up processes in savanna grasslands results in significant variability in structure within the biome, which we are still mostly unable to resolve (Sankaran et al., 2005; Bucini and Hanan, 2007). More explicit consideration of these interactions may help. Our current understanding of the ecology of fire and herbivory in savanna grassland suggests that the focus should be (1) on understanding how resource limitation affects tree and grass growth rates, which determine how those factors respond to chronic disturbances, and (2) on understanding the factors that control the intensity and temporal distribution of fire and especially herbivory.

This explicit emphasis on population-level processes represents a qualitative departure from most biome-level predictions. However, an accurate and predictive understanding of tree densities in savanna grasslands would be of immense value. Savanna grasslands contribute significantly to the global carbon cycle, and tree–grass ratios have profound – if poorly studied – effects on carbon and nutrient budgets in savanna grasslands (Scholes and Archer, 1997; Jackson et al., 2002). Within savanna grasslands, tree cover is changing as a direct consequence of rising CO_2, changing climate, and shifting fire management (Buitenwerf et al., 2011; Bond and Midgley, 2012). The distribution of savanna grassland may also shift significantly as rainfall and fire regimes respond to anthropogenic global change (Scheiter and Higgins, 2009; Higgins and Scheiter, 2012; see Chapter 14, this volume). Developing mechanistic models for tree and grass cover within savanna grassland that integrate top-down and bottom-up ecological processes represents a continuing challenge, but one that is fundamental to predicting savanna grassland responses to and feedbacks on global change.

References

Anderson, T. M. T., Hopcraft, J. G. C. J., Eby, S., et al. (2010). Landscape-scale analyses suggest both nutrient and antipredator advantages to Serengeti herbivore hotspots. *Ecology*, **91**, 1519–1529.

Archibald, S. and Bond, W. (2003). Growing tall vs growing wide: tree architecture and allometry of Acacia karroo in forest, savanna, and arid environments. *Oikos*, **102**, 3–14.

Archibald, S. and Scholes, R. J. (2007). Leaf green-up in a semi-arid African savanna – separating tree and grass responses to environmental cues. *Journal of Vegetation Science*, **18**, 583–594.

Archibald, S., Bond, W., Stock, W. and Fairbanks, D. (2005). Shaping the landscape: fire-grazer interactions in an African savanna. *Ecological Applications*, **15**, 96–109.

Archibald, S., Roy, D. P., Van Wilgen, B. W. and Scholes, R. J. (2009). What limits fire? An examination of drivers of burnt area in Southern Africa. *Global Change Biology*, **15**, 613–630.

Archibald, S., Staver, A. C. and Levin, S. A. (2012). Evolution of human-driven fire regimes in Africa. *Proceedings of the National Academy of Sciences of the USA*, **109**, 847–852.

Augustine, D. and McNaughton, S. (2004). Regulation of shrub dynamics by native browsing ungulates on East African rangeland. *Journal of Applied Ecology*, **41**, 45–58.

Bakker, E. S., Ritchie, M. E., Olff, H., Milchunas, D. G. and Knops, J. M. H. (2006). Herbivore impact on grassland plant diversity depends on habitat productivity and herbivore size. *Ecology Letters*, **9**, 780–788.

Balfour, D. and Howison, O. (2002). Spatial and temporal variation in a mesic savanna fire regime: responses to variation in annual rainfall. *African Journal of Range and Forage Science*, **19**, 45–53.

Barnes, M. (2001). Effects of large herbivores and fire on the regeneration of *Acacia erioloba* woodlands in Chobe National Park, Botswana. *African Journal of Ecology*, **39**, 340–350.

Beckage, B., Platt, W. J. and Gross, L. J. (2009). Vegetation, fire, and feedbacks: a disturbance-mediated model of savannas. *American Naturalist*, **174**, 805–818.

Beerling, D. J. and Osborne, C. P. (2006). The origin of the savanna biome. *Global Change Biology*, **12**, 2023–2031.

Belsky, A. J. (1992). Effects of grazing, competition, disturbance and fire on species composition and diversity in grassland communities. *Journal of Vegetation Science*, **3**, 187–200.

Bhattachan, A., Tatlhego, M., Dintwe, K., et al. (2012). evaluating ecohydrological theories of woody root distribution in the Kalahari. *PLoS One*, **7**, e33996.

Bond, W. (2005). Large parts of the world are brown or black: a different view on the "Green World" hypothesis. *Journal of Vegetation Science*, **16**, 261–266.

Bond, W. J. (2008). What limits trees in C-4 grasslands and savannas? *Annual Review of Ecology, Evolution and Systematics*, **39**, 641–659.

Bond, W. J. (2010). Do nutrient-poor soils inhibit development of forests? A nutrient stock analysis. *Plant Soil*, **334**, 47–60.

Bond, W. and Keeley, J. (2005). Fire as a global "herbivore": the ecology and evolution of flammable ecosystems. *Trends in Ecology and Evolution*, **20**, 387–394.

Bond, W. and Loffell, D. (2001). Introduction of giraffe changes acacia distribution in a South African savanna. *African Journal of Ecology*, **39**, 286–294.

Bond, W. J. and Midgley, G. F. (2012). Carbon dioxide and the uneasy interactions of trees and savannah grasses. *Philosophical Transactions of the Royal Society Series B*, **367**, 601–612.

Bond, W., Woodward, F. and Midgley, G. (2005). The global distribution of ecosystems in a world without fire. *New Phytologist*, **165**, 525–537.

Bond, W. J., Silander, J. A., Jr, Ranaivonasy, J. and Ratsirarson, J. (2008). The antiquity of Madagascar's grasslands and the rise of C4 grassy biomes. *Journal of Biogeography*, **35**, 1743–1758.

Brashares, J. S., Garland, T. and Arcese, P. (2000). Phylogenetic analysis of coadaptation in behavior, diet, and body size in the African antelope. *Behavioral Ecology*, **11**, 452–463.

Briggs, J. M., Knapp, A. K., Blair, J. M., et al. (2005). An ecosystem in transition: causes and consequences of the conversion of mesic grassland to shrubland. *Bioscience*, **55**, 243–254.

Briske, D. D. (1991). Developmental morphology and physiology of grasses. In *Grazing Management: An Ecological Perspective*, ed. R. K. Heitschmidt and J. W. Stuth. Portland, OR: Timber Press, pp. 85–108.

Bucini, G. and Hanan, N. P. (2007). A continental-scale analysis of tree cover in African savannas. *Global Ecology and Biogeography*, **16**, 593–605.

Buis, G. M., Blair, J. M., Burkepile, D. E., et al. (2009). Controls of aboveground net primary production in mesic savanna grasslands: an inter-hemispheric comparison. *Ecosystems*, **12**, 982–995.

Buitenwerf, R., Bond, W. J., Stevens, N. and Trollope, W. S. W. (2011). Increased tree densities in South African savannas: >50 years of data suggests CO_2 as a driver. *Global Change Biology*, **18**, 675–684.

Bustamante, M., Martinelli, L., Silva, D., et al. (2004). N-15 natural abundance in woody plants and soils of central Brazilian savannas (cerrado). *Ecological Applications*, **14**, S200–S213.

Bustamante, M. M. C., Medina, E., Asner, G. P., Nardoto, G. B. and Garcia-Montiel, D. C. (2006). Nitrogen cycling in tropical and temperate savannas. *Biogeochemistry*, **79**, 209–237.

Caylor, K. K., Scanlon, T. M. and Rodriguez-Iturbe, I. (2009). Ecohydrological optimization of pattern and processes in water-limited ecosystems: a trade-off-based hypothesis. *Water Resources Research*, **45**, W08407.

Cech, P. G., Kuster, T., Edwards, P. J. and Olde Venterink, H. (2008). Effects of herbivory, fire and N2-fixation on nutrient limitation in a humid African savanna. *Ecosystems*, **11**, 991–1004.

Chesson, P. (2000a). Mechanisms of maintenance of species diversity. *Annual Review of Ecology and Systematics*, **31**, 343–366.

Chesson, P. (2000b). General theory of competitive coexistence in spatially-varying environments. *Theoretical Population Biology*, **58**, 27–27.

Coetsee, C., Bond, W. J. and February, E. C. (2010). Frequent fire affects soil nitrogen and carbon in an African savanna by changing woody cover. *Oecologia*, **162**, 1027–1034.

Collins, S. L. (1992). Fire frequency and community heterogeneity in tallgrass prairie vegetation. *Ecology*, **73**, 2001–2006.

Collins, S. L., Glenn, S. M. and Gibson, D. J. (1995). Experimental-analysis of intermediate disturbance and initial floristic composition: decoupling cause and effect. *Ecology*, **76**, 486–492.

Coughenour, M. B. (1985). Graminoid responses to grazing by large herbivores: adaptations, exaptations, and interacting processes. *Annals of the Missouri Botanical Garden*, 852–863.

Cramer, M. D., Chimphango, S. B. M., van Cauter, A., Waldram, M. S. and Bond, W. J. (2007). Grass competition induces N-2 fixation in some species of African acacia. *Journal of Ecology*, **95**, 1123–1133.

Cramer, M. D., van Cauter, A. and Bond, W. J. (2010). Growth of N2-fixing African savanna *Acacia* species is constrained by below-ground competition with grass. *Journal of Ecology*, **98**, 156–167.

Cromsigt, J. P. G. M. and Olff, H. (2008). Dynamics of grazing lawn formation: an experimental test of the role of scale-dependent processes. *Oikos*, **117**, 1444–1452.

Dalgleish, H. J. and Hartnett, D. C. (2006). Belowground bud banks increase along a precipitation gradient of the North American Great Plains: a test of the meristem limitation hypothesis. *New Phytologist*, **171**, 81–89.

D'odorico, P., Caylor, K., Okin, G. S. and Scanlon, T. M. (2007). On soil moisture-vegetation feedbacks and their possible effects on the dynamics of dryland ecosystems. *Journal of Geophysical Research: Biogeosciences*, **112**, G04010.

Dobson, A. (1995). The ecology and epidemiology of rinderpest virus in Serengeti and Ngorongoro Conservation Area. *Serengeti II: Dynamics, Management, and Conservation of an Ecosystem*, **2**, 485.

Eby, S. and Ritchie, M. E. (2013). The impacts of burning on Thomson's gazelles, *Gazella thomsonii*, vigilance in Serengeti National Park, Tanzania. *African Journal of Ecology*, **51**, 337–342.

Elser, J. J., Bracken, M. E. S., Cleland, E. E., et al. (2007). Global analysis of nitrogen and phosphorus limitation of primary producers in freshwater, marine and terrestrial ecosystems. *Ecology Letters*, **10**, 1135–1142.

Facelli, J. M., Chesson, P. and Barnes, N. (2005). Differences in seed biology of annual plants in arid lands: a key ingredient of the storage effect. *Ecology*, **86**, 2998–3006.

February, E. C. and Higgins, S. I. (2010). The distribution of tree and grass roots in savannas in relation to soil nitrogen and water. *South African Journal of Botany*, **76**, 517–523.

February, E. C., Higgins, S. I., Bond, W. J. and Swemmer, L. (2013). Influence of competition and rainfall manipulation on the growth responses of savanna trees and grasses. *Ecology*, **94**, 1155–1164.

Fensham, R. J., Fairfax, R. J. and Ward, D. P. (2009). Drought-induced tree death in savanna. *Global Change Biology*, **15**, 380–387.

Fornara, D. A. and Toit Du, J. T. (2007). Browsing lawns? Responses of *Acacia nigrescens* to ungulate browsing in an African savanna. *Ecology*, **88**, 200–209.

Frank, D. A. and Groffman, P. M. (1998). Ungulate vs. landscape control of soil C and N processes in grasslands of Yellowstone National Park. *Ecology*, **79**, 2229–2241.

Gignoux, J., Clobert, J. and Menaut, J. (1997). Alternative fire resistance strategies in savanna trees. *Oecologia*, **110**, 576–583.

Gillson, L. (2006). A "large infrequent disturbance" in an East African savanna. *African Journal of Ecology*, **44**, 458–467.

Good, S. P. and Caylor, K. K. (2011). Climatological determinants of woody cover in Africa. *Proceedings of the National Academy of Sciences of the USA*, **108**, 4902–4907.

Govender, N., Trollope, W. S. W. and Van Wilgen, B. W. (2006). The effect of fire season, fire frequency, rainfall and management on fire intensity in savanna vegetation in South Africa. *Journa of Applied Ecology*, **43**, 748–758.

Grange, S., Owen-Smith, N., Gaillard, J.-M., et al. (2012). Changes of population trends and mortality patterns in response to the reintroduction of large predators: the case study of African ungulates. *Acta Oecologica*, **42**, 16–29.

Greve, M., Lykke, A. and Fagg, C. (2012). Continental-scale variability in browser diversity is a major driver of diversity patterns in acacias across Africa. *Journal of Ecology*, **100**, 1093–1104.

Hanan, N. P., Sea, W. B., Dangelmayr, G. and Govender, N. (2008). Do fires in savannas consume woody biomass? A comment on approaches to modeling savanna dynamics. *American Naturalist*, **171**, 851–856.

Harris, G., Thirgood, S., Hopcraft, J., Cromsight, J. and Berger, J. (2009). Global decline in aggregated migrations of large terrestrial mammals. *Endangered Species Research*, **7**, 55–76.

Hartnett, D. C., Hickman, K. R. and Walter, L. E. F. (1996). Effects of bison grazing, fire, and topography on floristic diversity in tallgrass prairie. *Journal of Range Management*, **49**, 413–420.

Hartshorn, A. S., Coetsee, C. and Chadwick, O. A. (2009). Pyromineralization of soil phosphorus in a South African savanna. *Chemical Geology*, **267**, 24–31.

Hawkins, B. A., Field, R., Cornell, H. V., et al. (2003). Energy, water, and broad-scale geographic patterns of species richness. *Ecology*, **84**, 3105–3117.

Hennenberg, K., Fischer, F., Kouadio, K., et al. (2006). Phytomass and fire occurrence along forest–savanna transects in the Comoé National Park, Ivory Coast. *Journal of Tropical Ecology*, **22**, 303–311.

Higgins, S. I. and Scheiter, S. (2012). Atmospheric CO_2 forces abrupt vegetation shifts locally, but not globally. *Nature*, **488**, 209–212.

Higgins, S., Bond, W. and Trollope, W. (2000). Fire, resprouting and variability: a recipe for grass–tree coexistence in savanna. *Journal of Ecology*, **88**, 213–229.

Higgins, S. I., Delgado-Cartay, M. D., February, E. C. and Combrink, H. J. (2011). Is there a temporal niche separation in the leaf phenology of savanna trees and grasses? *Journal of Biogeography*, **38**, 2165–2175.

Hillebrand, H., Gruner, D. S., Borer, E. T., et al. (2007). Consumer versus resource control of producer diversity depends on ecosystem type and producer community structure. *Proceedings of the National Academy of Sciences of the USA*, **104**, 10904–10909.

Hirota, M., Holmgren, M., Van Nes, E. H. and Scheffer, M. (2011). Global resilience of tropical forest and savanna to critical transitions. *Science*, **334**, 232–235.

Hoetzel, S., Dupont, L., Schefuss, E., Rommerskirchen, F. and Wefer, G. (2013). The role of fire in Miocene to Pliocene C-4 grassland and ecosystem evolution. *Nature Geoscience*, **6**, 1027–1030.

Hoffmann, W. and Moreira, A. (2003). The role of fire in population dynamics of woody plants. *Cerrados of Brazil*, 159–177.

Hoffmann, W., Orthen, B. and Do Nascimento, P. (2003). Comparative fire ecology of tropical savanna and forest trees. *Functional Ecology*, **17**, 720–726.

Hoffmann, W., Adasme, R., Haridasan, M. T., et al. (2009). Tree topkill, not mortality, governs the dynamics of savanna-forest boundaries under frequent fire in central Brazil. *Ecology*, **90**, 1326–1337.

Hoffmann, W. A., Geiger, E. L., Gotsch, S. G., et al. (2012). Ecological thresholds at the savanna–forest boundary: how plant traits, resources and fire govern the distribution of tropical biomes. *Ecology Letters*, **15**, 759–768.

Holdo, R. M., Holt, R. D. and Fryxell, J. M. (2009a). Grazers, browsers, and fire influence the extent and spatial pattern of tree cover in the Serengeti. *Ecological Applications*, **19**, 95–109.

Holdo, R. M., Sinclair, A. R. E., Dobson, A. P., et al. (2009b). A disease-mediated trophic cascade in the Serengeti and its implications for ecosystem C. *PLoS Biology*, **7**, e1000210.

Holdridge, L. R. (1947). Determination of world plant formations from simple climatic data. *Science*, **105**, 367–368.

Hopcraft, J. G. C., Olff, H. and Sinclair, A. R. E. (2010). Herbivores, resources and risks: alternating regulation along primary environmental gradients in savannas. *Trends in Ecology and Evolution*, **25**, 119–128.

Hopcraft, J. G. C., Anderson, T. M., Pérez-Vila, S., Mayemba, E. and Olff, H. (2011). Body size and the division of niche space: food and predation differentially shape the distribution of Serengeti grazers. *Journal of Animal Ecology*, **81**, 201–213.

Horn, H. S. and Macarthur, R. H. (1972). Competition among fugitive species in a harlequin environment. *Ecology*, **53**, 749–752.

Huenneke, L. F., Hamburg, S. P., Koide, R., Mooney, H. A. and Vitousek, P. M. (1990). Effects of soil resources on plant invasion and community structure in Californian serpentine grassland. *Ecology*, **71**, 478–491.

Hutchinson, G. E. (1959). Homage to Santa Rosalia or why are there so many kinds of animals? *American Naturalist*, **93**, 145–159.

Illius, A. W. and O'Connor, T. G. (2000). Resource heterogeneity and ungulate population dynamics. *Oikos*, **89**, 283–294.

Illius, A. W., Gordon, I. J., Elston, D. A. and Milne, J. D. (1999). Diet selection in goats: a test of intake-rate maximization. *Ecology*, **80**, 1008–1018.

Jackson, R. B., Banner, J. L., Jobb Aacute Gy, E. G., Pockman, W. T. and Wall, D. H. (2002). Ecosystem carbon loss with woody plant invasion of grasslands. *Nature*, **418**, 623.

Johnson, L. C. and Matchett, J. R. (2001). Fire and grazing regulate belowground processes in tallgrass prairie. *Ecology*, **82**, 3377–3389.

Keeley, J. E. and Rundel, P. W. (2005). Fire and the Miocene expansion of C4 grasslands. *Ecology Letters*, **8**, 683–690.

Kirkman, K. P., Collins, S. L., Smith, M. D., et al. (2014). Responses to fire differ between South African and North American grassland communities. *Journal of Vegetation Science*, **25**, 793–804.

Knapp, A. K. and Seastedt, T. R. (1986). Detritus accumulation limits productivity of tallgrass prairie. *Bioscience*, **36**, 662–668.

Knapp, A. K., Briggs, J. M., Blair, J. M. and Turner, C. L. (1998). Patterns and controls of

aboveground net primary production in tallgrass prairie. In *Grassland Dynamics: Long-term Ecological Research in Tallgrass Prairie*, ed. A. K. Knapp, J. M. Briggs, D. C. Hartnett and S. L. Collins. Oxford: Oxford University Press, pp. 193–221.

Knapp, A. K., Blair, J. M., Briggs, J. M., et al. (1999). The keystone role of bison in North American tallgrass prairie: bison increase habitat heterogeneity and alter a broad array of plant, community, and ecosystem processes. *Bioscience*, **49**, 39–50.

Knapp, A. K., Briggs, J. M. and Koelliker, J. K. (2001). Frequency and extent of water limitation to primary production in a mesic temperate grassland. *Ecosystems*, **4**, 19–28.

Koerner, S. E. and Collins, S. L. (2014). Interactive effects of grazing, drought, and fire on grassland communities in North America and South Africa. *Ecology*, **95**, 98–109.

Koerner, S. E., Burkepile, D. E., Fynn, R. W. S., et al. (2014). Plant community response to loss of large herbivores differs between North American and South African savanna grasslands. *Ecology*, **95**, 808–816. DOI: 10.1890/13-1828.1.

Krook, K., Bond, W. J. and Hockey, P. A. (2013). The effect of grassland shifts on the avifauna of a South African savanna. *Ostrich*, **78**, 271–279.

Kulmatiski, A. and Beard, K. H. (2012). Root niche partitioning among grasses, saplings, and trees measured using a tracer technique. *Oecologia*, **171**, 25–37.

Laca, E. A., Sokolow, S., Galli, J. R. and Cangiano, C. A. (2010). Allometry and spatial scales of foraging in mammalian herbivores. *Ecology Letters*, **13**, 311–320.

Ladwig, L. M., Collins, S. L., Swann, A. L., et al. (2012). Above- and belowground responses to nitrogen addition in a Chihuahuan Desert grassland. *Oecologia*, **169**, 177–185.

Lloyd, J., Bird, M. I., Vellen, L., et al. (2008). Contributions of woody and herbaceous vegetation to tropical savanna ecosystem productivity: a quasi-global estimate. *Tree Physiology*, **28**, 451–468.

Loarie, S. R., Tambling, C. J. and Asner, G. P. (2013). Lion hunting behaviour and vegetation structure in an African savanna. *Animal Behaviour*, **85**, 899–906.

Macarthur, R. and Levins, R. (1967). The limiting similarity, convergence, and divergence of coexisting species. *American Naturalist*, **101**, 377–382.

McNaughton, S. (1984). Grazing lawns: animals in herds, plant form, and coevolution. *American Naturalist*, **124**, 863–886.

McNaughton, S. J. (1985). Ecology of a grazing ecosystem: The Serengeti. *Ecological Monographs*, **55**, 259–294.

Michel, A. L., Bengis, R. G., Keet, D. F., et al. (2006). Wildlife tuberculosis in South African conservation areas: implications and challenges. *Veterinary Microbiology*, **112**, 91–100.

Midgley, J. J., Lawes, M. J. and Chamaillé-Jammes, S. (2010). TURNER REVIEW No. 19. Savanna woody plant dynamics: the role of fire and herbivory, separately and synergistically. *Australian Journal of Botany*, **58**, 1–11.

Miller, D., Archer, S., Zitzer, S. and Longnecker, M. (2001). Annual rainfall, topoedaphic heterogeneity and growth of an arid land tree (*Prosopis glandulosa*). *Journal of Arid Environments*, **48**, 23–33.

Moncrieff, G. R., Chamaillé-Jammes, S., Higgins, S. I., O'Hara, R. B. and Bond, W. J. (2011). Tree allometries reflect a lifetime of herbivory in an African savanna. *Ecology*, **92**, 2310–2315.

Naito, A. T. and Cairns, D. M. (2011). Patterns and processes of global shrub expansion. *Progress in Physical Geography*, **35**, 423–442.

Nardoto, G. B., da Cunha Bustamante, M. M., Pinto, A. S. and Klink, C. A. (2006). Nutrient use efficiency at ecosystem and species level in savanna areas of Central Brazil and impacts of fire. *Journal of Tropical Ecology*, **22**, 191–201.

Nippert, J. B., Knapp, A. K. and Briggs, J. M. (2005). Intra-annual rainfall variability and grassland productivity: can the past predict the future? *Plant Ecology*, **184**, 65–74.

O'Connor, T. G. (1993). The influence of rainfall and grazing on the demography of some African savanna grasses: a matrix modelling approach. *Journal of Applied Ecology*, **30**, 119–132.

O'Connor, T. G. (1995). Acacia karroo invasion of grassland: environmental and biotic effects influencing seedling emergence and establishment. *Oecologia*, **103**, 214–223.

O'Kane, C. A. J., Duffy, K. J., Page, B. R. and Macdonald, D. W. (2012). Heavy impact on seedlings by the impala suggests a central role in woodland dynamics. *Journal of Tropical Ecology*, **28**, 291–297.

Olff, H. and Ritchie, M. E. (1998). Effects of herbivores on grassland plant diversity. *Trends in Ecology and Evolution*, **13**, 261–265.

Olff, H. H., Ritchie, M. E. M. and Prins, H. H. T. H. (2002). Global environmental controls of diversity in large herbivores. *Nature*, **415**, 901–904.

Osem, Y., Perevolotsky, A. and Kigel, J. (2002). Grazing effect on diversity of annual plant communities in a semi-arid rangeland: interactions with small-scale spatial and temporal variation in primary productivity. *Journal of Ecology*, **90**, 936–946.

Paruelo, J. M., Jobbagy, E. G., Sala, O. E., Lauenroth, W. K. and Burke, I. C. (1998). Functional and structural convergence of temperate grassland and shrubland ecosystems. *Ecological Applications*, **8**, 194–206.

Peterson, D. and Reich, P. (2001). Prescribed fire in oak savanna: fire frequency effects on stand structure and dynamics. *Ecological Applications*, **11**, 914–927.

Phillips, O. L., Aragao, L. E. O. C., Lewis, S. L., et al. (2009). Drought sensitivity of the Amazon Rainforest. *Science*, **323**, 1344–1347.

Pianka, E. R. (1966). Latitudinal gradients in species diversity: a review of concepts. *American Naturalist*, **100**, 33–46.

Prins, H. and van der Jeugd, H. P. (1993). Herbivore population crashes and woodland structure in East Africa. *Journal of Ecology*, **81**, 305–314.

Proulx, M. and Mazumder, A. (1998). Reversal of grazing impact on plant species richness in nutrient-poor vs. nutrient-rich ecosystems. *Ecology*, **79**, 2581–2592.

Reich, P. B., Grigal, D. F., Aber, J. D. and Gower, S. T. (1997). Nitrogen mineralization and productivity in 50 hardwood and conifer stands on diverse soils. *Ecology*, **78**, 335–347.

Reich, P., Peterson, D., Wedin, D. and Wrage, K. (2001). Fire and vegetation effects on productivity and nitrogen cycling across a forest-grassland continuum. *Ecology*, **82**, 1703–1719.

Reichmann, L. G., Sala, O. E. and Peters, D. P. (2013). Precipitation legacies in desert grassland primary production occur through previous-year tiller density. *Ecology*, **94**, 435–443.

Riginos, C. C. and Grace, J. B. J. (2008). Savanna tree density, herbivores, and the herbaceous community: bottom-up vs. top-down effects. *Ecology*, **89**, 2228–2238.

Riginos, C., Grace, J. B., Augustine, D. J. and Young, T. P. (2009). Local versus landscape-scale effects of savanna trees on grasses. *Journal of Ecology*, **97**, 1337–1345.

Ripley, B. S., Donald, G., Osborne, C. P., Abraham, T. and Martin, T. (2010). Experimental investigation of fire ecology in the C_3 and C_4 subspecies of *Alloteropsis semialata*. *Journal of Ecology*, **98**, 1196–1203.

Ripple, W. J. and Beschta, R. L. (2004). Wolves and the ecology of fear: can predation risk structure ecosystems? *BioScience*, **54**, 755–766.

Rossatto, D. R., da Silveira Lobo Sternberg, L. and Franco, A. C. (2012). The partitioning of water uptake between growth forms in a Neotropical savanna: do herbs exploit a third water source niche? *Plant Biology*, **15**, 84–92.

Sala, O. E., Parton, W. J., Joyce, L. A. and Lauenroth, W. K. (1988). Primary production of the central grassland region of the United States. *Ecology*, **69**, 40–45.

Sala, O. E., Gherardi, L. A., Reichmann, L., Jobbagy, E. and Peters, D. (2012). Legacies of precipitation fluctuations on primary production: theory and data synthesis. *Philosophical Transactions of the Royal Society of London B*, **367**, 3135–3144.

Sankaran, M., Ratnam, J. and Hanan, N. P. (2004). Tree-grass coexistence in savannas revisited – insights from an examination of assumptions and mechanisms invoked in existing models. *Ecology Letters*, **7**, 480–490.

Sankaran, M., Hanan, N., Scholes, R., et al. (2005). Determinants of woody cover in African savannas. *Nature*, **438**, 846–849.

Scheiter, S. and Higgins, S. I. (2009). Impacts of climate change on the vegetation of Africa: an adaptive dynamic vegetation modelling approach. *Global Change Biology*, **15**, 2224–2246.

Scholes, R. and Archer, S. (1997). Tree-grass interactions in savannas. *Annual Review of Ecological Systems*, **28**, 517–544.

Scholes, R. J. and Walker, B. H. (1993). *An African Savanna: Synthesis of the Nylsvley Study*. Cambridge: Cambridge University Press.

Schutz, A. E. N., Bond, W. J. and Cramer, M. D. (2009). Juggling carbon: allocation patterns of a dominant tree in a fire-prone savanna. *Oecologia*, **160**, 235–246.

Sensenig, R. L., Demment, M. W. and Laca, E. A. (2010). Allometric scaling predicts preferences for burned patches in a guild of East African grazers. *Ecology*, **91**, 2898–2907.

Sharam, G., Sinclair, A. R. E. and Turkington, R. (2006). Establishment of broad-leaved thickets in Serengeti, Tanzania: the influence of fire, browsers, grass competition, and elephants. *Biotropica*, **38**, 599–605.

Sinclair, A. R. E., Mduma, S. and Brashares, J. S. (2003). Patterns of predation in a diverse predator–prey system. *Nature*, **425**, 288–290.

Smit, I. P., Asner, G. P., Govender, N., et al. (2010). Effects of fire on woody vegetation structure in African savanna. *Ecological Applications*, **20**, 1865–1875.

Smit, I. P. J., Smit, C. F., Govender, N., Linde, M. V. D. and MacFadyen, S. (2013). Rainfall, geology and landscape position generate large-scale spatiotemporal fire pattern heterogeneity in an African savanna. *Ecography*, **36**, 447–459.

Smith, M. D., van Wilgen B. W., Burns C. E., et al. (2012). Long-term effects of fire frequency and season on herbaceous vegetation in savannas of the Kruger National Park, South Africa. *Journal of Plant Ecology*, **6**, 71–83.

Staver, A. C. and Bond, W. J. (2014). Is there a 'browse trap'? Dynamics of herbivore impacts on trees and grasses in an African savanna. *Journal of Ecology*, **102**, 595–602.

Staver, A. C. and Levin, S. A. (2012). Integrating theoretical climate and fire effects on savanna and forest systems. *American Naturalist*, **180**, 211–224.

Staver, A. C., Bond, W. J., Stock, W. D., van Rensburg, S. J. and Waldram, M. S. (2009). Browsing and fire interact to suppress tree density in an African savanna. *Ecological Applications*, **19**, 1909–1919.

Staver, A. C., Archibald, S. and Levin, S. A. (2011a). The global extent and determinants of savanna and forest as alternative biome states. *Science*, **334**, 230–232.

Staver, A. C., Archibald, S. and Levin, S. A. (2011b). Tree cover in sub-Saharan Africa: rainfall and fire constrain forest and savanna as alternative stable states. *Ecology*, **92**, 1063–1072.

Staver, A. C., Bond, W. J. and February, E. C. (2011c). History matters: tree establishment variability and species turnover in an African savanna. *Ecosphere*, **2**, art49.

Staver, A. C., Bond, W. J., Cramer, M. D. and Wakeling, J. L. (2012). Top-down determinants of niche structure and adaptation among African acacias. *Ecology Letters*, **15**, 673–679.

Sternberg, L. (2001). Savanna-forest hysteresis in the tropics. *Global Ecology and Biogeography*, **10**, 369–378.

Tilman, D. and Pacala, S. (1994). The maintenance of species richness in plant communities. In *Species Diversity in Ecological Communities*, ed. R. E. Ricklefs and

D. Schluter. Chicago: University of Chicago Press, pp. 13–25.

Tilman, D., Wedin, D. and Knops, J. (1996). Productivity and sustainability influenced by biodiversity in grassland eco-systems. *Nature*, **379**, 718–720.

Trollope, W. S. W. and Tainton, N. M. (1986). Effect of fire intensity on the grass and bush components of the Eastern Cape thornveld. *Journal of the Grassland Society of Southern Africa*, **3**, 37–42.

Turner, W. C., Imologhome, P., Havarua, Z., et al. (2013). Soil ingestion, nutrition and the seasonality of anthrax in herbivores of Etosha National Park. *Ecosphere*, **4**, art13.

Viani, R. A. G., Rodrigues, R. R., Dawson, T. E. and Oliveira, R. S. (2011). Savanna soil fertility limits growth but not survival of tropical forest tree seedlings. *Plant Soil*, **349**, 341–353.

Visser, V., Woodward, F. I., Freckleton, R. P. and Osborne, C. P. (2012). Environmental factors determining the phylogenetic structure of C4 grass communities. *Journal of Biogeography*, **39**, 232–246.

Wakeling, J. L., Cramer, M. D. and Bond, W. J. (2012). The savanna-grassland "treeline": why don't savanna trees occur in upland grasslands? *Journal of Ecology*, **100**, 381–391.

Waldram, M. S., Bond, W. J. and William D Stock. (2008). Ecological engineering by a mega-grazer: White rhino impacts on a South African savanna. *Ecosystems*, **11**, 101–112.

Walter, H. (1971). *Ecology of Tropical and Subtropical Vegetation*. Edinburgh: Oliver and Boyd.

Watson, I., Westoby, M. and McR, A. (1997). Continuous and episodic components of demographic change in arid zone shrubs: models of two Eremophila species from Western Australia compared with published data on other species. *Journal of Ecology*, **85**, 833–846.

Werner, P. A. and Prior, L. D. (2013). Demography and growth of subadult savanna trees: interactions of life history, size, fire season, and grassy understory. *Ecological Monographs*, **83**, 67–93.

Whittaker, R. H. (1975). *Communities and Ecosystems*. 2nd edn. New York: Macmillan.

Wiegand, K., Jeltsch, F. and Ward, D. (2004). Minimum recruitment frequency in plants with episodic recruitment. *Oecologia*, **141**, 1–10.

Williams, R., Duff, G., Bowman, D. and Cook, G. (1996). Variation in the composition and structure of tropical savannas as a function of rainfall and soil texture along a large-scale climatic gradient in the Northern Territory, Australia. *Journal of Biogeography*, **23**, 747–756.

Yeo, J. J. (2005). Effects of grazing exclusion on rangeland vegetation and soils, East Central Idaho. *Western North American Naturalist*, **65**, 91–102.

Young, H. S., McCauley, D. J., Helgen, K. M., et al. (2013). Effects of mammalian herbivore declines on plant communities: observations and experiments in an African savanna. *Journal of Ecology*, **101**, 1030–1041.

Young, T. (1994). Natural die-offs of large mammals: implications for conservation. *Conservation Biology*, **8**, 410–418.

CHAPTER FIVE

Bottom-up and top-down forces shaping wooded ecosystems: lessons from a cross-biome comparison

DRIES P. J. KUIJPER

Polish Academy of Sciences, Białowieża, Poland

MARISKA TE BEEST

Umeå University, Umeå, Sweden

MARCIN CHURSKI

Polish Academy of Sciences, Białowieża, Poland

and

JORIS P. G. M. CROMSIGT

Swedish University of Agricultural Sciences, Umeå, Sweden; and Nelson Mandela Metropolitan University, South Africa

Introduction

Different climatic regions across the globe are associated with biomes that differ in their cover of woody plants, such as grasslands, savannas, and forests (Whitaker, 1962). For a long time, researchers have assumed that abiotic factors control the spatial distribution of woody plant-dominated ecosystems or biomes. According to this idea, rainfall and temperature determine the transitions from deserts to grasslands to savannas and eventually to forests (e.g., Prentice et al., 1992). However, we increasingly realize that biomes may be far less fixed entities than previously assumed. An alternative view for many regions might be that of "ecosystems uncertain," which Whittaker (1975) defined as zones "in which either grassland or one of the types dominated by woody plants" may occur under the same climatic conditions. As Bond (2005) discusses, many of these "ecosystems uncertain" may be seen as "consumer-controlled ecosystems" where plant consumers, such as herbivores and fire, prevent a closed forest from developing and are a major determinant of the ecosystem state. Bond (2005) showed that such "ecosystems uncertain" may in fact cover a very large part of the world (Fig. 5.1). More recently, several global analyses confirmed that

Trophic Ecology: Bottom-Up and Top-Down Interactions across Aquatic and Terrestrial Systems, eds T. C. Hanley and K. J. La Pierre. Published by Cambridge University Press. © Cambridge University Press 2015.

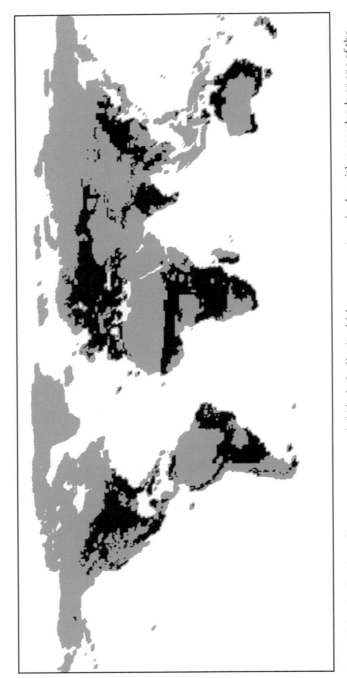

Figure 5.1 Distribution of "ecosystems uncertain" (dark shading), which are areas categorized as either grasslands or one of the biomes dominated by woody plants in the Whittaker biome diagram (reprinted with permission from Bond, 2005).

across a large part of the global land surface, tree cover is indeed bimodal (Staver et al., 2011) or even trimodal (Scheffer et al., 2012). This means that under the same climatic conditions, a system may be in a treeless, savanna, or forest state; this pattern has been described for (sub)tropical (Hirota et al., 2011; Staver et al., 2011), as well as boreal (Scheffer et al., 2012), parts of the world. From this, we can conclude that, rather than being purely controlled from the bottom up, ecosystems, and even biomes, are shaped by interacting bottom-up and top-down factors (Polis and Strong, 1996). There is general agreement that both bottom-up and top-down factors affect plant communities (Polis and Strong, 1996), but the question remains what the relative strengths of such top-down and bottom-up processes are and whether we can find general spatial and temporal patterns in their effects (Gripenberg and Roslin, 2007). Thus, here we aim to explore the relative role of bottom-up and top-down factors and assess their differences and/or similarities among woody-dominated ecosystems across the globe.

An important gradient along which woody plant communities across the globe can be arranged is productivity potential, which is largely determined by water availability (rainfall and temperature) and soil fertility. Productivity gradients also underlie one of the most influential theories predicting the relative importance of top-down processes, the *exploitation ecosystems hypothesis* (EEH from now onwards). EEH predicts that the importance of top-down processes as determinants of ecosystem structure increases with increasing productivity at a geographical scale. At low productivity, plant biomass is too low to support a herbivore population and plants will be regulated mainly by bottom-up effects, such as nutrient availability (Fretwell, 1977; Oksanen et al., 1981; DeAngelis, 1992). With increasing productivity, herbivore density will grow and regulate plant biomass from the top down. However, with further productivity increases, plants are predicted to escape from herbivore control as herbivores themselves come under top-down regulation by carnivores (Oksanen and Oksanen, 2000). Hence, top-down control of plant biomass by herbivores should be strongest at intermediate levels of productivity.

In this chapter we review the relative importance of bottom-up and top-down control of wooded ecosystems along a productivity gradient. We chose to focus on four different ecosystems that differ in potential plant productivity: boreal forest, savanna, temperate lowland forest, and tropical rainforest (ordered from low to high productivity according to Lieth and Whittaker, 1975; Table 5.1). For bottom-up processes, we consider factors which affect the resource supply to woody plants, whereas top-down processes are considered to be factors that result in plant biomass removal. Top-down and bottom-up factors can obviously influence many aspects of ecosystems and it is, therefore, important to specify the response variables of interest. We decided to focus on three characteristics of wooded ecosystems for which bottom-up and top-down control has been studied extensively: tree recruitment, tree species composition, and overall tree cover.

Table 5.1 *Ordering of forest systems based on Lieth and Whittaker (1975)*

	Net primary production (g m^{-2} year^{-1})	
	Range	Average
Equatorial tropical rainforest	1000–3500	2200
Equatorial seasonal rainforest	1000–2500	1600
Temperate forest – evergreen	600–2500	1300
Temperate forest – deciduous	600–2500	1200
Savanna	200–2000	900
Taiga (boreal forest and coniferous)	400–2000	800
Temperate steppe	200–1500	600
Boreal tundra and alpine	10–400	140

After reviewing the four systems, we synthesize the results and discuss similarities and differences in the relative role of top-down control in these systems.

The interaction of bottom-up and top-down controls of woody plants in boreal forest

Boreal forests are found in the northern hemisphere, covering large parts of Canada, northern Scandinavia, and Russia. Climatic conditions, and the factors associated with them (e.g., permafrost), set important limits to the productivity in these systems. Moreover, as the angle of the sun above the horizon decreases with latitude, solar energy input is lower, strengthened by the absence of sunlight during part of the year. As a result, high altitude boreal forests (at *c.* 45° and 65° N latitude) generally consists of two strata: a canopy layer and a herbaceous vegetation layer. In contrast, temperate forest systems are typically composed of three strata and tropical forests may be composed of up to five strata (Terborgh, 1985). In addition, the crown structures of the dominant tree shapes are related to latitude. Whereas typical boreal tree species are strongly inclined (e.g., Spruce, *Picea abies*), the crowns of canopy trees in tropical forest are generally flat or shallowly domed (Terborgh, 1985; Kuuluvainen, 1992). Together, low productivity and the change in angular radiation with altitude favor narrow-crowned evergreen conifers above broad-leaved and round-crowned deciduous trees in the boreal zone (Kuuluvainen, 1992).

Bottom-up and top-down processes controlling tree recruitment rate and species composition

Because soil nutrient supply, especially nitrogen, often limits the productivity of boreal forest vegetation (including woody species) and primary productivity (e.g., Van Cleve et al., 1983; Bonan and Shugart, 1989), nutrient cycling studies have been a prominent component of ecosystem research in boreal forests (Van Cleve et al., 1991; Nams et al., 1993; Pastor et al., 1993; Turkington et al., 2002). However, mammalian herbivores strongly interact with bottom-up nutrient availability and affect the structure and functioning of boreal ecosystems, especially by differentially affecting the recruitment of different tree species (Pastor et al., 1988; Brandner et al., 1990; McInnes et al., 1992; Pastor and Naiman, 1992; Pastor et al., 1993). As a result, the long-term effect of browsing appears to be the replacement of palatable deciduous tree species with unpalatable conifers or deciduous species (Bryant and Chapin, 1986; McInnes et al., 1992; Kielland and Bryant, 1998; Danell et al., 2003). These shifts in tree species composition due to browsing lead to strong indirect cascading effects on ecosystem functioning. A shift toward unpalatable species leads to a decrease in litter quality, because unpalatable species contain relatively high amounts of secondary plant metabolites (phenolics, terpenes) and structural components (cellulose and lignin). Reduced litter quality slows down microbial activity and leads to slower carbon and nutrient turnover (see also, Chapters 9 and 10 of this volume). When mineralization rates, particularly of nitrogen, are sufficiently depressed through these indirect feedbacks, the subsequent recovery of the ecosystem may be severely limited and long-term patterns of tree recruitment may be affected (Pastor et al., 1993). Such feedbacks cause the effects of herbivores on ecosystems to persist even after the herbivores are no longer present (Pastor and Naiman, 1992). This clearly illustrates the importance of interactions between bottom-up and top-down effects in these systems. Interestingly, conifer-dominance as a result of selective moose browsing can only be altered by alternative top-down factors, such as fire, disease, or insect herbivory (see *Bottom-up and top-down processes controlling tree cover* section below). These factors may reset the system to earlier successional stages where palatable, deciduous tree species again dominate and nutrient cycling rates are again increased (Pastor et al., 1993). Fire and insects are strong drivers in many boreal forest ecosystems, often interacting to affect succession and forest species composition (McCullough et al., 1998).

Bottom-up and top-down processes controlling tree cover

Fire and insects are natural disturbance agents in many boreal forest ecosystems, often interacting to affect succession, nutrient cycling, and forest species composition (McCullough et al., 1998). Typical for boreal systems is the role that bark beetles (e.g., *Ips typhograpus* in Eurasia and *Dendroctonus ponderosae* and *D. rufipennis* in North America), in addition to defoliators and fungi, play in driving cycles of mortality and recruitment of several coniferous species (see Malmström and Raffa, 2000 for a recent overview), thereby contributing to gap dynamics. The insects kill the trees, after which storms take down the dead trees and create forest gaps that may be tens to hundreds of hectares in size. Top-down insect control hence depends on abiotic conditions. Insect herbivore top-down effects also strongly interact with bottom-up effects in other ways.

For example, outbreaks of insects and pathogens in the boreal forest are controlled partly by weather conditions. Extended periods of warm weather can
favor the development of insect populations. Fire and mammalian herbivores
further modify the effect of insects as they may slow down gap regeneration
(McCullough et al., 1998). The relative importance of top-down control by fire
versus insects depends on the abiotic conditions. Insects play a larger role under
relatively wet conditions, while the importance of fire increases under drier
conditions (Malmström and Raffa, 2000).

Concluding, top-down factors clearly drive gap dynamics in boreal forests,
but interact strongly with bottom-up factors, perhaps in similar ways as they
do in tropical rain forests and temperate forests. However, top-down factors
are generally believed not to influence the ecosystem-level structure and woody
cover of boreal forests; i.e., they do not induce shifts from forest into a more open
savanna-like ecosystem, rather only affecting tree species composition (but see
Scheffer et al., 2012). While they may slow down gap regeneration, herbivores
and fire do not seem to prevent canopy closure in the long run. In contrast,
bottom-up factors, such as unproductive peat lands or permafrost, do lead to
reduced woody cover and more open forest structure.

Higher trophic levels affecting ungulate top-down effects in boreal forests

The classic study on the boreal forest of the Isle Royale in Lake Superior has
revealed that higher trophic levels may strongly regulate interactions between
herbivores and trees. In this system, wolves regulate moose population cycles
(Peterson et al., 1984), and, hence, indirectly affect top-down effects of moose
on patterns of tree regeneration (McLaren and Peterson, 1994). With an increase
in moose population size, regeneration of their principal food plant, Balsam
fir (*Abies balsamea*), is reduced, which eventually leads to decreased food supply.
However, the main factor determining herbivore density is predation. Increased
wolf numbers precede a decrease in moose density, indicating that this tri-
trophic system is dominated by top-down forces. Plant growth only reacted to
annual changes in productivity (bottom-up factors) when released from moose
herbivory due to more intense wolf predation (McLaren and Peterson, 1994).
While arguing for the primacy of top-down control in the Isle Royale food chain,
McLaren and Peterson (1994) acknowledged that large-scale forest disturbance
was nevertheless an important bottom-up stochastic influence.

The interaction of bottom-up and top-down controls of woody plants in savanna

Savannas are defined by the coexistence of trees and C4 grasses, where the
grass layer is continuous and tree cover may vary widely, ranging from dense
woodland to open grassland communities (Scholes and Archer, 1997). Although
there is ongoing debate about which factors are the main drivers of woody

plants in savannas, both bottom-up (soil properties, rainfall) and top-down (fire and herbivory) factors play an important role and interactions may be complex (Bond, 2008). Here, we provide a short summary of bottom-up and top-down processes that control tree recruitment, species composition, and overall tree cover in savannas. For a detailed description of these bottom-up and top-down processes in savannas, refer to Chapter 4 of this volume.

Bottom-up and top-down processes controlling tree recruitment and species composition

In savannas tree recruitment is often episodic and linked to periods with high rainfall, low fire frequencies (Bond and Van Wilgen, 1996; Skowno et al., 1999; Higgins et al., 2007), and/or low herbivore numbers (Prins and van der Jeugd, 1993; Holdo et al., 2009). Frequent fires in savannas can constrain tree recruitment to the extent that certain trees never reach adult heights – the so-called "fire trap" (Bond and Van Wilgen, 1996; Skowno et al., 1999). These suppressed individuals are only able to escape from this fire trap in periods of low fire frequencies, and even then only the individuals with the highest maximum growth rates will be able to grow above the flame zone (Wakeling et al., 2011). Fire exclusion experiments show unequivocally that woody species recruitment is constrained by fire (Higgins et al., 2007). However, the frequency of burning is a delicate balance. Infrequent fires will enhance woody establishment in grasslands, while too frequent fires will not give the grass enough time to recover and also enhance woody establishment (Roques et al., 2001; Wigley et al., 2010). Mammalian herbivory can have a large effect on tree recruitment in savannas as well. The direct effects of browsers on tree recruitment include reduction of seedling densities and keeping trees in smaller size classes through repeated browsing, creating a "browse trap" analogue to the fire trap (Sankaran et al., 2013). Exclosure experiments suggest that direct effects of browsers on tree recruitment may be large (Sankaran et al., 2013), something which is further exemplified by the increase in woody species following ungulate population crashes (Prins and Van der Jeugd, 1993). Interestingly, the effects of grazing ungulates may contrast with those of browsers. Grazers may increase woody recruitment by reducing cover of the competitive grasses, whereas browsers generally do the opposite. However, when intense grazing creates so-called grazing lawns (Cromsigt and Olff, 2008), tree recruitment is also heavily constrained (Van der Waal et al., 2011). Overall, the combined effect of grazers and browsers on tree recruitment in savannas is not well understood.

Bottom-up and top-down factors may regulate different aspects of tree recruitment. For example, a herbivore exclosure study by Staver et al. (2009) showed that bottom-up effects, including soil nutrients and moisture, were important in determining tree growth rates, whereas top-down disturbances from fire and herbivory limited tree densities. Moreover, several studies have addressed the interactions between bottom-up and top-down processes on savanna tree recruitment. Among the most striking examples is the effect of increasing CO_2 levels, which may reduce the role of top-down factors such as fire and herbivory by allowing trees to grow faster and more easily escape the fire and browse traps (Bond and Midgley, 2000). This phenomenon is suggested to drive the great increases in woody cover that are being observed in savannas worldwide (Buitenwerf et al., 2012). At a larger scale, it is clear that soil fertility and rainfall determine woody species composition in savannas. On fertile soils and under drier conditions, species from the Mimosaceae often dominate, leading to so-called fine-leaved savanna, while broad-leaved trees dominate on infertile soils (Scholes, 1999). Such patterns are also visible at relatively small scales, e.g., along catenal gradients, where broad-leaved species tend to dominate the nutrient-poor hill crests, while mostly fine-leaved species are found along the footslopes (Baldeck et al., 2014). On smaller scales, browsers may also drive shifts in woody species composition through differential effects on mortality rates of different tree species (Bond and Loffell, 2001; Sankaran et al., 2013). However, these effects of herbivores depend on bottom-up factors, with stronger changes in species composition in less productive environments (e.g., Young et al., 2013).

Bottom-up and top-down processes controlling tree cover

Both bottom-up and top-down factors are important in controlling tree cover and ultimately determine tree–grass ratios in savannas (Bond, 2008). Herbivores may influence tree density and cover in savannas, but the magnitude and direction of their effect depend on the underlying productivity (soil fertility and rainfall) of the system (Pringle et al., 2007; Asner et al., 2009). Sankaran et al. (2005) show that the importance of top-down control of woody cover increases with rainfall. Below 650 mm, woody cover is limited by and increases with rainfall; but above 650 mm, top-down factors such as fire are necessary to prevent canopy closure. Recent studies confirmed that at intermediate levels of rainfall, areas can either be in a grassland, savanna, or forest state depending on the intensity of fire or level of herbivory (Hirota et al., 2011; Staver et al., 2011).

A special feature of African savanna systems, in contrast to the other biomes we discuss, is the presence of megaherbivores (Owen-Smith, 1988). Due to their size of > 1000 kg, megaherbivores escape predator control and can have a disproportionally large effect on savanna structure and functioning. Most studies on megaherbivore impacts have focused on African elephants. Elephants can be a strong driver of savanna structure and may change closed woodland savannas into open grasslands by removing adult trees (Kerley et al., 2008). The effects of elephants, however, may vary depending on rainfall and elephant density (Guldemond and van Aarde, 2008). Similarly, white rhino, as a megagrazer, can have strong effects on savanna structure by creating grazing lawns (Cromsigt and Olff, 2008; Cromsigt and Te Beest, 2014), thereby impacting woody plant dynamics by affecting fire frequency (Waldram et al., 2008) and tree–grass interactions (Van der Waal et al., 2011). Similar to elephant, effects of rhino seem to depend on productivity, with stronger effects on vegetation structure in areas with higher rainfall (Waldram et al., 2008).

Higher trophic levels affect ungulate top-down effects in African savannas

Savannas contain some of the world's most diverse and abundant predator populations, but we have surprisingly little data on their top-down control. Although there is increasing evidence from savanna systems that carnivores influence herbivore space use and behavior (e.g., Valeix et al., 2009), we lack studies that link such behavioral responses to cascading effects on vegetation structure and function. Similarly, there are almost no empirical studies that link ecosystem responses to a numerical response of herbivores to predation (see Sinclair et al., 2010).

A main aspect that distinguishes African savannas from many temperate systems is the diversity of carnivores. Carnivore species have different hunting strategies and might drive different behaviors in their prey (Thaker et al., 2011). Moreover, not all herbivores are equally affected by predation and it has been shown that small herbivores are under stronger top-down regulation, whereas larger herbivores, especially megaherbivores, are mostly bottom-up regulated (Hopcraft et al., 2010). To date it is unclear what the summed effect of a diverse guild of predators is on a diverse guild of herbivores in shaping savanna ecosystems (Sinclair et al., 2010).

In addition to the top-down effects of predators on herbivores, pathogens can also have large effects on herbivore populations and, hence, tree recruitment. An excellent example of this is the episodic recruitment of *Acacia tortilis* and an increase in other woody species following crashes in impala numbers due to anthrax and rinderpest epidemics in Lake Manyare NP (Prins and Van der Jeugd, 1993). Similarly, eradication of rinderpest led to a strong increase in wildebeest numbers, whose intense grazing reduced fire frequency, resulting in an increase in woody cover (Holdo et al., 2009).

The interaction of bottom-up and top-down controls of woody plants in temperate forest: the Białowieża Primeval Forest as a case study

In the temperate zone, few examples exist of forest systems that have experienced little impact by humans (Hannah et al., 1995). Whereas in prehistoric times, large portions of this biome supposedly consisted of closed forest systems (Svenning, 2002; Birks, 2005; Mitchell, 2005), nowadays very little forest in a natural state remains (Hannah et al., 1995; Bengtsson et al., 2000). Temperate forests belong to one of the most threatened biomes of the world. It has been estimated that only 0.2% of central European deciduous forests remains in a relatively natural state (Hannah et al., 1995). A prime example of a temperate forest system that has been little affected by humans is the Białowieża Primeval Forest (BPF). The BPF (1450 km²), situated in eastern Poland (52′45′ N, 23′50′ E) and western Belarus, is a large continuous forest composed of multispecies tree stands, both deciduous (especially *Quercus robur*, *Tilia cordata*, and *Carpinus betulus*) and coniferous (*Pinus sylvestris* and *Picea abies*). In the Polish part, the Białowieża National Park (BNP) comprises the best preserved central part of the BPF, which experienced minimal human impact (Jędrzejewska et al., 1997; Samojlik et al., 2007). The BNP consists of a 47.5 km² area of strictly protected old growth forest in which no human intervention (forestry activities, hunting) has been allowed since 1921 and tourist access is only permitted with a guide. Before 1921, human impact in this region was very limited (Samojlik et al., 2007). Another unique aspect of BPF is that it is one of the very few European (lowland) forests still hosting complete native ungulate and carnivore assemblages (Fig. 5.4). Five ungulate species occur throughout the forest system (from most to least abundant based on Jędrzejewski et al., 1997: red deer (*Cervus elaphus*), wild boar (*Sus scrofa*), roe deer (*Capreolus capreolus*), European bison (*Bison bonasus*), and moose (*Alces alces*)). These ungulates have not been hunted inside the BNP for over 80 years and co-occur with their natural predators, wolf (*Canis lupus*) and lynx (*Lynx lynx*).

Bottom-up and top-down processes controlling tree recruitment and species composition in the Białowieża Primeval Forest

In this dense old-growth forest (< 1% of the area is devoid of tree cover in the BNP; Michalczuk, 2001), the ultimate factor leading to the regeneration of trees is gap formation (Runkle, 1981; Bobiec et al., 2000; Bobiec, 2007). However, increased light levels at the forest soil will not only stimulate the germination and growth of tree seedlings. Herbaceous vegetation is the first to react and creates increased competition with tree seedlings, hampering seedling growth or establishment (Modry et al., 2004; Van den Berghe et al., 2006; Kuijper et al., 2010a). Larger size classes (saplings up to 50 cm) are not related to herbaceous vegetation cover (Kuijper et al., 2010a), as they have outgrown and escaped competition with herbaceous plants. Still, their abundance is mainly determined by abiotic bottom-up factors, with soil fertility most strongly affecting their abundance (Kuijper et al., 2010a). There is a large shift in the importance of bottom-up versus top-down control when tree saplings reach the size of 50 cm (Fig. 5.2). In these size classes, ungulate browsing is the dominant factor affecting the recruitment process (Kuijper et al., 2013), with ungulates determining both the number and species composition of the regenerating trees (Kuijper et al., 2010a, Fig. 5.2).

Ungulates in this forest preferentially browse tree saplings between *c.* 50 and 150 cm (Kuijper et al., 2013) and may keep individuals of many tree species in this height class, creating a similar browse trap to that in savannas. Suppressed individuals are only able to escape from the browse trap once they have

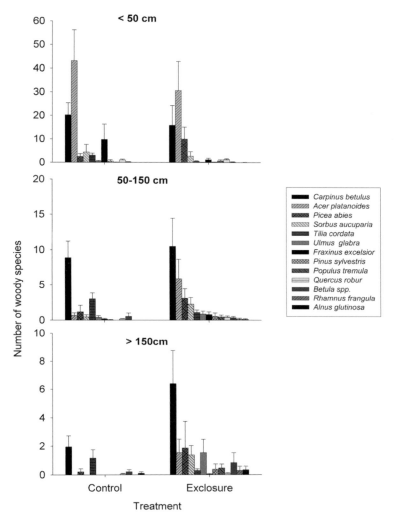

Figure 5.2 Different stages of tree recruitment on experimental plots with and without herbivores inside the Białowieża Primeval Forest. Number of trees inside exclosures (without ungulates) and controls (with ungulates) in 49 m² plots. Graphs depict the number of trees per species in different size classes 11 years after the start of the experiment. Ungulates clearly affect recruitment into size classes taller than 50 cm, whereas smaller size classes are not affected. For further details of the experimental set-up of this study, see Kuijper et al., 2010a.

reached a height to escape herbivore control (Kuijper et al., 2013) or profit from periods when herbivore numbers are low (Kuijper et al., 2010b), similar to what is observed in African systems following a population crash of browsing ungulates (Prins and Van der Jeugd, 1993; Chapter 4, this volume). As ungulates browse virtually all woody plant species that are present, only the most browse-tolerant tree species escape the browse trap. In our system, hornbeam (*Carpinus betulus*) is one of the most tolerant species. Both experimental (see Fig. 5.2) and long-term correlative data show that ungulate browsing leads to tree stands with a high proportion of this species (Kuijper et al., 2010a; 2010b; Kuijper, 2011). Tolerance traits (such as fast regrowth after browsing; see also, Chapter 8 of this volume) that enable these tree species to cope with high ungulate browsing pressure also make them palatable for ungulates. Interestingly, resistance traits, either physical (e.g., thorns) or chemical (e.g., secondary plant compounds), which play an important role in boreal forest systems, seem less important in this system. Whereas in boreal forests, ungulate browsing generally leads to the replacement of palatable species by unpalatable ones, in the temperate forest of Białowieża the opposite occurs. Here, ungulate browsing results in a dominance of palatable, browse-tolerant tree species (Kuijper et al., 2010a; 2010b; Cromsigt and Kuijper, 2011). This selection for tolerance traits by ungulate browsing might be a common feature of highly productive temperate forest systems (Cromsigt and Kuijper, 2011).

Herbivore top-down effects in these dense forests strongly interact with bottom-up factors, such as light availability. Due to elevated light levels at the forest soil, gaps in the tree canopy are characterized by strongly enhanced growth of tree saplings and herbaceous vegetation (Modry et al., 2004). The higher food availability also attracts ungulates and causes ungulate browsing to be strongly concentrated in forest gaps (Kuijper et al., 2009). Tree growth, depending on the tree species, is between 3- and 40-fold higher when ungulate browsing is excluded inside forest gaps, whereas the effects of ungulate browsing are much less pronounced (1- to 6-fold increase in tree height) under a closed canopy (Fig. 5.3).

Ungulates in this forest browse until a height of *c.* 200 cm, and trees have to reach this size class to escape herbivore control (Kuijper et al., 2013). Once trees escape the browse trap, bottom-up factors determine the final stage of the recruitment into the tree canopy. Although little research on this stage is available, plant–plant competition, light availability, and soil characteristics play a key role (Szwagrzyk et al., 2012). Hence during the recruitment process from a tree seedling to a canopy tree, there are switches between the relative importance of bottom-up and top-down effects. Bottom-up effects (light availability, soil fertility) regulate the seedling stage, while ungulate top-down effects control the intermediate sapling stages. Bottom-up effects (light availability, tree–tree competition) determine the final stage of large saplings above 2 m recruiting into the canopy.

Bottom-up and top-down processes controlling tree cover in the Białowieża Primeval Forest

In general, ungulate browsing reduces tree regeneration and affects tree species composition in this temperate forest system (Kuijper et al., 2010a; 2010b), but these top-down herbivore effects are not equally distributed in the area. Inside

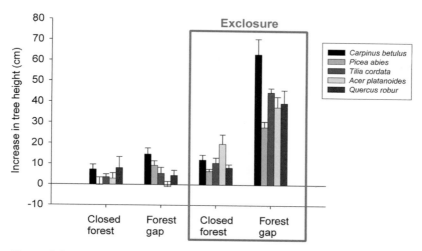

Figure 5.3 Increase in tree sapling height of five trees species in the height class of 50–100 cm, planted in April 2008 on experimental plots manipulating herbivory (exclosures versus controls) and light conditions (under closed tree canopy and in forest gaps). Six individuals of each tree species were planted per plot. This fully factorial design was replicated six times (for details on experimental set-up see Kuijper et al., 2009). Average increases in tree height between April 2008 and October 2009 per plot per species (±SE) are depicted. Besides significant effects of treatments (herbivory and light), both factors strongly interacted in all but one species (*Acer*: P = 0.011, *Carpinus*: P < 0.001, *Picea*: P = 0.053, *Quercus*: P = 0.003), illustrating that the effects of herbivory are more pronounced inside forest gaps.

forest gaps, where the tree recruitment process starts, herbivores exert much stronger effects on trees than outside forest gaps (Fig. 5.3). Despite the concentration of ungulates and the elevated strength of their top-down effects on tree saplings, ungulates are not able to prevent tree recruitment from happening in these highly productive forest gaps. Trees are regenerating with ungulates present, even during peaks in ungulate population density (Kuijper et al., 2010a; 2010b). Forest gaps are typically closed in by tall trees within 2–3 decades and no gaps that are kept in an open state beyond this time period exist without help from humans (Samojlik and Kuijper, 2013). Hence, ungulates in this forest only retard the regeneration of trees, but seem to be unable to maintain patches of open landscape.

Higher trophic levels affecting ungulate top-down effects in the Białowieża Primeval Forest

To understand the ungulate–tree interactions, processes at higher trophic levels cannot be ignored. Wolf and lynx are the two apex predators in the system and influence ungulate–tree interactions both directly (by influencing the number of ungulates) and indirectly (by influencing ungulate behavior) (Fig. 5.4). Considering the direct effects, predation slows down population growth rates,

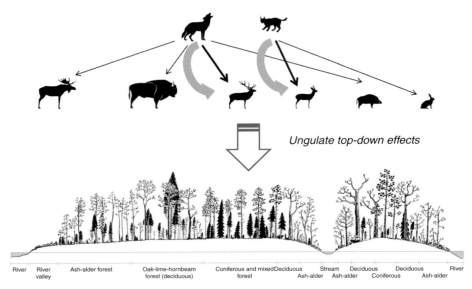

Figure 5.4 Wolf and lynx directly influence ungulate numbers, especially of their main prey red deer and roe deer, while large prey such as moose and bison are hardly affected due to low predation pressure (straight arrows). Spatial patterns of browsing by red deer are also modified by wolf, and lynx likely exerts similar indirect effects on its main prey (gray arrows). Via these density- and behaviorally mediated effects, apex predators influence ungulate top-down effects in the Białowieża Primeval Forest (tree stands drawn by T. Samojlik).

especially of red (predated mainly by wolf) and roe deer (predated mainly by lynx), and limits ungulate numbers (Jędrzejewska and Jędrzejewski, 2005). However, populations of ungulates are not regulated by predation, as the predation rate is inversely density-dependent or does not change with ungulate density (Jędrzejewski et al., 2002; Jędrzejewska and Jędrzejewski, 2005). During periods with increasing ungulate density, the predation rate (the proportion of prey consumed) declines. This illustrates that bottom-up factors (climate) cause fluctuations in ungulate numbers (regulation), and predators limit this number below the carrying capacity set by climatic conditions. The presence of apex predators, in this way, reduces the strength of top-down herbivore effects on tree regeneration by reducing their numbers.

In addition to these direct density-mediated effects, several studies from American temperate systems show that predators may indirectly influence ungulates by changing their distribution toward less risky habitat types, changing their movement patterns, or shifting their activity toward less risky times (Fortin et al., 2005; Mao et al., 2005; Creel et al., 2008). Most of these studies are from mixed-habitat systems with grassland as well as forested patches. In BPF, we are beginning to explore how behaviorally mediated effects of predators might operate in densely forested systems (Kuijper et al., 2013; 2014). In areas where wolf presence is higher (the core area of their territory), browsing ungulates avoid foraging close to objects that can act as escape impediments or block the view. Coarse woody debris (CWD; fallen logs, large branches, etc.) plays an important role in creating patches that are perceived by ungulates as being risky (Kuijper et al., 2013). Browsing intensity on tree saplings is reduced up to 20% when the amount of CWD increases in the direct vicinity. These effects are in line with findings from other studies illustrating that ungulate species avoid and are more vigilant in habitats with low visibility (Underwood, 1982; Halofsky and Ripple, 2008; Valeix et al., 2009) and high incidences of escape impediments (Halofsky and Ripple, 2008). Moreover it shows how environmental factors (presence of CWD) can affect ungulate top-down effects via higher trophic level interactions.

Whereas the direct effects of predators result in an overall lower browsing intensity in the BPF (by reducing the number of browsing ungulates), indirect effects of predators change the behavior of browsing ungulates and create a fine-scale pattern of areas with increased (perceived) predation risk and reduced browsing intensity. These locations with reduced browsing intensity offer windows of opportunity for tree saplings to escape herbivore control more quickly than areas with low predation risk (Kuijper et al., 2013).

The interaction of bottom-up and top-down controls of woody plants in tropical forest

Bottom-up and top-down processes controlling tree recruitment and species composition

Although there is plenty of discussion about which factors are the main drivers of tropical rainforest tree regeneration, evidently gap dynamics and top-down factors either caused by predators on herbivores or herbivores on the plant layer are highly important in these systems.

Whereas gap formation is also the ultimate factor regulating tree recruitment in tropical forest, abiotic factors seem to play a minor role in determining which tree species successfully recruit. Gap formation events are often highly unpredictable and seeds of trees tend to be patchily distributed. For these reasons the predictability of gap-phase recruitment is low and recruitment of different tree species in species-rich tropical and temperate forest systems often follows the principles of "lottery recruitment" (Busing and Brokaw, 2002). This assumes ecological equivalence of tree species, leading to equal chances of regeneration of different species. The hypothesis of ecological equivalence, which states that species have similar life history strategies, although they can still differ in many ways, is the cornerstone of the *unified neutral theory* (Hubbell, 2005). Species which are at least approximately ecologically equivalent are expected to arise frequently in fine-grained environments in species-rich communities that are strongly dispersal- and recruitment-limited. In these highly diverse tropical forest environments, different species are predicted to evolve adaptations to the most common environmental conditions, leading to converging life history strategies (Hubbell, 2006). This implies that at sites offering favorable abiotic conditions (e.g., in a gap) different tree species have equal chances to regenerate and it is not competitive exclusion that determines the outcome of this process. The neutral theory has also often been contested and criticized; see e.g., Terborgh et al. (1996).

A large number of studies stress the importance of top-down effects from herbivores, but also diseases or pathogens, in affecting tree growth or recruitment. This is, for example, expressed in the Janzen–Connell hypothesis (JC; Janzen, 1970; Connell, 1971), which is a widely accepted explanation for the maintenance of tree species biodiversity in tropical rainforests. According to their hypothesis, there is a density- or distance-dependent factor in the recruitment of seedlings from adults of tropical tree species due to host-specific herbivores or pathogens. These herbivores or pathogens that specifically target a species make the areas directly surrounding the seed-producing parent tree inhospitable for the survival of seedlings. In other words, specialist herbivores (often referred to as predators in tropical forest studies), pests, and pathogens keep key tree species rare enough to reduce their competitive ability, allowing space for many other species to co-occur. Similar to the neutral theory, the JC hypothesis is under continuous discussion (see, e.g., Hyatt et al., 2003; Carson et al., 2008).

Interestingly, compared to the other systems we discussed, there has been much attention on top-down effects of insect herbivores (Dyer et al., 2010) or larger invertebrates, such as land crabs (Green et al., 2008), and small invertebrates, such as amphibians (Beard et al., 2002), in tropical forests. Few studies have addressed or shown top-down effects imposed by large vertebrate herbivores, especially ungulates, in this system (but see, peccaries: Silman et al., 2003; browsing herbivores including deer and tapir: Dirzo and Miranda, 1991). Rather than direct effects of browsing by vertebrate herbivores, effects of rodents on tree recruitment of large-seeded trees have been widely documented. Central American agoutis (such as *Dasyproctapunctata*) scatter-hoard large seeds across their 2 to 3 ha home ranges as food reserves for the low-fruit season (Jansen et al., 2012). This behavior leads to the dispersal of seeds toward locations with lower conspecific tree densities, thus facilitating the escape of seeds from natural enemies. This could profoundly affect the spatial structure of plant communities (Hirsch et al., 2012). Scatter-hoarding rodents have been suggested to be a substitute for extinct megafaunal seed dispersers of tropical large-seeded trees (Jansen et al., 2012). Moreover, top-down effects have been shown to often operate indirectly, via insectivorous ants, birds, mammals, or bats affecting the density of insect herbivores and reducing herbivory (e.g., Dyer and Letourneau, 1999; Van Bael and Brawn, 2005; Dunham, 2008; Kalka et al., 2008).

Bottom-up and top-down processes controlling tree cover of tropical rainforests

The traditional view is that the high plant biomass and closed canopy cover of tropical rainforests is determined by the very high rainfall these forests receive (Hirota et al., 2011; Staver et al., 2011). This leads to very high rates of plant productivity, which cannot be controlled by top-down factors. Hence, tree cover in tropical rainforests is determined by abiotic control. However, a new hypothesis presents a revolutionarily different view on the relationship between forest cover and abiotic control (Sheil and Murdiyarso, 2009; Makarieva and Gorshkov, 2010). According to the "biotic pump" hypothesis, tropical rainforests (and other extensive tracts of closed canopy forest) strongly determine rainfall patterns themselves through their evaporation behavior. Makarieva and Gorshkov (2010), in fact, predict that clearing substantial parts of the rainforest might lead to a staggering 95% decline in rainfall, switching the system from wet to arid. Such arid conditions would make it very difficult for the forest to re-establish.

Higher trophic levels affecting top-down effects in tropical rainforests

There is also continuing debate on the role that large carnivores play in affecting plants. Several studies (e.g., Terborgh, 1988) suggest that large predators, such as cats (puma, jaguar, and ocelot) and raptors, collectively limit mid-sized terrestrial mammals, which would alter forest regeneration without their control. This view has been supported by studies on predator-free islands created by a hydroelectric impoundment in Venezuela. The absence of predators leads to higher herbivore intensity (mainly by leaf-cutter ants, *Atta* spp. and *Acromyrmex* sp., and howler monkeys, *Alouatta seniculus*) resulting in increased tree sapling and seedling mortality and reduced tree recruitment (Terborgh et al., 2006). According to the authors, these results showed the operation of strong top-down forces that negatively impacted nearly every plant species present, implying that predators maintained community stability (Terborgh et al., 2001; 2006). Other studies have contested this view. Wright et al. (1994) showed that contemporary mammal abundances at sites with and without large felids failed to support the hypothesis that felids control prey abundances. They argued that the impact of the subtle increases in herbivorous mammals that might follow release from felid predation is not likely to be dramatic. Although it is still a debated question whether large mammalian predators (Terborgh, 1988) or the small invertebrates (Wilson, 1987) rule these tropical ecosystems, clearly top-down effects (either by herbivores or carnivores) play an important role in maintaining tree diversity.

Synthesis: what can we learn from a cross-biome comparison of wooded ecosystems?

Theoretical models of trophic interactions predict that systems that are ordered along a gradient from low to high primary productivity should be increasingly consumer controlled (Fretwell, 1977; Oksanen et al., 1981; DeAngelis, 1992), with top-down herbivore control being strongest at intermediate levels of productivity. After reviewing the four different woody-dominated ecosystems (ordered from low to high productivity as: boreal forest, savanna, temperate lowland forest, and tropical rainforest; Lieth and Whittaker, 1975), can we observe similarities and differences in the relative role of top-down control in these systems?

It seems that the answer to this question depends on what ecosystem characteristic you talk about: woody plant recruitment, species composition, or overall tree cover. Looking at woody plant recruitment and species composition, our overview suggests that top-down effects play a similar role in all four ecosystems. In all these systems, top-down forces (herbivores or fire) can alter tree recruitment rates and tree species composition. Moreover, in all systems top-down factors strongly interact with bottom-up factors, such as rainfall and soil fertility. However, the direction of the effect of herbivory on species composition clearly differs between more productive systems, such as temperate forests and certain savannas, and less productive systems, such as boreal forest. In relatively unproductive boreal forest systems, interactions between browsing and woody plants are largely mediated by secondary plant metabolites, and this resistance strategy seems to dominate in plant communities in these ecosystems (Bryant et al., 1983; 1991; 1992). In more productive systems, the tolerance strategy seems more common as many woody plant species use fast growth to escape mammalian herbivory and to compensate biomass removal (Coley et al., 1985). As a result, in unproductive systems mammalian herbivory pushes the plant community toward more unpalatable plant species (with resistance traits) and decelerates nutrient cycling, while in more productive systems herbivory increases the proportion of palatable species (with tolerance traits) and accelerates nutrient cycling (Cromsigt and Kuijper, 2011; Du Toit and Olff, 2014).

Looking at overall woody cover, the influence of top-down factors seems to vary between ecosystems. In boreal, temperate, and tropical forests top-down factors play a role in gap dynamics, but at the ecosystem level woody cover seems to be driven mostly by bottom-up factors (water availability and soils). Top-down factors can delay, but not prevent, canopy closure in the long run. In contrast, top-down factors in savannas may actually shift the ecosystem between states varying from open grassland to savanna and even to closed canopy woodland.

Although in savanna systems the strength of top-down factors also strongly depends on abiotic conditions, especially rainfall. At low (and high) levels of water availability, maximum woody cover is again bottom-up regulated. These patterns indicate that at the more extreme ends of the productivity gradient, bottom-up factors determine woody cover, while for savannas at intermediate productivity, the balance between bottom-up and top-down processes is more variable and alternative stable states can more easily occur. As explained in the introduction, these patterns also relate to the view of "ecosystems uncertain" (Whittaker, 1975; Bond, 2005), and recent studies that illustrate that at zones with intermediate rainfall and mild seasonality, tree cover shows a bimodal pattern with either an open savanna or closed canopy tropical forest (Hirota et al., 2011; Staver et al., 2011). Similarly, for the more benign parts of the boreal biome, alternative modes in boreal tree cover have been shown, suggesting that they too may represent alternate states (Scheffer et al., 2012). Bond (2005) argues that consumer control is a key component of these "uncertain" ecosystems, where open habitats would be forest without consumers.

Only less than a quarter of the Earth's ice-free land can still be considered to be natural with minor alteration by humans, of which 20% are forests (Ellis and Ramankutty, 2008). The current global extent, duration, and intensity of human transformations of ecosystems caused a variety of novel ecological patterns and processes to emerge (Ellis, 2011). The temperate zone of the northern hemisphere has experienced the most anthropogenic changes and for the longest time (Fig. 3 in Bond and Keeley, 2005), and only small remnants of forests with a pristine or primeval character can be found (Hannah et al., 1995). This long history of human disturbance has strongly altered the presence and potential impact of top-down forces. First, there is a large-scale loss of apex carnivores through direct persecution and habitat change (Woodroffe, 2000; Laliberte and Ripple, 2004). Even in areas where large carnivores are currently still present, their population sizes have often been minimized to an ecologically ineffective size (Soulé et al., 2003; Nores et al., 2008). Was the temperate zone more strongly top-down controlled by carnivores before humans came along? We can only speculate about the answer, but certainly large carnivores had stronger effects on prey populations in these systems when their numbers were higher and diversity larger. Due to these impoverished carnivore communities, we may currently overestimate the role ungulates play in many areas. The reported overabundance of deer and their strong effects on woody plant communities (Côté et al., 2004), especially in temperate regions of North America, is at least partly related to the lack of natural predators. In addition, although in many of those areas humans hunt ungulates, it is very questionable if and how humans replace the top-down effects of carnivores, especially the indirect effects of carnivores on their prey (Cromsigt et al., 2013; see Chapter 14 of this volume for further discussion on this point). Second, humans have impoverished large herbivore communities through hunting and habitat destruction, especially in the temperate zone (Laliberte and Ripple, 2004). Moose disappeared from most European systems where it used to be present, and species like American bison and pronghorn antelope follow similar patterns in North America. Aurochs (*Bos taurus*, in the 17th century) and Tarpan (*Equus ferus*, in the 18th century), which once occurred throughout Eurasia, went extinct only recently. Remaining large herbivores, such as European bison (*Bison bonasus*), have been faced with large range reductions and now likely live in sub-optimal habitats (Kerley et al., 2012). As a result, ungulate community composition in temperate systems is now dominated by browsers (deer species) and large grazers have been lost. It is these large grazers who are suggested to play a key role in temperate systems, where habitats may shift among grassland, shrubland, and forest habitat under the influence of grazing (Olff et al., 1999). Hence, before humans started to alter temperate systems extensively, ungulate densities in some areas may have been high enough to have profound effects on woody plant cover, especially in the supposed preferred habitat of extinct grazers, such as fertile riverine areas (Hall, 2008). However, several studies indicate that even in prehistoric times, large herbivores were not able to open up closed forests on a large scale and that in large parts of Europe closed forests prevailed (Svenning, 2002; Birks, 2005; Mitchell, 2005).

Finally, what is true for mammalian herbivory is also true for another major top-down factor, fire. Although fire can have an enormous effect on the species composition and structure of wooded ecosystems (Bond and Keeley, 2005), fire has been effectively excluded across large parts of the temperate zone to minimize the risks to humans. Hence, we currently study vegetation patterns in areas in the temperate zones that are the result of centuries of fire exclusion. Considering the greatly impoverished large mammal communities and exclusion of fire across the temperate zone in the northern hemisphere, we have to be aware that studies carried out in these clearly human-impacted areas might not accurately reflect the importance of top-down control relative to bottom-up control.

To conclude, can we observe that woody plant communities ordered from low to high productivity show an increasing importance of top-down factors shaping these systems? When looking at the influence on tree recruitment and tree species composition, top-down forces seem to play an equally significant role in all biomes and strongly interact with bottom-up factors. When looking at ecosystem-level woody cover, bottom-up forces prevail in boreal, temperate, and tropical forest. In these systems, top-down factors generally cannot prevent canopy closure in the long run. In contrast, in savanna systems, top-down factors can control woody plant cover in the more benign parts of the climate gradients that occur in these regions. This illustrates that trying to simply fit forest biomes into categories of mainly top-down and mainly bottom-up controlled seems problematic and ignores much of the interactions between biotic and abiotic factors, which may potentially lead to alternative outcomes within the biomes. However, due to the loss of large mammals (both large herbivores and large carnivores) and suppression of fire in many systems, we may currently be underestimating the importance of consumer control in many regions, especially in the more densely populated areas.

Acknowledgments

The work of DPJK was supported by funding by the Polish Ministry of Science and Higher Education (grant no. 2012/05/B/NZ8/01010). MtB acknowledges support from the Nordic Centre of Excellence TUNDRA, funded by the Norden Top-Level Research Initiative "Effect Studies and Adaptation to Climate Change." JPGMC was supported by the Swedish thematic research programme Wildlife and Forestry. Additionally, MC received funding by the Polish Ministry of Science and Higher Education (grants no. N309 137 335).

References

Asner, G. P., Levick, S. R., Kennedy-Bowdoin, T., et al. (2009). Large-scale impacts of herbivores on the structural diversity of African savannas. *Proceedings of the National Academy of Sciences of the USA*, **106**, 4947–4952.

Baldeck, C. A, Colgan, M. S., Féret, J.-B., et al. (2014). Landscape-scale variation in plant community composition of an African savanna from airborne species mapping. *Ecological Applications*, **24**, 84–93.

Beard, K. H., Vogt, K. A. and Kulmatiski, A. (2002). Top-down effects of a terrestrial frog on forest nutrient dynamics. *Oecologia*, **133**, 583–593.

Bengtsson, J., Nilsson, S. G., Franc, A. and Menozzi, P. (2000). Biodiversity, disturbances, ecosystem function and management of European forests. *Forest Ecology and Management*, **132**, 39–50.

Birks, H. J. B. (2005). Mind the gap: how open were European primeval forests? *Trends in Ecology and Evolution*, **20**,154–156.

Bobiec, A. (2007). The influence of gaps on tree regeneration: a case study of the mixed lime-hornbeam (*Tilio-Carpinetum* Tracz. 1962) communities in the Białowieża Primeval Forest. *Polish Journal of Ecology*, **55**, 441–455.

Bobiec, A., van der Burgt, H., Meijer, K., et al. (2000). Rich deciduous forests in Białowieża as a dynamic mosaic of developmental phases: premises for nature conservation and restoration management. *Forest Ecology and Management*, **130**, 159–175.

Bonan, G. A. B. and Shugart, H. H. (1989). Environmental factors and ecological processes in boreal forests. *Annual Review of Ecology and Systematics*, **20**, 1–28.

Bond, W. J. (2005). Large parts of the world are brown or black: a different view on the 'Green World' hypothesis. *Journal of Vegetation Science*, **16**, 261–266.

Bond, W. J. (2008). What limits trees in C-4 grasslands and savannas? *Annual Review of Ecology, Evolution, and Systematics*, **39**, 641–659.

Bond, W. J. and Keeley, J. E. (2005). Fire as a global 'herbivore': the ecology and evolution of flammable ecosystems. *Trends in Ecology and Evolution*, **20**, 387–394.

Bond,W. J. and Loffell, D. (2001). Introduction of giraffe changes acacia distribution in a South African savanna. *African Journal of Ecology*, **39**, 286–294.

Bond, W. J. and Midgley, G. F. (2000). A proposed CO_2-controlled mechanism of woody plant invasion in grasslands and savannas. *Global Change Biology*, **6**, 865–869.

Bond, W. J. and van Wilgen, B. W. (1996). *Fire and Plants*. London: Chapman and Hall.

Brandner, T. A., Peterson, R. O. and Risenhoover, K. L. (1990). Balsam fir on Isle Royale: effects of moose herbivory and population density. *Ecology*, **71**, 155–164.

Bryant, J. P. and Chapin, F. S. III. (1986). Browsing-woody plant interactions during boreal forest plant succession. In *Forest Ecosystems in the Alaskan Taiga: A Synthesis of Structure and Function – Ecological Studies*, Volume 57, ed. K. Van Cleve, F. S. Chapin, P. W. Flanagan, L. A. Viereck and C. T. Dyrness. Berlin: Springer-Verlag, pp. 213–225.

Bryant, J. P., Provenza, F. D., Pastor, J., et al. (1991). Interactions between woody plants and browsing mammals mediated by secondary metabolites. *Annual Review of Ecology and Systematics*, **22**, 431–446.

Bryant, J. P., Reichardt, P. B. and Clausen, T. P. (1992). Chemically mediated interactions between woody plants and browsing mammals. *Journal of Range Management*, **45**, 18–24.

Buitenwerf, R., Bond, W. J., Stevens, N. and Trollope, W. S. W. (2012). Increased tree densities in South African savannas: > 50 years of data suggests CO_2 as a driver. *Global Change Biology*, **18**, 675–684.

Busing, R. T. and Brokaw, N. (2002). Tree species diversity in temperate and tropical forest gaps: the role of lottery recruitment. *Folia Geobotanica*, **37**, 33–43.

Carson, W. P., Anderson, J. T., Leigh, E. G. Jr. and Schnitzer, S. A. (2008). Challenges associated with testing and falsyfying the Janzen-Connell hypothesis: a review and critique. In *Tropical Forest Ecology*, ed. W. P. Carson and S. A. Schnitzer. Oxford: Wiley-Blackwell, pp. 210–241.

Coley, P. D., Bryant, J. P. and Chapin, F. S. III. (1985). Resource availability and plant antiherbivore defense. *Science*, **230**, 895–899.

Connell, J. H. (1971). On the role of natural enemies in preventing competitive exclusion in some marine animals and in rain forest trees. In *Dynamics of populations*, ed. P. J. den Boer and G. R. Gradwell. Wageningen: Centre for Agricultural Publications and Documentation, pp. 298–310.

Côté, S. D., Rooney, T. P., Trembley, J.-P., Dussault, C. and Waller, D. M. (2004). Ecological impacts of deer overabundance. *Annual Review of Ecology, Evolution, and Systematics*, **35**, 113–147.

Creel, S., Winnie Jr., J. A., Christianson, D. and Liley, S. (2008). Time and space in general models of antipredator response: test with wolves and elk. *Animal Behaviour*, **76**, 1139–1146.

Cromsigt, J. P. G. M. and Kuijper, D. P. J. (2011). Revisiting the browsing lawn concept: evolutionary interactions or pruning herbivores? *Perspectives in Plant Ecology, Evolution and Systematics*, **13**, 207–215.

Cromsigt, J. P. G. M. and Olff, H. (2008) Dynamics of grazing lawn formation: an experimental test of the role of scale-dependent processes. *Oikos*, **117**, 1444–1452.

Cromsigt, J. P. G. M. and Te Beest, M. (2014). Restoration of a megaherbivore – landscape-level impacts of white rhinoceros in Kruger National Park. *Journal of Ecology*. DOI: 10.1111/1365–2745.12218.

Cromsigt, J. P. G. M., Kuijper, D. P. J., Adam, M., et al. (2013). Hunting for fear: innovating management of human–wildlife conflicts. *Journal of Applied Ecology*, **50**, 544–549.

Danell, K., Bergström, R., Edenius, L. and Ericsson, G. (2003). Ungulates as drivers of tree population dynamics at module and genet levels. *Forest Ecology and Management*, **181**, 67–76.

DeAngelis, D. L. (1992). *Dynamics of Nutrient Cycling and Food Webs*. London: Chapman and Hall.

Dirzo, R. and Miranda, A. (1991). Altered patterns of herbivory and diversity in the forest understory: a case study of the possible consequences of contemporary defaunation. In *Plant–Animal Interactions: Evolutionary Ecology in Tropical and Temperate Regions*, ed. P. W. Price, P. W. Lewinsohn, G. W.

Fernandes and W. W. Benson. New York: Wiley, pp. 273–287.

Du Toit, J. T. and Olff, H. (2014). Generalities in grazing and browsing ecology: using across-guild comparisons to control contingencies. *Oecologia*. DOI: 10.1007/s00442-013-2864-8.

Dunham, A. E. (2008). Above and below ground impacts of terrestrial mammals and birds in a tropical forest. *Oikos*, **117**, 571–579.

Dyer, L. A. and Letourneau, D. K. (1999). Relative strengths of top-down and bottom-up forces in a tropical forest community. *Oecologia*, **119**, 265–274.

Dyer, L. A., Letourneau, D. K., Chavarria, G. V. and Amorett, D. S. (2010). Herbivores on a dominant understory shrub increase local plant diversity in rain forest communities. *Ecology*, **91**, 3707–3718.

Ellis, E. C. (2011). Anthropogenic transformation of the terrestrial biosphere. *Philosophical Transactions of the Royal Society A*, **369**, 1010–1035.

Ellis, E. C. and Ramankutty, N. (2008). Putting people in the map: anthropogenic biomes of the world. *Frontiers in Ecology and Environments*, **6**, 439–447.

Fortin, D., Beyer, H. L., Boyce, M. S., et al. (2005). Wolves influence elk movements: Behavior shapes a trophic cascade in Yellowstone National Park. *Ecology*, **86**, 1320–1330.

Fretwell, S. D. (1977). The regulation of plant communities by the food chains exploiting them. *Perspectives in Biology and Medicine*, **20**, 169–185.

Green, P. T., O'Dowd, D. J. and Lake, P. S. (2008). Recruitment dynamics in a rainforest seedling community: context-independent impact of a keystone consumer. *Oecologia*, **156**, 373–385.

Gripenberg, S. and Roslin, T. (2007). Up or down in space? Uniting the bottom-up versus top-down paradigm and spatial ecology. *Oikos*, **116**, 181–188.

Guldemond, R. and Van Aarde, R. (2008). A meta-analysis of the impact of African

elephants on savanna vegetation. *Journal of Wildlife Management*, **72**, 892–899.

Hall, S. J. G. (2008). A comparative analysis of the habitat of the extinct aurochs and other prehistoric mammals in Britain. *Ecography*, **31**, 187–190.

Halofsky, J. S. and Ripple, W. J. (2008). Fine-scale predation risk on elk after wolf reintroduction in Yellowstone National Park, USA. *Oecologia*, **155**, 869–877.

Hannah, L., Carr, J. L. and Landerani, A. (1995). Human disturbance and natural habitat – a biome level analysis of a global data set. *Biodiversity and Conservation*, **4**, 128–155.

Higgins, S. I., Bond, W. J., February, E. C., et al. (2007). Effects of four decades of fire manipulation on woody vegetation structure in savanna. *Ecology*, **88**, 1119–1125.

Hirota, M., Holmgren, M., Van Nes, E. H. and Scheffer, M. (2011). Global resilience of tropical forest and savanna to critical transitions. *Science*, **334**, 232–235.

Hirsch, B. T., Kays, R., Pereira, V. E. and Jansen, P. A. (2012). Directed seed dispersal toward areas with low conspecific tree density by a scatter-hoarding rodent. *Ecology Letters*, **15**, 1423–1429.

Holdo, R. M., Sinclair, A. R. E., Dobson, A. P., et al. (2009). A disease-mediated trophic cascade in the Serengeti and its implications for ecosystem C. *PLoS Biology*, **7**, e1000210.

Hopcraft, J. G. C., Olff, H. and Sinclair, A. R. E. (2010). Herbivores, resources and risks: alternating regulation along primary environmental gradients in savannas. *Trends in Ecology and Evolution*, **25** (2), 119–128.

Hubbell, S. P. (2005). Neutral theory in community ecology and the hypothesis of functional equivalence. *Functional Ecology*, **19**, 166–172.

Hubbell, S. P. (2006). Neutral theory and the evolution of ecological equivalence. *Ecology*, **87**, 1387–1398.

Hyatt, L. A., Rosenberg, M. S., Howard, T. G., et al. (2003). The distance dependence predictions of the Janzen-Conell hypothesis: a meta-analyis. *Oikos*, **103**, 590–602.

Jansen, P. A., Hirsch, B. T., Emsens, W. J., et al. (2012). Thieving rodents as substitute dispersers of megafaunal seeds. *Proceedings of the National Academy of Sciences of the USA*, **109**, 12610–12615.

Janzen, D. H. (1970). Herbivores and number of tree species in tropical forests. *American Naturalist*, **104**, 501–528.

Jędrzejewska, B. and Jędrzejewski, W. (2005). Large carnivores and ungulates in European temperate forest ecosystems: bottom-up and top-down control. In *Large Carnivores and the Conservation of Biodiversity*, ed. J. C. Ray, K. H. Redford, R. S. Steneck and J. Berger. Washington: Island Press, pp. 230–245.

Jędrzejewska, B., Jędrzejewski, W., Bunevich, A., N., Miłkowski, L. and Krasiński, Z. A. (1997). Factors shaping population densities and increased rates of ungulates in Białowieża Primeval Forest (Poland and Belarus) in the 19th and 20th century. *Acta Theriologica*, **42**, 399–451.

Jędrzejewski, W., Schmidt, K., Theuerkauf, J., et al. (2002). Kill rates and predation by wolves on ungulate populations in Białowieża primeval forest (Poland). *Ecology*, **83**, 1341–1356.

Kalka, M. B., Smith, A. R. and Kalko, E. K. V. (2008). Bats limit arthropods and herbivory in a tropical forest. *Science*, **320**, 71–71.

Kerley, G. I. H., Landman, M., Kruger, L., et al. (2008). Effects of elephants on ecosystems and biodiversity. In *Assessment of South African Elephant Management*, ed. R. J. Scholes and K. G. Mennell. Johannesburg: Witwatersrand University Press, pp. 146–205.

Kerley, G. I. H., Kowalczyk, R. and Cromsigt, J. P. G. M. (2012). Conservation implications of the refugee species concept and the European bison: king of the forest or refugee in a marginal habitat? *Ecography*, **35**, 519–529.

Kielland, K. and Bryant, J. P. (1998). Moose herbivory in taiga: effects on

biogeochemistry and vegetation dynamics in primary succession. *Oikos*, **82**, 377–383.

Kuijper, D. P. J. (2011). Lack of natural control mechanisms increases wildlife-forestry conflict in managed temperate European forest systems. *European Journal of Forest Research*, **130**, 895–909

Kuijper, D. P. J., Cromsigt, J. P. M. G., Churski, M., et al. (2009). Do ungulates preferentially feed in forest gaps in European temperate forests? *Forest Ecology and Management*, **258**, 1528–1535.

Kuijper, D. P. J., Cromsigt, J. P. G. M., Jędrzejewska, B., et al. (2010a). Bottom-up versus top-down control of tree regeneration in the Białowieża Primeval Forest, Poland. *Journal of Ecology*, **98**, 888–899.

Kuijper, D. P. J., Jędrzejewska, B., Brzeziecki, B., et al. (2010b). Fluctuating ungulate density shapes tree recruitment in natural stands of the Białowieża Primeval forest, Poland. *Journal of Vegetation Science*, **21**, 1082–1098.

Kuijper, D. P. J., de Kleine, C., Churski, M., et al. (2013). Landscape of fear in Europe: wolves affect spatial patterns of ungulate browsing in Białowieża Primeval Forest, Poland. *Ecography* **36**, 1263–1275. DOI: 10.1111/j.1600–0587.2013.00266.x.

Kuijper, D. P. J., Verwijmeren M., Churski, M., et al. (2014). What cues do ungulates use to assess predation risk in dense temperate forests? *PLoS One* **9**(1): e84607. DOI: 10.1371/journal.pone.0084607.

Kuuluvainen, T. (1992). Tree architecture adapted to efficient light utilization: is there a basis for latitudinal gradients? *Oikos*, **65**, 275–284.

Laliberte, A. S. and Ripple, W. J. (2004). Range contractions of North American carnivores and ungulates. *BioScience*, **54**, 123–138.

Lieth, H. and Whittaker, R. H. (eds.) (1975). *Primary Productivity of the Biosphere: Ecological Studies Vol. 14*. New York: Springer Verlag.

Makarieva, A. M. and Gorshkov, V. G. (2010). The biotic pump: condensation, atmospheric dynamics and climate. *International Journal of Water*, **5**, 365–385.

Malmström, C. M. and Raffa, K. F. (2000). Biotic disturbance agents in the boreal forest: considerations for vegetation change models. *Global Change Biology*, **6**, 35–48.

Mao, J. S., Boyce, M. S., Smith, D. W., et al. (2005). Habitat selection by elk before and after wolf reintroduction in Yellowstone National Park. *Journal of Wildlife Management*, **69**, 1691–1707.

McCullough, D. G., Werner, R. A. and Neumann, D. (1998). Fire and insects in northern and boreal forest ecosystems of North America. *Annual Review of Entomology*, **43**, 107–127.

McInnes, P., Naiman, R. J., Pastor, J. and Cohen, Y. (1992). Effects of moose browsing on vegetation and litter of the boreal forest, Isle Royale, Michigan, USA. *Ecology*, **73**, 2059–2075.

McLaren, B. E. and Peterson, R. O. (1994). Wolves, moose, and tree rings on Isle Royale. *Science*, **266**, 1555–1558.

Michalczuk, C. (2001). Forest habitats and tree stands of the Białowieża National Park. In *Phytocoenosis 13, Supplementum Cartographiae Geobotanicae 13*, Warszawa – Białowieża: Białowieża Geobotanical Station of Warsaw University.

Mitchell, F. J. G. (2005). How open were European primeval forests? Hypothesis testing using palaeoecological data. *Journal of Ecology*, **93**, 168–177.

Modry, M., Hubeny, D. and Rejsek, K. (2004). Differential response of naturally regenerated European shade tolerant tree species to soil type and light availability. *Forest Ecology and Management*, **188**, 185–195.

Nams, V. O., Folkard, N. F. G. and Smith, J. N. M. (1993). Effects of nitrogen fertilization on several woody and non woody boreal forest species. *Canadian Journal of Botany*, **71**, 93–97.

Nores, C., Llaneza, L. and Alvarez, M. A. (2008). Wild boar *Sus scrofa* mortality by hunting and wolf *Canis lupus* predation: an example in northern Spain. *Wildlife Biology*, **14**, 44–51.

Oksanen, L. and Oksanen, T. (2000). The logic and realism of the hypothesis of exploitation ecosystems. *The American Naturalist*, **155**, 703–723.

Oksanen, L., Fretwell, S. D., Arruda, J. and Niemelä. P. (1981). Exploitation ecosystems in gradients of primary productivity. *The American Naturalist*, **118**, 240–261.

Olff, H., Vera, F. W. M., Bokdam, J., et al. (1999). Shifting mosaics in grazed woodlands driven by the alternation of plant facilitation and competition. *Plant Biology*, **1**, 127–137.

Owen-Smith, R. N. (1988). *Megaherbivores: the Influence of Very Large Body Size on Ecology*. Cambridge, UK: Cambridge University Press.

Pastor, J. and Naiman, R. J. (1992). Selective foraging and ecosystem processes in boreal forests. *American Naturalist*, **139**, 690–705.

Pastor, J., Naiman, R. J., Dewey, B. and McInnes, P. (1988). Moose, microbes and the boreal forest. *BioScience*, **38**, 770–777.

Pastor, J., Dewey, B., Naiman, R. J., McInnes, P. and Cohen, Y. (1993). Moose browsing and soil fertility in the boreal forests of Isle Royale National Park. *Ecology*, **74**, 467–480.

Peterson, R. O., Page, R. E. and Dodge, K. M. (1984).Wolves, moose, and the allometry of population cycles. *Science*, **224**, 1350–1352.

Polis, G. A. and Strong, D. R. (1996). Food web complexity and community dynamics. *The American Naturalist*, **147**, 813–846.

Prentice, I. C., Cramer, W., Harrison, W. P., et al. (1992). A global biome model based on plant physiology and dominance, soil properties and climate. *Journal of Biogeography*, **19**, 117–134.

Pringle, R. M., Young, T. P., Rubenstein, D. I. and McCauley, D. J. (2007). Herbivore-initiated interaction cascades and their modulation by productivity in an African savanna. *Proceedings of the National Academy of Sciences of the USA*, **104**, 193–197.

Prins, H. T. and Van der Jeugd, H. P. (1993). Herbivore population crashes and woodland structure in East Africa. *Journal of Ecology*, **81**, 305–314.

Roques, K. G., O'Connor, T. G. and Watkinson, A. R. (2001). Dynamics of shrub encroachment in an African savanna: relative influences of fire, herbivory, rainfall and density dependence. *Journal of Applied Ecology*, **38**, 268–280.

Runkle, J. R. (1981). Gap regeneration in some old-growth forests of the eastern United States. *Ecology*, **62**, 1041–1051.

Samojlik, T. and Kuijper, D. P. J. (2013). Grazed wood pasture versus browsed high forests: impact of ungulates on forest landscapes from the perspective of the Białowieża Primeval Forest. In *Trees, Forested Landscapes and Grazing Animals: A European Perspective on Woodlands and Grazed Treescapes*, ed. I. D. Rotherham. London/New York: Routledge, pp. 143–162.

Samojlik, T., Jędrzejewska, B., Krasnodębski, D., Dulinicz, M. and Olczak, H. (2007). Man in the ancient forest. *Academia, The Magazine of the Polish Academy of Sciences*, **4**(12), 36–37.

Sankaran, M., Hanan, N. P, Scholes, R. J., et al. (2005). Determinants of woody cover in African savannas. *Nature*, **438**, 846–849.

Sankaran M, Augustine, D. J. and Ratnam J. (2013). Native ungulates of diverse body sizes collectively regulate long-term woody plant demography and structure of a semi-arid savanna. *Journal of Ecology*, **101**, 1389–1399.

Scheffer, M., Hirotaa, M., Holmgren, M., Van Nes, E. H. and Chapin III, F. S. (2012). Thresholds for boreal biome transitions. *Proceedings of the National Academy of Sciences of the USA*, **109**, 21384–21389.

Scholes, R. J. (1999). Savannas. In *The Vegetation of Southern Africa*, eds. R. Cowling, D. Richardson and S. Pierce. Cambridge, UK: Cambridge University Press, pp. 258–277.

Scholes, R. J. and Archer, S. R. (1997). Tree-grass interactions in sanannas. *Annual Review of Ecology, Evolution, and Systematics*, **28**, 517–544.

Sheil, D. and Murdiyarso, D. (2009). How forests attract rain: an examination of a new hypothesis. *BioScience*, **59**, 341–347.

Silman, M. R., Terborgh, J. W. and Kiltie, R. A. (2003). Population regulation of a dominant rain forest tree by a major seed predator. *Ecology*, **84**, 431–438.

Sinclair, A. R. E., Metzger, K., Brashares, J. S., et al. (2010). Trophic cascades in African savanna: Serengeti as a case study. In *Trophic Cascades: Predators, Prey, and the Changing Dynamics of Nature*, ed. J. A. Estes and J. Terborgh. Washington: Island Press, pp. 255–274.

Skowno, A. L., Midgley, J. J., Bond, W. J. and Balfour, D. (1999). Secondary succession in *Acacia nilotica* (L.) savanna in the Hluhluwe Game Reserve, South Africa. *Plant Ecology*, **145**, 1–9.

Soulé, M. E., Estes, J. A., Berger, J. and Martinez del Rio, C. (2003). Ecological effectiveness: conservation goals for interactive species. *Conservation Biology*, **17**, 1238–1250.

Staver, C. A., Bond, W. J., Stock, W. D., van Rensburg, S. J. and Waldram, M. S. (2009). Browsing and fire interact to suppress tree density in an African savanna. *Ecological Applications*, **19**, 1909–1919.

Staver, C. A., Archibald, S. and Levin. S. A. (2011). The global extent and determinants of savanna and forest as alternative biome states. *Science*, **334**, 230–232.

Svenning, J.-C. (2002). A review of natural vegetation openness in north-western Europe. *Biological Conservation*, **104**, 133–148.

Szwagrzyk, J., Szewczyk, J. and Maciejewski, Z. (2012). Shade-tolerant tree species from temperate forests differ in their competitive abilities: a case study from Roztocze, south-eastern Poland. *Forest Ecology and Management*, **282**, 28–35.

Terborgh, J. (1985). The vertical component of plant species diversity in temperate and tropical forests. *The American Naturalist*, **126**, 760–776.

Terborgh, J. (1988). The big things that run the world – a sequel to E. O. Wilson. *Conservation Biology*, **2**, 402–403.

Terborgh, J., Foster, R. B. and Nunez, P. (1996). Tropical tree communities: a test of the nonequilibrium hypothesis. *Ecology*, **77**, 561–567.

Terborgh, J., Lopez, L., Nuñez, V. P., et al. (2001). Ecological meltdown in predator-free forest fragments. *Science*, **294**, 1923–1926.

Terborgh, J., Feeley, K., Silman, M., Nuñez, P. and Balukjian, B. (2006). Vegetation dynamics of predator-free land-bridge islands. *Journal of Ecology*, **94**, 253–263.

Thaker, M., Vanak, A. T., Owen, C. R., et al. (2011). Minimizing predation risk in a landscape of multiple predators: effects on the spatial distribution of African ungulates. *Ecology*, **92**, 398–407.

Turkington, R., John, E., Watson, S. and Seccombe-Hett, P. (2002). The effects of fertilization and herbivory on the herbaceous vegetation of the boreal forest in north-western Canada: a 10-year study. *Journal of Ecology*, **90**, 325–337.

Underwood, R. (1982). Vigilance behaviour in grazing African antelopes. *Behaviour*, **79**, 82–107.

Valeix, M., Loveridge, A. J., Chamaillé-Jammes, S., et al. (2009). Behavioral adjustments of African herbivores to predation risk by lions: spatiotemporal variations influence habitat use. *Ecology*, **90**, 23–30.

Van Bael, S. A. and Brawn, J. D. (2005). The direct and indirect effects of insectivory by birds in two contrasting Neotropical forests. *Oecologia*, **143**, 106–116.

Van Cleve, K., Oliver, L., Schlentner, R., Viereck, L. A. and Dyrness, C. T. (1983). Productivity and nutrient cycling in taiga forest ecosystems. *Canadian Journal of Forest Research*, **13**, 747–766.

Van Cleve, K., Chapin, F. S. III, Dyrness, C. T. and Viereck, L. A. (1991). Elemental cycling in taiga forests: state-factor control. *BioScience*, **41**(2), 78–88.

Van den Berghe, C., Frelechoux, F., Gadallah, F. and Butler, A. (2006). Competitive effects of herbaceous vegetation on tree seedling emergence, growth and survival: does gap size matter? *Journal of Vegetation Science*, **17**, 481–488.

Van der Waal, C., Kool, A., Meijer, S. S., et al. (2011). Large herbivores may alter vegetation structure of semi-arid savannas through soil nutrient mediation. *Oecologia*, **165**, 1095–1107.

Wakeling, J. L., Staver, A. C. and Bond, W. J. (2011). Simply the best: the transition of savanna saplings to trees. *Oikos*, **120**, 1448–1451.

Waldram, M., Bond, W. and Stock, W. (2008). Ecological engineering by a mega-grazer: white rhino impacts on a South African savanna. *Ecosystems*, **11**, 101–112.

Whittaker, R. H. (1962). Classification of natural communities. *Botanical Review*, **28**, 1–239.

Whittaker, R. H. (1975). *Communities and Ecosystems*, 2nd edn. London: Collier MacMillan.

Wigley, B. J. Bond, W. J. and Hoffman, M. T. (2010). Thicket expansion in a South African savanna under divergent land use: local vs. global drivers? *Global Change Biology*, **16**, 964–976.

Wilson, E. O. (1987). The little things that run the world (The importance and conservation of invertebrates). *Conservation Biology*, **1**, 344–346.

Wright, S. J., Gompper, M. E. and Deleon, B. (1994). Are large predators keystone species in Neotropical forests – the evidence from Barro Colorado Island. *Oikos*, **71**, 279–294.

Woodroffe, R. (2000). Predators and people: using human densities to interpret declines of large carnivores. *Animal Conservation*, **3**, 165–173.

Young, H. S., McCauley, D. J., Helgen, K. M., et al. (2013). Effects of mammalian herbivore declines on plant communities: observations and experiments in an African savanna. *Journal of Ecology*, **101**, 1030–1041.

CHAPTER SIX

Dynamic systems of exchange link trophic dynamics in freshwater and terrestrial food webs

JOHN L. SABO

Arizona State University, Phoenix, AZ, USA

and

DAVID HOEKMAN

National Ecological Observatory Network, Boulder, CO; University of Wisconsin, Madison, WI; and Southern Nazarene University, Bethany, OK, USA

Introduction

Trophic dynamics and the exchange of materials across ecosystem boundaries are topics that have each been treated in detail, but separately, in freshwater ecosystems. The observation that common food web motifs occur – if not flourish – in spite of a boundary between water and air, which divides realms that impose fundamentally different physiological demands, would suggest that the forest doesn't end at the lake, and vice versa. This is an important observation because it further suggests that conservation and restoration cannot be successful without considering these linkages explicitly. Further, the watershed food web provides a rich framework for a more comprehensive management paradigm. In this chapter, we review some of the few, but salient, examples of cascading trophic interactions across the land–water interface. Many of these examples are "incomplete" cascades that include strong interactions across two, but not more, trophic levels. We also discuss how a dynamic systems framework can be used to conceptualize and organize the diverse array of trophic interactions in which exchange of organic matter and mobile animals across the land–water boundary may instigate strong top-down dynamics.

Aquatic–terrestrial exchange of carbon, organisms, and energy

Energy exchange between land and freshwater ecosystems has a long history in stream and lake ecology. Terrestrial organic matter is the dominant source of carbon (C) to streams in many temperate catchments (Fisher and Likens, 1973)

Trophic Ecology: Bottom-Up and Top-Down Interactions across Aquatic and Terrestrial Systems, eds T. C. Hanley and K. J. La Pierre. Published by Cambridge University Press. © Cambridge University Press 2015.

and its removal has consequences for abundances of primary consumers (detritivores) and their predators (Wallace et al., 1999). Strong energetic dependence by rivers on terrestrial forests is the basis for the River Continuum Concept (RCC; Vannote et al., 1980). Similarly, in tropical floodplain rivers, terrestrial plants – especially fruit – provide key nutrition for fish. This observation forms the basis for the Flood Pulse Concept (FPC; Junk et al., 1989). The RCC and the FPC continue to be influential concepts and productive lines of empirical pursuit to this day (Cross et al., 2011; 2013; Jardine et al., 2012a; 2012b).

In lake ecosystems, autochthonous resources – phytoplankton and benthic diatoms – provide the bulk of the food base to higher trophic levels. The footprint of land on lake food webs was largely ignored until very recently (Pace et al., 2004; Carpenter et al., 2005; Cole et al., 2006; 2011). This relatively new recognition of the importance of terrestrial organic matter (OM) in lakes has stirred a very lively debate about the relative importance of large fluxes of dissolved OM of terrestrial origin and higher quality autochthonous resources to lacrustrine consumers (Brett et al., 2009; 2012).

In contrast to a much longer tradition acknowledging the importance of terrestrial fluxes of OM to aquatic ecosystems, the reciprocal flux (from freshwater to land) has only recently been addressed (Polis et al., 1997a; Nakano and Murakami, 2001; Bastow et al., 2002; Sabo and Power, 2002a; 2002b; Sanzone et al., 2003; Power et al., 2004), and this work and that of many others (reviewed by Baxter et al., 2005) has broadened our definition of river food webs. In river ecosystems, there is a plethora of empirical evidence and recent theory to suggest that the influence of riverine algae, insects, and fish on terrestrial consumer abundance and community composition is large, both in terms of effect size and spatial extent (reviewed in Naiman et al., 2002; Marczak et al., 2007; Sabo and Hagen, 2012; Muehlbauer et al., 2013). Similarly, in lake ecosystems, there is growing evidence that aquatic insects influence abundance and community composition of terrestrial plants and animals. Terrestrial productivity of plant and arthropod communities near productive lakes is bolstered by emerging aquatic insects (Jonsson and Wardle, 2009; Hoekman et al., 2011; Dreyer et al., 2012; Hoekman et al., 2012). **In summary, freshwater ecosystems – streams and rivers in particular – have been one of the most important demonstration sites of the significance of energy flow between spatially structured food webs.**

Trophic dynamics in freshwater ecosystems

The study of trophic dynamics also has a rich tradition in lake and river food webs (Carpenter et al., 1985; 2001; 2008; Power, 1990; Wootton et al., 1996; Pace et al., 1999; Power et al., 2008); in fact, the first observation of a trophic cascade was in European lakes (Hrbacek et al., 1961) and these results may have stimulated the development of green world theory (HSS; Hairston et al.,

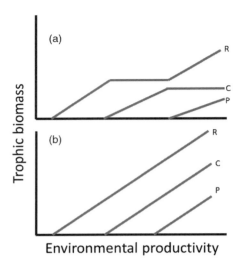

Figure 6.1 Illustration of green world theory (HSS) or top-down control (a) and resource or bottom-up control (b). Trophic level biomass (ordinate) increases with environmental productivity (abscissa) (b), but top-down control in (a) by consumers decouples biomass accrual from production in a stair-step fashion. Symbols for trophic levels follow equations 6.1: R is plant, C is herbivore (consumer), and P is predator.

1960) and later the exploitation ecosystems hypothesis (EEH; Oksanen et al., 1981; Oksanen, 1983; Oksanen and Oksanen, 2000). In green world theory (HSS), Hairston et al. (1960) argued that ecosystems maintain green vegetation because predators hold herbivores in check; HSS is the root of the notion that predators exert "top-down" control on ecosystem patterns of biomass. More broadly, Fretwell (1977) generalized the cascade concept across a productivity gradient (Fig. 6.1a; *sensu* Mittelbach et al., 1988). Specifically, as plant *productivity* increases, plant *biomass* initially increases, until enough plant energy is produced by the system to support herbivores. Further increases in *plant* productivity then increase *herbivore* biomass – but not plant biomass, which remains constant until enough herbivore energy is produced by the system to support carnivores. Further increases in *plant* productivity increase plant *and* carnivore biomass – but not herbivore biomass. Many of the textbook examples of trophic cascades have been observed from elegant experiments in rivers (Power, 1990) and lakes (Carpenter et al., 1985; 2001; Pace et al., 1999).

Green world theory (HSS) and the top-down control paradigm have been challenged by observations of a number of phenomena that influence trophic dynamics from the bottom-up. First, and foremost, plant defenses to herbivores and the inedibility of plant structural tissues (e.g., wood) may more strongly limit herbivory than predation pressure (Murdoch, 1966; Stachowicz and Hay, 1999; Shurin et al., 2006; also, see Chapter 8). Second, where plant defenses are not potent, herbivores may antagonize each other or consumers may hoard resources in territories leading to trophic inefficiency at the plant–herbivore link (Arditi and Saiah, 1992; Ginzburg and Akcakaya, 1992). Third, there is a rich theoretical debate about the role of species richness and food web link density (i.e., trophic complexity) on the occurrence of cascades (Hairston and Hairston, 1993; 1997; Polis and Strong, 1996; also, see Chapter 12), and vice versa the role of cascades in determining the structural stability of complex food webs (Neutel et al., 2002; Bascompte and Melian, 2005; Allesina et al., 2008). Briefly, simple "linear" food chains may be more amenable to cascades than "reticulate' or complex food webs (Polis and Strong, 1996), and complex webs might be structured in a way to minimize strong chains of interactions and cascades (Neutel et al., 2002; Bascompte and Melian, 2005; McCann, 2012). Coincidentally, many of the best examples of cascades come from freshwater ecosystems with low species diversity, prompting Strong to conjecture that "all cascades are wet" or from freshwater ecosystems (Strong, 1992). Several recent syntheses of trophic cascades across marine, freshwater, and terrestrial ecosystems have added some depth and nuance to this conjecture. First, while cascades do appear to be stronger in aquatic than terrestrial ecosystems, variation in the strength of top-down control appears to be as great within each type of ecosystem as between them (Shurin et al., 2002). Moreover, there are now many well-documented examples of trophic cascades in terrestrial ecosystems (Pace et al., 1999). Second, predators appear to have stronger effects on the variability of herbivore than plant biomass, and this top-down induced herbivore variability is stronger in lentic and pelagic ecosystems (Halpern et al., 2005). Finally, physiological and taxonomic differences among predators explain more of the variation in the strength of cascades across ecosystems than the distinction between aquatic and terrestrial ecosystem types (Borer et al., 2005). These and other complexities (McQueen et al., 1986) support an alternate view to HSS – one in which biomass patterns are determined more strongly by resource supply, or "bottom-up" rather than top-down (HSS) forces (Fig. 6.1b; *sensu* Mittelbach et al., 1988). **In summary, freshwater systems have been one of the most important demonstration sites of the importance of top predators in structuring patterns of trophic biomass in food webs and there continues to be debate about why freshwater cascades appear to be more robust and commonly observed.**

Overview of dynamic systems approach to river/lake–watershed exchange and trophic cascades

The purpose of this chapter is to review and synthesize empirical evidence for cross-system trophic cascades between freshwater and land. *Here, we define cross-system trophic cascades as regulation of the biomass of one or more trophic levels by predators but where cross-boundary movement of predators, their prey, or non-living resources influences the strength of these top-down dynamics.* To do this, we will use a dynamic systems (DS) approach based on McCann (2012). DS approaches have been central to the development of food web theory and continue to provide deep insights into understanding the dynamics of small clusters of strongly interacting species within larger, more complex webs. Below, we define some relevant DS terms and concepts that we will then apply to synthesize what is known about river–land and lake–land cross-system trophic cascades.

Dynamic systems models are based on coupled dynamic (time-dependent) equations that represent the population growth of a small subset (2–6) of strongly interacting species. A classic coupled dynamic system is the MacArthur–Rosenzweig consumer-resource (MRCR) model (following the notation of McCann (2012)):

$$\frac{dR}{dt} = R\,(growth - loss), \quad \text{or} \quad \frac{dR}{dt} = R\left\{r\left(1 - \frac{R}{K}\right) - \frac{a_C\,C}{R + R_0}\right\} \tag{6.1a}$$

$$\frac{dC}{dt} = C\,(growth - loss), \quad \text{or} \quad \frac{dC}{dt} = C\left\{\frac{e a_C\,R}{R + R_0} - m\right\} \tag{6.1b}$$

where R and C are the abundance, biomass, or density of the "resource" (here, plant) and "consumer" (here, herbivore). The growth term for the plant is density dependent or logistic growth. The loss term for the plant is consumption by the herbivore, and search or attacks by the herbivore are represented as a saturating function or Type II functional response (Oaten and Murdoch, 1975a; 1975b). The growth term for the herbivore is the loss term of the plant multiplied by a constant that is equivalent to the trophic transfer efficiency of the consumer (Lindeman, 1942). Finally, the loss term for the herbivore is a simple density-independent death rate. A more detailed description of the model can be found in McCann (2012).

McCann's treatment of dynamic systems covers several key concepts that are relevant to our review of cross-system trophic cascades. First, McCann distinguishes between excitable and non-excitable systems. A non-excitable system when perturbed will exhibit population trajectories that approach a steady state (where C and R in Eqns 6.1a–b do not change with time) without oscillations. By contrast, an excitable system responds to perturbation with population trajectories for C and R that approach steady state values with oscillations and may even exhibit an oscillatory steady state. The MRCR is inherently excitable and oscillates when the "coupling strength" between consumers and resources $\left(\frac{a_{max}C}{R+R_0}\right)$ is high relative to the consumer "loss term" (m) and resource self-regulation $\left(1 - \frac{R}{K}\right)$. As a rule of thumb, dynamic systems (not restricted to MRCR) are stabilized when losses/self-regulation are high relative to coupling strength. Examples of stabilizing biological processes that reduce coupling strength relative to losses are density dependence, inedibility, anti-herbivore defenses, and consumer interference. In the context of HSS, green world dynamics and trophic cascades arise in excitable systems and can be counterbalanced by high relative loss rates and/or density dependence of the resource population. How do these concepts relate to cross-system trophic cascades?

A trophic cascade involves inverse biomass patterns across a chain of at least three trophic levels: plant (R), herbivore (C), and predator (P), so this would involve a third equation describing coupling between C–P and loss rates for the predator, and modifications to the herbivore dynamics such that the losses reflect predation, viz:

$$\frac{dC}{dt} = C \, (growth - loss), \quad \text{or} \quad \frac{dC}{dt} = C \left\{ \frac{ea_C R}{R + R_0} - a_P P \right\} \tag{6.2a}$$

$$\frac{dP}{dt} = P \, (growth - loss), \quad \text{or} \quad \frac{dP}{dt} = P \left\{ ea_P C - m \right\} \tag{6.2b}$$

In such a system (Eqns 6.1a and 6.2a–b), strong coupling between C–R and P–C is almost unanimously destabilizing (excitable) and can be stabilized by weakening the coupling strength between the predator and herbivore relative to that of the herbivore and plant (McQueen et al., 1986; Lafontaine and McQueen, 1991; DeMelo et al., 1992; Shurin et al., 2002; McCann, 2012). Three types of dynamics – all relevant to cross-system cascades – increase coupling strength between P–C, enhance cascading trophic interactions, and thereby destabilize the food chain. Paradoxically, the first of these is donor-controlled resource inputs to consumers at either trophic level. Here, resource fluxes from a donor ecosystem to a consumer in a recipient ecosystem inflate the recipient consumer's equilibrium abundance without feedback on the donor resource population. Such "no cost consumption" increases the coupling strength of the consumer on *in situ* ecosystem resources via consumer biomass leading to cascading dynamics in the recipient ecosystem. The second type of destabilizing dynamic is the movement (dispersal or migration) of predators across ecosystem boundaries. If the magnitude of movement is high, P abundance in the recipient ecosystem is inflated above limits set by resource supply, leading to stronger coupling strength between the cosmopolitan predator and recipient resource. Finally, apparent competition can be destabilizing in many circumstances. Apparent competition is defined as mutual indirect negative effects of competitors (i.e., resources in this context) mediated through a shared consumer; this interaction can be asymmetrical (indirect amensalism). Apparent competition is ubiquitous and if the lack of diet specialization by P is at no cost (i.e., switching is easy), apparent competition is moderately destabilizing. These destabilizing effects can be muted if resources also compete directly.

In this chapter, we contend that all three of these destabilizing effects are prominent features of food webs that cross the land–water interface. Passive, donor-controlled fluxes of organic matter (detritus) are commonplace, especially from land to freshwater; and these fluxes have been shown by extensive empirical work to inflate consumer abundance and biomass (e.g., Polis et al., 1997; Wallace et al., 1997; 1999). Similarly, some predators (e.g., adult dragonflies, frogs and some salamanders, snakes, fishing birds, and otters) move with ease between aquatic and terrestrial habitats and feed on unique resources in each ecosystem. Finally, apparent competition is a common species interaction in cross-system studies of food webs. Invertebrates in one ecosystem enhance

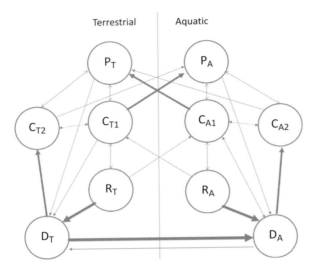

Figure 6.2 A dynamic system of trophic interactions between predators (P), herbivores (C with 1 subscript), detritivores (C with 2 subscript), living plants (R), and detritus (D) in aquatic (A subscript) and terrestrial (T subscript) ecosystems. Arrows point to effects (i.e., a node with changed abundance or biomass) and arrow weight indicates effect magnitude.

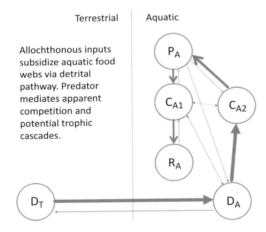

Allochthonous inputs subsidize aquatic food webs via detrital pathway. Predator mediates apparent competition and potential trophic cascades.

Figure 6.3 Allochthonous inputs motif (Motif 1) from dynamic system in Fig. 6.2 highlighting the effects of detrital inputs from terrestrial ecosystems to the base of freshwater food webs and subsequent altered trophic dynamics in the aquatic food web. Symbols as in Fig. 6.2.

predator (e.g., fish, lizard, and spider) abundances in the recipient system at "no cost," with the consequence of increased coupling strength between recipient predators and prey. Moreover, the potentially dampening effects of direct competition between alternate prey (i.e., aquatic and terrestrial invertebrates in the example above) do not occur because of spatial segregation of population dynamics of these two resources.

Below, we review trophic dynamics across the river–land and lake–land interface and classify empirical findings in terms of a generalized watershed food web (Fig. 6.2) and three food web motifs that involve subsets of the five general actors included in the complete web (Figs. 6.3 to 6.5): generalist consumers feeding on both herbivores in the green channel and detritivores in the detrital channel,

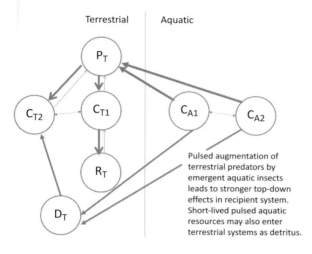

Terrestrial | Aquatic

Pulsed augmentation of terrestrial predators by emergent aquatic insects leads to stronger top-down effects in recipient system. Short-lived pulsed aquatic resources may also enter terrestrial systems as detritus.

Figure 6.4 Resource exchange motif (Motif 2) from dynamic system in Fig. 6.2 highlighting the effects of fluxes of aquatic resource species (herbivores and detritivores) to terrestrial consumers and subsequent altered trophic dynamics in the terrestrial food web. Note that this motif can be inverted such that terrestrial resources drive aquatic trophic dynamics. Symbols as in Fig. 6.2.

herbivores, detritivores, and detritus. Three general motifs appear common in both lake and river settings:

- Motif 1: **Detrital inputs** (Fig. 6.3): Donor-controlled inputs of terrestrial organic matter (i.e., detritus) to freshwater food webs, bottom-up detrital chain dynamics, no cost inflation of aquatic predator abundance, and destabilization of aquatic green channel (trophic cascade)
- Motif 2: **Resource exchange** (Fig. 6.4): Donor-controlled flows of the resource population (i.e., herbivores in Eqns 6.1a, 6.2a–b) across land–water boundaries to generalist top predators leading to increased coupling strength between these predators and herbivores and plants in the recipient ecosystem
- Motif 3: **Amphibious predation** (Fig. 6.5): Movement of mobile top predators across the land–water interface – local resources inflate consumer density causing immigration and increasing coupling strength between the consumer and resources in neighboring systems

Empirical observations of watershed food web motifs in lakes and rivers

We used five recent reviews (Baxter et al., 2005; Marczak et al., 2007; Bartels et al., 2011a; Sabo and Hagen, 2012; Muehlbauer et al., 2013) and a snowball sampling methodology based on cited literature and citations of these papers to identify empirical case studies for our review of river case studies. In choosing papers, we selected only those *field* studies that both measured fluxes of OM or organisms across the land–water boundary *and* quantified the impact of these fluxes on food web dynamics in the donor or recipient system. We found only nine articles that fit these criteria. This subset of papers may not be exhaustive, but we think it is representative.

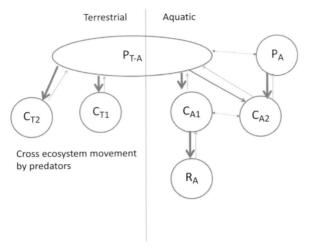

Figure 6.5 Amphibious predation motif (Motif 3) from dynamic system in Fig. 6.2 highlighting the effects of a mobile terrestrial predator that exerts control of resource populations and alters trophic dynamics in the aquatic food web. Note that this predator could also be aquatic, with similar effects in a terrestrial ecosystem. Symbols as in Fig. 6.2.

In general, our case studies can be classified according to our three DS motifs (Table 6.1). For each motif, there are two additional complexities to consider. First, a cross-system flux of species or energy may alter trophic dynamics in the *recipient* system, or a trophic cascade in the *donor* system may modify the magnitude of a flux – we call these recipient- and donor-based trophic cascades, respectively. Second, subsidized trophic dynamics can span three or more trophic levels (a "complete" trophic cascade module). Alternatively, subsidized dynamics that include only two trophic levels can be conceptualized as apparent competition modules. We treat these apparent competition modules as "incomplete cascades" (i.e., dynamics don't extend to plants), as these sorts of case studies were much more commonly observed than complete cascades. A rich set of examples of food webs linked across the land–river boundary exist, including three elegant examples of complete trophic cascades.

Empirical observation of detrital inputs (Motif 1)

Allochthonous inputs have a long tradition in stream – but not lake – ecology (Fisher and Likens, 1973). Not surprisingly, there are a number of reports of large-scale manipulations of terrestrial detritus (wood and leaves), most notably from the Coweeta study site in the southeastern US. These studies unanimously support the notion that terrestrial detritus fuels production of aquatic detritivores, and that this energy fuels production and abundance of aquatic predators (Wallace et al., 1997; 1999; Johnson and Wallace, 2005). Unfortunately, the effects of inflated predator densities on the green channel (i.e., living plants) were not evaluated in these studies; in theory, predators supported at no cost by terrestrial detritus could exert runaway consumption in this green channel (i.e., on aquatic herbivores).

Table 6.1 *Summary of empirical work reviewed in this paper in the context of cross-boundary dynamic systems*

Description	Donor system	Recipient system	Citation	Effect	Effect type
Motif 1. Allochthonous inputs					
Terrestrial leaf litter inputs to stream food webs	Land	River	Wallace et al., 1997; 1999; Johnson and Wallace, 2005	Large-scale removal of leaf litter led to reductions in abundance and biomass of all trophic levels in stream food webs (detritivores and predators); bottom-up effects demonstrated forcefully, but impact on herbivores was not examined manipulatively	Bottom-up energy flow through whole recipient food web module
Experimental N and P stream enrichment	River	Land	Davis et al., 2011	Experimental N and P addition increased productivity of stream food web and increased biomass export to terrestrial system. No response of spider biomass, perhaps due to shifts in body-size distribution of emergent insects	Bottom-up energy flow
Deposition of salmon carcasses in riparian areas	River/ocean	Land	Bartz and Naiman, 2005; Helfield and Naiman, 2001	Salmon-borne nutrients enrich soil and alter riparian vegetation	Bottom-up energy flow
Experimental DIC addition to lake food web	Land	Lake	Pace et al., 2004	Addition of inorganic dissolved 13C increases productivity of pelagic food web and demonstrates terrestrial subsidy to lake food web	Bottom-up energy flow

(cont.)

Table 6.1 (cont.)

Description	Donor system	Recipient system	Citation	Effect	Effect type
Experimental POC (as opposed to DOC) addition to benthic lake food web	Land	Lake	Bartels et al., 2012	Experimental POC (C4 corn starch with distinct isotopic signature) addition resulted in an isotopic shift of benthic herbivorous and predatory invertebrates, as well as zooplankton in the pelagic food web. Demonstrates importance of benthic pathway for incorporating allochthonous inputs into lake food webs	Bottom-up energy flow through whole recipient food web module and spatially adjacent (pelagic) module
Motif 2. Resource exchange					
Terrestrial insect fluxes to fish	Land	River	Nakano et al., 1999	Experimentally reduced terrestrial consumer inputs to aquatic top predator cause increases in consumption and decreases in abundance of herbivorous invertebrates in aquatic system and increases in algal biomass	Trophic cascade
Invasion of a predator modifies flux	Land	River	Baxter et al., 2004	Invasion of non-native aquatic predator usurps terrestrial consumer fluxes and causes native predator to have inflated effects on herbivores and plants. Reduced *in situ* prey also reduces back flux of consumers from aquatic systems to spiders, and reduction of spider densities	Trophic cascade

Aquatic insect fluxes to lizards	River	Land	Sabo and Power, 2002a; 2002b	Experimentally reduced aquatic–terrestrial fluxes of aquatic insects to lizards which elicited an increase in abundance of lizards and decrease in abundance of some groups of terrestrial insects (prey subsidies stabilize the terrestrial food chain module)	Apparent mutualism
Aquatic insect fluxes to birds	River	Land	Murakami and Nakano, 2002	Experimental manipulation of aquatic insect subsidies to neighboring forests reveal that birds recruit to areas with high aquatic insect abundance and have increased effects on terrestrial herbivores (prey subsidies destabilize the terrestrial food chain module)	Apparent competition
Experimental flux of aquatic insect carcasses to heathland food webs	Lake	Land	Hoekman et al., 2011; 2012	When midge carcasses are deposited on experimental plots, plants respond positively to the nutrient addition, arthropod densities increase, and isotopic evidence demonstrates that midge C enters all levels of the recipient food web	Bottom-up energy flow to terrestrial consumers

(cont.)

Table 6.1 (cont.)

Description	Donor system	Recipient system	Citation	Effect	Effect type
Aquatic insect fluxes to terrestrial arthropod food webs	Lake	Land	Dreyer et al., 2012; Gratton et al., 2008	Mensurative study at lakes in northeast Iceland: in areas of high aquatic insect (midge) density (closer to shore vs. farther and high- vs. low-midge lakes), virtually all taxonomic groups of terrestrial arthropods were more abundant, including herbivores, detritivores, and predators. Isotopic evidence in Gratton et al. (2008) confirms the consumption of midges in terrestrial food webs. These studies highlight the multiple trophic pathways by which midges are incorporated into terrestrial food webs	Bottom-up energy flow to terrestrial consumers
Aquatic insect fluxes to spiders	Lake	Land	Jonsson and Wardle, 2009	Observational study of terrestrial invertebrates on islands in a boreal lake. Freshwater subsidy declined with increasing distance from shore and was most pronounced on smaller islands. Web-building spiders responded positively to subsidies (aquatic insect abundance)	Bottom-up energy flow to terrestrial consumers

Description	Donor system	Recipient system	Citation	Observation	Type of effect
Invasion of a predator modifies flux	Lake	Land	Epanchin et al., 2010	Introduced aquatic predators (trout in alpine lakes) consume mayflies, cutting off cross-ecosystem subsidy from reaching nesting passerine birds. Fishless lakes contained 98% more mayflies and 5.9 times more rosy-finches	Bottom-up energy flow diverted by introduced predator
Biodiversity loss	River (artificial stream)	Land	Wesner, 2012	Reductions in fish diversity (3 to 1 species) reduce aquatic insect biomass and spider abundance	Apparent competition
Motif 3. Amphibious predation					
Fishing birds and the dynamics of a strong interaction	Land	River	Power, 1984	Fishing birds cause large herbivorous fish to move to deep water, releasing algae in shallow pool habitats. Fear of predation by terrestrial bird causes two-level cascades in deep water	Trophic cascade
Fish facilitate terrestrial plants by suppressing dragonfly densities	Lake/pond	Land	Knight et al., 2005	Comparing fishless ponds to ponds that contain fish, they found that fish suppress dragonfly densities, resulting in fewer adult dragonflies around fish-containing ponds. Because adult dragonflies consume pollinators and change their behavior, fewer dragonflies resulted in plants receiving more pollinator visits and being less pollen-limited	Trophic cascade

The table is organized around the motif studied (see text for description). For each example, the donor system (i.e., which system supplies resources or consumers), the recipient system, where effects are typically observed), the citation, a brief description of the observations (effect), and the type of effect observed (i.e., bottom-up versus top-down and apparent competition or mutualism versus full cascade) are listed.

DIC = dissolved inorganic carbon; POC = particulate organic carbon.

The interconnected nature of lakes and their surrounding landscape has been demonstrated by studies focused in both aquatic and terrestrial systems (Vander Zanden and Gratton, 2011). In lakes, experimental addition to simulate terrestrial runoff and deposition has demonstrated the importance of terrestrial subsidies to aquatic systems. Pace et al. (2004) added dissolved inorganic C to two lakes and found that it was readily incorporated into pelagic food webs, including zooplankton and fish. Similarly, Bartels et al. (2012) added particulate organic C to the benthos in experimental enclosures in a lake and tracked its movement through the benthic and pelagic food web compartments. While these and other examples demonstrate a flux of C into lakes, they do not quantify effects on trophic dynamics in recipient systems.

Empirical observation of resource exchange (Motif 2)

Resource exchange is the most commonly documented motif in watershed food webs; there are examples of complete and incomplete cascades either elicited by cross-system exchange (recipient-based cascade) or causing it (donor-based cascade). In Japan, Nakano et al. (1999) executed an impressive field experiment in which they modified terrestrial insect fluxes to stream fish using plastic greenhouse barriers built over the stream. Reductions in the donor-controlled flux of terrestrial invertebrates caused stream fish to increase their consumption of stream herbivores. Increased coupling strength between P and C led to lower abundance of stream herbivores and increased plant biomass. This is an example of a recipient-based aquatic cascade driven by inputs (or experimental removal of inputs) of terrestrial herbivorous and carnivorous invertebrates. Note that this interaction is actually apparent *mutualism* because the presence of the flux reduces runaway consumption in the aquatic system.

Similarly, Baxter et al. (2004) illustrate an elegant web of unexpected indirect interactions in a Japanese river. Specifically, non-native trout outcompete native char for terrestrial invertebrates (donor-controlled inputs). Due to reduced (terrestrial) prey for native char, these native predators consume more aquatic herbivores, thereby releasing algae from herbivory. Reductions in aquatic herbivores lead to a concomitant reduction in emergent aquatic insects and lower densities of spiders. Finally, a cascade through herbivores and plants in the aquatic system results in lower fluxes of aquatic herbivores and lower densities of spiders on land, completing the aquatic–terrestrial loop. This set of dynamics illustrates both a donor-based cascade (increased coupling strength of char and aquatic herbivores mediated by non-native trout) and a recipient-based cascade (char effects on aquatic biomass patterns lowers flux of emergent insects from river back to land).

There is a similar number of examples of incomplete cascades in our case studies, falling under the umbrella of apparent competition/mutualism. Two examples illustrate how the effects of aquatic invertebrates on terrestrial predators can have contrasting effects on the abundance of terrestrial invertebrates. In Japan, aquatic insect fluxes elicit a strong numerical response of birds, which in turn lowers the abundance of terrestrial invertebrates (Murakami and Nakano, 2002). By contrast, Sabo and Power (2002a) showed that aquatic insect fluxes similarly elicited a strong numerical response of a terrestrial consumer (a lizard), but that strong preference for aquatic prey released terrestrial prey from predation pressure. Birds mediate apparent competition and lizards mediate apparent mutualism in these examples, both in the recipient (terrestrial) system.

In lakes, Epanchin et al. (2010) showed that introduced aquatic predators (trout in alpine lakes) consume mayflies, cutting off cross-ecosystem subsidies from reaching nesting passerine birds. Fishless lakes contained 98% more mayflies and 5.9 times more rosy-finches. Unfortunately this study did not quantify the consequences for rosy-finch reproduction or their alternative prey sources. More generally, movement of organisms (and OM from insect fallout after mass emergence) out of lakes has been less studied, but some recent examples have demonstrated a high potential for effects in nearshore terrestrial communities (Gratton and Vander Zanden, 2009; Bartrons et al., 2013). The importance of aquatic resource inputs to terrestrial food webs has been demonstrated in an experimental setting where insect deposition was simulated by manually depositing insect carcasses (Hoekman et al., 2011; 2012). Nutrient addition increased plant productivity when carcasses were deposited on experimental plots, and the abundance of a wide variety of arthropods increased, including herbivores, detritivores, and predators. Stable isotopes revealed that aquatic insect C entered all levels of the recipient terrestrial food web, demonstrating bottom-up energy flow. Mensurative studies have shown that aquatic insect inputs to terrestrial systems can be substantial and decline with distance from shore (Gratton et al., 2008; Jonsson and Wardle, 2009; Dreyer et al., 2012). Jonsson and Wardle (2009) showed that smaller islands have higher aquatic insect densities and also a greater abundance of spiders. Similarly, Gratton et al. (2008) found isotopic evidence that aquatic C was present in terrestrial arthropods and Dreyer et al. (2012) found that densities of arthropods at all trophic levels responded positively to aquatic insect density (inputs). Predator densities may have increased via aggregation or reproduction and their presence can result in apparent competition between aquatic insects and *in situ* herbivores (Dreyer et al., in review).

Empirical observation of amphibious predation (Motif 3)

Aquatic birds, mammals, and reptiles, as well as aquatic organisms with complex life histories like amphibians and odonates, can cross the aquatic–terrestrial boundary, allowing them to capitalize on production in both ecosystems. Fishing birds in Panamanian streams cause large-bodied herbivorous catfish to move to deep water, and algae achieve higher biomass in the shallower portions of these streams (Power, 1984). In this case, mobile birds – which may also eat terrestrial prey like small mammals, snakes, and lizards – cause a recipient-based trophic cascade in rivers. Predators can respond to consumers that cross the aquatic–terrestrial boundary and in other cases, predators themselves cross the lake–land margin. Knight et al. (2005) compared fishless ponds to ponds that contain fish, and found that fish suppress dragonfly densities, resulting in fewer predacious adult dragonflies around fish-containing ponds. Because adult dragonflies consume pollinators and change their behavior, aquatic predators (fish) can initiate a cascade of indirect interactions, thereby influencing the number of pollinator visits and amount of pollen limitation of the plants surrounding ponds (by consuming sub-adult dragonflies). Fishless versus fish-stocked alpine lakes provide another example where cascading effects in the donor system influenced the magnitude of the cross-ecosystem flux.

Conclusion

Most studies of tropic dynamics at the land–freshwater interface have focused on stream–riparian connections, though lakes can also be very important at the landscape level, and future work should include both lentic– and lotic–terrestrial exchanges. Here we have identified three general motifs in which trophic cascades and other dynamics can occur across the land–freshwater boundary (Figs. 6.3 to 6.5). These contexts are motifs that are commonly observed in larger food webs and have well-studied theoretical properties in a dynamic systems framework (McCann, 2012). There are at least three variations in the dynamics of these motifs (Table 6.1). First, trophic dynamics in a donor system can alter the exchange of OM, consumers, and predators across ecosystem boundaries, reducing the flux to recipient systems. By contrast, the movement of OM, consumers, or predators from a donor system can alter the strength of cascades in the recipient system. We call these two dynamics donor- and recipient-based cascades. Second, many, if not the majority of our case studies find strong cross-system dynamics that end at the predator–herbivore (P–C) link. We found only a few examples of full trophic cascades; our review provides a framework for future studies to build on. Finally, experimental manipulation of fluxes of consumers at intermediate trophic levels between freshwater and land can produce results consistent with either greater or lower consumption of recipient resources by *in situ* predators; these results are called apparent competition or mutualism, respectively. This distinction is likely a consequence of countervailing effects of the numerical and functional responses of recipient predators to donor resources and the relative preference of the predator for donor or recipient prey. Strong apparent competition and potentially cascades should be expected where donor resource fluxes augment predator density (numerical response) to the extent that subsidized densities offset any preference and switching of these predators from recipient to donor resources.

The next step is to explore the ecosystem consequences of some of the trophic effects noted above. We have highlighted the responses of a few simple food web motifs to the movement of detritus, prey, and predators, all with potentially different effects (Marczak et al., 2007). For example, when generalist predators receive resource inputs from adjacent ecosystems, a broad array of trophic effects can result. Generalist predators can mediate apparent competition between cross-ecosystem fluxes and *in situ* herbivores and detritivores. What are the consequences of bolstered predator populations on ecosystem processes like decomposition rates or plant productivity? How might herbivory in the recipient system be affected? In general, future studies should build on the demonstrated strong links between aquatic and terrestrial systems to develop a more thorough understanding of both the trophic and ecosystem consequences of terrestrial–freshwater fluxes in both space and time.

Space: the final frontier

Understanding the complexities of the food web modules described above (Figs. 6.3 to 6.5) and their dynamics takes us one critical step toward creating a more general knowledge of the way food webs function across landscape boundaries. Articulating the spatial relevance (see Chapter 11) and impact of cross-system exchange and trophic dynamics in a landscape setting is the next step. Watersheds provide the appropriate spatial construct for linking the terrestrial realm to both lake and river ecosystems. The central question is: "How extensive are cross-system cascades relative to the spatial scale of the recipient ecosystem?" We propose that future studies not only seek to observe trophic dynamics across that land–water boundary, but also quantify spatial variation in the boundaries of these dynamics deep into the recipient ecosystem. In this way, we can learn whether exchange and the trophic consequences of this exchange manifest across only a small portion or a majority of the recipient ecosystem. Wide reaching impacts of exchange imply more strongly coupled food web dynamics and underscore effective boundaries for ecosystems. Several studies have presented theoretical constructs for this (Gratton and Vander Zanden, 2009; Sabo and Hagen, 2012) and one recent study did explicitly estimate emergent insect fluxes to land and the extent of the landscape influenced by these cross-ecosystem movements at the watershed and regional scales (Bartrons et al., 2013). We propose that the landscape context must be considered in all future studies of cross-system trophic dynamics.

Time: the other final frontier

Resource pulses have a rich set of effects on communities and ecosystems (Ostfeld and Keesing, 2000; McClain et al., 2003). Cross-system exchanges are typically seasonal and, hence, pulsed by nature (Polis et al., 1997). Leaves fall in the fall and insects emerge in the spring. How the amplitude, timing, and frequency of pulsed exchange across the land–water interface affects populations of consumers and dynamics in recipient food webs is largely an unanswered question. One system that lends itself to these questions is the pulsed emergence of insects into nearshore terrestrial systems (Gratton et al., 2008). An experiment that simulated pulsed aquatic inputs by adding midge carcasses has shown strong responses in recipient systems, some immediate (small-bodied detritivores) and others lagged (plants and predators), some persistent (plant litter) and others ephemeral (aquatic C isotopes) (Hoekman et al., 2011; 2012). But long-term measurements of food web responses to resource pulses are few and many questions remain to be answered. Are the effects of strong pulses of donor-controlled resources lagged for some consumers but not for others? Is next year's brood bigger after a bonanza of pulsed resources? How do these lagged effects play out in a food web where the on–off nature of resource pulses is boom or bust and also unpredictable? These are important questions to address as they likely have strong effects on the stability and resilience of food webs coupled across landscape boundaries.

In summary, a simple food web characterizes a watershed ecosystem (Fig. 6.2) and the empirical evidence suggests that three motifs within this food web (Figs. 6.3 to 6.5) are commonly observed. These observations suggest that the boundary – the lake or river edge – has been a boundary of convenience for ecologists. More importantly, these common motifs suggest that the watershed food web can be a useful tool for conservation planners and ecosystem restoration. Biodiversity portfolios must contain meaningful species interactions – not just species – and the watershed food web is a critical framework for defining these important interactions on land and in water.

Acknowledgments

We thank Kevin McCann for encouragement and reading an earlier version of this manuscript. This article is dedicated to Gary Polis, Masahiko Higashi, Takuya Abe, Shigeru Nakano, and Michael Rose, who died in pursuit of knowledge about dynamic ecological systems, and who continue to this day to inspire us to do theoretically motivated empirical ecology.

References

Allesina, S., Alonso, D. and Pascual, M., 2008. A general model for food web structure. *Science*, **320**(5876), 658–661.

Arditi, R. and Saiah, H. (1992). Empirical evidence of the role of heterogeneity in ratio-dependent consumption. *Ecology*, **73**(5), 1544.

Bartels, P., Cucherousset, J., Steger, K., et al. (2011). Reciprocal subsidies between freshwater and terrestrial ecosystems structure consumer resource dynamics. *Ecology*, **93**(5), 1173–1182.

Bartels, P., Cucherousset, J., Gudasz, C., et al. (2012). Terrestrial subsidies to lake food webs: an experimental approach. *Oecologia*, **168**(3), 807–818.

Bartrons, M., Papes, M., Diebel, M. W., Gratton, C. and Vander Zanden, M. J. (2013). Regional-level inputs of emergent aquatic insects from water to land. *Ecosystems*, **16**, 1353–1363.

Bartz, K. and Naiman, R. (2005). Effects of salmon-borne nutrients on riparian soils and vegetation in southwest Alaska. *Ecosystems*, **8**, 529–545.

Bascompte, J. and Melian, C. J. (2005). Simple trophic modules for complex food webs. *Ecology*, **86**(11), 2868–2873.

Bastow, J. L., Sabo, J. L., Finlay, J. C. and Power, M. E. (2002). A basal aquatic-terrestrial trophic link in rivers: algal subsidies via shore-dwelling grasshoppers. *Oecologia*, **131**(2), 261–268.

Baxter, C. V., Fausch, K. D., Murakami, M. and Chapman, P. L. (2004). Fish invasion restructures stream and forest food webs by interrupting reciprocal prey subsidies. *Ecology*, **85**(10), 2656–2663.

Baxter, C. V., Fausch, K. D. and Saunders, W. C. (2005). Tangled webs: reciprocal flows of invertebrate prey link streams and riparian zones. *Freshwater Biology*, **50**(2), 201–220.

Borer, E. T., Seabloom, E. W., Shurin, J. B., et al. (2005). What determines the strength of a trophic cascade? *Ecology*, **86**(2), 528–537.

Brett, M. T., Kainz, M. J., Taipale, S. J. and Seshan, H. (2009). Phytoplankton, not allochthonous carbon, sustains herbivorous zooplankton production. *Proceedings of the National Academy of Sciences of the USA*, **106**(50), 21197–21201.

Brett, M. T., Arhonditsis, G. B., Chandra, S. and Kainz, M. J. (2012). Mass flux calculations show strong allochthonous support of freshwater zooplankton production is unlikely. *PLoS One*, **7**(6), p. e39508.

Carpenter, S. R., Kitchell, J. F. and Hodgson, J. R. (1985). Cascading trophic interactions and lake productivity. *BioScience*, **35**, 634–639.

Carpenter, S. R., Cole, J. J., Hodgson, J. R., et al. (2001). Trophic cascades, nutrients, and lake productivity: whole-lake experiments. *Ecological Monographs*, **71**(2), 163–186.

Carpenter, S. R., Cole, J. J., Pace, M. L., et al. (2005). Ecosystem subsidies: terrestrial support of aquatic food webs from C-13 addition to contrasting lakes. *Ecology*, **86**(10), 2737–2750.

Carpenter, S. R., Brock, W. A., Cole, J. J., Kitchell, J. F. and Pace, M. L. (2008). Leading indicators of trophic cascades. *Ecology Letters*, **11**(2), 128–138.

Cole, J. J., Carpenter, S. R., Pace, M. L., et al. (2006). Differential support of lake food webs by three types of terrestrial organic carbon. *Ecology Letters*, **9**(5), 558–568.

Cole, J. J., Carpenter, S. R., Kitchell, J., et al. (2011). Strong evidence for terrestrial support of zooplankton in small lakes based on stable isotopes of carbon, nitrogen, and hydrogen. *Proceedings of the National Academy of Sciences of the USA*, **108**(5), 1975–1980.

Cross, W. F., Baxter, C. V., Donner, K. C., et al. (2011). Ecosystem ecology meets adaptive management: food web response to a controlled flood on the Colorado River, Glen Canyon. *Ecological Applications*, **21**(6), 2016–2033.

Cross, W. F., Baxter, C. V., Rosi-Marshall, E. J., et al. (2013). Food-web dynamics in a large river discontinuum. *Ecological Monographs*, **83**(3), 311–337.

Davis, J. M., Rosemond, A. D. and Small, G. E. (2011). Increasing donor ecosystem productivity decreases terrestrial consumer reliance on a stream resource subsidy. *Oecologia*, **167**, 821–834.

DeMelo, R., France, R. and McQueen, D. J. (1992). Biomanipulation: hit or myth? *Limnology and Oceanography*, **37**(1), 192–207.

Dreyer, J., Hoekman, D. and Gratton, C. (2012). Lake-derived midges increase abundance of shoreline terrestrial arthropods via multiple trophic pathways. *Oikos*, **121**(2), 252–258.

Epanchin, P., Knapp, R. and Lawler, S. (2010). Nonnative trout impact an alpine-nesting bird by altering aquatic-insect subsidies. *Ecology*, **91**(8), 2406–2415.

Fisher, S. G. and Likens, G. E. (1973). Energy flow in Bear Brook, New Hampshire: an integrative approach to stream ecosystem metabolism. *Ecological Monographs*, **43**(4), 421–439.

Fretwell, S. D. (1977). Regulation of plant communities by food-chains exploiting them. *Perspectives in Biology and Medicine*, **20**(2), 169–185.

Ginzburg, L. R. and Akcakaya, H. R. (1992). Consequences of ratio-dependent predation for steady-state properties of ecosystems. *Ecology*, **73**(5), 1536.

Gratton, C. and Vander Zanden, M. (2009). Flux of aquatic insect productivity to land: comparison of lentic and lotic ecosystems. *Ecology*, **90**(10), 2689–2699.

Gratton, C., Donaldson, J. and Vander Zanden, M. J. (2008). Ecosystem linkages between lakes and the surrounding terrestrial landscape in northeast Iceland. *Ecosystems*, **11**(5), 764–774.

Hairston, N. G. and Hairston, N. G. (1993). Cause-effect relationships in energy-flow, trophic structure, and interspecific interactions. *American Naturalist*, **142**(3), 379–411.

Hairston, N. G. J. and Hairston, N. G. S. (1997). Does food web complexity eliminate trophic-level dynamics? *American Naturalist*, **149**(5), 1001–1007.

Hairston, N. G., Smith, F. E. and Slobodkin, L. B. (1960). Community structure, population control, and competition: Paper 17. *Foundations of Ecology*, **94**(879), 421–425.

Halpern, B. S., Borer, T., Seabloom, E. W. and Shurin, J. B. (2005). Predator effects on herbivore and plant stability. *Ecology Letters*, **8**(2), 189–194.

Helfield, J. M. and Naiman, R. J. (2001). Effects of salmon-derived nitrogen on riparian forest growth and implications for stream productivity. *Ecology*, **82**, 2403–2409.

Hoekman, D., Dreyer, J., Jackson, R., Townsend, P. and Gratton, C. (2011). Lake to land subsidies: experimental addition of aquatic insects increases terrestrial arthropod densities. *Ecology*, **92**(11), 2063–2072.

Hoekman, D., Bartrons, M. and Gratton, C. (2012). Ecosystem linkages revealed by experimental lake-derived isotope signal in heathland food webs. *Oecologia*, 1–9.

Hrbacek, J., Dvoráková, V., Korínek, V., et al. (1961). Demonstration of the effect of the fish stock on the species composition of zooplankton and the intensity of metabolism of the whole plankton association. *Verhandlungen der Internationalen Vereinigung für Theoretische und Angewandte Limnologie*, **14**, 192–195.

Jardine, T. D., Pettit, N. E., Warfe, D. M., et al. (2012a). Consumer–resource coupling in wet–dry tropical rivers. *Journal of Animal Ecology*, **81**(2), 310–322.

Jardine, T. D., Pusey, B. J., Hamilton, S. K., et al. (2012b). Fish mediate high food web connectivity in the lower reaches of a tropical floodplain river. *Oecologia*, **168**(3), 829–838.

Johnson, B. R. and Wallace, J. B. (2005). Bottom-up limitation of a stream salamander in a detritus-based food web. *Canadian Journal of Fisheries and Aquatic Sciences*, **62**(2), 301–311.

Jonsson, M. and Wardle, D. A. (2009). The influence of freshwater-lake subsidies on invertebrates occupying terrestrial vegetation. *Acta Oecologica*, **35**(5), 698–704.

Junk, W. J., Bayley, P. B. and Sparks, R. E. (1989). The flood pulse concept in river-floodplain systems. *Canadian Special Publication of Fisheries and Aquatic Sciences*, **106**(1), 110–127.

Knight, T. M., McCoy, M. W., Chase, J. M., McCoy, K. A. and Holt, R. D. (2005). Trophic cascades across ecosystems. *Nature*, **437**(7060), 880–883.

Lafontaine, N. and McQueen, D. J. (1991). Contrasting trophic level interactions in Lake St. George and Haynes Lake (Ontario, Canada). *Canadian Journal of Fisheries and Aquatic Sciences*, **48**(3), 356–363.

Lindeman, R. L. (1942). The trophic-dynamic aspect of ecology: Paper 7. *Foundations of Ecology*, **23**, 399–418.

Marczak, L. B., Thompson, R. M. and Richardson, J. S. (2007). Meta-analysis: Trophic level, habitat, and productivity shape the food web effects of resource subsidies. *Ecology*, **88**(1), 140–148.

McCann, K. S. (2012). *Food Webs*. Princeton, NJ: Princeton University Press.

McClain, M. E., Boyer, E. W., Dent, C. L., et al. (2003). Biogeochemical hot spots and hot moments at the interface of terrestrial and aquatic. *Ecosystems*, **6**(4), 301–312.

McQueen, D. J., Post, J. R. and Mills, E. L. (1986). Trophic relationships in fresh-water pelagic ecosystems. *Canadian Journal of Fisheries and Aquatic Sciences*, **43**(8), 1571–1581.

Mittelbach, G. G., Osenberg, C. W. and Leibold, M. A. (1988). Trophic relations and ontogenetic niche shifts in aquatic ecosystems. In *Size-Structured Populations*, ed. D. B. Ebenman and D. L. Persson. Berlin, Heidelberg: Springer, pp. 219–235. Available at: http://link.springer.com/chapter/10.1007/978–3–642–74001–5_15 (Accessed December 6, 2013).

Muehlbauer, J. D., Collins, S. F., Doyle, M. W. and Tockner, K. (2013). How wide is a stream? Spatial extent of the potential "stream signature" in terrestrial food webs using meta-analysis. *Ecology*. Available at: http://www.esajournals.org/doi/abs/10.1890/12--1628.1 (Accessed December 6, 2013).

Murakami, M. and Nakano, S. (2002). Indirect effect of aquatic insect emergence on a terrestrial insect population through predation by birds. *Ecology Letters*, **5**(3), 333–337.

Murdoch, W. W. (1966). Community structure, population control, and competition – a Critique. *American Naturalist*, **100**(912), 219–226.

Naiman, R. J., Bilby, R. E., Schindler, D. E. and Helfield, J. M. (2002). Pacific salmon, nutrients, and the dynamics of freshwater and riparian ecosystems. *Ecosystems*, **5**(4), 399–417.

Nakano, S. and Murakami, M. (2001). Reciprocal subsidies: dynamic interdependence between terrestrial and aquatic food webs. *Proceedings of the National Academy of Sciences of the USA*, **98**(1), 166–170.

Nakano, S., Miyasaka, H. and Kuhara, N. (1999). Terrestrial–aquatic linkages: riparian arthropod inputs alter trophic cascades in a stream food web. *Ecology*, **80**(7), 2435–2441.

Neutel, A.-M., Heesterbeek, J. A. P. and de Ruiter, P. C. (2002). Stability in real food webs: weak links in long loops. *Science*, **296**(5570), 1120–1123.

Oaten, A. and Murdoch, W. W. (1975a). Functional response and stability in predator-prey systems. *The American Naturalist*, **109**, 289–298.

Oaten, A. and Murdoch, W. W. (1975b). Switching, functional response, and stability in predator-prey systems. *The American Naturalist*, **109**, 299–318.

Oksanen, L. (1983). Trophic exploitation and arctic phytomass patterns. *The American Naturalist*, **122**, 45–52.

Oksanen, L. and Oksanen, T. (2000). The logic and realism of the hypothesis of exploitation ecosystems. *The American Naturalist*, **155**(6), 703–723.

Oksanen, L., Fretwell, S. D., Arruda, J. and Niemela, P. (1981). Exploitation ecosystems in gradients of primary productivity. *The American Naturalist*, **118**(2), 240–261.

Ostfeld, R. S. and Keesing, F. (2000). Pulsed resources and community dynamics of consumers in terrestrial ecosystems. *Trends in Ecology and Evolution*, **15**(6), 232–237.

Pace, M. L., Cole, J. J., Carpenter, S. R. and Kitchell, J. F. (1999). Trophic cascades revealed in diverse ecosystems. *Trends in Ecology and Evolution*, **14**(12), 483–488.

Pace, M. L., Cole, J. J., Carpenter, S. R., et al. (2004). Whole-lake carbon-13 additions reveal terrestrial support of aquatic food webs. *Nature*, **427**(6971), 240–243.

Polis, G. A. and Strong, D. R. (1996). Food web complexity and community dynamics. *American Naturalist*, **147**(5), 813–846.

Polis, G. A, Hurd, S. D., Jackson, C. T. and Pinero, F. S. (1997a). El Niño effects on the dynamics and control of an island ecosystem in the Gulf of California. *Ecology*, **78**(6), 1884–1897.

Polis, G. A., Anderson, W. B. and Holt, R. D. (1997b). Toward an integration of landscape and food web ecology: the dynamics of spatially subsidized food webs. *Annual Review of Ecology and Systematics*, **28**(1), 289–316.

Power, M. E. (1984). Depth distributions of armored catfish: predator-induced resource avoidance? *Ecology*, **65**(2), 523.

Power, M. E. (1990). Effects of fish in river food webs. *Science*, **250**(4982), 811–814.

Power, M. E., Rainey, W. E., Parker, M. S., et al. (2004). River-to-watershed subsidies in an old-growth conifer forest. In *Food Webs at the Landscape Level*, ed. A. Polis, M. E. Power, and G. R. Huxel. Chicago, IL: University of Chicago Press, pp. 387–409.

Power, M. E., Parker, M. S. and Dietrich, W. E. (2008). Seasonal reassembly of a river food web: floods, droughts, and impacts of fish. *Ecological Monographs*, **78**(2), 263–282.

Sabo, J. L. and Hagen, E. M. (2012). A network theory for resource exchange between rivers and their watersheds. *Water Resources Research*, **48**(4). Available at: http://www.agu.org/journals/wr/wr1204/2011WR010703/ (Accessed December 7, 2013).

Sabo, J. L. and Power, M. E. (2002a). Numerical response of lizards to aquatic insects and short-term consequences for terrestrial prey. *Ecology*, **83**(11), 3023–3036.

Sabo, J. L. and Power, M. E. (2002b). River-watershed exchange: effects of riverine subsidies on riparian lizards and their terrestrial prey. *Ecology*, **83**(7), 1860–1869.

Sanzone, D. M., Meyer, J. L., Marti, E., et al. (2003). Carbon and nitrogen transfer from a desert stream to riparian predators. *Oecologia*, **134**(2), 238–250.

Shurin, J. B., Borer, T., Seabloom, E. W., et al. (2002). A cross-ecosystem comparison of the strength of trophic cascades. *Ecology Letters*, **5**(6), 785–791.

Shurin, J. B., Gruner, D. S. and Hillebrand, H. (2006). All wet or dried up? Real differences between aquatic and terrestrial food webs. *Proceedings of the Royal Society B: Biological Sciences*, **273**(1582), 1–9.

Stachowicz, J. J. and Hay, M. E. (1999). Reducing predation through chemically mediated camouflage: indirect effects of plant defenses on herbivores. *Ecology*, **80**(2), 495–509.

Strong, D. R. (1992). Are trophic cascades all wet? Differentiation and donor-control in speciose ecosystems. *Ecology*, **73**(3), 747–754.

Vander Zanden, M. J. and Gratton, C. (2011). Blowin' in the wind: Reciprocal airborne carbon fluxes between lakes and land. *Canadian Journal of Fisheries and Aquatic Sciences*, **68**(1), 170–182.

Vannote, R. L., Minshall, G. W., Cummins, K. W., Sedell, J. R. and Cushing, C. E. (1980). River Continuum Concept. *Canadian Journal of Fisheries and Aquatic Sciences*, **37**(1), 130–137.

Wallace, J., Eggert, S., Meyer, J. and Webster, J. (1997). Multiple trophic levels of a forest stream linked to terrestrial litter inputs. *Science*, **277**(5322), 102–104.

Wallace, J. B., Eggert, S., Meyer, J. and Webster, J. (1999). Effects of resource limitation on a detrital-based ecosystem. *Ecological Monographs*, **69**(4), 409–442.

Wesner, J. S. (2012). Predator diversity effects cascade across an ecosystem boundary. *Oikos*, **121**, 53–60.

Wootton, J. T., Parker, M. S. and Power, M. E. (1996). Effects of disturbance on river food webs. *Science*, **273**(5281), 1558–1560.

CHAPTER SEVEN

Bottom-up and top-down interactions in coastal interface systems

JAN P. BAKKER*
University of Groningen, The Netherlands

KARINA J. NIELSEN*
Sonoma State University, Rohnert Park, USA

JUAN ALBERTI
Universidad Nacional de Mar del Plata (UNMDP) – Consejo Nacional
de Investigaciones Científicas y Técnicas (CONICET), Argentina

FRANCIS CHAN
Oregon State University, Corvallis, USA

SALLY D. HACKER
Oregon State University, Corvallis, USA

OSCAR O. IRIBARNE
Universidad Nacional de Mar del Plata (UNMDP) – Consejo Nacional
de Investigaciones Científicas y Técnicas (CONICET), Argentina

DRIES P. J. KUIJPER
Polish Academy of Sciences, Białowieża, Poland

BRUCE A. MENGE
Oregon State University, Corvallis, USA

MAARTEN SCHRAMA
University of Manchester, UK

and

BRIAN R. SILLIMAN
Duke University, Beaufort, North Carolina, USA

General introduction of rocky intertidal and salt marsh systems

The land–sea margin encompasses a variety of hard and soft-bottom habitats where organisms are exposed to a dynamic range of aquatic and atmospheric

* Both authors contributed equally.

Trophic Ecology: Bottom-Up and Top-Down Interactions across Aquatic and Terrestrial Systems, eds T. C. Hanley and K. J. La Pierre. Published by Cambridge University Press. © Cambridge University Press 2015.

conditions dependent on a rhythm set by the tides. In this chapter, we focus on rocky intertidal and salt marsh ecosystems, which have been extensively studied on many continents. Both rocky shore and salt marsh communities exhibit strong and consistent patterns of intertidal zonation over relatively compressed spatial scales, making them excellent systems for understanding the context-dependency of species interactions. Hard-bottomed rocky intertidal communities are dominated by marine macroalgae and sessile marine invertebrates extending their reach to the furthest edge of the influence of sea spray, while soft-bottomed salt marsh communities are anchored by terrestrial plants with adaptations or tolerance to inundation by salty and brackish waters. Rocky shore communities may be battered by the full force of large ocean waves or gently lapped with seawater on more protected shorelines. In contrast, salt marshes are restricted to quiet waters where sediment accretion by plants is the main mechanism for habitat creation. Both communities may experience very large tidal excursions or only minimal ones, depending on the local dynamics of the tides, with corresponding consequences for the spatial extent of these communities across the shoreline. The steep environmental gradients and distinctive biological zonation patterns that characterize both rocky shore and salt marsh ecosystems (Fig. 7.1) have provided ecologists with accessible and highly tractable ecosystems for investigating the role of bottom-up and top-down factors along environmental gradients.

Bottom-up and top-down interactions in rocky intertidal systems

Introduction to rocky intertidal systems

Rocky intertidal communities have been the subject of intensive study worldwide, especially at temperate latitudes. The typically broad tidal range and relatively moderate atmospheric conditions create a wide zone of intertidal habitat that is generally hospitable to rocky intertidal species, while also readily accessible to investigators for hours at a time during periods of low tide and calm sea state. The effects of consumers, or top-down factors, have been documented over decades (Robles and Desharnais, 2002), while the role of nutrients and other influences on the base of the food web have only more recently been a focus of exploration (Menge, 2000). The marine invertebrates and macroalgae that dominate this ecosystem have adapted to living on hard substrates that are periodically exposed to air during low tides, while surfgrasses (*Phyllospadix* spp.), the only true plants found in the rocky intertidal, have adaptations that allow them to thrive while submerged in saltwater. Rocky intertidal communities display distinct patterns of zonation that are associated with steep environmental gradients, typically over scales of a few centimeters to several meters (Fig. 7.1). Tidal range varies dramatically in different regions of the world, which is a primary constraint on the extent of the intertidal zone (together with the steepness

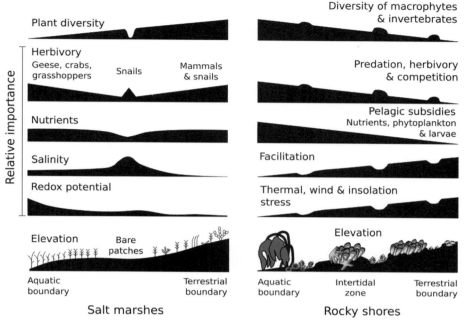

Figure 7.1 Schematic representation of the effects of top-down (consumers) and bottom-up (nutrients) and other ecologically important factors on rocky intertidal systems and salt marshes. Typical zonation of salt marsh and rocky intertidal organisms along an elevational tidal gradient is shown at the bottom, the relative importance of abiotic and biotic factors that influence community structure in both systems is represented above it, and the relative diversity of plants (salt marshes), and macrophytes and sessile invertebrates (rocky intertidal) is depicted at the top. Depressions along the elevation gradient in the rocky intertidal represent tidepools.

of the shoreline). The character (i.e., diurnal, semidiurnal, mixed) and timing of tidal cycles with respect to daily, seasonal, and even celestial cycles are also important to local species distributions (Denny and Paine, 1998). These tidal rhythms set the stage for the degree of exposure to potentially stressful atmospheric conditions (e.g., intense light, high or low temperatures, wind, rain, ice), especially for those organisms living higher on the shore. Even organisms living in tidal pools experience a wider range of physical conditions (e.g., temperature, pH, salinity) than is typical for marine organisms, as they can be isolated from the ocean for periods ranging from hours to days, depending on the tidal height of the pool.

Most rocky intertidal species have complex life histories that include propagules (larvae or spores) that are dispersed by ocean currents. Dispersal distances may be very short (1–10 cm) or long (100s of km), depending on pelagic larval duration (ranging from minutes to months), their behavior, and prevailing

ocean currents (Kinlan and Gaines, 2003). As a result, the scale at which intertidal communities are considered determines whether interacting species can be
thought of as sharing open or closed population dynamics (Kinlan and Gaines,
2003). For example, over ≤ 1 km scales, individual subpopulations (or "local"
populations) of barnacles on a rocky shoreline are open (i.e., offspring may not
return to their population of origin; instead, they disperse and populate a different population). However, over scales of 100s of km, they can be viewed as
part of a larger (closed) metapopulation (Hanski and Gilpin, 1991). These char-
acteristically diverse life histories and scaling of population dynamics distinguish marine communities from terrestrial communities and have important,
and sometimes surprising, consequences for conceptualizing and modeling the
influence of top-down and bottom-up processes on rocky intertidal populations,
communities, and ecosystems.

Rocky intertidal communities occur on the most wave-exposed headlands of
the open coast, where the intertidal zone is broadened by the reach of wave run-
up and splash, and on the low energy shores found in wave-protected coves of
the open coast or straits and bays, where the intertidal zone is more compressed
in space (for shores of similar slope). Disturbances by waves that free up limited
space on open-coast rocky intertidal zones have a predictable, seasonal rhythm.
These disturbances enhance diversity by creating a mosaic of patches in different
stages of succession (Paine and Levin, 1981). The life histories of some species are
even tuned to take advantage of the seasonal disturbance pattern (e.g., Paine,
1979). Organisms attached to sedimentary rocks are more easily dislodged by
waves than those attached to igneous rocks, thus the type of rock itself can have a
strong influence on community susceptibility to disturbance. Episodic scouring
or burial by sand or ice is another common disturbance in this ecosystem.
Seasonal ice scour is a common disturbance at higher latitudes (e.g., Wethey,
1985; Scrosati and Heaven, 2007). Large rocky boulders and smaller benches (flat,
narrow, wave-cut areas often at the base of a seaside cliff) are often interspersed
in a mostly sandy shoreline, resulting in so-called psammophilic (sand associated
or tolerant) species assemblages (e.g., Díaz-Tapia et al., 2013).

Historical development of top-down and bottom-up perspectives

Early and influential field experiments from rocky intertidal ecosystems
(Connell, 1961; Paine, 1966; Dayton, 1975; Menge, 1976; Lubchenco, 1978)
closely paralleled, and contributed to, emerging ecological theories on the
role(s) of energy flow and species interactions in determining community structure (Bertness et al., 2014). Initially, zonation of rocky intertidal communities
was thought to be determined by the limitations of organismal physiology and
biomechanics to atmospheric conditions and wave forces, respectively (Colman,

1933; Stephenson and Stephenson, 1949; Lewis, 1964). However, field experiments revealed that competition and consumption often determined the lower limits of species distributions, while physical factors prevailed in controlling their upper distributional limits (Connell, 1961; Paine, 1966).

Paralleling these mechanistic insights, early trophic and energetic theories of ecology (Lotka, 1925; Volterra, 1926; Elton, 1927; Lindeman, 1942) were being integrated into models of community structure and dynamics (Hairston et al., 1960; Oksanen et al., 1981; Fretwell, 1987; also, see Chapter 1). Importantly, these models also provided insights into where competition or trophic interactions should be expected to regulate species abundances. Menge and Sutherland (1976), using evidence from rocky shores, expanded on these ideas to explain patterns of species diversity. Niche theory, previously thought to be the dominant process defining the structure of communities (reviewed by Vandermeer, 1972), was largely abandoned by rocky intertidal ecologists as more compelling evidence emerged from field experiments clearly demonstrating the importance of predation and disturbance (Robles and Desharnais, 2002).

Focused research on the influence of consumers further revealed the importance of indirect effects (e.g., Wootton, 1993; Menge, 1995) and the context-dependency of species interactions (e.g., Menge and Sutherland, 1987; Menge and Olson, 1990). Variation in the ability of rocky intertidal consumers to influence the abundance of their prey and community structure began to be seen in the context of open rather than closed population models (e.g., Gaines and Roughgarden, 1985; Menge, 1991). Most rocky intertidal predators have open (local) populations with complex life histories and dispersing larvae, but some are direct developers (e.g., the whelks *Nucella* spp. and *Acanthina* spp.) or brooders with closed populations (like the sea stars *Leptasterias* spp.). The omnipresent mussels and barnacles also have dispersing larvae. Incorporating variation in the scaling of demographic processes is an important consideration for modeling marine community dynamics in general, and is explicitly considered in marine conservation models (i.e., networks of marine protected areas; e.g., Gaines et al., 2010).

Models incorporating open population dynamics have focused on space-occupying invertebrates and their predators (e.g., Wieters et al., 2008), sometimes treating sessile heterotrophs as the base of the food web and disregarding phytoplankton. In contrast, investigations of benthic autotrophs and consumers continued to emphasize the relative rates of algal growth and consumption by invertebrate herbivores (e.g., Cubit, 1984). As argued by Menge (1992), the potential role of nutrient supply, or bottom-up factors, had been largely overlooked by benthic marine ecologists, in striking contrast to the perspective held by pelagic marine ecologists (i.e., biological oceanographers) that nutrient dynamics structured pelagic ecosystems.

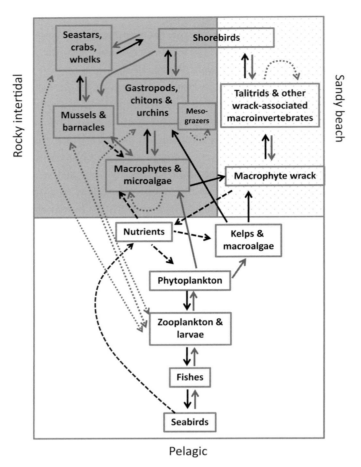

Rocky intertidal

Sandy beach

Pelagic

Figure 7.2 Connectivity and subsidization for rocky intertidal, pelagic, and sandy beach ecosystems. Black arrows indicate energy flow (organic matter), gray arrows indicate species interactions, dotted arrows indicate life history cycles, and dashed arrows indicate nutrient flows.

Since the early 1990s, a more balanced appreciation of the interplay between bottom-up and top-down processes in rocky intertidal communities has emerged, but studies typically focus on one or the other of the two major food webs: (1) phytoplankton, sessile, suspension-feeding herbivores, and mobile predators, or (2) benthic algae, mobile grazers, and mobile predators (Fig. 7.2). Below we discuss each of these "sub-webs" along with more recent work that integrates the roles of nutrients (bottom-up factors), consumers (top-down factors), and scales of connectivity among ecosystems (i.e., meta-ecosystem dynamics, *sensu* Loreau et al., 2003) across both to shape the structure and dynamics of rocky intertidal ecosystems. But first, we consider the roles of consumers and nutrients generally.

Consumers

Top-down effects of consumers are commonly observed in rocky intertidal ecosystems (Menge, 2000). These may result from consumption, consistent with the mathematical foundation of classic food chain models, or through non-consumptive effects whereby prey behavior or traits are altered by limiting foraging excursions or inducing the production of defensive structures or chemicals (e.g., Raimondi et al., 2000). If these modified behaviors or traits influence additional species, they are referred to as trait-mediated indirect effects (of consumers). For example, waterborne predator cues may inhibit the activity of lower level consumers, reducing their consumption of prey (e.g., Raimondi et al., 2000; Trussell et al., 2002). Omnivory is not unusual among rocky intertidal consumers. The adults of benthic predators and even so-called herbivores often feed on more than one trophic level, while larval stages range from planktotrophic to non-feeding. Ontogenetic and adult omnivory may also augment the influence and effectiveness of invertebrate consumers in structuring rocky intertidal communities (Menge and Sutherland, 1987). Furthermore, highly mobile, vertebrate consumers from adjacent ecosystems such as surf zone fishes (e.g., Paine and Palmer, 1978; Menge and Lubchenco, 1981; Ojeda and Muñoz, 1999; Taylor and Schiel, 2010; Vinueza et al., 2014) and shorebirds (e.g., Marsh, 1986; Wootton, 1997; Ellis et al., 2007) forage during high and low tides, respectively. Conversely, the effects of consumers on the abundance of lower trophic levels may be minimized when prey recruitment rates are high, or when their activities are curtailed by environmental stress (Menge and Sutherland, 1987).

Direct exploitation of intertidal animals and plants by humans is ubiquitous, but the intensity of top-down effects varies geographically and historically with cultural and economic context. Prehistoric middens filled with the hard remains of intertidal organisms are common worldwide (e.g., Hockey, 1994; Moreno, 2001; Fa, 2008). Mussels, abalone, and other gastropods, chitons, barnacles, crabs, urchins, and tunicates are all gathered and eaten. Macroalgae, including the rhodophytes *Porphyra* spp. (e.g., nori in Japan, laver in Wales, luche in Chile) and *Chondrus chrispus* (Irish moss), many kelps (e.g., *Laminaria, Undaria, Alaria, Durvillaea,* and *Postelsia*), and sea lettuce (a chlorophyte) are all eaten, and some intertidal algae (including *Fucus, Ascophyllum, Lessonia,* and *Chondrus*) are used industrially or agriculturally (Thompson et al., 2010). Human impacts can be substantial (e.g., Castilla and Duran, 1985; Roy et al., 2003; Salomon et al., 2007), but have been treated within the more applied contexts of ecosystem-based management and marine conservation biology, rather than as consumers in natural ecosystems (McLeod and Leslie, 2009; also, see Chapter 14).

Nutrients

Bottom-up inputs of inorganic nutrients that fuel the growth of benthic macrophytes (macroalgae and surfgrasses) and nearshore phytoplankton production provide a direct connection between intertidal communities and adjacent coastal oceans. These inputs generally originate from physical oceanographic processes such as coastal upwelling and tidal mixing. For some systems, anthropogenic nutrient inputs via river runoff, groundwater discharge, and atmospheric deposition can further represent important contributions. Macroalgal uptake rates can be very high, and this is accentuated high on the shore where nutrient uptake is limited to periods of high tide inundation (Bracken et al., 2011). Nutrient uptake occurs over the entire macroalgal thallus and primarily via leaves in surfgrasses (Terrados and Williams, 1997), instead of, as in terrestrial systems, being taken up by roots embedded in a matrix of soil where microbial processes and physical conditions affect rates of nutrient cycling. In rocky intertidal systems, recycling of nutrients is very limited relative to physically mediated inputs, but ammonium excreted by mussels can be important in isolated tidepools and for macrophytes in close association with mussel beds (Bracken and Nielsen, 2004; Aquilino et al., 2009; Pather et al., 2014). Excretion by small macroinvertebrates may also affect surfgrass growth (Moulton and Hacker, 2011). Interestingly, despite the very high nutrient uptake capacity of macrophytes, local nutrient depletion is rare (in contrast to soil nutrient pools in terrestrial ecosystems), with the exception of isolated tidepools or environments with low water flow (Hurd, 2000; Nielsen, 2003). Due to the typically high seawater flow rates in this ecosystem, nutrient availability is largely independent of the uptake capacity of macrophytes and instead reflects the dynamics of the pelagic subsidy. This stands in sharp juxtaposition to the well-known phenomenon of nutrient draw-down by pelagic phytoplankton, the other major source of primary production for intertidal ecosystems. Phytoplankton may thus reduce (or pre-empt) pelagic nutrients before they reach intertidal macrophytes, and be nutrient-limited themselves. In summary, rocky intertidal ecosystems rely almost exclusively on allochthonously sourced nutrients (from adjacent ocean and terrestrial ecosystems) with almost no internal (re)cycling or local nutrient pool depletion. These are not standard assumptions of classic food chain models (Oksanen et al., 1981; Oksanen and Oksanen, 2000), and this has consequences for the influence of bottom-up processes on higher trophic levels, as illustrated below (also, see Chapter 1).

Effects of consumers and subsidies on food webs

Sub-web #1: effects of consumers and pelagic subsidies on sessile invertebrates
The effects of predators have been intensively studied within the most experimentally tractable sub-web of the community: mussels and barnacles, and the sea stars and carnivorous snails that prey on them. In this sub-web, the effects of predation on community structure are consistently evident (Navarrete and Castilla, 2003; Menge and Menge, 2013), except where recruitment rates of prey

are very high or predation rates are very low (e.g., Menge, 1976; Menge and Menge, 2013), or where environmental stress, such as desiccation from winds, is extreme (Bertness et al., 2006). The effects of individual predators (on a per capita or per population basis) can vary substantially both within and among sites, and among species (Navarrete and Menge, 1996; Navarrete and Castilla, 2003). Variation in pelagic subsidies, in the form of juvenile mussels and barnacles and their primary food source, phytoplankton, can strongly influence rates of predation (i.e., individuals eaten per day), the pace of succession, and ultimately community structure (Fig. 7.3; Menge et al., 2003; Menge and Menge, 2013). For example, in a synthesis of many studies from the west coast of North America and New Zealand's South Island, Menge and Menge (2013) found that increases in phytoplankton availability were strongly associated with increases in rates of predation (primarily by sea stars) on, competition among, and abundances of sessile invertebrate herbivore populations (mussels and barnacles). One interpretation is that this represents an interaction between bottom-up and top-down forces (Menge et al., 1997). An alternate interpretation is that physical oceanographic processes that mediate the supply of ecological subsidies (i.e., propagules, nutrients, and organic matter) over larger spatial scales play the major role in structuring this ecosystem, rather than local-scale species interactions (Menge et al., in press).

Nonetheless, even in the face of substantial variation in mussel recruitment among sites, the sea star *Pisaster ochraceus* is consistently able to exclude mussels from the low zone (Paine, 1974; Menge et al., 1994; Robles et al., 1995). Predation can also dampen the effects of variable prey recruitment on community succession and final prey abundance, as long as prey recruitment rates do not swamp consumption rates. This variable dampening effect was clearly illustrated for subordinate predatory whelks on both barnacles and mussels during community succession in mussel bed patches originating in different years (Berlow, 1997). In a larger analysis of the effects of total predation among many sites, Menge and Menge (2013) likewise found that despite high variation in predation rates among sites, the late-successional (final) effects of predators on prey abundance were fairly consistent, except where predation rates were very low, perhaps reflecting low predator recruitment rates. The rate of predation (by sea stars and whelks) and its total effect across different stages of succession can increase as phytoplankton abundance and prey recruitment rates increase (Menge and Menge, 2013). These differences in subsidies, and the associated variation in community dynamics, generate substantial differences in community structure among sites (Fig. 7.3; Menge et al., 1997; Connolly and Roughgarden, 1998).

Importantly though, high inputs at the base of the food web, although correlated with the abundance of predators (Figs. 7.3, 7.4), do not necessarily translate into increased predator populations through food chain mechanisms in the

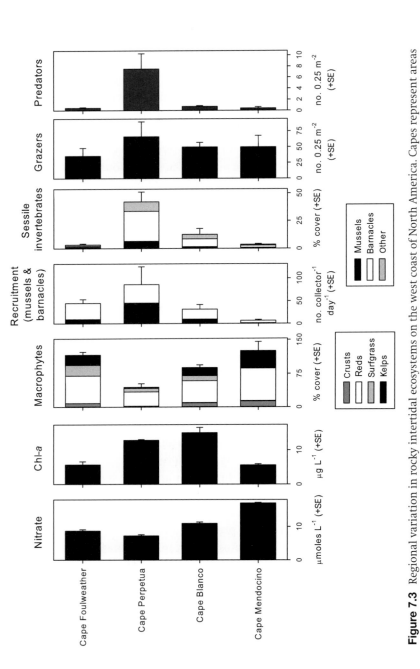

Figure 7.3 Regional variation in rocky intertidal ecosystems on the west coast of North America. Capes represent areas under the influence of a major oceanographic cell, typically a region that has characteristic nearshore circulation features that influence nutrient availability and transport of propagules (error bars = 1 SE of site level averages). Each cape includes data from three to four sites surveyed over 3 to 4 years (cover and density are from annual surveys of 0.25 m² quadrats (n = 10–30), recruitment data are annual averages of monthly estimates (n = 3–5), and nutrients and chlorophyll a (chl-a) data are annual averages of water samples collected monthly during spring and summer months (n = 3)). Capes are arranged geographically from north to south: Cape Mendocino sites are located in northern California while the remaining sites are in central and southern Oregon. Macrophytes include macroalgae and surfgrasses; here, broken down into crustose forms (crusts), kelps and other Phaeophyceae (browns), Chlorophyta (greens), Rhodophyta (reds), and surfgrasses. Greens and other browns are too scarce to be visualized. Sessile invertebrates include mussels, barnacles, anemones, sponges, bryozoans, tunicates, hydroids, and tube-dwelling polychaetes. Grazers include mobile invertebrates that are primarily herbivorous. Figures based on data presented in Menge et al. (in press).

intertidal, with the exception of predators with closed (local) populations (consistent with predictions of simple demographic models) (Wieters et al., 2008). Instead, in direct contradiction of a food chain mechanism, mussel and barnacle abundances are positively correlated with the abundance of predators on the west coast of North America (Fig. 7.4d). In contrast, in a synthetic study of barnacles and mussels and their predators on the west coasts of North and South America, Wieters et al. (2008) found that predatory whelks with crawl-away juveniles do track the subsidies (recruitment rates) of their (demographically open) prey, but are not correlated with their abundances. In their analysis, abundance of sea stars and muricid gastropod predators with open populations did not track prey recruitment rates (at a local scale), as expected for open populations (Wieters et al., 2008), although on the northwest coasts of North America and the South Island of New Zealand, predator consumption rates track prey recruitment (Menge and Menge, 2013). This may lead to differential reproductive output back into the regional, pelagic larval pool among sites (Wieters et al., 2008). Empirical evidence of this phenomenon is scarce for rocky intertidal organisms. However, per capita reproductive output of *Pisaster* does respond to regional variation in prey recruitment, as predicted. In Oregon, where prey recruitment is relatively high, per capita reproductive output of *Pisaster* was on average higher (Sanford and Menge, 2007) than in northern California (Wood, 2008), where prey recruitment is much lower (Fig. 7.3; Connolly et al., 2001; Broitman et al., 2008).

The inherently open population dynamics in the phytoplankton, sessile suspension-feeding invertebrates, predators sub-web, and the cross-ecosystem spatial subsidies of phytoplankton and prey from pelagic ecosystems do not conform to simple, food chain model assumptions (Oksanen et al., 1981; Menge, 2000; also, see Chapter 1). Thus, it is not terribly surprising that the predicted relationships among trophic levels are not observed when examined across sites that vary substantially in phytoplankton availability (proxied by chlorophyll *a* concentration (chl-*a*)) (Fig. 7.4). In this sub-web, phytoplankton and predator abundances should be positively correlated among sites, while neither should be correlated with the abundance of suspension feeders. Instead, among sites we observe: (1) a strong positive correlation between the abundances of predators and herbivores (Fig. 7.4d); (2) a weaker (and non-linear) positive relationship between herbivores and chl-*a* (Fig. 7.4g) and (3) no correlation between chl-*a* and predators (Fig. 7.4f). Initially though, intertidal researchers framed these spatial gradients in phytoplankton abundance and prey recruitment (e.g., Fig. 7.3) as "bottom-up" factors. High variation in phytoplankton among sites was conceptualized as a gradient in productivity that co-varied with larval recruitment rates, possibly due to pelagic ecosystem food chain effects (Menge et al., 1997). Thus the effects of variation in prey recruitment on predation rates were viewed as the interaction of bottom-up and top-down processes (Menge et al.,

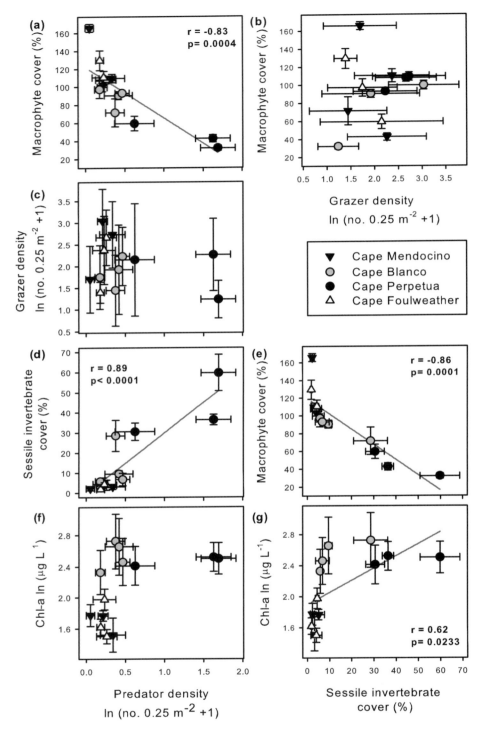

Figure 7.4 Correlations among trophic levels and space occupiers in rocky intertidal communities. Geographical regions and number of surveys are as in Fig. 7.3. Data are annual averages for each site (error bars = 1 SE of annual averages). Correlations

1997; Menge, 2000). However, Menge (2000) also recognized the limitations of simple food chain models due, in part, to the prevalence of omnivory in most rocky intertidal food webs.

Emerging meta-ecosystem theory (Loreau et al., 2003; Leroux and Loreau, 2008) seems to hold more promise for modeling the dynamics of (local scale) communities that are so inextricably dependent on the exchange of life history stages, as well as energetic and nutrient subsidies from adjacent ecosystems (Fig. 7.2). These subsidies are often driven by regional- rather than local-scale processes (Loreau et al., 2003; Menge et al., unpublished data). The scale(s) of demographic processes for animals with pelagic larval stages is a function of the interaction of larval behavior and duration with oceanographic transport (Kinlan and Gaines, 2003; Shanks, 2009; Morgan and Fisher, 2010). Models explicitly incorporating spatial subsidies and cross-scale interactions, such as meta-ecosystem models, may more accurately represent the dynamics shaping the structure of this rocky intertidal sub-web. Additionally, it may be critically important to include horizontal interactions (i.e., competition and facilitation) with the second major sub-web in this ecosystem, as we discuss below.

Sub-web #2: effects of consumers and pelagic subsidies on benthic macrophytes
Benthic macroalgae and surfgrasses form the base of the second main sub-web, and are the other major space-occupying organisms, besides sessile invertebrates, in this ecosystem; they also interact with adjacent ecosystems in terms of nutrient subsidies, larval dispersal, and predators (Fig. 7.2). Larger kelps, rockweeds, and surfgrasses form algal canopies which are important refuge habitats for many taxa, especially in the high intertidal zones where the potential for

---

(except e) are grouped within the two major sub-webs of rocky intertidal communities in accordance with expected relationships if simple food chain dynamics influence the relative abundance of trophic levels. In the macrophyte–grazer–predator sub-web, if top-down effects are strong enough to influence the base of the food chain then:
(a) predator and macrophyte abundances should be indirectly and positively related;
(b) grazer and macrophyte abundances should be unrelated (however, if predators are not effective, then grazer and macrophyte abundances should be positively related); and
(c) predator and grazer abundances should not be related. Similarly, in the phytoplankton–sessile invertebrate–predator sub-web: (f) predator and phytoplankton abundances should be indirectly and positively related; (g) sessile invertebrate and phytoplankton abundances should be unrelated (however, if predators are not effective, then sessile invertebrate and phytoplankton abundances should be positively related); and (d) predator and sessile invertebrate abundances should not be related. Sessile organisms compete for limited space, thus (e) sessile invertebrates and macrophytes should be negatively correlated. Figures based on data presented in Menge et al., (in press).

desiccation and thermal stress is high (e.g., Bertness et al., 1999; Burnaford, 2004; Moulton and Hacker, 2011). Urchins, chitons, limpets, and other gastropods, as well as a suite of meso-grazers (e.g., isopods, gammarid amphipods, small gastropods) feed directly on benthic macroalgae, the smaller epiphytic algae that colonize them, and the microalgae, algal spores, and cyanobacteria that form thin films on otherwise unoccupied rock surfaces. Urchins, chitons, and limpets (especially the larger Patellogastropoda, Fissurelidae, and Siphonariidae) can have strong negative effects on the abundance and diversity of intertidal macroalgae (e.g., Paine and Vadas, 1969; Duggins and Dethier, 1985; Duran and Castilla, 1989; Nielsen, 2003). Smaller limpets, if abundant, have strong effects on early successional stages, but can be swamped if macrophyte growth rates are high (e.g., Dethier and Duggins, 1984; Freidenburg et al., 2007), as their ability to feed on the upright portions of larger, attached algae is limited.

Surfgrasses (*Phyllospadix* spp.) only occur in the North Pacific (Short et al., 2007). *Phyllospadix scouleri* is a dominant, late-succession species (Turner, 1983; Moulton and Hacker, 2011). It is relatively invulnerable to direct herbivory due to a combination of chemical and structural traits, but some specialized limpets feed on the leaves (Fishlyn and Phillips, 1980). Surfgrasses are vulnerable, however, to seed predation by small crustaceans (Holbrook et al., 2000). Low intertidal zone meadows of surfgrass are an important, and sometimes dominant, seascape element that provides habitat for dozens of species of macroinvertebrates (Moulton and Hacker, 2011) and nearshore fishes (Galst and Anderson, 2008).

In the macrophyte-based sub-web, there are fewer allochthonous inputs of organic matter than in the sessile, suspension feeder sub-web. Drift algae from adjacent subtidal or intertidal rocky habitat is the most common subsidy to this sub-web. In South Africa, drift of subtidal kelps supports limpets that live at such high densities they occupy most of the substratum, pre-empting other organisms (Bustamante et al., 1995). Once these limpets exceed 5 cm in length, they escape their primary predators, the oystercatcher *Haematopus moquini* and the giant clingfish *Choriscochismus dentex* (Bustamante et al., 1995). When limpets are denied access to drift kelp, they starve to death. This subsidy probably not only intensifies herbivory on other algae within the foraging range of the limpets, but also increases competition for space between limpets and algae. This exemplifies the strong influence cross-ecosystem subsidies can have on simple food chains, but subsidy effects of this magnitude are not commonly observed in this sub-web. Most invertebrate herbivores rely on attached benthic micro- and macroalgae, including, importantly, algal gametophytes and sporelings (Duggins and Dethier, 1985; Paine, 1992). Survivorship of kelp spores can be facilitated by other organisms such as the dense, calcified algal turfs that are resistant to grazers (Milligan, 1998; Menge et al., unpublished data).

As noted previously, intertidal macrophytes are largely supported by exogenous nutrient inputs. In some instances, nutrients recycled within intertidal habitats can augment nutrient supply (Aquilino et al., 2009). Where seabirds are abundant guano may contribute to the nutrient budget, but this is also an exogenous source and its effects can be variable (e.g., Bosman and Hockey, 1986; Wootton, 1991; Kolb et al., 2010). Variation in macroalgal standing crop due to differences in nutrient loading is often apparent only at the lowest levels of herbivory, at the small spatial scales of experimental plots (Nielsen, 2001; Guerry et al., 2009). Where nutrients are plentiful, such as in upwelling regions, they may only be limiting high on the shore, in areas where flow rates are low, or during periods of reduced availability (e.g., El Niño years) (Wootton et al., 1996; Nielsen, 2003; Bracken and Nielsen, 2004). However, experimental manipulations of nutrients may not be sufficient in duration or magnitude to elicit a response (Nielsen, 2003; Kraufvelin et al., 2006). In contrast, and despite evidence that herbivory is generally high in intertidal ecosystems (Poore et al., 2012), algal abundances are often elevated where regional nutrient loading is naturally higher, including near centers of localized coastal upwelling (Fig. 7.3 (Capes Blanco and Mendocino), Bosman et al., 1987; Broitman et al., 2001; Nielsen and Navarrete, 2004; Bustamante et al., 1995; Menge et al., in press), although light availability may also be important (Kavanaugh et al., 2009). At larger spatial scales, correlations among multiple sites in western North America reveal that (new) nutrients are positively related to macrophyte abundance, but the relationship is stronger for macroalgae alone (details as in Fig. 7.4, data not shown), consistent with data from the west coast of South America (Nielsen and Navarrete, 2004; but see Bustamante et al., 1995 for contrasting results in South Africa).

In food chain models with two trophic levels (i.e., no predators), algal standing crop is predicted to remain unchanged in the face of increasing nutrients, but herbivore abundances should increase with nutrients (Oksanen et al., 1981). In small-scale experiments, this outcome has not been observed (Nielsen, 2001; Guerry et al., 2009). Larger herbivores probably forage over scales greater than the typical experimental replicate (usually $< 1 \, m^2$) and may not aggregate to feed in high productivity plots. However, chitons and limpets foraging in areas with increased availability of microalgae are often larger and more fecund (Dethier and Duggins, 1984; Bosman et al., 1987), as are intertidal urchins near regions of coastal upwelling where nutrients and drift kelp from subtidal ecosystems are abundant (Lester et al., 2007). Local populations with enhanced reproductive output may make a disproportionate contribution to a regional larval pool. But, population responses to localized nutrient enhancement at small spatial scales would not be expected for the most common herbivores enumerated and manipulated in intertidal ecosystems (i.e., urchins and molluscs) due to their open populations at this scale. Interestingly, the few experiments where increased nutrients did increase herbivore abundances involved species with closed populations. Small herbivorous arthropod or "mesograzers" with short generation times (or brooders) – such as gammarid amphipods, isopods, chironomids, and the snail *Littorina saxatilis* – all show positive responses to experimentally increased nutrient levels (Wootton et al., 1996; Worm et al., 2000). However, consistent with the results above for larger herbivores, observational data from multiple rocky shore sites in western North America show no correlations between either herbivore or predator abundances and nutrient availability or macrophyte cover (Fig. 7.4b and c).

Although there is scant evidence of food chain-mediated nutrient effects on benthic herbivores, predators such as crabs and birds can have strong top-down impacts on intertidal herbivores (Lubchenco, 1978; Ellis et al., 2007). In addition to direct effects, they can alter the behavior of their prey through risk cues, indirectly enhancing predator effects on lower trophic levels (Trussell et al., 2004). On northeastern rocky shores of North America, the subtidal crab *Cancer borealis* forages in the intertidal zone during high tides on two other predators (the green crab *Carcinus maenus* and the whelk *Nucella lapillus*) in addition to the herbivore *Littorina littorea*. However, large-scale experiments where gulls were scared away from 50 m stretches of shore (by researchers during low tides) demonstrated that gulls initiate a trophic cascade by reducing the abundance of foraging crabs (Ellis et al., 2007). In separate experiments, Trussell et al. (2004) demonstrated that the green crab can have both trait- and density-mediated indirect effects on the algae in this ecosystem. Thus, gulls have the potential to initiate a trophic cascade through five trophic levels (Ellis et al., 2007).

Synthesis of the two rocky intertidal zone sub-webs: a meta-ecosystem perspective

Until this point we have focused on the separate effects of the two sub-webs without considering how they interact or the potential for horizontal interactions (competition and facilitation). A central dogma of rocky intertidal ecology is that mussels are competitive dominants that exclude macroalgae and other low zone organisms in the absence of predators (Paine, 1966; 1974). Yet observations across many sites suggest that the base of one or the other sub-webs (sessile invertebrates or macrophytes) typically dominates low zone seascapes (Menge, 1992; Menge et al., 1997; Broitman et al., 2001). Recent research suggests that these differences appear to be driven by meta-ecosystem dynamics mediated through regional oceanographic influences on nutrient subsidies and plankton transport (including larvae and phytoplankton) (Menge and Menge, 2013; Menge et al., in press). Macrophytes dominate where nutrient loading rates are high but phytoplankton levels are low (coinciding with narrow continental shelves) (Fig. 7.2). Here, alongshore and offshore currents carry abundant phytoplankton blooms, generated by nutrient-rich and well-lit upwelled water, away from these upwelling centers. The inevitable time lag between the initial phytoplankton "seeding" of newly upwelled waters and a phytoplankton population growth response (Dugdale et al., 1990) effectively decouples locations of high phytoplankton and high benthic algal production (Broitman and Kinlan, 2006). High mussel and barnacle growth and recruitment rates often co-occur with high phytoplankton concentrations on wider continental shelves, or where upwelled waters are detained or entrained (e.g., mesoscale oceanographic gyres, headland–lee complexes, etc.). Furthermore, evidence from field experiments and observations indicate growth rates of intertidal kelps can be light-limited in regions where phytoplankton flourish (or accumulate), pre-empting light before it reaches the benthos (Kavanaugh et al., 2009). Sessile invertebrates should be strong competitors for space in these locations. Interestingly though, recent field experiments (across multiple sites and years on the west coast of North America) provide a different perspective (Hacker et al., unpublished data). The effect (interaction strength) of sessile invertebrates on low zone macrophytes is consistently very small or zero, while macrophytes can have substantial negative effects on sessile invertebrates, except where phytoplankton abundance is very high. Thus, meta-ecosystem dynamics driven by oceanographic regimes appear to also influence the horizontal interactions (in this case, competition for space) that help shape the structure of rocky intertidal communities.

The use of the conceptual terms "top-down" and "bottom-up" to refer to nutrient availability and trophic cascades, respectively, as originally conceived by Hunter and Price (1992) was not explicitly or implicitly intended to encompass the kinds of demographic and spatial complexities that typify rocky intertidal ecosystems. However, the broader conceptualization of bottom-up and top-down processes (and their interaction) that ensued encouraged rocky intertidal ecologists to re-examine their assumptions about the primacy of top-down processes. The exchange of energy, nutrients, and propagules between pelagic and rocky intertidal ecosystems, exploitation of prey by terrestrial and pelagic consumers, and virtually complete spatial segregation of the "brown web" and its associated role in nutrient recycling to other marine ecosystems have a substantial influence on the way bottom-up (e.g., nutrients and light) and top-down processes are manifested at the meta-ecosystem scale and thus determine the structure of rocky intertidal communities. This more mature appreciation for the complexity and cross-ecosystem connectivity of rocky intertidal systems, along with the increasing prevalence of conducting research in larger interdisciplinary teams, provides an excellent foundation for investigating the influence of these factors (including meta-ecosystem models that explicitly incorporate them) on top-down and bottom-up process.

Bottom-up and top-down interactions in salt marsh systems

Introduction to salt marsh systems

Coastal salt marshes are regularly inundated by tidal flooding. Marshes harbor plant communities that show zonation depending on the elevational gradient. Characteristic plant species growing in these environments have to cope with salt stress, and waterlogged and anoxic soil conditions. Their nutrient supply depends on the substrate – peat, sand, clay, or pebbles – and input from marine sources. In addition to abiotic conditions, top-down forces, in the form of invertebrate (e.g., snails, crabs, grasshoppers, and stem-borer and sucking insects) and vertebrate (e.g., wild guinea pigs, geese, hares, and cattle) herbivores and predators (e.g., blue crabs), can have a large effect on plants. The relative importance of bottom-up and top-down forces varies within salt marsh communities; for example, in temperate European salt marshes, predators have not been identified as important actors (Kuijper and Bakker, 2005). In addition, once plants have died and transformed into detritus, detritivores play an important role in this system. We will discuss the conditions that determine the importance of these abiotic and biotic (bottom-up and top-down) factors in shaping the community composition of salt marshes.

Atlantic salt marshes

Here, we consider salt marshes at either side of the Atlantic Ocean, which have been studied in great detail over the last 50 years. In terms of structure, they differ in vertical accretion – namely, plant parts, dead organic material, and plant litter in northwestern (North America) and southwestern (South America) marshes compared to silt and sand in northeastern (Europe) marshes. Overall, northwestern and northeastern Atlantic salt marshes have been intensively modified for centuries by different human activities (i.e., livestock grazing, hay cutting, and agriculture), while southwestern Atlantic marshes have remained more pristine.

Northeastern (Europe) Atlantic salt marshes (as well as western Atlantic salt marshes) mainly include the interaction between plants, sediment/nutrients, and (in)vertebrates. The lack of naturally occurring large herbivores on salt marshes in Europe implies that the effects of large herbivores are restricted to livestock grazing. In fact, livestock grazing has been the most common land use of European salt marshes in the last millennia (Bakker et al., 2005a; Davy et al., 2009).

Northwestern (North America) and southwestern (South America) Atlantic salt marshes

Northwestern (North America) and southwestern (South America) Atlantic marshes naturally occur on relatively sheltered soft-bottom intertidal areas. In most northwestern Atlantic marshes the vertical accretion (between 0.9 and 17.8 mm/year) mainly depends on the organic contribution of decaying plant litter and roots (Turner et al., 2002). In addition, these marshes exhibit natural drainage systems with meandering creeks and levees. However, during the 20th century, many northwestern Atlantic marshes were artificially ditched to control mosquito populations, which dramatically altered drainage, and consequently affected plant communities (Bromberg Gedan et al., 2009b). Main plant species in the southwestern Atlantic include *Spartina alterniflora*, which dominates the low marsh, and *Spartina densiflora* and *Sarcocornia perennis* (usually acting as a pioneer species) that typically occur at higher elevations. In the northwestern Atlantic, *S. alterniflora* (often described as a foundation species; i.e., Bruno, 2000) also dominates the low marsh, *Spartina patens* dominates the seaward edge of the high marsh, *Phragmites australis*, *Juncus gerardii*, and *Juncus roemerianus* are more abundant at higher elevations, and *Distichlis spicata* can be found throughout the high marsh following a fugitive strategy (Pennings and Bertness, 2001; Pennings et al., 2005; Isacch et al., 2006). The relative abundance of these species also varies with environmental conditions. *Spartina alterniflora*, *S. perennis*, and *D. spicata* are more abundant in higher salinity sites, while the remaining species prefer lower salinities (Pennings et al., 2005; Isacch et al., 2006). Thus, the abiotic context plays a key role in determining small- and large-scale patterns of plant species dominance.

Northeastern (Europe) Atlantic marshes

Northeastern (Europe) Atlantic marshes emerge along the coast with sufficient elevation, shelter against energy by streaming and wave action, and enough supply of suspended sediment and seeds or plant parts. On such intertidal flats, the first plant species such as *Salicornia* spp. or *Spartina anglica* can establish. During succession, the marshes become older and higher by vertical accretion, developing into suitable habitat for late-successional species, such as the tall grass *Elytrigia atherica* on the high marsh and shrub *Atriplex portulacoides* on the low marsh (Olff et al., 1997). Thus, in these marshes, sedimentation strongly determines zonation and successional patterns.

In Europe, two types of salt marshes can be distinguished based on their development: back-barrier and foreland salt marshes. Back-barrier marshes establish in the lee of a sand barrier. They show a natural drainage system with meandering creeks and levees with higher elevation than the adjacent depressions. They have a relatively thin layer of clay (up to 0.5 m). In contrast, foreland marshes develop without the shelter of a sandy barrier. They have a thick layer of clay, which can amount to several meters. These salt marshes either have an extensive natural creek system, or are located within man-made sedimentation fields with a drainage system of ditches and are grazed by livestock or left fallow after previous grazing (Bakker et al., 2005a). These marshes are minerogenic, as their vertical accretion mainly depends on the input from tidal flooding. In the case of back-barrier marshes, the elevation gradient runs from the upper marsh at the foot of a dune; in foreland marshes, the elevation gradient runs from the foot of the seawall along the foreland coast to the intertidal flats. This elevational gradient influences the rate of sedimentation, which is the main driver of plant succession. Sedimentation rates differ largely between back-barrier and foreland marshes due to wave activity, which is artificially reduced on foreland marshes to enhance sedimentation. The rate of sediment input on northeastern Atlantic salt marshes varies from < 5 mm/year on sandy back-barrier marshes to up to 20 mm/year on marshes with sedimentation fields (Bakker et al., 2005a). As a result of these differences in development, successional trajectories markedly differ between these two types of salt marshes. Whereas high and low marshes feature their own successional pathway on back-barrier marshes (Leendertse et al., 1997), the low marsh transforms into high marsh on foreland marshes (De Leeuw et al., 1993). In the high marsh on back-barrier marshes and on foreland marshes, the tall grass *Elytrigia atherica* will ultimately dominate the vegetation, and its occurrence and dominance are positively related to elevation and accretion rate (Suchrow et al., 2012).

Spatial variation in salt marshes

Elevation determines the duration and frequency of flooding, and thus is the best predictor of species distribution along the elevational gradient (Fig. 7.1). Among the physical stressors that covary with elevation, salinity and redox potential (mediated by sediment type) have been shown to greatly influence plant productivity and zonation (Hacker and Bertness, 1995a). Additionally, salt marshes around the world are nutrient limited, and N additions increase plant production (e.g., Kiehl et al., 1997; Van Wijnen and Bakker, 1999; Silliman and Zieman, 2001; Alberti et al., 2010b) and arthropod numbers (Vince et al., 1981; Levine et al., 1998), as well as regulating both competitive hierarchies and zonation patterns (i.e., subordinate plants like *S. alterniflora* become dominant after nutrient additions and move their upper zonation limit upwards; Emery et al., 2001).

Experimental studies also revealed that both invertebrates and vertebrates can exert strong control on salt marsh plant production and zonation (geese: Esselink et al., 1997; crabs: Bortolus and Iribarne, 1999; geese and hares: Kuijper and Bakker, 2005; snails: Silliman et al., 2005; grasshoppers: Bertness et al., 2008; small aboveground rodents: Bromberg Gedan et al., 2009a; belowground rodents: Kuijper and Bakker, 2012). Moreover, human activities in European and North American salt marshes have removed a significant amount of plant production through livestock grazing and hay cutting (Bakker et al., 1993; Bos et al., 2005; Bromberg Gedan et al., 2009b).

Elevation (and its effects on oxygen availability and salinity)
The importance of trophic interactions is expected to decrease as environmental stress increases (usually toward lower elevations) because higher trophic levels are more sensitive to stressful conditions (Menge and Sutherland, 1987). Given

that salt marshes usually exhibit marked environmental differences across elevation and thus tidal duration and frequency, it is assumed that the relative importance of top-down and bottom-up factors would vary with elevation. This prediction is supported by results from field experiments. For example, the abundance of different crab predators (*Callinectes sapidus*: Silliman and Bertness, 2002; *Panopeus herbstii*: Silliman et al., 2004) increases with decreasing elevation in many salt marshes located between Georgia and Delaware, USA. These crabs strongly control grazer snails (*Littoraria irrorata*), which in turn control *S. alterniflora* biomass, and thus, crabs indirectly release otherwise herbivore-controlled plants (Silliman et al., 2004), particularly at low elevations (Silliman and Bertness, 2002). Even more, the loss of top predators (most often crabs and fishes in these salt marsh systems) due to human activities has been hypothesized to be a fundamental cause of massive die-offs of *S. alterniflora* in northwestern Atlantic salt marshes, because herbivore populations grow wildly due to relaxed predation (Altieri et al., 2012).

In addition, experiments conducted in a southwestern Atlantic salt marsh manipulating nutrients and crab herbivores revealed that results are strongly variable across elevations (Alberti et al., 2010b). At the lowest reaches of the marsh, the relative importance of crab (*Neohelice granulata*) herbivory is highest and, jointly with nutrients, controls plant biomass. At intermediate elevations, the relative importance of crab herbivory and nutrients is greatly reduced and hypersalinity turns into the major driver of plant performance (Alberti et al., 2010b). At the highest reaches of the marsh, physical stressors (salinity, anoxia, and flooding frequency) are greatly reduced, leading to increased species diversity (see next section), and, as in the opposite end of the tidal gradient, both herbivory (by the wild guinea pigs, *Cavia aperea*) and nutrients interact to determine plant biomass, diversity, and overall species assemblage (Alberti et al., 2011). This is analogous to the results described above for the northwestern Atlantic where herbivorous crabs exert a strong control on the low marsh, while snails can reduce plant biomass on the high marsh. Studies from European back-barrier marshes illustrated that intermediate-sized, vertebrate herbivores (geese and hares) have pronounced effects on the successional pathway of plant species at low marsh elevation, whereas their effects were less pronounced on the high marsh (Kuijper and Bakker, 2005).

Besides affecting trophic interactions, elevation generally affects the relative importance of physical stress. More and longer flooding often creates saline, but also waterlogged conditions, with lower oxygen levels in the soil resulting in reduced redox potentials and leading to increased production of aerenchymatous roots (air-filled channels that allow transfer of gases between the shoot and the root) to cope with these anoxic conditions (Burdick and Mendelssohn, 1987). Oxygen limitation diminishes the production of salt marsh plants such as *Spartina* (Linthurst and Seneca, 1981; Castillo et al., 2000), but crab burrowing can ameliorate this stressful scenario by oxygenating the sediment (*Uca pugnax*: Bertness, 1985; *Neohelice granulata*: Daleo et al., 2007). Indeed, the positive effect of crab burrowing on plant production is then translated to higher trophic levels, benefiting rodents (*Akodon azarae* and *Oligoryzomys flavescens*: Canepuccia et al., 2008) and stem-borer moths (*Haimbachia* sp.: Canepuccia et al., 2010a). Similarly, bioturbation by a small crustacean (*Orchestia gammarellus*) on northern European Atlantic marshes has recently been hypothesized to be important for the establishment of anoxia-intolerant plant species such as *Elytrigia atherica* (Schrama et al., 2012). Finally, a rush, *Juncus gerardii*, in New England salt marshes has been shown to increase species diversity through its ability to increase oxygen conditions and decrease salinity for neighboring plants, which serve as the base of the food chain for insects and their predators (Hacker and Bertness, 1995b; 1996; Hacker and Gaines, 1997). Thus, facilitator species, particularly in the salt marsh, represent a clear interaction of bottom-up and top-down forces that has important implications for community and ecosystem processes.

Effects of elevation and sediment redox potential have been distinguished from each other in a United Kingdom salt marsh. Anoxic conditions occurred at lower elevation, with redox potential generally increasing with elevation, resulting in oxic sediments at higher elevation. However, sediment oxygenation at any given elevation was variable, particularly at intermediate levels in the tidal range. This imperfect correlation between elevation and sediment redox allowed quantification of their independent effects on species distributions (Davy et al., 2011). Some species were affected by both elevation and redox potential (*Elytrigia atherica*), while other species were more affected by redox potential than by elevation (*Suaeda maritima*, *Atriplex portulacoides*).

Among many other abiotic factors that regulate marsh functioning, salinity is probably one of the most intensively studied in salt marshes (Odum, 1988). High salinities inhibit nitrogen uptake by plant roots and can impose serious restrictions on plant vigor and growth, masking the otherwise positive effects of increased nutrient availability (Linthurst and Seneca, 1981; Hacker and Bertness, 1995a). The impact of salinity does not only affect plants; it can also affect habitat use, and densities and numbers of herbivores. For example, in many southwestern Atlantic salt marshes, salinity increases during dry years, preventing wild guinea pigs (*Cavia aperea*, a terrestrial herbivore) from using the marsh due to reduced plant quality (Canepuccia et al., 2010b). In addition, estuarine salinity gradients influence plant assemblages in the northwestern Atlantic. Only a few species can tolerate the most saline extreme of the gradient, whereas they are outcompeted by other plant species at locations with lower salinity (Odum, 1988; Crain et al., 2004). Moreover, at the lower end of the salt gradient, the impact of herbivorous rodents on plant biomass (Taylor and Grace, 1995) and diversity (Bromberg Gedan et al., 2009a) is more pronounced. Along the elevational gradient of a northeastern Atlantic salt marsh, Brent geese (*Branta bernicla*) forage on the relatively saline, low marsh, whereas Barnacle geese (*Branta leucopsis*) mainly forage on the higher, less saline marsh. This difference was confirmed by experimental addition of salt water to the high marsh. Brent geese have a bigger salt gland than Barnacle geese (Stahl et al., 2002), which enables them to forage in different portions of the marsh.

However, the negative effect of salinity on plant vigor can also act synergistically with herbivory, imposing severe and persistent reductions in salt marsh plant cover and biomass. For example, drought-induced increases in salinity in northwestern Atlantic marshes caused plant decay and formation of snail fronts that propagated through healthy marsh, leading to massive die-offs (Silliman et al., 2005). These results led to a refinement of the community organization models originally proposed by Menge and Sutherland (1987; i.e., the importance of negative biotic interactions decreases as physical stress increases), predicting that, in these scenarios, plant cover would rapidly decrease as physical stress increases (Silliman et al., 2013). Collectively this evidence reveals that physical stress can negatively affect plants while negatively *or* positively affecting their consumers, depending on the system.

Sediment type

Sediment type can modify the aforementioned gradients driven by elevation. Salt marshes throughout the world develop on very different substrates including cobble/pebble, sand, peat, and silt/clay (peat: Bertness et al., 2009; clay: Van Wijnen and Bakker, 1999; clay – various substrate types: Daleo and Iribarne, 2009). As well as other environmental factors, sediment type can play a key role in marsh functioning, determining the relative importance of top-down and bottom-up interactions. For example, massive die-offs of *S. alterniflora* marshes have been occurring in Cape Cod (Massachusetts, USA), driven by elevated herbivorous crab densities (Altieri et al., 2012). However, this extremely negative effect of crabs on salt marsh plants depends on sediment type. Crabs recruit and burrow in peat but not in sandy marshes, so substrate type thus determines the potential for occurrence of this key grazer and its control of the marsh die-off (Bertness et al., 2009).

A similar influence of sediment type on plant–animal interactions occurs in southwestern Atlantic salt marshes. These salt marshes occur across a gradient of silt/clay-dominated to cobble/pebble-dominated sediments, which closely parallel a gradient of increasing oxygen availability in the sediment. On marshes with larger grain size, oxygen limitation is less common and, consequently, the relative importance of the positive effect of crab burrows on plant production is outweighed by the negative effects of crab herbivory (Daleo and Iribarne, 2009). Sediment type therefore regulates plant productivity and, consequently, the relative strength of positive and negative impacts of grazers on plants and the overall net effect of top-down and bottom-up control.

Nutrients

Salt marshes often exhibit gradients in nutrient availability (particularly N) that are correlated with plant density and biomass (Valiela et al., 1978; Alberti et al., 2010b), and can even lead to distinctive growth forms such as the short and tall forms of *S. alterniflora* (Valiela et al., 1978). Additionally, salt marshes are threatened by increasing N inputs due to human activities (sewage discharge, runoff from agricultural fields, and atmospheric deposition; Adam, 2002). Persistent ecosystem scale N enrichment can have severe consequences for salt marshes given that the globally observed increase in aboveground production (e.g., Van Wijnen and Bakker, 1999; Silliman and Zieman, 2001) occurs jointly with a reduction in belowground biomass that leads to marsh loss by creek-bank collapse (Deegan et al., 2012). Many experimental studies that jointly manipulated nutrients and herbivores have found that both factors regulate plant biomass and/or diversity (Silliman and Zieman, 2001; Bertness et al., 2008; Alberti et al., 2010b; 2011; Kuijper and Bakker, 2012). Nutrient additions increase overall plant production, while the effects of herbivores range from moderate to high enough to consume all the extra plant biomass produced under nutrient-enriched scenarios. In particular, the consumption of extra plant biomass promoted by nutrient additions (e.g., Bertness et al., 2008) likely occurs due to increased herbivore densities in fertilized plots. The varying impact of herbivores depends on their densities (Abraham et al., 2005), the presence of their predators (Silliman and Bertness, 2002), and the abiotic context (Bromberg Gedan et al., 2009a).

However, rather than acting in isolation, nutrients interact with herbivores in marshes (as in other systems); the relative importance of these forces in regulating plant production is likely context-dependent. As salt marshes often occur in very dynamic habitats (e.g., exposed to tidal wave action and extreme weather events), the alternation of bottom-up and top-down factors dominating trophic interactions might be a common feature of these unique systems. Whereas herbivores might play an important role in influencing plant species composition, their effects could be overruled when extreme events change the abiotic conditions and reset the successional clock to earlier stages (Kuijper and Bakker, 2012).

Short-term succession across environmental gradients in salt marshes

Natural disturbance by floating plant debris (wrack), sediment deposition, herbivores, fire, and ice can all play an essential role in marsh functioning by creating unvegetated areas and eliciting short-term secondary succession (Pennings and Bertness, 2001). For example, goose herbivory in the Canadian Arctic can be intense enough to create unvegetated areas that can remain unchanged for years due to changes in edaphic conditions (Abraham et al., 2005). Recovery is usually driven by sexual or asexual colonization of bare patches and can be modulated by herbivores and bioturbators regulating sediment dynamics. Indeed, as salinity is reduced from salt to freshwater marshes, plant species diversity increases (Odum, 1988), as well as the importance of herbivory during secondary succession in northwestern Atlantic marshes (mostly initiated by wrack deposition), and thus, full recovery after disturbance takes longer if herbivores are present (Bromberg Gedan et al., 2009a). Similarly, recovery in southwestern Atlantic salt marshes occurs very slowly due to the impact of herbivores at different intertidal heights (Alberti et al., 2008; 2010a; Daleo et al., 2011; 2014). In contrast, sediment dynamics seem to play a greater role in regulating succession in northeastern Atlantic marshes, even mediating the impact of herbivores. It is suggested for northeastern Atlantic marshes that herbivory and bioturbation by the ragworm *Nereis diversicolor* may account for a loss of salt marsh vegetation from the pioneer zone by foraging and burial of seeds and seedlings. It is also possible that abiotic conditions such as mobility of the top layer during winter periods may cause removal of seeds (Wolters et al., 2005). Experiments in the field and in mesocosms revealed that the establishment of small *S. anglica* plants in the pioneer zone is inhibited by the lugworm *Arenicola marina* because of low sediment stability induced by the lugworms. In turn, *Arenicola* establishment in *Spartina*-dominated patches is limited by high silt content and compactness and dense rooting of the sediment by *Spartina* (Van Wesenbeeck et al., 2007), illustrating the potential for negative interactions to determine community composition. Similarly, Greylag geese (*Anser anser*) feed on aboveground biomass of grasses and dicotyledons, as well as on rhizomes and winterbuds of *S. anglica* and tubers of *Scirpus maritimus*. Exploitation of juvenile plants of *Spartina* by geese prevented new establishment on bare soil. After removal of tuber biomass by geese, recovery took 2 years in an exclosure experiment. Without exclosures, bare soil remained (Esselink et al., 1997). Overall, these examples suggest that herbivores alone, through modulation of sediment dynamics, generally retard secondary succession after disturbance. This highlights the importance of top-down forces and their potential interaction with bottom-up processes in short-term succession of salt marsh ecosystems.

Long-term succession: a case study from European back-barrier salt marshes

Changes in abiotic conditions and biotic interactions during salt marsh succession
An inherent problem of studying succession is the time required to study an entire successional sequence. Space for time substitutions provide a useful tool to deal with the problem of lack of successional studies over many decades. The back-barrier salt marsh of the Dutch island of Schiermonnikoog represents a rare, clear chronosequence. The establishment of vegetation in different parts

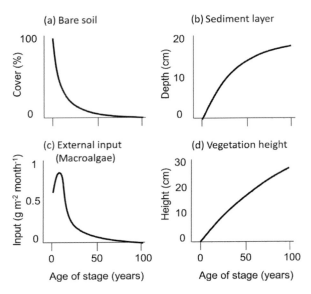

Figure 7.5 Changes in abiotic and biotic parameters over the chronosequence of the back-barrier salt marsh on Schiermonnikoog, The Netherlands: (a) percentage of bare soil decreases strongly with age of succession, (b) sediment layer increases with succession to about 16 cm, (c) external input of macroalgae is only present in the early stages of succession, and (d) vegetation height increases strongly toward late-successional stages (modified after Schrama et al., 2012).

of the salt marsh can be derived from aerial photographs (Olff et al., 1997). The chronosequence including very young marsh (from 0 years onwards) to older marsh (up to 100 years) allows an inference of succession of soil, vegetation, vertebrate, and invertebrate animals in the same ecosystem and the opportunity to look at the relative importance of bottom-up and top-down forces during succession.

Along this chronosequence, the cover of bare soil decreases, whereas the depth of deposited sediment increases with age of the salt marsh (Fig. 7.5). Deposited sediment contributes to surface elevation change, and it also contains N. The N pool of the rooting zone of 50 cm is positively correlated with the thickness of the clay layer on back-barrier marshes (Olff et al., 1997). As N mineralization is positively related to the N pool (Bakker et al., 2005b), plant production increases and vegetation grows taller with an increase in clay layer thickness. Hence, the chronosequence represents a productivity gradient.

Early in succession, external input from marine sources such as macroalgae plays an important role (Fig. 7.6a). In this early stage, large numbers of the dipteran *Fucellia maritima* occur under decaying algae, as well as microbivores and their predators (Schrama et al., 2012). This stage thus harbors a food web dominated by detritivores (i.e., a brown food web) and a low plant standing biomass. Stable isotope analyses revealed that marine-derived N is not only found in the various layers of the food web, but also in the pioneer plants at this stage (Schrama et al., 2013a). Nutrient flows from the marine ecosystem to the terrestrial ecosystem thus drive early successional communities in the salt marsh from the bottom-up.

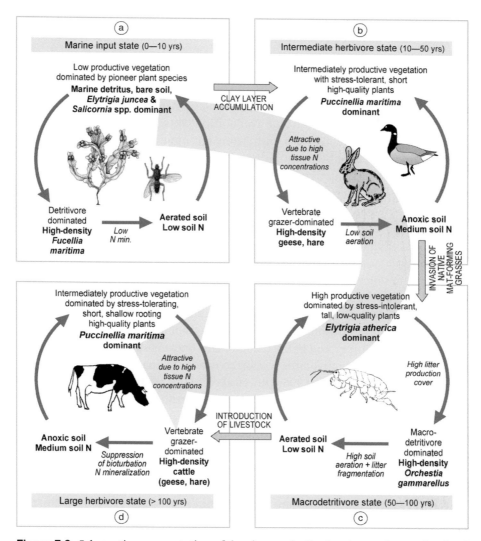

Figure 7.6 Schematic representation of the changes in the dominant plant and animal species along the chronosequence of the barrier island of Schiermonnikoog, The Netherlands. Gray arrows depict dominant top-down and bottom-up drivers that maintain the status quo in each of the four different states. Thick open arrows represent the bottom-up or top-down drivers that force the system from one stage to the next.

Effects of plants on intermediate-sized vertebrate herbivores (bottom-up control)
Intermediate successional stages are characterized by a strong increase in above-ground standing plant biomass. This is accompanied by a strong increase in the biomass of small (invertebrates) and intermediate-sized herbivores (e.g., geese and hares; Fig. 7.6b). Along the productivity gradient, densities of herbivores, such as migratory Arctic-breeding geese (e.g., Brent geese (*Branta bernicla*

bernicla), Barnacle geese (*Branta leucopsis*)), sedentary European brown hares (*Lepus europaeus*), and European rabbits (*Oryctolagus cuniculus*), initially increase to an optimum at intermediate productivity, but decline at sites with high productivity (Van de Koppel et al., 1996; Bakker et al., 2009). According to the exploitation ecosystem hypothesis (EEH), at sites with low productivity, plant biomass is too low to support herbivore populations, and plant growth will be regulated by bottom-up effects such as nutrient availability (Oksanen and Oksanen, 2000). With increasing productivity, a shift from bottom-up to top-down effects is expected to occur. Top-down regulation of plant biomass occurs at sites of intermediate levels of productivity, with herbivore populations regulated by top-down processes (namely, carnivores) at high productivity (Oksanen and Oksanen, 2000). However, in the absence of carnivores, bottom-up effects play an important role even at highly productive sites. Forage quality declines at sites of high biomass and tall canopy (Van Der Wal et al., 2000a), featuring a decreasing leaf:stem ratio. As a result, herbivore density at high biomass sites can decrease even in the absence of carnivores in these systems because intake rate and foraging efficiency of geese level off or decline with plant biomass above a certain threshold (Van Der Wal et al., 1998; Bos et al., 2004; Van Der Graaf et al., 2006) and hare foraging conditions are negatively affected by increasing biomass (Kuijper et al., 2008; Kuijper and Bakker, 2008). This bottom-up control of herbivore density at highly productive sites is referred to as the "quality threshold hypothesis" (Van De Koppel et al., 1996; Olff et al., 1997). This hypothesis states that without top predators, plant quality can exert strong bottom-up effects that regulate herbivore density at highly productive successional stages.

There is good evidence that the aforementioned patterns of herbivore density are related to succession and in this way linked to dietary quality and foraging conditions for intermediate-sized herbivores. Geese numbers were estimated at young, intermediate, and older parts of the salt marsh on Schiermonnikoog between 1971 and 1997. In the late 1970s, Brent geese numbers were high in the old marsh. However, geese numbers declined significantly in the following 20 years (Van Der Wal et al., 2000b). This decrease was not related to a decrease in size of the area, as the surface area increased over the years as a result of ongoing vegetation succession. Geese numbers increased in the intermediate-aged salt marsh, followed by a slight but significant decrease toward 1997. In the young salt marsh, geese numbers only increased. Furthermore, the development of new young marsh led to an eastward movement of geese on this salt marsh (Van Der Wal et al., 2000b). Hence, the reduction of the community with preferred food plants in the western parts was compensated for by an increase in this community in newly developed eastern parts of the salt marsh. We observed that ongoing plant succession pushed the geese eastward and geese had to follow the changing vegetation. Thus, vegetation succession evicted spring staging geese (Van Der Wal et al., 2000b). Similarly, the increasing abundance of the unpalatable, tall plants during salt marsh succession reduces the grazing intensity of non-migratory hares (Kuijper et al., 2008; Kuijper and Bakker, 2008). As a result, hare numbers also decrease with increasing salt marsh age (Van De Koppel et al., 1996) and bottom-up factors control the vegetation at the older, productive successional stages (Bakker et al., 2005a).

While the biomass of herbivores decreases toward late succession, biomass of detritivores increases (Fig. 7.6c). This coincides with a stable biomass of live plants toward late-successional stages, and a continuous increase in dead biomass. The crustacean macrodetritivore beach hopper *Orchestia gammarellus* becomes dominant in the final stages of succession (Schrama et al., 2012). It creates oxic conditions by digging ("bioturbating") in the upper layer of the sediment. This increases net N mineralization and promotes dominance of the tall grass sea couch, *Elytrigia atherica*. The spreading of this tall grass during salt marsh succession is a phenomenon of natural succession on back-barrier marshes (Veeneklaas et al., 2013).

Effects of intermediate-sized herbivores on plants (top-down control)
Are intermediate-sized herbivores only a victim of plant succession? The effects of intermediate-sized herbivores on vegetation were tested in long-term field experiments. At four sites along the chronosequence on the island of Schiermonnikoog, exclosures were established. After 7 years, top-down effects of herbivores on the vegetation were most pronounced in the low salt marsh. Plant species of later successional stages increased in cover inside full exclosures, especially at the youngest, least productive marshes. These experiments revealed that herbivory retarded the establishment and spread of late-successional species (i.e., *Atriplex portulacoides* and *Elytrigia atherica*) on the low salt marsh. Hares are estimated to retard vegetation succession for at least 20–25 years (Van Der Wal et al., 2000a; Kuijper and Bakker, 2005). The open vegetation in the young, least productive marshes offers the opportunity for establishment of late-successional species, as long as selective grazing by herbivores is absent. Once late-successional species have established, they will spread more rapidly in the absence of herbivores, indicating that establishment is actually the limiting factor in this invasion, but that herbivory can retard further spread (Kuijper et al., 2004). In the absence of herbivores, late-successional species can directly invade, during the "window of opportunity" in young marshes, and will dominate the vegetation at an earlier stage. Hence, the top-down effects of the herbivores combined with the bottom-up effects of the vegetation can retard vegetation succession in these salt marsh systems for several decades (Kuijper and Bakker, 2005) until herbivores themselves are evicted by vegetation succession.

Importantly, *Elytrigia atherica* is not always the final successional stage. Sites far away from the intertidal flats receive little or no sediment. Hence, they are building a sedimentation deficit due to continuous sea-level rise, and hence get wetter. The result is that *Elytrigia* stands decline and become replaced by the native *Phragmites australis* (Veeneklaas et al., 2013). This implies that successional pathways to climax communities are governed by bottom-up and top-down control at young successional stages, but that in some cases, bottom-up control may play a dominant role in determining the final successional stage. It appears that herbivorous snails in eastern Atlantic marshes take a similar top-down position to geese and hares in northwestern Atlantic salt marshes (Silliman et al., 2005).

Effects of large herbivores on plants and smaller herbivores (top-down control)
Natural large herbivores are nowadays lacking in European salt marshes. However, these natural grasslands likely were important foraging sites for extinct large grazers such as aurochs (*Bos taurus*, 17th century) and Tarpan (*Equus ferus*, 18th century), which once occurred throughout Eurasia (Hall, 2008). Northwestern European salt marshes have often experienced millennia of livestock grazing (Bakker et al., 2005a; Davy et al., 2009), which can be seen as a substitute for the processes induced once by natural large herbivores. Grazing by large herbivores causes soil compaction by trampling, with subsequent increased bulk density, increased water content, reduced redox potential, and hence reduced rates of net N mineralization on grazed sites compared to ungrazed sites (Fig. 7.6d). This coincides with a strong decline in the abundance of *Orchestia gammarellus* and a resulting decrease in bioturbation. Such differences were found on clay marsh with fine-grained sediment, and not at the foot of dunes with sandy coarse-grained sediment, and thus depend on the type of sediment. This suggests that grazing by large herbivores affects the ecosystem more by habitat destruction (i.e., trampling) than by removing biomass and hence competition for light (Schrama et al., 2013b).

Livestock grazing in the final stage of succession in the chronosequence can set back the successional clock toward earlier stages in the vegetation (Bos et al., 2002; Schrama et al., 2013b). Such plant communities pave the way for intermediate-sized herbivores that would otherwise have been evicted by vegetation succession. Geese were found to depend on the facilitation by livestock (Bos et al., 2005), and hares also profit from the better foraging conditions created by livestock on older salt marshes (Kuijper et al., 2008). High densities of livestock, hares, geese, and herbivorous invertebrates turn the brown food web back to a green food web (Schrama et al., 2012). It seems that both top-down and bottom-up factors simultaneously control plant species composition when grazing by large herbivores occurs on old, productive back-barrier salt marshes.

Summary of salt marsh systems
Nutrients, physical stressors, plant communities, and herbivore pressure vary within and between marshes largely because of changes in the elevational gradient. Rather than acting in isolation, these factors interact in many different ways

and their balance is strongly spatially and temporally context-dependent. For example, while nutrients generally have positive short-term impacts on plant production, they might also promote herbivore densities and grazing pressure, as well as marsh erosion in the long term. Analogously, physical stress usually has negative impacts on plant biomass (e.g., inhibiting nutrient uptake), and it can also promote or deter herbivory, depending on whether herbivores are affected by that stressor. Both nutrients and herbivores influence plant communities, but in addition, changes in plant communities over time influence herbivores, determining the relative importance of bottom-up or top-down forces. Overall changes in a natural salt marsh succession reveal a transition from a brown food web in the earliest stages, to a green food web during intermediate successional stages, followed again by a brown food web at the oldest, most productive stages of salt marsh succession (Schrama et al., 2012). Along this successional gradient, there are clear shifts in bottom-up and top-down processes that co-occur with changes in food web configuration. Plant and animal communities in very early and late-successional stages are mainly regulated by bottom-up processes. The availability of nutrients, either from marine sources or from oxygenated sediment, as well as the local abiotic conditions, largely determine plant and animal species composition. In both cases, the food web is largely dominated by brown web trophic groups (Schrama et al., 2012). However, in intermediate stages of succession, trophic groups of the green web (e.g., intermediate-sized herbivores, such as geese and hares) reach their highest biomass. Here, they have the biggest impact on the vegetation, thus top-down processes are playing an important role in shaping plant community composition. In late-successional stages, plant–plant interactions play an important role, when the geese and hares can no longer cope with the high plant production, and the tall grass *Elytrigia atherica* outcompetes low-statured plant species. Importantly, the introduction of livestock in the final stage of succession can set back the successional clock toward earlier stages in plant and animal communities. These results show how the balance between top-down and bottom-up forces varies across environmental gradients, highlighting the strong spatial and temporal context-dependency of these interactions.

Overall summary of rocky intertidal and salt marsh systems

The two coastal interface ecosystems we have focused on in this chapter share many characteristics. First and foremost, both experience the rhythm of the tides, which sets the scene for most of the abiotic conditions and biotic factors that influence these ecosystems. In these systems, community structure and the relative importance of top-down and bottom-up factors result from differences in oceanographic and climatological conditions, geographical location, substrate type, and local and regional species' pools.

A striking difference between rocky shore and salt marsh ecosystems is the opposed orientation of their major vertical stress gradients: abiotic stress on rocky shores is strongly linked to the duration and conditions of atmospheric exposure during low tides, whereas stressful conditions on the salt marsh arise from the duration of inundation by seawater (Fig. 7.1). Rocky shores generally become less physiologically stressful for macrophytes and sessile invertebrates lower on the shore, where periods of inundation are longer, whereas abiotic stress for salt marsh plants generally decreases with increasing salt marsh elevation. For rocky shore macrophytes, winds and insolation result in desiccation, and thermal and light stresses are greater higher on the shore, where nutrients are also less readily available. In contrast, the soils of salt marshes are highly heterogeneous with respect to salinity, nutrients, and redox potential, and these factors, absent from rocky shores, are important determinants for plant community composition in salt marshes. However, mobile marine consumers in both ecosystems are negatively influenced by atmospheric conditions at higher elevations, and both rocky intertidal and salt marsh communities are vulnerable to terrestrial consumers. On rocky shores, these are primarily birds, whereas mammals and birds are common on salt marshes.

Further, one of the most fundamental differences between salt marsh and rocky shore ecosystems is the difference in substrates, noted above. Detritus does not accumulate in rocky intertidal ecosystems and contribute to the formation of soil; instead, it is largely exported to adjacent beaches (Fig. 7.2) and other benthic marine ecosystems. As a result, there is limited opportunity for consumers to influence nutrient recycling or ecosystem nutrient supply rates. Rather, exchange or inputs of nutrients and energy with adjacent ecosystems are both common and more influential. In contrast, salt marsh plants and animals can influence nutrient dynamics and sediment properties through uptake and sequestration of nutrients in plant biomass, bioturbation and excreta (e.g., pseudofeces of mussels at the edge of the marsh), and the brown food web. Top-down and bottom-up processes differ markedly between these systems. Trophic interactions on rocky shores can be strong, but also vary as a function of environmental stress and oceanographically mediated subsidies of energy, materials, and propagules from adjacent ecosystems. Some of these processes mediate bottom-up effects (e.g., oceanographic processes that affect nutrient supply or light availability), but many are spatially structured subsidies transported among ecosystems by ocean currents. Adjacent pelagic ecosystems serve as "nurseries" that also provide avenues of dispersal for larvae. Additionally, the marine macro- and microalgae that are the dominant primary producers on rocky shores have lower C:N ratios and thus higher forage quality than the vascular plants that dominate salt marshes. Although this makes rocky shore autotrophs more palatable and nutritious, these macrophytes often go uneaten, and in some places, especially where nutrients and light are plentiful, they can dominate the shore. Mobile rocky shore herbivores seem to have the greatest impact early in succession before macrophytes grow large, escaping vulnerability. Suspension-feeding invertebrates do not make much of a dent in nearshore phytoplankton concentrations, yet phytoplankton abundance fuels the growth and fecundity of suspension feeders, and of their invertebrate predators. Thus, energy from cross-ecosystem subsidies plays a major role in supporting rocky shore consumer populations.

In comparison, the greatest impacts of herbivores in salt marshes occur at intermediate stages of succession in European salt marshes (Kuijper and Bakker, 2005), and during early and intermediate stages of succession in South American salt marshes (Alberti et al., 2008; 2010a; Daleo et al., 2014). Top-down forces in salt marshes often play a relatively small role in structuring communities. However, this strongly depends on the geographical range. On the vast low-lying western Atlantic marshes, herbivore top-down control can play a major role; consumer assemblages in these areas are often dominated by marine grazers (for example, snails and crabs), or a mix of terrestrial and aquatic grazers (such as aphids, stem-borer moths, nutrias, wild guinea pigs, and other rodents). In the extensive lower marshes, plants can be completely devoured by consumer fronts of snails, which overwhelm the vascular plants. Low-lying salt marshes in South America are also top-down controlled by invertebrate grazers (e.g., crabs). In contrast, in northeastern Atlantic marshes (Europe), which are dominated by high marsh, grazers are mostly terrestrial vertebrates, such as hares and geese. Grazing by these vertebrates, especially in early successional stages, can have a profound effect on vegetation development. In later stages of salt marsh development, only grazing by large vertebrate herbivores (livestock) can have a similar impact. In general, top-down control by carnivores is rare in salt marshes.

Ecosystems of the land–sea margin are characterized by their strong connectivity with adjacent pelagic, estuarine, shoreline, and terrestrial ecosystems (Menge et al., 1997; in press; Polis et al., 1997 (and references therein); Menge, 2000; Valiela et al., 2002; Dugan et al., 2003; Silliman and Bertness, 2004; Schrama et al., 2012; 2013a; Menge and Menge, 2013). Empirical studies in rocky shore and salt marsh (and other) land–sea margin ecosystems have made important contributions to our increasing appreciation of the interplay between bottom-up and top-down processes, and how they structure communities and influence ecosystem functioning. In addition, these systems highlight the importance of spatial and temporal scales (see Chapter 11) in determining the interaction of bottom-up and top-down processes at the land–sea boundary.

Acknowledgments

We thank Torrance Hanley and Kimberly La Pierre for insightful comments on the manuscript. We are grateful to Dick Visser for drawing Figs. 7.5 and 7.6. The work of DPJK was supported by funding from the Polish Ministry of Science and Higher Education (grant no. 2012/05/B/NZ8/01010). The work of OI was supported by grants from UNMDP, CONICET, and MINCyT. Research by KJN,

SDH, BAM, and FC was supported by the National Science Foundation (grant nos. OCE-0726983, OCE-0727611, OCE-1061233, 992 OCE-1061530), the Andrew W. Mellon Foundation, the David and Lucile Packard Foundation, the Gordon and Betty Moore Foundation, endowment funds from the Wayne and Gladys Valley Foundation, and startup funds to SDH from the College of Science at OSU.

References

Abraham, K. F., Jefferies, R. L. and Alisauskas, R. T. (2005). The dynamics of landscape change and snow geese in mid-continent North America. *Global Change Biology*, **11**, 841–855.

Adam, P. (2002). Saltmarshes in a time of change. *Environmental Conservation*, **29**, 39–61.

Alberti, J., Escapa, M., Iribarne, O., Silliman, B. and Bertness, M. (2008). Crab herbivory regulates plant facilitative and competitive processes in Argentinean marshes. *Ecology*, **89**, 155–164.

Alberti, J., Escapa, M., Daleo, P., Méndez Casariego, A. and Iribarne, O. (2010a). Crab bioturbation and herbivory reduce pre- and post-germination success of *Sarcocornia perennis* in bare patches of SW Atlantic salt marshes. *Marine Ecology Progress Series*, **400**, 55–61.

Alberti, J., Méndez Casariego, A., Daleo, P., et al. (2010b). Abiotic stress mediates top-down and bottom-up control in a Southwestern Atlantic salt marsh. *Oecologia*, **163**, 181–191.

Alberti, J., Canepuccia, A., Pascual, J., Pérez, C. and Iribarne, O. (2011). Joint control by rodent herbivory and nutrient availability of plant diversity in a salt marsh-salty steppe transition zone. *Journal of Vegetation Science*, **22**, 216–224.

Altieri, A. H., Bertness, M. D., Coverdale, T. C., Herrmann, N. C. and Angelini, C. (2012). A trophic cascade triggers collapse of a salt marsh ecosystem with intensive recreational fishing. *Ecology*, **93**, 1402–1410.

Aquilino, K. M., Bracken, M. E., Faubel, M. N. and Stachowicz, J. J. (2009). Local-scale nutrient regeneration facilitates seaweed growth on wave-exposed rocky shores in an upwelling system. *Limnology and Oceanography*, **54**, 309–317.

Bakker, J. P., de Leeuw, J., Dijkema, K. S., et al. (1993). Salt marshes along the coast of The Netherlands. *Hydrobiologia*, **265**, 73–95.

Bakker, J. P., Bunje, J., Dijkema, K. S., et al. (2005a). Salt marshes. In *Wadden Sea Quality Status Report 2004. Wadden Sea Ecosystem No 19*, ed. K. Essink, C. Dettmann, H. Farke, G. Lüerssen, H. Marencic, and W. Wiersinga. Wilhelmshaven, Germany: Trilateral Monitoring and Assessment Group, Common Wadden Sea Secretariat, pp. 163–179.

Bakker, J. P., Bouma, T. J. and Van Wijnen, H. J. (2005b). Interactions between microorganisms and intertidal plant communities. In *Interactions Between Macro- and Microorganisms in Marine Sediments: Coastal and Estuarine Studies 60*, ed. K. Kristensen, J. E. Kostka and R. R. Haese. Washington: American Geophysical Union, pp. 179–198.

Bakker, J. P., Kuijper, D. P. J. and Stahl, J. (2009). Community ecology and management of salt marshes. In *Community Ecology Processes, Models and Applications*, ed. H. A. Verhoef and P. J. Morin. Oxford: Oxford University Press, pp. 131–147.

Berlow, E. L. (1997). From canalization to contingency: historical effects in a successional rocky intertidal community. *Ecological Monographs*, **67**, 435–460.

Bertness, M. D. (1985). Fiddler crab regulation of *Spartina alterniflora* production on a New England salt marsh. *Ecology*, **66**, 1042–1055.

Bertness, M. D., Leonard, G. H., Levine, J. M. and Bruno, J. F. (1999). Climate-driven interactions among rocky intertidal organisms caught between a rock and a hot place. *Oecologia*, **120**, 446–450.

Bertness, M. D., Crain, C. M., Silliman, B. R., et al. (2006). The community structure of western Atlantic Patagonian rocky shores. *Ecological Monographs*, **76**, 439–460.

Bertness, M. D., Crain, C., Holdredge, C. and Sala, N. (2008). Eutrophication and consumer control of New England salt marsh primary productivity. *Conservation Biology*, **22**, 131–139.

Bertness, M. D., Holdredge, C. and Altieri, A. H. (2009). Substrate mediates consumer control of salt marsh cordgrass on Cape Cod, New England. *Ecology*, **90**, 2108–2117.

Bertness, M. D., Bruno, J. F., Silliman, B. R. and Stachowicz, J. J. (2014). A short history of marine community ecology. In *Marine Community Ecology and Conservation*, ed. M. D. Bertness, J. F. Bruno, B. R. Silliman and J. J. Stachowicz. Sunderland, MA: Sinauer Associates, pp. 2–8.

Bortolus, A. and Iribarne, O. (1999). Effects of the SW Atlantic burrowing crab *Chasmagnathus granulata* on a *Spartina* salt marsh. *Marine Ecology Progress Series*, **178**, 79–88.

Bos, D., Bakker, J. P., De Vries, Y. and Van Lieshout, S. (2002). Long-term vegetation changes in experimentally grazed and ungrazed back-barrier marshes in the Wadden Sea. *Applied Vegetation Science*, **5**, 45–54.

Bos, D., Van De Koppel, J. and Weissing, F. J. (2004). Dark-bellied Brent geese aggregate to cope with increased levels of primary production. *Oikos*, **107**, 485–496.

Bos, D., Loonen, M., Stock, M., et al. (2005). Utilisation of Wadden Sea salt marshes by geese in relation to livestock grazing. *Journal for Nature Conservation*, **15**, 1–15.

Bosman, A. L. and Hockey, P. A. R. (1986). Seabird guano as a determinant of rocky intertidal community structure. *Marine Ecology Progress Series*, **32**, 247–257.

Bosman, A. L., Hockey, P. A. R. and Siegfried, W. R. (1987). The influence of coastal upwelling on the functional structure of rocky intertidal communities. *Oecologia*, **72**, 226–232.

Bracken, M. E. and Nielsen, K. J. (2004). Diversity of intertidal macroalgae increases with nitrogen loading by invertebrates. *Ecology*, **85**, 2828–2836.

Bracken, M. E., Jones, E. and Williams, S. L. (2011). Herbivores, tidal elevation, and species richness simultaneously mediate nitrate uptake by seaweed assemblages. *Ecology*, **92**, 1083–1093.

Broitman, B. R. and Kinlan, B. P. (2006). Spatial scales of benthic and pelagic producer biomass in a coastal upwelling ecosystem. *Marine Ecology Progress Series*, **327**, 15–25.

Broitman, B. R., Navarrete, S. A., Smith, F. and Gaines, S. D. (2001). Geographic variation of southeastern Pacific intertidal communities. *Marine Ecology Progress Series*, **224**, 21–34.

Broitman, B. R., Blanchette, C. A., Menge, B. A., et al. (2008). Spatial and temporal patterns of invertebrate recruitment along the west coast of the United States. *Ecological Monographs*, **78**, 403–421.

Bromberg Gedan, K., Crain, C. M. and Bertness, M. D. (2009a). Small-mammal herbivore control of secondary succession in New England tidal marshes. *Ecology*, **90**, 430–440.

Bromberg Gedan, K., Silliman, B. R. and Bertness, M. D. (2009b). Centuries of human-driven change in salt marsh ecosystems. *Annual Review of Marine Science*, **1**, 117–141.

Bruno, J. F. (2000). Facilitation of cobble beach plant communities through habitat modification by *Spartina alterniflora*. *Ecology*, **81**, 1179–1192.

Burdick, D. M. and Mendelssohn, I. A. (1987). Waterlogging responses in dune, swale and marsh populations of *Spartina patens* under field conditions. *Oecologia*, **74**, 321–329.

Burnaford, J. L. (2004). Habitat modification and refuge from sublethal stress drive a marine plant-herbivore association. *Ecology*, **85**, 2837–2849.

Bustamante, R. H., Branch, G. M. and Eekhout, S. (1995). Maintenance of an exceptional intertidal grazer biomass in South Africa: subsidy by subtidal kelps. *Ecology*, **76**, 2314–2329.

Canepuccia, A. D., Fanjul, M. S., Fanjul, E., Botto, F. and Iribarne, O. O. (2008). The intertidal burrowing crab *Neohelice* (= *Chasmagnathus*) *granulata* positively affects foraging of rodents in south western Atlantic salt marshes. *Estuaries and Coasts*, **31**, 920–930.

Canepuccia, A. D., Alberti, J., Daleo, P., Farina, J. L. and Iribarne, O. O. (2010a). Ecosystem engineering by burrowing crabs increases cordgrass mortality caused by stem-boring insects. *Marine Ecology Progress Series*, **404**, 151–159.

Canepuccia, A. D., Alberti, J., Pascual, J., et al. (2010b). ENSO episodes modify plant/terrestrial-herbivore interactions in a southwestern Atlantic salt marsh. *Journal of Experimental Marine Biology and Ecology*, **396**, 42–47.

Castilla, J. C. and Duran, L. R. (1985). Human exclusion from the rocky intertidal zone of central Chile: the effects on *Concholepas concholepas* (Gastropoda). *Oikos*, **45**, 391–399.

Castillo, J. M., Fernández-Baco, L., Castellanos, E. M., et al. (2000). Lower limits of *Spartina densiflora* and *S. maritima* in a Mediterranean salt marsh determined by different ecophysiological tolerances. *Journal of Ecology*, **88**, 801–812.

Colman, J. (1933). The nature of the intertidal zonation of plants and animals. *Journal of the Marine Biological Association of the United Kingdom*, **18**, 435–476.

Connell, J. H. (1961). The influence of interspecific competition and other factors on the distribution of the barnacle *Chthamalus stellatus*. *Ecology*, **42**, 710–723.

Connolly, S. R. and Roughgarden, J. (1998). A latitudinal gradient in northeast Pacific intertidal community structure: evidence for an oceanographically based synthesis of marine community theory. *The American Naturalist*, **151**, 311–326.

Connolly, S. R., Menge, B. A. and Roughgarden, J. (2001). A latitudinal gradient in recruitment of intertidal invertebrates in the northeast Pacific Ocean. *Ecology*, **82**, 1799–1813.

Crain, C. M., Silliman, B. R., Bertness, S. L. and Bertness, M. D. (2004). Physical and biotic drivers of plant distribution across estuarine salinity gradients. *Ecology*, **85**, 2539–2549.

Cubit, J. D. (1984). Herbivory and the seasonal abundance of algae on a high intertidal rocky shore. *Ecology*, **65**, 1904–1917.

Daleo, P. and Iribarne, O. (2009). Beyond competition: the stress-gradient hypothesis tested in plant–herbivore interactions. *Ecology*, **90**, 2368–2374.

Daleo, P., Fanjul, E., Méndez Casariego, A., et al. (2007). Ecosystem engineers activate mycorrhizal mutualism in salt marshes. *Ecology Letters*, **10**, 902–908.

Daleo, P., Alberti, J. and Iribarne, O. (2011). Crab herbivory regulates re-colonization of disturbed patches in a southwestern Atlantic salt marsh. *Oikos*, **120**, 842–847.

Daleo, P., Alberti, J., Pascual, J., Canepuccia, A. and Iribarne, O. (2014). Asexual reproduction, herbivory and disturbance recovery of SW Atlantic salt marsh plant communities. *Oecologia*, **175**, 335–343.

Davy, A. J., Bakker, J. P. and Figueroa, M. E. (2009). Human modification of European salt marshes. In *Human Impacts on Salt Marshes: A Global Perspective*, ed. B. R. Silliman, M. D. Bertness and D. Strong. California: University of California Press.

Davy, A., Brown, M. J. H., Mossman, H. L. and Grant, A. (2011). Colonization of a newly developing salt marsh: disentangling independent effects of elevation and redox potential on halophytes. *Journal of Ecology*, **99**, 1350–1357.

Dayton, P. K. (1975). Experimental evaluation of ecological dominance in a rocky intertidal algal community. *Ecological Monographs*, **45**, 137–159.

De Leeuw, J., De Munck, W., Olff, H. and Bakker, J. P. (1993). Does zonation reflect the succession of salt marsh vegetation? A comparison of an estuarine and a coastal bar island marsh in the Netherlands. *Acta Botanica Neerlandica*, **42**, 435–445.

Deegan, L. A., Johnson, D. S., Warren, R. S., et al. (2012). Coastal eutrophication as a driver of salt marsh loss. *Nature*, **490**, 388–392.

Denny, M. W. and Paine, R. T. (1998). Celestial mechanics, sea-level changes, and intertidal ecology. *The Biological Bulletin*, **194**, 108–115.

Dethier, M. N. and Duggins, D. O. (1984). An "indirect commensalism" between marine herbivores and the importance of competitive hierarchies. *The American Naturalist*, **124**, 205–219.

Díaz-Tapia, P., Bárbara, I. and Díez, I. (2013). Multi-scale spatial variability in intertidal benthic assemblages: differences between sand-free and sand-covered rocky habitats. *Estuarine, Coastal and Shelf Science*, **133**, 97–108.

Dugan, J. E., Hubbard, D. M., McCrary, M. D. and Pierson, M. O. (2003). The response of macrofauna communities and shorebirds to macrophyte wrack subsidies on exposed sandy beaches of southern California. *Estuarine, Coastal and Shelf Science*, **58**, 25–40.

Dugdale, R. C., Wilkerson, F. P. and Morel, A. (1990). Realization of new production in coastal upwelling areas: a means to compare relative performance. *Limnology and Oceanography*, **35**, 822–829.

Duggins, D. O. and Dethier, M. N. (1985). Experimental studies of herbivory and algal competition in a low intertidal habitat. *Oecologia*, **67**, 183–191.

Duran, L. R. and Castilla, J. C. (1989). Variation and persistence of the middle rocky intertidal community of central Chile, with and without human harvesting. *MarineBiology*, **103**, 555–562.

Ellis, J. C., Shulman, M. J., Wood, M., Witman, J. D. and Lozyniak, S. (2007). Regulation of intertidal food webs by avian predators on New England rocky shores. *Ecology*, **88**, 853–863.

Elton, C. S. (1927). *Animal Ecology*. Chicago, IL: University of Chicago Press.

Emery, N. C., Ewanchuk, P. J. and Bertness, M. D. (2001). Competition and salt-marsh plant zonation: stress tolerators may be dominant competitors. *Ecology*, **82**, 2471–2485.

Esselink, P., Helder, G. J. F., Aerts, B. A. and Gerdes, K. (1997). The impact of grubbing greylag geese *Anser anser* on vegetation dynamics of a tidal marsh. *Aquatic Botany*, **55**, 261–279.

Fa, D. A. (2008). Effects of tidal amplitude on intertidal resource availability and dispersal pressure in prehistoric human coastal populations: the Mediterranean–Atlantic transition. *Quaternary Science Reviews*, **27**, 2194–2209.

Fishlyn, D. A. and Phillips, D. W. (1980). Chemical camouflaging and behavioral defenses against a predatory seastar by three species of gastropods from the surfgrass *Phyllospadix* community. *The Biological Bulletin*, **158**, 34–48.

Freidenburg, T. L., Menge, B. A., Halpin, P., Webster, M. A. and Sutton-Grier, A. (2007). Cross-scale variation in top-down and bottom-up control of algal abundance. *Journal of Experimental Marine Biology and Ecology*, **347**, 8–29.

Fretwell, S. D. (1987). Food chain dynamics: the central theory of ecology? *Oikos*, **50**, 291–301.

Gaines, S. and Roughgarden, J. (1985). Larval settlement rate: a leading determinant of structure in an ecological community of the marine intertidal zone. *Proceedings of the National Academy of Sciences of the USA*, **82**, 3707–3711.

Gaines, S. D., White, C., Carr, M. H. and Palumbi, S. R. (2010). Designing marine reserve networks for both conservation and fisheries management. *Proceedings of the National Academy of Sciences of the USA*, **107**, 18286–18293.

Galst, C. A. and Anderson, T. W. (2008). Fish–habitat associations and the role of disturbance in surfgrass beds. *Marine Ecology Progress Series*, **365**, 177–186.

Guerry, A. D., Menge, B. A. and Dunmore, R. A. (2009). Effects of consumers and enrichment on abundance and diversity of benthic algae in a rocky intertidal community. *Journal of Experimental Marine Biology and Ecology*, **369**, 155–164.

Hacker, S. D. and Bertness, M. D. (1995a). A herbivore paradox: why salt marsh aphids live on poor-quality plants. *American Naturalist*, **145**, 192–210.

Hacker, S. D. and Bertness, M. D. (1995b). Morphological and physiological consequences of a positive plant interaction. *Ecology*, **76**, 2165–2175.

Hacker, S. D. and Bertness, M. D. (1996). Trophic consequences of a positive plant interaction. *The American Naturalist*, **148**, 559–575.

Hacker, S. D. and Gaines, S. D. (1997). Some implications of direct positive interactions for community species diversity. *Ecology*, **78**, 1990–2003.

Hairston, N. G., Smith, F. E. and Slobodkin, L. B. (1960). Community structure, population control, and competition. *The American Naturalist*, **100**, 421–425.

Hall, S. J. G. (2008). A comparative analysis of the habitat of the extinct aurochs and other prehistoric mammals in Britain. *Ecography*, **31**, 187–190.

Hanski, I. and Gilpin, M. (1991). *Metapopulation Dynamics: Empirical and Theoretical Investigations*. San Diego, CA: Academic Press.

Hockey, P. A. R. (1994). Man as a component of the littoral predator spectrum: a conceptual overview. In *Rocky Shores: Exploitation in Chile and South Africa*, ed. W. R. Siegfried. Berlin: Springer, pp. 17–31.

Holbrook, S. J., Reed, D. C., Hansen, K. and Blanchette, C. A. (2000). Spatial and temporal patterns of predation on seeds of the surfgrass *Phyllospadix torreyi*. *Marine Biology*, **136**, 739–747.

Hunter, M. D. and Price, P. W. (1992). Playing chutes and ladders: heterogeneity and the relative roles of bottom-up and top-down forces in natural communities. *Ecology*, **73**, 723–732.

Hurd, C. L. (2000). Water motion, marine macroalgal physiology, and production. *Journal of Phycology*, **36**, 453–472.

Isacch, J. P., Costa, C. S. B., Rodríguez-Gallego, L., et al. (2006). Distribution of salt marsh plant communities associated with environmental factors along a latitudinal gradient on the South-West Atlantic coast. *Journal of Biogeography*, **33**, 888–900.

Kavanaugh, M. T., Nielsen, K. J., Chan, F. T., et al. (2009). Experimental assessment of the effects of shade on an intertidal kelp: do phytoplankton blooms inhibit growth of open-coast macroalgae? *Limnology and Oceanography*, **54**, 276–288.

Kiehl, K., Esselink, P. and Bakker, J. P. (1997). Nutrient limitation and plant species composition in temperate salt marshes. *Oecologia*, **111**, 325–330.

Kinlan, B. P. and Gaines, S. D. (2003). Propagule dispersal in marine and terrestrial environments: a community perspective. *Ecology*, **84**, 2007–2020.

Kolb, G. S., Ekholm, J. and Hambäck, P. A. (2010). Effects of seabird nesting colonies on algae and aquatic invertebrates in coastal waters. *Marine Ecology Progress Series*, **417**, 287–300.

Kraufvelin, P., Salovius, S., Christie, H., et al. (2006). Eutrophication-induced changes in benthic algae affect the behaviour and fitness of the marine amphipod *Gammarus locusta*. *Aquatic Botany*, **84**, 199–209.

Kuijper, D. P. J. and Bakker, J. P. (2005). Top-down control of small herbivores on salt-marsh vegetation along a productivity gradient. *Ecology*, **86**, 914–923.

Kuijper, D. P. J. and Bakker, J. P. (2008). Unpreferred plants affect patch choice and spatial distribution of brown hares. *Acta Oecologica*, **4**, 339–344.

Kuijper, D. P. J. and Bakker, J. P. (2012). Vertebrate below- and above-ground herbivory and abiotic factors alternate in shaping salt-marsh plant communities.

Journal of Experimental Marine Biology and Ecology, **432–433**, 17–28.

Kuijper, D. P. J., Nijhoff, D. J. and Bakker, J. P. (2004). Herbivory and competition slow down invasion of a tall grass along a productivity gradient. *Oecologia*, **141**, 452–459.

Kuijper, D. P. J., Beek, P., Van Wieren, S. E. and Bakker, J. P. (2008). Time-scale effects in the interaction between a large and a small herbivore. *Basic and Applied Ecology*, **9**, 126–134.

Leendertse, P. C., Roozen, A. J. M. and Rozema, J. (1997). Long-term changes (1953–1990) in the salt marsh vegetation at the Boschplaat on Terschelling in relation to sedimentation and flooding. *Plant Ecology*, **132**, 49–58.

Leroux, S. J. and Loreau, M. (2008).Subsidy hypothesis and strength of trophic cascades across ecosystems. *Ecology Letters*, **11**, 1147–1156.

Lester, S. E., Gaines, S. D. and Kinlan, B. P. (2007). Reproduction on the edge: large-scale patterns of individual performance in a marine invertebrate. *Ecology*, **88**, 2229–2239.

Levine J. M., Hacker, S. D., Harley, C. D. G. and Bertness, M. D. (1998). Nitrogen effects on an interaction chain in a salt marsh community. *Oecologia*, **117**, 266–272.

Lewis, J. R. (1964). *The Ecology of Rocky Shores*. London, UK: English University Press.

Lindeman, R. L. (1942). The trophic-dynamic aspect of ecology. *Ecology*, **23**, 399–417.

Linthurst, R. A. and Seneca, E. D. (1981). Aeration, nitrogen and salinity as determinants of *Spartina alterniflora* Loisel growth response. *Estuaries*, **4**, 53–63.

Loreau, M., Mouquet, N. and Holt, R. D. (2003). Meta-ecosystems: a theoretical framework for a spatial ecosystem ecology. *Ecology Letters*, **6**, 673–679.

Lotka, A. J. (1925). *Elements of Physical Biology*. Baltimore: Williams and Wilkins.

Lubchenco, J. (1978). Plant species diversity in a marine intertidal community: importance of herbivore food preference and algal competitive abilities. *The American Naturalist*, **112**, 23–39.

Marsh, C. P. (1986). Rocky intertidal community organization: the impact of avian predators on mussel recruitment. *Ecology*, **67**, 771–786.

McLeod, K. and Leslie, H. (eds.). (2009). *Ecosystem-Based Management for the Oceans*. Washington: Island Press, pp. 3–6.

Menge, B. A. (1976). Organization of the New England rocky intertidal community: role of predation, competition, and environmental heterogeneity. *Ecological Monographs*, **46**, 355–393.

Menge, B. A. (1991). Relative importance of recruitment and other causes of variation in rocky intertidal community structure. *Journal of Experimental Marine Biology and Ecology*, **146**, 69–100.

Menge, B. A. (1992). Community regulation: under what conditions are bottom-up factors important on rocky shores? *Ecology*, **73**, 755–765.

Menge, B. A. (1995). Indirect effects in marine rocky intertidal interaction webs: patterns and importance. *Ecological Monographs*, **65**, 21–74.

Menge, B. A. (2000). Top-down and bottom-up community regulation in marine rocky intertidal habitats. *Journal of Experimental Marine Biology and Ecology*, **250**, 257–289.

Menge, B. A. and Lubchenco, J. (1981). Community organization in temperate and tropical rocky intertidal habitats: prey refuges in relation to consumer pressure gradients. *Ecological Monographs*, **51**, 429–450.

Menge, B. A. and Menge D. N. L. (2013). Dynamics of coastal meta-ecosystems: the intermittent upwelling hypothesis and a test in rocky intertidal regions. *Ecological Monographs*, **83**, 283–310.

Menge, B. A. and Olson, A. M. (1990).Role of scale and environmental factors in regulation of community structure. *Trends in Ecology and Evolution*, **5**, 52–57.

Menge, B. A. and Sutherland, J. P. (1976). Species diversity gradients: synthesis of the roles of predation, competition, and temporal heterogeneity. *The American Naturalist*, **110**, 351–369.

Menge, B. A. and Sutherland, J. P. (1987). Community regulation: variation in disturbance, competition, and predation in relation to environmental stress and recruitment. *The American Naturalist*, **130**, 730–757.

Menge, B. A., Berlow, E. L., Blanchette, C. A., Navarrete, S. A. and Yamada, S. B. (1994). The keystone species concept: variation in interaction strength in a rocky intertidal habitat. *Ecological Monographs*, **64**, 249–286.

Menge, B. A., Daley, B. A., Wheeler, P. A., et al. (1997). Benthic–pelagic links and rocky intertidal communities: bottom-up effects on top-down control? *Proceedings of the National Academy of Sciences of the USA*, **94**, 14530–14535.

Menge, B. A., Lubchenco, J., Bracken, M. E. S., et al. (2003). Coastal oceanography sets the pace of rocky intertidal community dynamics. *Proceedings of the National Academy of Sciences of the USA*, **100**, 12229–12234.

Menge, B. A., Gouhier, T. C., Hacker, S. D., et al. (in press). Are meta-ecosystems organized hierarchically? A model and test in rocky intertidal habitats. *Ecological Monographs*.

Milligan, K. L. D. (1998). Effects of wave-exposure on an intertidal kelp species *Hedophyllum sessile* (C. Agardh) Setchell: demographics and biomechanics. PhD dissertation, University of British Columbia, Vancouver.

Moreno, C. A. (2001). Community patterns generated by human harvesting on Chilean shores: a review. *Aquatic Conservation: Marine and Freshwater Ecosystems*, **11**, 19–30.

Morgan, S. G. and Fisher, J. L. (2010). Larval behavior regulates nearshore retention and offshore migration in an upwelling shadow and along the open coast. *Marine Ecology Progress Series*, **404**, 109–126.

Moulton, O. M. and Hacker, S. D. (2011). Congeneric variation in surfgrasses

and ocean conditions influence macroinvertebrate community structure. *Marine Ecology Progress Series*, **433**, 53–63.

Navarrete, S. A. and Castilla, J. C. (2003). Experimental determination of predation intensity in an intertidal predator guild: dominant versus subordinate prey. *Oikos*, **100**, 251–262.

Navarrete, S. A. and Menge, B. A. (1996). Keystone predation and interaction strength: interactive effects of predators on their main prey. *Ecological Monographs*, **66**, 409–429.

Nielsen, K. J. (2001). Bottom-up and top-down forces in tide pools: test of a food chain model in an intertidal community. *Ecological Monographs*, **71**, 187–217.

Nielsen, K. J. (2003). Nutrient loading and consumers: agents of change in open-coast macrophyte assemblages. *Proceedings of the National Academy of Sciences of the USA*, **100**, 7660–7665.

Nielsen, K. J. and Navarrete, S. A. (2004). Mesoscale regulation comes from the bottom-up: intertidal interactions between consumers and upwelling. *Ecology Letters*, **7**, 31–41.

Odum, W. E. (1988). Comparative ecology of tidal freshwater and salt marshes. *Annual Review of Ecology and Systematics*, **19**, 147–176.

Ojeda, F. P. and Muñoz, A. A. (1999). Feeding selectivity of the herbivorous fish *Scartichthys viridis*: effects on macroalgal community structure in a temperate rocky intertidal coastal zone. *Marine Ecology Progress Series*, **184**, 219–229.

Oksanen, L. and Oksanen, T. (2000). The logic and realism of the hypothesis of exploitation ecosystems. *The American Naturalist*, **155**, 703–723.

Oksanen, L., Fretwell, S. D., Arruda, J. and Niemela, P. (1981). Exploitation ecosystems in gradients of primary productivity. *The American Naturalist*, **118**, 240–261.

Olff, H., De Leeuw, J., Bakker, J. P., et al. (1997). Vegetation succession and herbivory on a salt marsh: changes induced by sea level rise and silt deposition along an elevational gradient. *Journal of Ecology*, **85**, 799–814.

Paine, R. T. (1966). Food web complexity and species diversity. *The American Naturalist*, **100**, 65–75.

Paine, R. T. (1974). Intertidal community structure: experimental studies on the relationship between a dominant competitor and its principal predator. *Oecologia*, **15**, 93–120.

Paine, R. T. (1979). Disaster, catastrophe, and local persistence of the sea palm *Postelsia palmaeformis*. *Science*, **205**, 685–687.

Paine, R. T. (1992). Food-web analysis through field measurement of per capita interaction strength. *Nature*, **355**, 73–75.

Paine, R. T. and Levin, S. A. (1981). Intertidal landscapes: disturbance and the dynamics of pattern. *Ecological Monographs*, **51**, 145–178.

Paine, R. T. and Palmer, A. R. (1978). *Sicyases sanguineus*: a unique trophic generalist from the Chilean intertidal zone. *Copeia*, **1978**, 75–81.

Paine, R. T. and Vadas, R. L. (1969). The effects of grazing by sea urchins, *Strongylocentrotus* spp., on benthic algal populations. *Limnology and Oceanography*, **14**, 710–719.

Pather, S., Pfister, C. A., Post, D. M. and Altabet, M. A. (2014). Ammonium cycling in the rocky intertidal: remineralization, removal, and retention. *Limnography and Oceanography* **59**, 361–372.

Pennings, S. C. and Bertness, M. D. (2001). Salt marsh communities. In *Marine Community Ecology*, ed. M. D. Bertness, S. D. Gaines and M. Hay. Sunderland, MA: Sinauer Associates, pp. 289–316.

Pennings, S. C., Grant, M. B. and Bertness, M. D. (2005). Plant zonation in low-latitude salt marshes: disentangling the roles of flooding, salinity and competition. *Journal of Ecology*, **93**, 159–167.

Polis, G. A., Anderson, W. B. and Holt, R. D. (1997). Toward an integration of landscape and food web ecology: the dynamics of spatially subsidized food webs. *Annual Review of Ecology and Systematics*, **28**, 289–316.

Poore, A. G., Campbell, A. H., Coleman, R. A., et al. (2012). Global patterns in the impact of marine herbivores on benthic primary producers. *Ecology Letters*, **15**, 912–922.

Raimondi, P. T., Forde, S. E., Delph, L. F. and Lively, C. M. (2000). Processes structuring communities: evidence for trait-mediated indirect effects through induced polymorphisms. *Oikos*, **91**, 353–361.

Robles, C. and Desharnais, R. (2002). History and current development of a paradigm of predation in rocky intertidal communities. *Ecology*, **83**, 1521–1536.

Robles, C., Sherwood-Stephens, R. and Alvarado, M. (1995). Responses of a key intertidal predator to varying recruitment of its prey. *Ecology*, **76**, 565–579.

Roy, K., Collins, A. G., Becker, B. J., Begovic, E. and Engle, J. M. (2003). Anthropogenic impacts and historical decline in body size of rocky intertidal gastropods in southern California. *Ecology Letters*, **6**, 205–211.

Salomon, A. K., Tanape Sr, N. M. and Huntington, H. P. (2007). Serial depletion of marine invertebrates leads to the decline of a strongly interacting grazer. *Ecological Applications*, **17**, 1752–1770.

Sanford, E. and Menge, B. A. (2007). Reproductive output and consistency of source populations in the sea star *Pisaster ochraceus*. *Marine Ecology Progress Series*, **349**, 1–12.

Schrama, M., Berg, M. P. and Olff, H. (2012). Ecosystem assembly rules: the interplay of green and brown webs during salt marsh succession. *Ecology*, **93**, 2353–2364.

Schrama, M., Jouta, J., Berg, M. P. and Olff, H. (2013a). Food web assembly at the landscape scale: using stable isotopes to reveal changes in trophic structure during succession. *Ecosystems*, **16**, 627–638.

Schrama, M. J. J., Heijing, P., Van Wijnen, H. J., et al. (2013b). Herbivore trampling as an alternative pathway for explaining differences in nitrogen mineralization in moist grasslands. *Oecologia*, **172**, 231–243.

Scrosati, R.and Heaven, C. (2007). Spatial trends in community richness, diversity, and evenness across rocky intertidal environmental stress gradients in eastern Canada. *Marine Ecology Progress Series*, **342**, 1–14.

Shanks, A. L. (2009). Pelagic larval duration and dispersal distance revisited. *The Biological Bulletin*, **216**, 373–385.

Short, F., Carruthers, T., Dennison, W. and Waycott, M. (2007). Global seagrass distribution and diversity: a bioregional model. *Journal of Experimental Marine Biology and Ecology*, **350**, 3–20.

Silliman, B. R. and Bertness, M. D. (2002). A trophic cascade regulates salt marsh primary production. *Proceedings of the National Academy of Sciences of the USA*, **99**, 10500–10505.

Silliman, B. R. and Bertness, M. D. (2004). Shoreline development drives invasion of *Phragmites australis* and the loss of plant diversity on New England salt marshes. *Conservation Biology*, **18**, 1424–1434.

Silliman, B. R. and Zieman, J. C. (2001). Top-down control of *Spartina alterniflora* production by periwinkle grazing in a Virginia salt marsh. *Ecology*, **82**, 2830–2845.

Silliman, B. R., Layman, C. A., Geyer, K. and Zieman, J. C. (2004). Predation by the black-clawed mud crab, *Panopeus herbstii*, in Mid-Atlantic salt marshes: further evidence for top-down control of marsh grass production. *Estuaries*, **27**, 188–196.

Silliman, B. R., Van De Koppel, J., Bertness, M. D., Stanton, L. E. and Mendelssohn, I. A. (2005). Drought, snails and large-scale die-off of Southern U.S. salt marshes. *Science*, **310**, 1803–1806.

Silliman, B. R., McCoy, M. W., Angelini, C., et al. (2013). Consumer fronts, global change, and runaway collapse in ecosystems. *Annual Review of Ecology, Evolution, and Systematics*, **44**, 503–538.

Stahl, J., Bos, D. and Loonen, M. J. J. E. (2002). Foraging along a salinity gradient – the effect of tidal inundation on site choice by Brent and barnacle geese. *Ardea*, **90**, 201–212

Stephenson, T. A. and Stephenson, A. (1949). The universal features of zonation between tide-marks on rocky coasts. *The Journal of Ecology*, **37**, 289–305.

Suchrow, S., Pohlman, M., Stock, M. and Jensen, K. (2012). Long-term surface elevation change in German North Sea salt marshes. *Estuarine, Coastal and Shelf Science*, **98**, 75–83.

Taylor, D. I. and Schiel, D. R. (2010). Algal populations controlled by fish herbivory across a wave exposure gradient on southern temperate shores. *Ecology*, **91**, 201–211.

Taylor, K. L. and Grace, J. B. (1995). The effects of vertebrate herbivory on plant community structure in the coastal marshes of the Pearl River, Louisiana, USA. *Wetlands*, **15**, 68–73.

Terrados, J. and Williams, S. L. (1997). Leaf versus root nitrogen uptake by the surfgrass *Phyllospadix torreyi*. *Marine Ecology Progress Series*, **149**, 267–277

Thompson, S. A., Knoll, H., Blanchette, C. A. and Nielsen, K. J. (2010). Population consequences of biomass loss due to commercial collection of the wild seaweed *Postelsia palmaeformis*. *Marine Ecology Progress Series*, **413**, 17–31.

Trussell, G. C., Ewanchuk, P. J. and Bertness, M. D. (2002). Field evidence of trait-mediated indirect interactions in a rocky intertidal food web. *Ecology Letters*, **5**, 241–245.

Trussell, G. C., Ewanchuk, P. J., Bertness, M. D. and Silliman, B. R. (2004). Trophic cascades in rocky shore tide pools: distinguishing lethal and nonlethal effects. *Oecologia*, **139**, 427–432.

Turner, R. E., Swenson, E. M. and Milan, C. S. (2002). Organic and inorganic contributions to vertical accretion in salt marsh sediments. In *Concepts and Controversies in Tidal Marsh Ecology*, ed. M. P. Weinstein and

D. A. Kreeger. New York: Kluwer Academic Publishers, pp. 583–594

Turner, T. (1983). Facilitation as a successional mechanism in a rocky intertidal community. *The American Naturalist*, **121**, 729–738.

Valiela, I., Teal, J. M. and Deuser, W. G. (1978). The nature of growth forms in the salt marsh grass *Spartina alterniflora*. *The American Naturalist*, **112**, 461–470.

Valiela, I., Cole, M. L., McClelland, J., et al. (2002). Role of salt marshes as part of coastal landscapes. In *Concepts and Controversies in Tidal Marsh Ecology*, ed. M. P. Weinstein and D. A. Kreeger. New York: Kluwer Academic Publishers, pp. 23–36.

Van De Koppel, J., Huisman, J., Van Der Wal, R. and Olff, H. (1996). Patterns of herbivory along a productivity gradient: an empirical and theoretical investigation. *Ecology*, **77**, 736–745.

Van Der Graaf, A. J., Coehoorn, P. and Stahl, J. (2006). Sward height and bite size affect the functional response of *Branta leucopsis*. *Journal of Ornithology*, **147**, 479–484.

Van Der Wal, R., Van De Koppel, J. and Sagel, M. (1998). On the relation between herbivore foraging efficiency and plant standing crop: an experiment with barnacle geese. *Oikos*, **82**, 123–130.

Van Der Wal, R., Van Lieshout, S., Bos, D. and Drent, R. H. (2000a). Are spring staging Brent geese evicted by vegetation succession? *Ecography*, **23**, 60–69.

Van Der Wal, R., Van Wieren, S. E., Van Wijnen, H. J., Beucher, O. and Bos, D. (2000b). On facilitation between herbivores: how Brent geese profit from brown hares. *Ecology*, **81**, 969–980.

Van Wesenbeeck, B. K., Van De Koppel, J., Herman, P. M. J., Bakker, J. P. and Bouma, T. J. (2007). Biomechanical warfare in ecology: negative interactions between species by habitat modification. *Oikos*, **116**, 742–750.

Van Wijnen, H. J. and Bakker, J. P. (1999). Nitrogen and phosphorus limitation in a coastal barrier salt marsh: the implications for vegetation succession. *Journal of Ecology*, **87**, 265–272.

Vandermeer, J. H. (1972). Niche theory. *Annual Review of Ecology and Systematics*, **3**, 107–132.

Veeneklaas, R. M., Dijkema, K. S., Hecker, N. and Bakker, J. P. (2013). Spatio-temporal dynamics of the invasive salt-marsh plant species *Elytrigia atherica* on natural salt marshes. *Applied Vegetation Science*, **16**, 205–216.

Vince, S. W., Valiela, I. and Teal, J. M. (1981). An experimental study of the structure of herbivorous insect communities in a salt marsh. *Ecology*, **62**, 1662–1678.

Vinueza, L., Post, A., Guarderas, P., Smith, F. and Idrovo, F. (2014). Ecosystem-based management for rocky shores of the Galapagos Islands. In *The Galapagos Marine Reserve: A Dynamic Social-Ecological System*, ed. J. Denkinger and L. Vinueza. New York, NY: Springer International Publishing, pp. 81–107.

Volterra, V. (1926). Fluctuations in the abundance of a species considered mathematically. *Nature*, **118**, 558–560.

Wethey, D. S. (1985). Catastrophe, extinction, and species diversity: a rocky intertidal example. *Ecology*, **66**, 445–456.

Wieters, E. A., Gaines, S. D., Navarrete, S. A., Blanchette, C. A. and Menge, B. A. (2008). Scales of dispersal and the biogeography of marine predator–prey interactions. *The American Naturalist*, **171**, 405–417.

Wolters, M., Bakker, J. P., Bertness, M., Jefferies, R. L. and Möller, I. (2005). Salt marsh erosion and restoration in south-east England: squeezing the evidence requires realignment. *Journal of Applied Ecology*, **42**, 844–851.

Wood, M. (2008). *Reproductive output of a keystone predator and its preferred prey: the differential influence of oceanographic regime and local habitat*. MS thesis, Sonoma State University, Rohnert Park, CA.

Wootton, J. T. (1991). Direct and indirect effects of nutrients on intertidal community structure: variable consequences of seabird guano. *Journal of Experimental Marine Biology and Ecology*, **151**, 139–153.

Wootton, J. T. (1993). Indirect effects and habitat use in an intertidal community: interaction chains and interaction modifications. *The American Naturalist*, **141**, 71–89.

Wootton, J. T. (1997). Estimates and tests of per capita interaction strength: diet, abundance, and impact of intertidally foraging birds. *Ecological Monographs*, **67**, 45–64.

Wootton, J. T., Power, M. E., Paine, R. T. and Pfister, C. A. (1996). Effects of productivity, consumers, competitors and El Niño events on food chain patterns in a rocky intertidal community. *Proceedings of the National Academy of Sciences of the USA*, **93**, 13855–13858.

Worm, B., Lotze, H. and Sommer, U. (2000). Coastal food web structure, carbon storage, and nitrogen retention regulated by consumer pressure and nutrient loading. *Limnology and Oceanography*, **45**, 339–349.

PART III

Patterns and Processes

Influence of plant defenses and nutrients on trophic control of ecosystems

KARIN T. BURGHARDT and OSWALD J. SCHMITZ

Yale University, New Haven, CT, USA

Introduction

Ecological systems are extraordinarily complex. Thus classical approaches to resolve ecosystem functioning have simplified analyses by conceptualizing ecosystems as being organized into trophic level compartments that contain organisms with similar feeding dependencies (e.g., producers, herbivores, carnivores) (Elton, 1927; Lindeman, 1942). Two competing worldviews on the regulation of ecosystem productivity emanated from such a conceptualization of ecosystem structure. The bottom-up view posits that the productivity of each trophic level is essentially limited by the one immediately below it (Lindeman, 1942; Feeny, 1968), while the top-down view recognizes that resource levels influence production, but contends that herbivore populations are mostly limited by predators rather than producer biomass (Hairston et al., 1960). Accordingly, predators can indirectly increase the productivity of a given system by reducing the negative effects of herbivores on plant biomass, resulting in a world that is green with plant material, rather than denuded by herbivory (Paine, 1969; Oksanen et al., 1981). Bottom-up theory countered that the world is green not because of predators, but instead due to variation in plant quality as a result of anti-herbivore defenses or weather patterns (Murdoch, 1966; Ehrlich and Birch, 1967; Scriber and Feeny, 1975; White, 1978; Feeny, 1991; Polis and Strong, 1996). This variation causes much of the "green" world to be inedible to herbivores; thus herbivores are still resource-limited.

The recognition of context-dependence in the degree of top-down or bottom-up control of ecosystems has resulted in gradual changes in how ecosystem functioning is envisioned. For instance, the "exploitation ecosystems" hypothesis (EEH) addresses context-dependence by combining elements of top-down and bottom-up concepts (Oksanen et al., 1981; Oksanen and Oksanen, 2000). At low levels of soil resource availability, plants are not productive enough to support herbivore populations and are thus bottom-up controlled (see Fig. 5.3).

Trophic Ecology: Bottom-Up and Top-Down Interactions across Aquatic and Terrestrial Systems, eds T. C. Hanley and K. J. La Pierre. Published by Cambridge University Press. © Cambridge University Press 2015.

At medium levels of soil resources, an ecosystem can support herbivore populations, which in turn control plant productivity, while carnivores enter the ecosystem and control the herbivore population at the highest resource availability, thus releasing plant productivity from herbivore control. As a result, there is now a general consensus that both top-down and bottom-up control can occur within the same ecosystem, but that their relative magnitude is context specific (Hunter and Price, 1992; Power, 1992; Chase et al., 2000b). Understanding of the basis for this context-dependence in strength remains incomplete: while explanations for cross-ecosystem differences have been offered (Shurin et al., 2002), explanations for spatial differences within ecosystems remain elusive. This chapter aims to begin resolving the basis for within ecosystem context-dependency in the strength of trophic control by focusing on one of the important mediating factors identified in early debates about top-down and bottom-up forcing within ecosystems: the expression of plant defensive traits. This focus is a natural extension of classic theory because the expression of plant defensive traits is also intimately tied to resource availability. We review here the interplay between resources, plant defenses, and top-down and bottom-up control strength in an effort to offer generalizable principles that extend to explain differences across terrestrial and aquatic ecosystems.

Strong (1992) suggested that aquatic and terrestrial ecosystems are controlled in fundamentally different ways, with top-down control more prevalent in aquatic ecosystems, due in part to differences in primary productivity. However, a recent meta-analysis of experimental evidence concludes that net primary productivity does not differ between aquatic and terrestrial habitats, and instead producer nutritional quality is a consistently better indicator of the importance of consumers for top-down control (Cebrian and Lartigue, 2004). This makes sense in light of the fact that plant defensive strategies directly interact with nutritional quality to determine plant palatability (Raubenheimer, 1992). Subsequent theory (Vos et al., 2004) and experimental work (Verschoor et al., 2004b) have demonstrated that defensive traits that limit the efficacy of consumers to impact plants can be an important determinant of the relative strength of top-down and bottom-up effects, ultimately mediating the presence of trophic cascades in ecosystems. That is, this integrative view of trophic control of ecosystems is beginning to be one of "control from the middle out" (*sensu* Trussell and Schmitz, 2012), rather than from the top-down or bottom-up.

We introduce and elaborate on why defensive traits may play a key role in moderating trophic control of ecosystems from the middle out. We begin by clarifying the terminology used throughout the chapter to refer to defensive traits and then introduce a trait-based framework for thinking about how plant defenses may impact trophic control. Next, we highlight the dominant defensive traits found within aquatic versus terrestrial systems and review how nutrient availability may impact the strength of individual plant defenses within a species through phenotypic plasticity (Cipollini et al., 2003) or through average community trait defense levels via filtering of species that perform well in particular nutrient environments (Uriarte, 2000) across terrestrial, freshwater, and marine ecosystems. We propose here that a trait-based approach offers greater opportunity for understanding context-dependency in the way defenses mediate trophic control than approaches that focus merely at the species level or lump all species into trophic groups. We then end with an exploration of the link between the expressed plant defense traits and ensuing food web interactions and ecosystem functioning.

Primary producer anti-herbivore defenses

Most plants lack the capability to actively move away from potential herbivores. Vascular plants in terrestrial or littoral systems are rooted in place and floating phytoplankton species in marine and pelagic systems lack directional escape from their consumers. However, none of these organisms are passive in their interactions with consumers. Thousands of plant species reduce herbivory by producing an arsenal of anti-herbivore defenses (Karban and Baldwin, 1997). These include structural defenses, such as thorns, spines, or tough tissues that are difficult to chew, as well as chemical defenses, such as toxic compounds. Chemical defenses can be qualitative, where the mode of action is to poison a herbivore, or quantitative, such as leaf toughness or digestion inhibitors that force a herbivore to consume a larger quantity of food in order to extract the same nutrients, thereby prolonging their exposure to potential predation or parasitism (Feeny, 1976). Defensive traits that decrease plant damage from herbivores or lower herbivore performance are collectively known as *resistance* traits.

A second general defensive strategy, known as *tolerance*, minimizes the negative impact of herbivory by enabling a plant to regrow quickly and thus regain lost photosynthetic capacity (Strauss and Agrawal, 1999). This strategy may include an increase in growth rate, utilization of stored reserves, activation of dormant meristems, or a decrease in allocation to structural tissue, which lowers leaf toughness and leaf mass per area (LMA) (Tiffin, 2000). These traits would seem to increase the palatability of plant tissue, thereby rendering them ineffective as a defense. But, if a plant is able to produce tissue faster than the herbivore can remove it, or if the herbivore completes its life cycle and leaves the plant, then tolerance can overcome herbivore impacts.

In addition, these defenses can be described as being either *constitutive* or *induced*. If the defenses are always produced within a plant regardless of the presence of a herbivore, they are constitutive. Defenses are considered induced if they are expressed after a herbivore begins to inflict damage (Agrawal and Karban, 1998). Inducible tolerance or resistance responses are a form of phenotypic (trait) plasticity that may be adaptive (Agrawal, 2001) and could impact community dynamics through increasing trait variation within populations

(Schmitz et al., 2003). The focus of this chapter will be on direct defenses, such as those described above; however, many plants also utilize indirect defenses, such as the release of plant volatiles that attract parasitoids and predators of the herbivore to the attacked plant (Arimura et al., 2005; Pohnert et al., 2007). Indirect defenses merit independent treatment and are described in more detail in Chapter 13 of this volume. Moreover, the efficacy of a plant defense is inherently tied to the environmental context in which it is expressed. A putative defense may not decrease a herbivore attack when it is expressed within a milieu of plants all expressing defense, but may work quite well if better quality, less defended plants are in the surrounding environment (Belovsky and Schmitz, 1994).

Historically plant defenses have been measured in isolation. However, terrestrial and aquatic plants may respond to herbivores through the simultaneous expression of several commonly co-occurring traits or "plant defense syndromes" (Agrawal and Fishbein, 2006; Ruehl and Trexler, 2013). Structural and chemical defense expression and tissue allocation are individual traits that cumulatively determine the overall tolerance or resistance of a particular plant. As such, we consider tolerance and resistance strategies (albeit not mutually exclusive; Mauricio et al., 1997) to represent two common "plant defense syndromes" with distinct trait expression levels that are nonetheless useful for exploring potentially different effects of plant defense on trophic cascades.

Conceptual framework

Mechanism Switching Hypothesis

The impact of a herbivore on plants will depend on the nature of herbivore resource limitation. Herbivores could be limited by *relative* resource supply if their per capita uptake rate of edible plant biomass is limited by the amount of time available to feed (Schmitz, 2008). In this case, there may be a surfeit of plants that herbivores cannot eat due to daily limitations on feeding imposed by the abiotic environment. Alternatively, herbivores could be limited by *absolute* resource supply if their per capita uptake of plant biomass is limited by the availability of total edible plant biomass (Schmitz, 2008). In this case, herbivores increase their per capita intake rate of edible plant biomass in direct proportion to the abundance of edible plant biomass. The nature of herbivore resource limitation also determines the extent to which predators can indirectly alleviate plant damage via direct interactions with herbivore prey. These ideas are encapsulated in the *Mechanism Switching Hypothesis* (MSH) of trophic control of ecosystems (Schmitz, 2008).

For example, consider a simple system of three trophic levels comprised of plants, herbivores, and predators. In the absence of predators, plant abundance is limited by consumption from herbivores. Predators can reduce herbivore

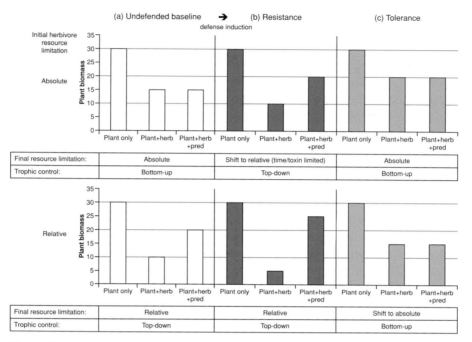

Figure 8.1 A conceptual framework for extending the Mechanism Switching Hypothesis of trophic control (Schmitz, 2008) to include plant defenses and their impact on herbivore resource limitation. Bars represent predicted outcomes of herbivore resource limitation and plant defense on plant biomass. (a) Undefended leaf tissue can be eaten by herbivores experiencing either absolute resource limitation leading to bottom-up control or relative resource limitation (e.g., temperature limitations on feeding time) leading to top-down control of plant biomass. (b) If plants induce a resistance response to herbivory (toxin or structural), the defenses impose relative resource limitation on herbivores because herbivores cannot increase feeding rate when a predator removes a herbivore (time or toxin limited feeding). (c) In contrast, induced tolerance traits impose absolute resource limitation on herbivores due to high-quality regrowth tissue. If a predator removes a herbivore, other herbivores will consume more, preventing a trophic cascade.

abundances, and thereby have an indirect effect on plants through cascading effects that alleviate plant damage – called a trophic cascade (Fig. 8.1a). However, this response will only occur if the herbivores that remain do not compensate and consume a larger per capita share of the plant biomass (i.e., herbivores experience relative resource limitation). If instead, herbivores experience absolute resource limitation, any remaining herbivores are able to increase their per capita uptake of plants, such that predators have no net indirect effect on plant damage. In this conceptualization, the interaction between resource limitation and predators determines whether top-down control emerges.

We suggest that the MSH offers the means to extend the consideration of trophic control of ecosystems to include plant defenses. In essence, plant defenses can determine whether herbivores become relative or absolute resource limited. For example, the presence of a structural "resistance" defensive trait may increase the amount of foraging time a herbivore requires to gain the same nutritional pay-off (Moran and Hamilton, 1980; Raubenheimer, 1992). This strategy also has the advantage of increasing the amount of time that a herbivore is exposed to predation. In addition, when a predator consumes a herbivore, the remaining herbivores on the plant cannot increase their per capita feeding rate because spines and structural defenses inhibit feeding rate. These herbivores are foraging time limited and experience relative resource limitation that leads to a trophic cascade and top-down control (Fig. 8.1b). If the resistance defense is a toxin rather than structural the same qualitative outcome occurs, but the mechanism differs. The herbivores experience toxin-limitation upon feeding. Despite perhaps having ample time to feed, herbivores can nonetheless only process a limited quantity of any toxin-containing tissue per unit time. Therefore, when a predator removes a herbivore from the plant, other herbivores cannot increase their per capita feeding rate, resulting again in a trophic cascade. This case of a herbivore experiencing relative resource limitation created by a toxin, rather than by time, is not a scenario included in the undefended world originally assumed by MSH.

If the herbivores were originally absolute resource limited before plant induction, the presence of resistance causes a switch in the nature of trophic control, relative to undefended plants, leading to a trophic cascade. If the herbivores were originally relative resource limited, then there is no switch in trophic control; however, through inducing a defense (bottom-up effect), a plant is able to exacerbate the positive direct effects of the defense through the help of predators (top-down effect) that prey on herbivores.

In contrast, the induction of tolerance traits (increased growth rate, thinner leaves) may lead to an overall increase in herbivory through absolute resource limitation of herbivores (Fig. 8.1c). If a predator removes a herbivore from a plant with tolerance traits, all other herbivores will increase their per capita feeding rate due to a lack of defended tissue. This will result in bottom-up control of primary production. If herbivores were relative resource limited in the presence of undefended tissue, the induction of tolerance traits would then shift them to absolute resource limitation, removing top-down control.

Because plant defensive traits or herbivore behaviors mediate the strength of trophic control over productivity, trophic control is from the middle out, rather than from the top-down or bottom-up. Moreover, the framework leads to an interesting new insight. While plants with resistance traits certainly derive a direct benefit by reducing herbivore feeding, plants expressing such traits gain a greater indirect benefit from predators through trophic cascades than would similar plant species that did not express such traits. While predators have been invoked before to explain low nutritive defenses that cause more damage to the plant through increased feeding requirements of the herbivore (Moran and Hamilton, 1980), the result here is more general and applies to toxin-based qualitative defenses as well as structural ones. In addition, while the quantity of primary production shifts in response to herbivores and plant defensive syndrome response (resistance versus tolerance), the traits of uneaten plant material are also impacted by these same factors. For example, plant litter in the absence of herbivores will be qualitatively different due to the lack of expressed defensive traits. Accordingly, the MSH can be extended to consider how these shifts in quality have the potential to impact community dynamics through nutrient cycling (see *Nutrient cycling links top-down and bottom-up effects* section below).

Functional trait-based approach

The MSH does not attempt to predict which plants will express which defensive traits in what environment (as do the plant defense or tolerance theories). Instead, given a defensive plant syndrome (resistance or tolerance), it predicts qualitatively whether bottom-up and top-down effects will prevail to impact community processes. Because it does not assume all individuals within a trophic level (or even species) have identical responses and traits, the MSH has the components of a trait-mediated approach for determining what regulates community processes (Schmitz et al., 2003; 2004; Duffy, 2009). This functional trait approach of resistance versus tolerance can be applied within communities, species, or genotypes. We propose this framework as a way to predict when plant defensive traits will impact top-down and bottom-up control in ecosystems. This approach may also be useful for better understanding the basis for the purported contingency in trophic control observed between and within ecosystem types, such as between aquatic and terrestrial ecosystems.

Dominant defense strategies in aquatic and terrestrial systems

Much previous work elucidating the differences between terrestrial and aquatic systems focused on the differences between the dominant primary producers in each system (Strong, 1992; Chase, 2000). Below, we summarize the known defenses of the primary producers within pelagic (open water), terrestrial, and littoral (nearshore) ecosystems to explore whether there are systematic differences among ecosystem types in defense expression. We do not provide an exhaustive treatment here, as recent reviews have already been completed for most systems (Pohnert, 2004; Hanley et al., 2007; Toth and Pavia, 2007; Van Donk et al., 2011).

Pelagic autotrophs

The dominant players in aquatic pelagic systems are unicellular and multicellular phytoplankton that allocate little to structural tissue, resulting in highly edible tissues due to low C:N ratios (Sardans et al., 2012). Phytoplankton must be small enough to remain suspended in the water column, yet can escape predation if they exceed an herbivore's gape limitation (Fogg, 1991). As a result, one common defense strategy is for groups of unicellular phytoplankton to join into colonies called coenobia, at the cost of an increased risk of sinking out of resource-rich surface waters and potential decreases in nutrient uptake due to lower surface area (Lürling and Beekman, 1999; Verschoor et al., 2004a). In contrast to terrestrial systems, phytoplankton are small relative to the zooplankton and other herbivores that eat them; an encounter with a herbivore often means a complete loss of fitness. Thus traditional tolerance strategies are not likely to be effective; instead, some phytoplankton and diatoms exude activated chemical defenses (secondary metabolites) into the water to deter herbivores from attacking or produce morphological structures, such as spines (Leibold, 1989; Leibold, 1999; Van Donk et al., 2011), in the presence of herbivores. Another strategy expressed at low resource availability in green algae is a tough morphology that allows some individuals to pass through the zooplankton digestive system unharmed (Van Donk, 1997).

Often plant defenses are induced, not by direct contact with the herbivore, but by the detection of chemical cues in the water column (kairomones) released by the herbivore (Pohnert et al., 2007). At high resource availability and in the presence of herbivores, some species are also able to induce changes in life history traits to speed up growth rates and generation times to outgrow herbivore species (Agrawal, 1998). While not referred to as such in the literature, we argue that changing life history traits in the presence of herbivores can be thought of as belonging to a "tolerance" defensive strategy, because the effect is that different induced plant traits are expressed within the system. Defense induction is a more ubiquitous response within freshwater pelagic systems than in marine systems (Lass and Spaak, 2003). In marine systems, induction is rare, but a few species of algal phytoplankton produce constitutive chemical resistance traits that can lead to toxic algal blooms and corresponding consumer die-offs (Pohnert, 2004).

Terrestrial autotrophs

Terrestrial plants tend to be vascular, relatively long-lived, and allocate more resources to plant structure than most aquatic plants. Overall, plant tissue quality is lower than in aquatic systems due to the increased presence of lignin and cellulose (Sardans et al., 2012). In addition, terrestrial plants produce a cornucopia of chemical defenses (Harborne et al., 1999; Kaplan et al., 2008; Arnason and Bernards, 2010). Some of these defenses, such as digestion inhibitors and structural defenses, force herbivores to consume more tissue to attain the same nutrition. These defenses are common in terrestrial plants, in part, because herbivores do not consume an entire plant at one time and can choose to move to a more palatable plant before causing plant mortality (Moran and Hamilton, 1980; Hanley et al., 2007). In contrast to pelagic systems, one encounter with a herbivore does not usually cause vascular plant mortality. Direct contact with a herbivore's salivary chemicals or characteristic damage patterns are usually required for induction in vascular plants, although recent evidence also points to neighbor induction by leaf volatiles through airborne plant/plant communication (Karban et al., 1999; 2000). The ability of terrestrial plants to avoid mortality when attacked enables tolerance to be a more viable strategy for them to deal with herbivores (Rosenthal and Kotanen, 1994). Some of the best examples of tolerance come from terrestrial systems with grazing herbivores; for instance, grasslands can be more productive in the presence of herbivory than without due to compensatory growth strategies (McNaughton, 1985).

Resistance traits also vary by plant functional group. The resistance traits of the closely related grasses are dominated by phenolics, nitrogen-containing defenses, toughness, and silica deposits in leaf tissue. While herbaceous and woody plants are derived from across the vascular plant phylogeny and express a wide range of resistance traits, there is a general pattern of greater inducibility and N-based defensive chemistry in herbaceous plants compared to woody species (Massad et al., 2011). Differences in functional group defense expression are manifest through succession, as perennial plants and then woody plants replace annual, herbaceous colonizers. As a result, resource-rich early successional systems are often dominated by tolerance responses and N-based defenses that shift toward toxic C-based defenses in late-successional, slow-growing species (Davidson, 1993).

Littoral and benthic autotrophs

Littoral and benthic autotrophs possess size, life history traits, and stoichiometric properties that are often intermediate between pelagic and terrestrial systems (Shurin et al., 2006). Communities consist of periphyton and macrophytes, including macroalgal species as well as vascular macrophytes (derived from terrestrial lineages), which root and access light in the photic zone. Often these systems are characterized by resource subsidy inputs from the terrestrial community (Nowlin et al., 2008).

Marine systems contain a diverse array of non-vascular macroalgae that are both free-living and part of benthic periphyton communities. Their tissue can become calcified which confers both structural and chemical defense (Hay et al., 1994). Many toxic resistance compounds (primarily phlorotannins in brown algae) are expressed as well (Hay and Fenical, 1988). However, few of these putative resistance compounds have been shown to provide effective defense against herbivores (*sensu* Karban and Baldwin, 1997). In addition, the lack of a vascular system in these plants would suggest a limited capacity for induction; however, recent work has demonstrated widespread induced resistance in response to small crustaceans and gastropods within this plant group, particularly in brown and green algae (Toth and Pavia, 2007). There is also within-plant variation in chemical defense expression (Cronin and Hay, 1996).

Historically, herbivores were considered unimportant to freshwater macroalgae, as herbivory rates were thought to be very low (Hutchinson, 1975). However, meta-analysis has shown that herbivory rates are higher on macrophytes than terrestrial plants (Cyr and Pace, 1993), suggesting that selection should favor defense expression in these plants. Although there is evidence of chemical resistance in macroalgae (Prusak et al., 2005), evidence of induction is rare (Camacho, 2008). While unusual in marine systems, vascular macrophytes dominate littoral zones in freshwater communities. They produce chemical defenses, such as alkaloids, that are also common in terrestrial plants due to derived ancestry from many terrestrial vascular plant lineages (Ostrofsky and Zettler, 1986; Chambers et al., 2008). In addition they produce structural defenses that lower plant palatability (Cronin and Lodge, 2003; Lamberti-Raverot and Puijalon, 2012). Tolerance traits are not very well studied in either freshwater littoral or marine benthic systems, but they have the potential to be quite important, particularly in systems dominated by large grazers (Burkepile and Hay, 2006; 2013; Nolet, 2004).

Grouping plant defense response by habitat or relatedness?

Most syntheses of trophic control in terrestrial and aquatic systems look for broad-brush similarities and differences and thus treat all species within a shared habitat type (e.g., pelagic) as though they are selected for and capable of expressing the same convergent, adaptive traits. This may not be appropriate to do. For example, macrophytes are found within seven plant divisions, resulting in Chlorophyta (green algae) macrophytes that are more closely related to green algal phytoplankton species than to any vascular macrophyte (only found within Pteridophyta and Spermatophyta divisions; Chambers et al., 2008). A result of macrophytes being spread across most of the plant phylogeny is that their trait expression may be constrained by the evolutionary history of the group from which they are derived.

For example, the molecular machinery necessary to produce many polyphenolic chemical defenses in terrestrial plants, such as tannins, flavonoids, and lignins, is thought to be a relic of evolutionary history, originally deployed to protect aquatic plants from damaging UV light as they gradually evolved to live on land (Rozema et al., 2002). These UV-activated defenses are therefore less prevalent in algal species that remained in aquatic environments, because water is much more effective at filtering UV rays. Therefore, chemical defenses (at least

UV-activated ones) are predicted to be of greater importance in terrestrial than aquatic systems. However, closely related vascular macrophytes that reinvaded aquatic environments from many terrestrial vascular lineages (at least 211 independent re-colonization events; Cook, 1999) should have molecular machinery more similar to terrestrial plants and thus produce these defenses (Rozema et al., 2002). Therefore, we argue for more finely resolved comparisons when exploring contingency among ecosystems, such as considering vascular land plants and littoral zone vascular macrophytes as equivalent and pelagic phytoplankton as being different. While rarely implemented in the aquatic literature, this approach would respect phylogenetic constraints on trait evolution in response to herbivores that may determine which potential plant defense strategies are available to an organism and perhaps explain some of the contingency in the outcomes across distantly related species.

Influence of nutrient availability on expressed defense strategies

MSH is incomplete in that it excludes a factor known to be important to plant defense expression: resource availability to plants. A shift in nutrient availability can change the absolute and relative costs of constitutive and induced defenses and potentially the outcome of plant competitive interactions (Cipollini et al., 2003). Thus the efficacy and selection for the plant defensive traits outlined above are influenced by the environmental context in which they are expressed (Belovsky and Schmitz, 1994). Classical ways of thinking about the interaction of resource availability and trophic control depict a static pool of resources (Oksanen et al., 1981). Another approach is to take a dynamic perspective of nutrient pools in ecosystems that allows for consideration of feedbacks between the abiotic nutrient pool and biotic responses such as plant defense traits and trophic interactions (Loreau, 2010; DeAngelis et al., 2012). In this section, we review a number of ways to approach how plant defense expression interacts with nutrient availability and then propose a more dynamic way of viewing interactions between primary producers and their environment.

Interspecific variation and community shifts

Environments with particular resource conditions may favor communities comprised of species with particular plant traits. Within the MSH framework previously outlined, at an interspecific level, defensive response can be thought of as an aggregate expression of functional traits of all members of a community – a so-called interspecific defense perspective. The growth/defense tradeoff hypothesis posits that at high nutrient levels, adapted plants grow so rapidly as to preclude investment in defense. At low nutrient levels, however, species are favored that grow slowly and have time to invest in defenses for their longer-lived more valuable leaves (Coley et al., 1985). In theory, therefore, if low nutrient availability filters out species that express tolerance traits and over-represents species with resistance traits, then we may expect to see trophic cascades in those systems.

While there are many evaluations of this interspecific defense theory for terrestrial systems (Fine et al., 2006), few tests have been performed in aquatic systems particularly within littoral habitats or between macroalgal species (Pavia and Toth, 2008). Because the goal of this chapter is to compare ecosystems on an equal footing, we will not focus on interspecific plant defense theory. Nevertheless, it is noteworthy that in planktonic algal systems, an interspecific growth/defense tradeoff is often invoked to explain community shifts due to herbivory or nutrients (Grover, 1995). Here edible phytoplankton with high growth rates are replaced by defended, but slow-growing species at low nutrient levels or high herbivory rates. The existence of such a growth–defense tradeoff was supported by meta-analysis, but size-selective grazing by zooplankton species complicates the effect on trophic cascades, with edible species still able to bloom in the presence of herbivores (Agrawal, 1998).

Intra-specific variation and phenotypic plasticity

While interspecific species turnover is more often invoked in aquatic systems, possibly due to the short lifespans of phytoplankton, plants can also exhibit genotypic and phenotypic variation in defense allocation to resistance or tolerance within a species or over a single individual's lifespan (Glynn et al., 2007). A recent meta-analysis of ontogenetic changes in plant defense allocation in terrestrial plants showed little influence of ontogeny on tolerance. However, herbaceous plants shifted from relying on induced chemical defenses when young to constitutive chemical defenses when old. Woody plants also exhibited an increase in constitutive defenses over time, with an initial reliance on chemical defenses in the seedling stage shifting to physical defenses during the juvenile stage, and then an overall decrease in defense allocation when mature (Barton and Koricheva, 2010). While untested, according to the MSH hypothesis extended in this chapter, these life-cycle stage shifts in defense expression in response to ontogenetically staged herbivory may result in different likelihoods of trophic cascades occurring throughout a growing season or plant's lifetime.

Resistance models

Plants show the bottom-up effect of nutrient gradients even in the absence of herbivores through variation in quality (nutrient content) and the level of constitutive defense allocation. For resistance traits, these relationships have been extensively investigated and formalized as plant defense theories, particularly for terrestrial systems (Herms and Mattson, 1992; Stamp, 2003; Wise and Abrahamson, 2007). There are competing views about how plant defense allocation is related to nutrient and other abiotic resource levels. According to these different views, peak defense allocation could happen at high (for nitrogenous-based defenses) (Bryant et al., 1987), low (Coley et al., 1985), or intermediate (Herms and Mattson, 1992) nutrient levels. Detailed treatment of resistance-based defense theory lies outside of the scope of this chapter and has been reviewed recently elsewhere (Koricheva, 2002; Stamp, 2003; Pavia and Toth, 2008). However, a review of recent studies that manipulated nutrients and measured constitutive defensive traits found increasing, decreasing, and no effect of nutrient supply on resistance trait expression across ecosystems (Table 8.1). This supports the view that no clear theory has yet emerged as a leading contender to explain resistance defense expression in terrestrial or aquatic systems (Stamp, 2003; Toth and Pavia, 2007).

Tolerance models

While many intra-specific theories of tolerance have been proposed and tested (e.g., the compensatory continuum hypothesis or the growth rate model), one recent approach integrates previous models to explain tolerance across resource conditions and may help predict where we might expect to see either tolerance or resistance traits dominating in ecosystems. The limiting resource model of tolerance (LRM), developed in terrestrial systems for vascular plants, uses a multistep dichotomous key to predict how changing the availability of a focal resource will impact tolerance by accounting for: (1) whether the focal abiotic resource is limiting plant fitness in the low-focal resource environment; (2) if the herbivore damage affects the use/acquisition of the focal resource or of an alternative resource; and (3) whether the herbivore damage causes the alternative resource to limit plant fitness (Wise and Abrahamson, 2005).

While complex, these three factors offer the flexibility needed to explain whether tolerance would be higher, lower, or equal at different nutrient levels. For example, imagine that nitrogen is the focal limiting resource for a plant species and a foliar herbivore primarily impacts carbon acquisition. If the addition of nitrogen does not cause carbon to become limiting, then the model predicts that the plant should exhibit equal tolerance in both high and low nitrogen environments (Wise and Abrahamson, 2005). When tested, the model accurately predicted the level of tolerance in 22 out of 24 cases of varying nutrient availability in terrestrial plants; 17 of these showed higher tolerance at lower nutrient availability (Wise and Abrahamson, 2007). This result may be generalizable to most terrestrial species. We know of only one study to apply the LRM to aquatic plants – which measured brown seaweed response to herbivory across different N environments (Hay et al., 2011) – and the prediction of the LRM of equal tolerance between high and low nitrogen environments in this system was supported. Clearly, further examination of this idea (and possible expansion to include herbivore-mediated linkages between resources; Bagchi and Ritchie, 2011), especially in non-terrestrial ecosystems, is needed.

Table 8.1 *Studies that manipulated a focal nutrient and measured the effect on constitutive plant defense expression*

Reference	Ecosystem	Zone	Primary producer	Species	Focal nutrient (FN)	Type of defense	Trait measured	Effect of ↑ in FN on trait
(Lundgren, 2010)	Marine	Pelagic	Phytoplankton	*Phaeocystis globosa*	N, P, N&P	Structural	Colony formation	↑
(O'Donnell et al., 2013)	Freshwater	Pelagic	Phytoplankton	*Scenedesmus acutus*	P	Structural	Colony formation	↑
(Gavis et al., 1979)	Freshwater	Pelagic	Phytoplankton	*Scenedesmus quadricauda*	Nitrate (N)	Structural	Colony formation	↑
(Trainor and Siver, 1983)	Freshwater	Pelagic	Phytoplankton	*Scenedesmus quadricauda*	Ammonium (N)	Structural	Colony formation	↑
(Lampert et al., 1994)	Freshwater	Pelagic	Phytoplankton	*Scenedesmus acutus*	Urea (N)	Structural	Colony formation	=
–	–	–	–	–	Ammonium (N)	Structural	Colony formation	=
(Wiltshire and Lampert, 1999)	Freshwater	Pelagic	Phytoplankton	*Scenedesmus obliquus*	Urea (N)	Structural	Colony formation	↑
(Van Donk, 1997)	Freshwater	Pelagic	Phytoplankton	*Scenedesmus* spp.	Mult. nutrients	Structural	Cell wall thickness	→
–	–	–	–	–	–	Structural	Size	→
(Cronin and Lodge, 2003)	Freshwater	Littoral	Vascular macrophyte	*Potamogeton amplifolius; Nuphar advena*	Mult. nutrients	Chemical	Phenols	↑
–	–	–	–	–	–	Growth	Growth rate	↑

(Lamberti-Raverot and Puijalon, 2012)	Freshwater	Littoral	Vascular macrophyte	*Myosotis scorpioides; Mentha aquatica*	Mult. nutrients	Structural	Breaking force	→
(Cronin and Hay, 1996)	Marine	–	Macroalgae	*Dictyota ciliolata; Sargassum filipendula*	–; Mult. nutrients	Structural; Chemical	Density; Terpenoids	→; =
(Van Alstyne, 2000)	Marine	–	Macroalgae	*Fucus gardneri*	P; –	Growth; Chemical	Growth rate; Phlorotannin	←; →
(Arnold, 1995)	Marine	–	Macroalgae	*Lobophora variegata*	N; –	Growth; Chemical	Growth rate; Phlorotannin	→; →
(Ilvessalo et al., 1989)	Marine	Littoral	Macroalgae	*Fucus vesiculosus*	N	Chemical	Phlorotannin	→
(Hemmi and Jormalainen, 2002)	Marine	Littoral	Macroalgae	*Fucus vesiculosus*	Mult. nutrients	Chemical	Phlorotannin	=
(Gowda et al., 2003)	Terrestrial	Forest	Woody	*Acacia tortilis*	–; Mult. nutrients	Structural; Structural	Toughness; Spine mass	→; ←
(Cash and Fulbright, 2005)	Terrestrial	Forest	Woody	*Acacia spp.*	Mult. nutrients	Structural	Spine mass	=
(Bazely et al., 1991)	Terrestrial	Grassland	Herb	*Rubus fruticosus*	Mult. nutrients	Structural	Spine density	→

(cont.)

Table 8.1 (cont.)

Reference	Ecosystem	Zone	Primary producer	Species	Focal nutrient (FN)	Type of defense	Trait measured	Effect of ↑ in FN on trait
(Hoffland et al., 2000)	Terrestrial	Grassland	Herb	*Lycopersicon esculentum*	Mult. nutrients	Structural	Trichome density	→
(Richardson et al., 1999)	Terrestrial	Fen	Herb	*Cladium jamaicense*	P	Chemical	Phenols	→
(Forkner and Hunter, 2000)	Terrestrial	Forest	Woody	*Quercus* spp.	Mult. nutrients	Chemical	Tannins, phenols	→
(Osier and Lindroth, 2001)	Terrestrial	Forest	Woody	*Populus* spp.	Mult. nutrients	Chemical	Tannins	→
(Cornelissen and Stiling, 2006)	Terrestrial	Forest	Woody	*Quercus* spp.	Mult. nutrients	Chemical	Tannin	=
–	–	–	–	–	–	Structural	Toughness	=
–	–	–	–	–	–	Growth	N content	←
(Wallace, 1989)	Terrestrial	Grassland	Herb	Var. Monocots	N	Structural	Silica	→
(Osier and Lindroth, 2004)	Terrestrial	Forest	Woody	*Populus* spp.	Mult. nutrients	Chemical	Phenolic glycosides, condensed tannins	=
(Cipollini and Bergelson, 2001)	Terrestrial	Greenhouse	Herb	*Brassica napus*	Mult. nutrients	Chemical	Protein-based trypsin inhibitors	←

While tolerance is rarely investigated under that terminology in aquatic systems, aquatic ecologists have thoroughly tested the Growth Rate Hypothesis (GRH), which links N and P usage within an individual via protein synthesis. Fast growth strategies require high P-allocation to synthesize ribosomal RNA (Sterner and Elser, 2002), thus environments with low N:P ratios favor species with fast growth rates. There is considerable empirical support for GRH from aquatic pelagic environments, but the model is rarely tested in terrestrial systems, where support is weak (Sardans et al., 2012). While not explicitly presented as an intra-specific tolerance model, the GRH meets the criteria for tolerance if a mitigation of fitness impact is produced within a species in response to herbivory and available resources, and is therefore complementary to the LRM, outlined above. The GRH and LRM represent an example where terrestrial and aquatic ecologists are wrestling with similar concepts, but with different jargon, leading to the incorrect perception that aquatic and terrestrial systems operate differently.

Induced defenses

Studies rarely explicitly investigate whether resource availability influences whether plants induce or continuously express anti-herbivore defenses. An intriguing recent study that quantified this with the phytoplankton *Scenedesmus acutus* showed that low P availability resulted in the induction of colony formation in the presence of herbivores, whereas under high P colony formation was constitutive (O'Donnell et al., 2013). In terrestrial systems, a similar kind of experiment found that the constitutive expression of protein-based trypsin inhibitors and the ability to induce them increased with nutrient availability (Cipollini and Bergelson, 2001). Future studies that manipulate both nutrient availability and herbivore presence are needed to resolve the general patterns among herbivory, nutrient availability, and defense induction across aquatic and terrestrial ecosystems.

Nutrient cycling links top-down and bottom-up effects

All classical plant defense theories (including EEH) view soil nutrient conditions as static and homogeneous. However, this may not be an accurate representation of nutrient dynamics. There is increasing recognition that species, especially consumers in higher trophic levels, play an important role in structuring nutrient environments through resource consumption, nutrient cycling, and translocation (Kitchell et al., 1979; Vanni, 2002; Pringle et al., 2010; Schmitz et al., 2010). Moreover, phenotypic variation in species traits may determine spatial heterogeneity in the nutrient environment as well (Norberg et al., 2001; Cornwell et al., 2008). Thus, while the nutrient environment certainly impacts the degree to which plant resources express tolerance and resistance traits, their expression may also feedback to influence nutrient cycling and hence change

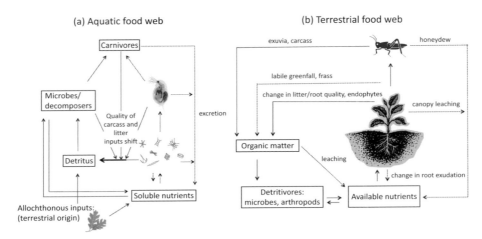

Figure 8.2 Potential pathways through which herbivores can influence nutrient cycling in (a) a generalized aquatic food web (adapted from Moore et al., 2004) and (b) a generalized terrestrial food web. Dashed lines indicate a fast-cycle pathway that has within season/generation effects on nutrient cycling. Solid lines represent slow-cycle pathways with primarily between season or generation effects. Induced plant defensive trait responses to herbivory have the potential to alter the relative magnitude of these pathways resulting in differential cycling rates. Clip art from Integration and Application Network, University of Maryland Center for Environmental Science (ian.umces.edu/imagelibrary/).

nutrient conditions. Whether a plant species utilizes a resistance or tolerance strategy against herbivores may thus have implications at both the community (Chase et al., 2000a) and ecosystem level by mediating bottom-up and top-down effects on nutrient cycling.

How defensive phenotypes (resistance versus tolerance) may alter ecosystem processes can be examined by expanding the linear trophic interaction chain perspective to include both above- and belowground linkages through nutrient cycling (Fig. 8.2). Nutrient cycling broadly encompasses several ecosystem processes, including production following nutrient uptake and decomposition leading to nutrient release (Deangelis, 1980; Deangelis et al., 1989; Moore et al., 2004; see Chapter 9 of this volume for more on nutrient cycling). Nutrients create a common currency for all trophic levels (Andersen et al., 2004). Moreover, linking above- and belowground processes reveals interesting reciprocal feedbacks between herbivores and the nutrient base through direct and indirect interactions (Van der Putten et al., 2001; Bardgett and Wardle, 2003; Schmitz, 2010).

This conception facilitates consideration of a dynamic nature of plant–herbivore interactions. For instance, herbivores not only influence productivity through direct consumption of plants, but also indirectly by influencing the way nutrient availability becomes altered via induced plant responses that can decrease or increase plant palatability (nutrient content) and thereby alter decomposition of organic matter by microbes or the release of inorganic waste by animals (Schmitz, 2010). Herbivore-induced responses by plants may impact slow-cycle inputs from uneaten organic plant litter (termed "after-life" effects), as well as fast-cycle inputs, such as inorganic materials from herbivore fecal output and canopy leaching (Hunter, 2001). These indirect effects on cycling (Fig. 8.2) are rarely quantified, particularly in terrestrial systems (Choudhury, 1988; Bardgett and Wardle, 2003; but see Frost and Hunter, 2008), but point to the potential importance of a plastic plant trait (defense allocation) for mediating the relative magnitudes of nutrients entering the slow- and fast-cycle pathways of ecosystems.

Can plant defenses affect how nutrients move through aquatic and terrestrial systems?

A classic idea of herbivore-mediated nutrient cycling is the acceleration hypothesis (McNaughton et al., 1989; Belovsky and Slade, 2000; Chapman et al., 2003), which proposes a positive feedback between herbivory and nutrient cycling. Herbivores consume a dominant species with highly nutritious leaf litter. These plants tolerate herbivory and by producing highly nutritious leaf regrowth cause herbivores to release large quantities of high quantity egesta, as well as facilitating plant canopy leaching and greenfall inputs. These factors collectively act to increase decomposition rates and ultimately increase the rate of nutrient supply to plants. In subsequent years, high resource supply favors the same dominant, nutritious plant species. In contrast, the deceleration hypothesis (Ritchie et al., 1998) posits that herbivores consume palatable plants selectively, thus shifting community composition toward less palatable species (Fig. 8.3). Litter from a community of unpalatable species decomposes more slowly than that from a palatable community because of a positive relationship between palatability and decomposability (Grime et al., 1996; but see Palkova and Leps, 2008; Ohgushi, 2008).

The acceleration hypothesis uses intra-specific changes in plant tolerance traits to predict an increase in nutrient cycling through herbivory, while the deceleration hypothesis relies on interspecific trait changes within a community. We propose that both deceleration and acceleration of nutrient cycling are viable outcomes at both the inter- or intra-specific levels depending on (1) the degree of intra-specific variation in plant traits (genotypic and phenotypic plasticity) and (2) the degree to which the plant community is dominated by a single plant defense syndrome. For example, uneaten litter from a plant (or plant community) that expresses structural or quantitative resistance defenses may be broken down more slowly by the microbial community than plants expressing tolerance traits, thereby impacting available nitrogen in the system (Schweitzer et al., 2008). Qualitative resistance defenses that persist in the environment may have a similar effect (Fig. 8.3). In contrast, plants that express tolerance traits

Trait- mediated herbivore impact on nutrient cycling

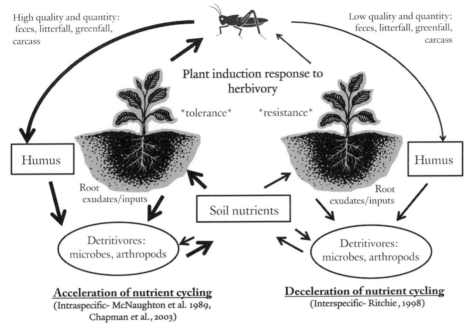

Figure 8.3 The defensive response trait (resistance versus tolerance) a plant produces in the face of herbivory may change the rate of nutrient cycling in a given system. Arrow line width represents the magnitude of nutrients moving through the pathway. Tolerance traits may result in an increase in herbivore egestion and high-quality litter entering the detrital food web. Resistance responses may decrease nutrient return to the soil through herbivory, as well as providing low-quality recalcitrant leaf tissue that is slowly broken down by the detrital food web, thus decreasing cycling rates.

produce high-quality litter that may be broken down rapidly by the microbial community, resulting in a larger available nitrogen pool (Fig. 8.3).

Few studies have looked for evidence of the impact of plant defense traits on nutrient cycling. However, it is clear that herbivores do have the potential to affect cycling rates across all systems. For example, in benthic kelp beds or pelagic lakes, consumers can increase net primary productivity (NPP) through increased nutrient cycling (Sterner et al., 1992; Steinberg, 1995; Vanni, 2002). Experiments also demonstrate that herbivores and plant traits can influence nutrient cycling in terrestrial systems. For example, pulses of cicada cadavers in northern temperate forests increase plant growth rates the following year (Yang, 2004). In addition, intra-specific variation in oak leaf phenotype influences fast- and slow-cycle litter decomposition (Madritch and Hunter, 2005), and recent meta-analyses indicated plant traits (e.g., LMA, lignin, and nutrient content) are the most important drivers of litter decomposition across global ecosystems (Cornelissen, 1996; Cornwell et al., 2008). Moreover, there is evidence that resource pulses move more quickly through aquatic than terrestrial systems (Nowlin et al., 2008). Whether this is due in part to differential expression of defensive traits, while plausible given our synthesis above, remains unknown.

Differences in herbivore feeding guilds

Aquatic algae (phytoplankton and reef periphyton) experience greater herbivory than vascular macrophytes, which experience greater herbivory than terrestrial plants, with median annual primary productivity removed of 79%, 30%, and 18%, respectively (Cyr and Pace, 1993). These differences in herbivory rates have often been cited as reasons for differences between top-down and bottom-up effects among ecosystems (Strong, 1992). However, plant responses may also be impacted by the functional group of the herbivores that consume them (Gruner and Mooney, 2013). Plant responses to herbivory in the grazing systems of the Serengeti may be more similar to marine kelp forests with extensive grazing by marine mammals than to other terrestrial ecosystem types (Burkepile, 2013). It is often assumed that herbivores are more specialized on land (insects) than in pelagic or littoral ecosystems (Newman and Rotjan, 2013). Specialized herbivores are likely to induce different plant defense responses than generalists (Feeny, 1976; Bernays, 2001; see also Chapter 13, this volume). Herbivore feeding guild and specialization is not currently explicitly incorporated into the MSH, but it is another trait-based approach that may be worthwhile to pursue in an examination of contingency in the interplay between plant defense and nutrients on trophic control of ecosystems.

Conclusions

Plants can produce both tolerance and resistance responses to herbivory and we see examples of each of these strategies across terrestrial and aquatic ecosystems. Chemical and structural resistance defenses tend to dominate terrestrial ecosystems, but play a smaller role in aquatic systems. The exception to this is terrestrial grazing ecosystems that are clearly dominated by plant tolerance responses to herbivory. In terrestrial systems, there is evidence that defense allocation is constrained to some degree by phylogenetic relationships (Armbruster, 1997; Ronsted et al., 2012; but see Haak et al., 2013), however this subject remains ripe for investigation within aquatic ecosystems. In particular, we suggest that a phylogenetic approach would be useful for understanding patterns within the phylogenetically diverse functional group of macrophytes. While tolerance responses are not often studied in aquatic systems under that terminology, we argue that induced changes in life history attributes that increase fitness in the presence of herbivory should be considered a tolerance trait and that tolerance traits may be very common yet overlooked in pelagic, benthic, and littoral communities. Plant defense theories are more refined and well tested in terrestrial systems than in aquatic systems. In aquatic systems the stoichiometrically based GRH accurately predicts higher growth rates in low N:P ratio environments. Which plant defense strategy (tolerance or resistance) a plant induces in response to herbivory has different ramifications for nutrient cycling, the coevolution of herbivores and plants, and community dynamics (Chase et al., 2000a).

Plant defense theory could advance through empirical tests among a broader range of ecosystem types, as well as benefiting from contextualizing a system not in terms merely of a plant–herbivore linkage, but instead in terms of a trophic chain with direct and indirect effects among soil nutrients, plants, herbivores, and predators. Tests could also benefit from more emphasis on the role of tolerance as a defensive trait, because it helps to unify thinking across ecosystem types once a common conceptual jargon is used. In general tolerance has been overlooked as an explanatory plant functional trait. For example, in Korcheiva's extensive meta-analysis on the cost of defensive traits, chemical, mechanical, and induced defenses were examined, but not tolerance traits (Koricheva, 2002). A recently proposed terrestrial-based model, LRM (Wise and Abrahamson, 2005), holds great promise for predicting tolerance traits across resource environments. We suggest that this model be tested broadly across ecosystems to determine whether it is generalizable.

The unresolved basis for wide variation in expression of resistance traits may stem from an incomplete conceptualization of the "system" and the context-dependent feedbacks that determine their expression. We suggest that taking a trait-based approach in the context of a food chain may help to resolve when and where these traits are expressed and how they impact trophic control of ecosystems. The MSH of trophic control may provide the basis for including plant defense traits (Schmitz, 2008). We predict that "resistance" traits (both structural and qualitative) will result in a trophic cascade through relative resource limitation of herbivores, while "tolerance" traits will invoke absolute resource limitation of herbivores, resulting in herbivore control of primary productivity. We realize that this framework does not yet consider important additional factors, such as plant volatiles, herbivore feeding guild, and ontological shifts in plant defense, but nonetheless view it as a useful starting point.

This conception may also help offer a complementary explanation for variation in the strength of top-down control across nutrient supply or productivity gradients implicit in the classic EEH of trophic control of ecosystems. This theory predicts that top-down control should be strongest at intermediate levels of productivity, which is attributed to predator satiation (Oksanen and Oksanen, 2000). This result, as well as the finding that herbivore and predator efficiency are important explanatory factors, was supported by meta-analysis (Borer et al., 2005). The MSH framework developed here suggests that plant defense traits may also account for the weakening of top-down control. The expression of tolerance regrowth traits at high nutrient levels could cause herbivores that were relative resource limited at lower nutrient levels to become absolute resource limited. In turn, predators would no longer have an indirect positive effect on productivity. At high nutrient levels, these tolerance traits may allow plants to escape their herbivores by outgrowing them. This outcome is not formalized within the EEH, but is consistent with the outcomes presented there.

The induction of resistance and tolerance traits in plant communities may also have important effects on nutrient cycling and future resource availability through "after-life" effects of plant defense or tolerance traits that remain in uneaten plant litter entering the detrital food web. While different rates of nutrient cycling have been predicted and recorded within aquatic and terrestrial systems (Nowlin et al., 2008), it remains to be seen whether taking this functional trait approach may explain some of the contingency found within and between aquatic and terrestrial ecosystems.

The lack of empirical investigation into these topics makes generalization difficult. However, as anthropogenic nitrogen inputs increase (Vitousek et al., 1997) and climate change increases herbivory and the potential induction of plant defenses (Ayres, 1993), it is increasingly important to understand how herbivory and nutrient context influence plant and herbivore populations across ecosystems. Tackling this question of whether and when plant defensive traits and nutrient availability modify trophic cascades within many ecosystem types is the first step. Only then will we be able to adequately address the question of whether defensive traits map to similar community responses in both aquatic and terrestrial ecosystems. This knowledge of how soil nutrient environment changes the expression of plant defensive traits and productivity may also be useful to agriculturists interested in lowering pesticide use while maximizing yield.

Acknowledgments

This research was supported by a GRFP under NSF Grant # DGE-1122492 to KTB and the Yale Institute for Biospheric Studies.

References

Agrawal, A. A. (1998). Algal defense, grazers, and their interactions in aquatic trophic cascades. *Acta Oecologica*, **19**, 331–337.

Agrawal, A. A. (2001). Phenotypic plasticity in the interactions and evolution of species. *Science*, **294**, 321–326.

Agrawal, A. A. and Fishbein, M. (2006). Plant defense syndromes. *Ecology*, **87**, S132–S149.

Agrawal, A. A. and Karban, R. (1998). *Why Induced Defenses May Be Favored over Constitutive Strategies in Plants*. Princeton, NJ: Princeton University Press.

Andersen, T., Elser, J. J. and Hessen, D. O. (2004). Stoichiometry and population dynamics. *Ecology Letters*, **7**, 884–900.

Arimura, G.-I., Kost, C. and Boland, W. (2005). Herbivore-induced, indirect plant defences. *Biochimica et Biophysica Acta (BBA) – Molecular and Cell Biology of Lipids*, **1734**, 91–111.

Armbruster, W. S. (1997). Exaptations link evolution of plant-herbivore and plant-pollinator interactions: a phylogenetic inquiry. *Ecology*, **78**, 1661–1672.

Arnason, J. T. and Bernards, M. A. (2010). Impact of constitutive plant natural products on herbivores and pathogens. *Canadian Journal of Zoology – Revue Canadienne De Zoologie*, **88**, 615–627.

Arnold, T. M. (1995). Phenotypic variation in polyphenolic content of the tropical brown alga *Lobophora variegata* as a function of nitrogen availability. *Marine Ecology Progress Series*, **123**, 177.

Ayres, M. P. (1993). Plant defense, herbivory, and climate change. *Biotic Interactions and Global Change*, **75**, 94.

Bagchi, S. and Ritchie, M. E. (2011). Herbivory and plant tolerance: experimental tests of alternative hypotheses involving non-substitutable resources. *Oikos*, **120**, 119–127.

Bardgett, R. D. and Wardle, D. A. (2003). Herbivore-mediated linkages between aboveground and belowground communities. *Ecology*, **84**, 2258–2268.

Barton, K. E. and Koricheva, J. (2010). The ontogeny of plant defense and herbivory: characterizing general patterns using meta-analysis. *The American Naturalist*, **175**, 481–493.

Bazely, D. R., Myers, J. H. and Burke Da Silva, K. (1991). The response of numbers of bramble prickles to herbivory and depressed resource availability. *Oikos*, **61**, 327–336.

Belovsky, G. E. and Schmitz, O. J. (1994). Plant defenses and optimal foraging by mammalian herbivores. *Journal of Mammalogy*, **75**, 816–832.

Belovsky, G. E. and Slade, J. B. (2000). Insect herbivory accelerates nutrient cycling and increases plant production. *Proceedings of the National Academy of Sciences of the USA*, **97**, 14412–14417.

Bernays, E. A. (2001). Neural limitations in phytophagous insects: implications for diet breadth and evolution of host affiliation. *Annual Review of Entomology*, **46**, 703–27.

Borer, E. T., Seabloom, E. W., Shurin, J. B., et al. (2005). What determines the strength of a trophic cascade? *Ecology*, **86**, 528–537.

Bryant, J. P., Chapin, F. S., Reichardt, P. B. and Clausen, T. P. (1987). Response of winter chemical defense in Alaska paper birch and green alder to manipulation of plant carbon nutrient balance. *Oecologia*, **72**, 510–514.

Burkepile, D. E. (2013). Comparing aquatic and terrestrial grazing ecosystems: is the grass really greener? *Oikos*, **122**, 306–312.

Burkepile, D. E. and Hay, M. E. (2006). Herbivore vs. nutrient control of marine primary producers: context-dependent effects. *Ecology*, **87**, 3128–3139.

Camacho, F. (2008). Macroalgal and cyanobacterial chemical defenses in freshwater communities. In *Algal Chemical Ecology*, ed. C. Amsler. Berlin Heidelberg Springer, pp. 105–120.

Cash, V. W. and Fulbright, T. E. (2005). Nutrient enrichment, tannins, and thorns: effects on browsing of shrub seedlings. *Journal of Wildlife Management*, **69**, 782–793.

Cebrian, J. and Lartigue, J. (2004). Patterns of herbivory and decomposition in aquatic and terrestrial ecosystems. *Ecological Monographs*, **74**, 237–259.

Chambers, P. A., Lacoul, P., Murphy, K. J. and Thomaz, S. M. (2008). Global diversity of aquatic macrophytes in freshwater. *Hydrobiologia*, **595**, 9–26.

Chapman, S. K., Hart, S. C., Cobb, N. S., Whitham, T. G. and Koch, G. W. (2003). Insect herbivory increases litter quality and decomposition: an extension of the acceleration hypothesis. *Ecology*, **84**, 2867–2876.

Chase, J. M. (2000). Are there real differences among aquatic and terrestrial food webs? *Trends in Ecology and Evolution*, **15**, 408–412.

Chase, J. M., Leibold, M. A. and Simms, E. (2000a). Plant tolerance and resistance in food webs: community-level predictions and evolutionary implications. *Evolutionary Ecology*, **14**, 289–314.

Chase, J. M., Leibold, M. A., Downing, A. L. and Shurin, J. B. (2000b). The effects of productivity, herbivory, and plant species turnover in grassland food webs. *Ecology*, **81**, 2485–2497.

Choudhury, D. (1988). Herbivore induced changes in leaf-litter resource quality: a neglected aspect of herbivory in ecosystem nutrient dynamics. *Oikos*, **51**, 389–393.

Cipollini, D. and Bergelson, J. (2001). Plant density and nutrient availability constrain constitutive and wound-induced expression of trypsin inhibitors in *Brassica napus*. *Journal of Chemical Ecology*, **27**, 593–610.

Cipollini, D., Purrington, C. B. and Bergelson, J. (2003). Costs of induced responses in plants. *Basic and Applied Ecology*, **4**, 79–89.

Coley, P. D., Bryant, J. P. and Chapin, F. S. (1985). Resource availability and plant antiherbivore defense. *Science*, **230**, 895–899.

Cook, C. D. K. (1999). The number and kinds of embryo-bearing plants which have become aquatic: a survey. *Perspectives in Plant Ecology, Evolution and Systematics*, **2**, 79–102.

Cornelissen, J. H. C. (1996). An experimental comparison of leaf decomposition rates in a wide range of temperate plant species and types. *Journal of Ecology*, **84**, 573–582.

Cornelissen, T. and Stiling, P. (2006). Does low nutritional quality act as a plant defence? An experimental test of the slow-growth, high-mortality hypothesis. *Ecological Entomology*, **31**, 32–40.

Cornwell, W. K., Cornelissen, J. H. C., Amatangelo, K., et al. (2008). Plant species traits are the predominant control on litter decomposition rates within biomes worldwide. *Ecology Letters*, **11**, 1065–1071.

Cronin, G. and Hay, M. (1996). Within-plant variation in seaweed palatability and chemical defenses: optimal defense theory versus the growth-differentiation balance hypothesis. *Oecologia*, **105**, 361–368.

Cronin, G. and Lodge, D. (2003). Effects of light and nutrient availability on the growth, allocation, carbon/nitrogen balance, phenolic chemistry, and resistance to herbivory of two freshwater macrophytes. *Oecologia*, **137**, 32–41.

Cyr, H. and Pace, M. L. (1993). Magnitude and patterns of herbivory in aquatic and terrestrial ecosystems. *Nature*, **361**, 148–150.

Davidson, D. W. (1993). The effects of herbivory and granivory on terrestrial plant succession. *Oikos*, **68**, 23–35.

Deangelis, D. L. (1980). Energy-flow, nutrient cycling, and ecosystem resilience. *Ecology*, **61**, 764–771.

Deangelis, D. L., Mulholland, P. J., Palumbo, A. V., et al. (1989). Nutrient dynamics and food-web stability. *Annual Review of Ecology and Systematics*, **20**, 71–95.

Deangelis, D., Ju, S., Liu, R., Bryant, J. and Gourley, S. (2012). Plant allocation of carbon to defense as a function of herbivory, light and nutrient availability. *Theoretical Ecology*, **5**, 445–456.

Duffy, J. E. (2009). Why biodiversity is important to the functioning of real-world ecosystems. *Frontiers in Ecology and the Environment*, **7**, 437–444.

Ehrlich, P. R. and Birch, L. C. (1967). The "Balance of Nature" and "Population Control". *The American Naturalist*, **101**, 97–107.

Elton, C. (1927). *Animal Ecology*. London: Sidgwick and Jackson.

Feeny, P. P. (1968). Effect of oak leaf tannins on larval growth of winter moth *Operophtera brumata*. *Journal of Insect Physiology*, **14**, 805–817.

Feeny, P. (1976). Plant apparency and chemical defense. In *Biochemical Interaction Between Plants and Insects*, ed. J. W. Wallace and R. L. Mansell. USA: Springer, pp. 1–40.

Feeny, P. (1991). Theories of plant-chemical defense: a brief historical survey. *Symposia Biologica Hungarica*, **39**, 163–175.

Fine, P. V. A., Miller, Z. J., Mesones, I., et al. (2006). The growth-defense trade-off and habitat specialization by plants in Amazonian forests. *Ecology*, **87**, S150–S162.

Fogg, G. E. (1991). Tansley Review No. 30. The phytoplanktonic ways of life. *New Phytologist*, **118**, 191–232.

Forkner, R. E. and Hunter, M. D. (2000). What goes up must come down? Nutrient addition and predation pressure on oak herbivores. *Ecology*, **81**, 1588–1600.

Frost, C. J. and Hunter, M. D. (2008). Insect herbivores and their frass affect *Quercus rubra* leaf quality and initial stages of subsequent litter decomposition. *Oikos*, **117**, 13–22.

Gavis, J., Chamberlin, C. and Lystad, L. (1979). Coenobial cell number in *Scenedesmus quadricauda* (Chlorophyceae) as a function of growth rate in nitrate-limited chemostats. *Journal of Phycology*, **15**, 273–275.

Glynn, C., Herms, D. A., Orians, C. M., Hansen, R. C. and Larsson, S. (2007). Testing the growth-differentiation balance hypothesis: dynamic responses of willows to nutrient availability. *New Phytologist*, **176**, 623–634.

Gowda, J. H., Albrectsen, B. R., Ball, J. P., Sjöberg, M. and Palo, R. T. (2003). Spines as a mechanical defence: the effects of fertiliser treatment on juvenile *Acacia tortilis* plants. *Acta Oecologica*, **24**, 1–4.

Grime, J. P., Cornelissen, J. H. C., Thompson, K. and Hodgson, J. G. (1996). Evidence of a causal connection between anti-herbivore defence and the decomposition rate of leaves. *Oikos*, **77**, 489–494.

Grover, J. P. (1995). Competition, herbivory, and enrichment: nutrient-based models for edible and inedible plants. *The American Naturalist*, **145**, 746–774.

Gruner, D. S. and Mooney, K. A. (2013). Green grass and high tides: grazing lawns in terrestrial and aquatic ecosystems (commentary on Burkepile 2013). *Oikos*, **122**, 313–316.

Haak, D., Ballenger, B. and Moyle, L. (2013). No evidence for phylogenetic constraint on natural defense evolution among wild tomatoes. *Ecology*, **95**, 1633–1641

Hairston, N. G., Smith, F. E. and Slobodkin, L. B. (1960). Community structure, population control, and competition. *American Naturalist*, **94**, 421–425.

Hanley, M. E., Lamont, B. B., Fairbanks, M. M. and Rafferty, C. M. (2007). Plant structural traits and their role in anti-herbivore defence. *Perspectives in Plant Ecology, Evolution and Systematics*, **8**, 157–178.

Harborne, J. B., Baxter, H. and Moss, G. P. (1999). *Phytochemical Dictionary: A Handbook of Bioactive Compounds From Plants*. London: Taylor and Francis.

Hay, K. B., Poore, A. G. B. and Lovelock, C. E. (2011). The effects of nutrient availability on tolerance to herbivory in a brown seaweed. *Journal of Ecology*, **99**, 1540–1550.

Hay, M. E. and Fenical, W. (1988). Marine plant-herbivore interactions: the ecology of chemical defense. *Annual Review of Ecology and Systematics*, **19**, 111–145.

Hay, M. E., Kappel, Q. E. and Fenical, W. (1994). Synergisms in plant defenses against herbivores: interactions of chemistry, calcification, and plant quality. *Ecology*, **75**, 1714–1726.

Hemmi, A. and Jormalainen, V. (2002). Nutrient enhancement increases performance of a marine herbivore via quality of its food alga. *Ecology*, **83**, 1052–1064.

Herms, D. A. and Mattson, W. J. (1992). The dilemma of plants: to grow or defend? *Quarterly Review of Biology*, **67**, 283–335.

Hoffland, E., Dicke, M., Van Tintelen, W., Dijkman, H. and Van Beusichem, M. (2000). Nitrogen availability and defense of tomato against two-spotted spider mite. *Journal of Chemical Ecology*, **26**, 2697–2711.

Hunter, M. (2001). Insect population dynamics meets ecosystem ecology: effects of herbivory on soil nutrient dynamics. *Agricultural and Forest Entomology*, **3**, 77–84.

Hunter, M. D. and Price, P. W. (1992). Playing chutes and ladders: heterogenity and the relative roles of bottom-up and top-down forces in natural communities. *Ecology*, **73**, 724–732.

Hutchinson, G. E. (1975). *A Treatise on Limnology: Limnological Botany*. New York, NY: Wiley.

Ilvessalo, H., Ilvessalo, J. and Tuomi (1989). Nutrient availability and accumulation of

phenolic compounds in the brown alga
Fucus vesiculosus. *Marine Biology*, **101**, 115–119.

Kaplan, I., Halitschke, R., Kessler, A., Sardanelli,
S. and Denno, R. F. (2008). Constitutive and
induced defenses to herbivory in above-
and belowground plant tissues. *Ecology*, **89**,
392–406.

Karban, R. and Baldwin, I. T. (1997). *Induced
Responses to Herbivory*. Chicago, IL: University
of Chicago Press.

Karban, R., Agrawal, A. A., Thaler, J. S. and Adler,
L. S. (1999). Induced plant responses and
information content about risk of herbivory.
Trends in Ecology and Evolution, **14**, 443–447.

Karban, R., Baldwin, I. T., Baxter, K. J., Laue, G.
and Felton, G. W. (2000). Communication
between plants: induced resistance in wild
tobacco plants following clipping of
neighboring sagebrush. *Oecologia*, **125**,
66–71.

Kitchell, J. F., Oneill, R. V., Webb, D., et al. (1979).
Consumer regulation of nutrient cycling.
Bioscience, **29**, 28–34.

Koricheva, J. (2002). Meta-analysis of sources of
variation in fitness costs of plant
antiherbivore defenses. *Ecology*, **83**, 176–190.

Lamberti-Raverot, B. and Puijalon, S. (2012).
Nutrient enrichment affects the mechanical
resistance of aquatic plants. *Journal of
Experimental Botany*, **63**, 6115–6123.

Lampert, W., Rothhaupt, K. O. and Vonelert, E.
(1994). Chemical induction of colony
formation in a green-alga *Scenedesmus acutus*
by grazers (Daphnia). *Limnology and
Oceanography*, **39**, 1543–1550.

Lass, S. and Spaak, P. (2003). Chemically induced
anti-predator defences in plankton: a
review. *Hydrobiologia*, **491**, 221–239.

Leibold, M. A. (1989). Resource edibility and the
effects of predators and productivity on the
outcome of trophic interactions. *American
Naturalist*, **134**, 922–949.

Leibold, M. A. (1999). Biodiversity and nutrient
enrichment in pond plankton communities.
Evolutionary Ecology Research, **1**, 73–95.

Lindeman, R. L. (1942). The trophic-dynamic
aspect of ecology. *Ecology*, **23**, 399–418.

Loreau, M. (2010). *From Populations to Ecosystems*.
Princeton, NJ: Princeton University Press.

Lundgren, V. (2010). Grazer-induced defense in
Phaeocystis globosa (Prymnesiophyceae):
influence of different nutrient conditions.
Limnology and Oceanography, **55**, 1965.

Lürling, M. and Beekman, W. (1999). Grazer-
induced defenses in *Scenedesmus*
(Chlorococcales; Chlorophyceae):
coenobium and spine formation. *Phycologia*,
38, 368–376.

Madritch, M. D. and Hunter, M. D. (2005).
Phenotypic variation in oak litter influences
short- and long-term nutrient cycling
through litter chemistry. *Soil Biology and
Biochemistry*, **37**, 319–327.

Massad, T., Fincher, R. M., Smilanich, A. and
Dyer, L. (2011). A quantitative evaluation of
major plant defense hypotheses, nature
versus nurture, and chemistry versus ants.
Arthropod-Plant Interactions, **5**, 125–139.

Mauricio, R., Rausher, M. D. and Burdick, D. S.
(1997). Variation in the defense strategies of
plants: are resistance and tolerance
mutually exclusive? *Ecology*, **78**, 1301–1311.

McNaughton, S. J. (1985). Ecology of a grazing
ecosystem – the Serengeti. *Ecological
Monographs*, **55**, 259–294.

McNaughton, S. J., Oesterheld, M., Frank, D. A.
and Williams, K. J. (1989). Ecosystem-level
patterns of primary productivity and
herbivory in terrestrial habitats. *Nature*, **341**,
142–144.

Moore, J. C., Berlow, E. L., Coleman, D. C., et al.
(2004). Detritus, trophic dynamics and
biodiversity. *Ecology Letters*, **7**, 584–600.

Moran, N. and Hamilton, W. D. (1980). Low
nutritive quality as defense against
herbivores. *Journal of Theoretical Biology*, **86**,
247–254.

Murdoch, W. W. (1966). Community structure,
population control, and competition: a
critique. *The American Naturalist*, **100**, 219–
226.

Newman, R. M. and Rotjan, R. D. (2013).
Re-examining the fundamentals of grazing:
freshwater, marine and terrestrial

similarities and contrasts (commentary on Burkepile 2013). *Oikos*, **122**, 317–320.

Nolet, B. (2004). Overcompensation and grazing optimisation in a swan-pondweed system? *Freshwater Biology*, **49**, 1391–1399.

Norberg, J., Swaney, D. P., Dushoff, J., et al. (2001). Phenotypic diversity and ecosystem functioning in changing environments: a theoretical framework. *Proceedings of the National Academy of Sciences of the USA*, **98**, 11376–11381.

Nowlin, W. H., Vanni, M. J. and Yang, L. H. (2008). Comparing resource pulses in aquatic and terrestrial ecosystems. *Ecology*, **89**, 647–659.

O'Donnell, D. R., Fey, S. B. and Cottingham, K. L. (2013). Nutrient availability influences kairomone-induced defenses in *Scenedesmus acutus* (Chlorophyceae). *Journal of Plankton Research*, **35**, 191–200.

Ohgushi, T. (2008). Herbivore-induced indirect interaction webs on terrestrial plants: the importance of non-trophic, indirect, and facilitative interactions. *Entomologia Experimentalis et Applicata*, **128**, 217–229.

Oksanen, L. and Oksanen, T. (2000). The logic and realism of the hypothesis of exploitation ecosystems. *The American Naturalist*, **155**, 703–723.

Oksanen, L., Fretwell, S. D., Arruda, J. and Niemela, P. (1981). Exploitation ecosystems in gradients of primary productivity. *American Naturalist*, **118**, 240–261.

Osier, T. L. and Lindroth, R. L. (2001). Effects of genotype, nutrient availability, and defoliation on aspen phytochemistry and insect performance. *Journal of Chemical Ecology*, **27**, 1289–1313.

Osier, T. L. and Lindroth, R. L. (2004). Long-term effects of defoliation on quaking aspen in relation to genotype and nutrient availability: plant growth, phytochemistry and insect performance. *Oecologia*, **139**, 55–65.

Ostrofsky, M. L. and Zettler, E. R. (1986). Chemical defences in aquatic plants. *Journal of Ecology*, **74**, 279–287.

Paine, R. T. (1969). *Pisaster-Tegula* interaction: prey patches, predator food preference, and intertidal community structure. *Ecology*, **50**, 950–961.

Palkova, K. and Leps, J. (2008). Positive relationship between plant palatability and litter decomposition in meadow plants. *Community Ecology*, **9**, 17–27.

Pavia, H. and Toth, G. (2008). Macroalgal models in testing and extending defense theories. In *Algal Chemical Ecology*, ed. C. Amsler. Berlin Heidelberg: Springer, pp. 147–172

Pohnert, G. (2004). Chemical defense strategies of marine organisms. In *The Chemistry of Pheromones and Other Semiochemicals I*, ed. S. Schulz. Berlin Heidelberg: Springer, pp. 180–219

Pohnert, G., Steinke, M. and Tollrian, R. (2007). Chemical cues, defence metabolites and the shaping of pelagic interspecific interactions. *Trends in Ecology & Evolution*, **22**, 198–204.

Polis, G. A. and Strong, D. R. (1996). Food web complexity and community dynamics. *American Naturalist*, **147**, 813–846.

Power, M. E. (1992). Top-down and bottom-up forces in food webs: do plants have primacy? *Ecology*, **73**, 733–746.

Pringle, R. M., Doak, D. F., Brody, A. K., Jocque, R. and Palmer, T. M. (2010). Spatial pattern enhances ecosystem functioning in an African savanna. *PLoS Biology*, **8**.

Prusak, A., O'Neal, J. and Kubanek, J. (2005). Prevalence of chemical defenses among freshwater plants. *Journal of Chemical Ecology*, **31**, 1145–1160.

Raubenheimer, D. (1992). Tannic acid, protein, and digestible carbohydrate: dietary imbalance and nutritional compensation in locusts. *Ecology*, **73**, 1012–1027.

Richardson, C. J., Ferrell, G. M. and Vaithiyanathan, P. (1999). Nutrient effects on stand structure, resorption efficiency, and secondary compounds in everglades sawgrass. *Ecology*, **80**, 2182–2192.

Ritchie, M. E., Tilman, D. and Knops, J. M. H. (1998). Herbivore effects on plant and nitrogen dynamics in oak savanna. *Ecology*, **79**, 165–177.

Ronsted, N., Symonds, M. R., Birkholm, T., et al. (2012). Can phylogeny predict chemical diversity and potential medicinal activity of plants? A case study of amaryllidaceae. *BMC Evolutionary Biology*, **12**, 182.

Rosenthal, J. P. and Kotanen, P. M. (1994). Terrestrial plant tolerance to herbivory. *Trends in Ecology and Evolution*, **9**, 145–148.

Rozema, J., Rozema, L. O., Björn, J. F., et al. (2002). The role of UV-B radiation in aquatic and terrestrial ecosystems: an experimental and functional analysis of the evolution of UV-absorbing compounds. *Journal of Photochemistry and Photobiology B: Biology*, **66**, 2–12.

Ruehl, C. B. and Trexler, J. C. (2013). A suite of prey traits determine predator and nutrient enrichment effects in a tri-trophic food chain. *Ecosphere*, **4**, art75.

Sardans, J., Rivas-Ubach, A. and Peñuelas, J. (2012). The elemental stoichiometry of aquatic and terrestrial ecosystems and its relationships with organismic lifestyle and ecosystem structure and function: a review and perspectives. *Biogeochemistry*, **111**, 1–39.

Schmitz, O. (2010). *Resolving Ecosystem Complexity*. Princeton, NJ: Princeton University Press.

Schmitz, O. J. (2008). Herbivory from individuals to ecosystems. *Annual Review of Ecology, Evolution, and Systematics*, **39**, 133–152.

Schmitz, O. J., Adler, F. R. and Agrawal, A. A. (2003). Linking individual-scale trait plasticity to community dynamics. *Ecology*, **84**, 1081–1082.

Schmitz, O. J., Krivan, V. and Ovadia, O. (2004). Trophic cascades: the primacy of trait-mediated indirect interactions. *Ecology Letters*, **7**, 153–163.

Schmitz, O. J., Hawlena, D. and Trussell, G. C. (2010). Predator control of ecosystem nutrient dynamics. *Ecology Letters*, **13**, 1199–1209.

Schweitzer, J., Madritch, M., Bailey, J., et al. (2008). From genes to ecosystems: the genetic basis of condensed tannins and

their role in nutrient regulation in a *Populus* model system. *Ecosystems*, **11**, 1005–1020.

Scriber, J. M. and Feeny, P. P. (1975). Growth form of host plant as a determinant of feeding efficiencies and growth-rates in Papilionidae and Saturniidae (Lepidoptera). *Journal of the New York Entomological Society*, **83**, 247–248.

Shurin, J., Shurin, E., Borer, E., et al. (2002). A cross-ecosystem comparison of the strength of trophic cascades. *Ecology Letters*, **5**, 785–791.

Shurin, J. B., Gruner, D. S. and Hillebrand, H. (2006). All wet or dried up? Real differences between aquatic and terrestrial food webs. *Proceedings of the Royal Society B: Biological Sciences*, **273**, 1–9.

Stamp, N. (2003). Out of the quagmire of plant defense hypotheses. *Quarterly Review of Biology*, **78**, 23–55.

Steinberg, P. D. (1995). Evolutionary consequences of food chain length in kelp forest communities. *Proceedings of the National Academy of Sciences of the USA*, **92**, 8145.

Sterner, R. W. and Elser, J. J. (2002). *Ecological Stoichiometry: The Biology of Elements from Molecules to Biosphere*. Princeton, NJ: Princeton University Press.

Sterner, R. W., Elser, J. J. and Hessen, D. O. (1992). Stoichiometric relationships among producers, consumers and nutrient cycling in pelagic ecosystems. *Biogeochemistry*, **17**, 49–67.

Strauss, S. Y. and Agrawal, A. A. (1999). The ecology and evolution of plant tolerance to herbivory. *Trends in Ecology and Evolution*, **14**, 179–185.

Strong, D. (1992). Are trophic cascades all wet? Differentiation and donor-control in speciose ecosystems. *Ecology*, **73**, 747.

Tiffin, P. (2000). Mechanisms of tolerance to herbivore damage: what do we know? *Evolutionary Ecology*, **14**, 523–536.

Toth, G. B. and Pavia, H. (2007). Induced herbivore resistance in seaweeds: a meta-analysis. *Journal of Ecology*, **95**, 425–434.

Trainor, F. and Siver, P. (1983). Effect of growth rate on unicell production in two strains of

Scenedesmus (Chlorophyta). *Phycologia*, **22**, 127–131.

Trussell, G. R. and Schmitz, O. J. (2012). Species functional traits, trophic control, and the ecosystem comsequences of adaptive foraging in the middle of food chains. In *Trait-mediated Indirect Interactions: Ecological and Evolutionary Perspectives*, ed. T. Ohgushi, O. Schmitz and R. D. Holt. New York, NY: Cambridge University Press, pp. 324–338.

Uriarte, M. (2000). Interactions between goldenrod (*Solidago altissima* L.) and its insect herbivore (*Trirhabda virgata*) over the course of succession. *Oecologia*, **122**, 521–528.

Van Alstyne, K. L. (2000). Effects of nutrient enrichment on growth and phlorotannin production in *Fucus gardneri* embryos. *Marine Ecology Progress Series*, **206**, 33–43.

Van Der Putten, W. H., Vet, L. E. M., Harvey, J. A. and Wäckers, F. L. (2001). Linking above- and belowground multitrophic interactions of plants, herbivores, pathogens, and their antagonists. *Trends in Ecology and Evolution*, **16**, 547–554.

Van Donk, E. (1997). Defenses in phytoplankton against grazing induced by nutrient limitation, UV-B stress and infochemicals. *Aquatic Ecology Series*, **31**, 53–58.

Van Donk, E., Ianora, A. and Vos, M. (2011). Induced defences in marine and freshwater phytoplankton: a review. *Hydrobiologia*, **668**, 3–19.

Vanni, M. J. (2002). Nutrient cycling by animals in freshwater ecosystems. *Annual Review of Ecology and Systematics*, **33**, 341–370.

Verschoor, A. M., Van Der Stap, I., Helmsing, N. R., Lürling, M. and Van Donk, E. (2004a).

Inducible colony formation within the Scenedesmaceae: adaptive responses to infochemicals from two different herbivore taxa. *Journal of Phycology*, **40**, 808–814.

Verschoor, A. M., Vos, M. and Van Der Stap, I. (2004b). Inducible defences prevent strong population fluctuations in bi- and tritrophic food chains. *Ecology Letters*, **7**, 1143–1148.

Vitousek, P., Vitousek, J., Aber, R., et al. (1997). Human alterations of the global nitrogen cycle: sources and consequences. *Ecological Applications*, **7**, 737–750.

Vos, M., Verschoor, A. M., Kooi, B. W., et al. (2004). Inducible defenses and their trophic structure. *Ecology*, **85**, 2783–2794.

Wallace, A. (1989). Relationships among nitrogen, silicon, and heavy metal uptake by plants. *Soil Science*, **147**, 457–460.

White, T. C. R. (1978). Importance of a relative shortage of food in animal ecology. *Oecologia*, **33**, 71–86.

Wiltshire, K. H. and Lampert, W. (1999). Urea excretion by *Daphnia*: A colony-inducing factor in *Scenedesmus*? *Limnology and Oceanography*, **44**, 1894–1903.

Wise, M. J. and Abrahamson, W. G. (2005). Beyond the compensatory continuum: environmental resource levels and plant tolerance of herbivory. *Oikos*, **109**, 417–428.

Wise, M. J. and Abrahamson, W. G. (2007). Effects of resource availability on tolerance of herbivory: a review and assessment of three opposing models. *American Naturalist*, **169**, 443–454.

Yang, L. H. (2004). Periodical cicadas as resource pulses in North American forests. *Science*, **306**, 1565–1567.

CHAPTER NINE

Interactive effects of plants, decomposers, herbivores, and predators on nutrient cycling

SARAH E. HOBBIE

University of Minnesota, St. Paul, MN, USA

and

SÉBASTIEN VILLÉGER

Université Montpellier 2, France

Context

The rates and pathways of nutrient cycling through ecosystems depend on interactions between both bottom-up forces, including the chemical characteristics of biomass that influence its decomposition and consumption by higher trophic levels, and top-down forces, such as the nutritional requirements and metabolic efficiencies of consumers and decomposers that influence their feeding and excretion. At the base of the food web, nutrient cycling is influenced by whether NPP becomes detritus, entering the so-called "brown" food web, or is consumed by herbivores before death or senescence. Here, we use "nutrient" to refer to essential elements other than carbon (C), such as nitrogen (N), phosphorus (P), and calcium (Ca). The fraction of NPP in ecosystems that is consumed by decomposers (primarily bacteria and fungi) versus herbivores is hugely variable among ecosystems, with a larger fraction being consumed by herbivores in aquatic than in terrestrial ecosystems on average (Cyr and Pace, 1993; Cebrian and Lartigue, 2004). This pattern arises because primary producers in aquatic environments have higher nutritional quality than terrestrial primary producers (Cebrian and Lartigue, 2004) and are largely unicellular, whereas those in terrestrial ecosystems are multicellular and structurally complex – requiring compounds, such as lignin, that are difficult to digest (Lindeman, 1942; Shurin et al., 2006).

In this chapter, we discuss bottom-up and top-down influences on nutrient cycling (Fig. 9.1), focusing first on decomposer food webs, and the characteristics of primary producers (bottom-up forces) and decomposers (top-down forces)

Trophic Ecology: Bottom-Up and Top-Down Interactions across Aquatic and Terrestrial Systems, eds T. C. Hanley and K. J. La Pierre. Published by Cambridge University Press. © Cambridge University Press 2015.

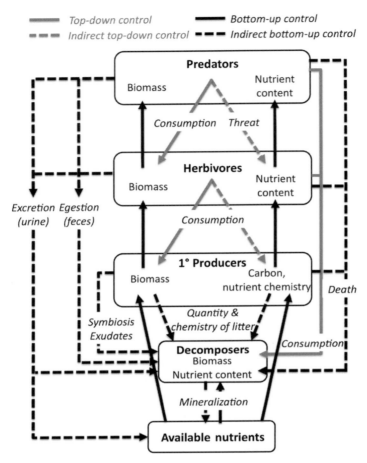

Figure 9.1 Conceptual diagram showing top-down and bottom-up effects on nutrient cycling. Solid black lines represent bottom-up control, whereby resources at one trophic level influence the biomass and nutrient content (and, for primary producers, carbon chemistry) of the next highest trophic level. Organisms can also exert indirect bottom-up control of decomposers and available nutrients (dashed black lines) through the production of excreta and through death (the production of cadavers and detritus). Primary producers also have indirect effects on nutrient cycling by supplying carbon to heterotrophic microbes, either by exuding soluble organic compounds into soil, sediment, or water that influence nutrient mineralization, or by supplying carbon to symbiotic organisms such as N-fixing microbes and mycorrhizae that influence nutrient availability. Solid gray lines represent top-down effects on nutrient cycling that occur through consumption effects on biomass or nutrient content of lower trophic levels (e.g., by changing nutrient content either within individuals or by changing species composition), or by inducing defenses. Higher trophic levels can also exert indirect top-down effects on nutrient cycling (dashed gray lines) by altering the feeding behavior (e.g., prey and location) of their prey.

that affect nutrient cycling. Next, we discuss herbivore-based food webs and highlight mechanisms by which herbivores can have top-down effects that either enhance or depress rates of nutrient cycling. Then, we extend that discussion to the influence of higher trophic levels on nutrient cycling. Finally, we discuss mechanisms by which trophic interactions mediate the transfer of nutrients among ecosystems.

Bottom-up and top-down influences on nutrient cycling in decomposer food webs

Bottom-up forces in decomposer food webs

Decomposers, primarily bacteria and fungi, but also detritivorous invertebrates and vertebrates, consume dead organic matter to obtain energy, and in the process of decomposition, break down macromolecules into smaller ones and excrete nutrients as waste products. The rate and stoichiometry of nutrients released during decomposition are influenced in part by interactions between bottom-up (i.e., primary producer detritus chemistry) and top-down (i.e., decomposer nutritional requirements and metabolic efficiency) factors, although bottom-up influences are better studied, at least in terrestrial ecosystems. For example, nutrient release from decomposing primary producer detritus is influenced by autotrophic tissue C:nutrient ratios. C:N and C:P ratios can vary among species in both aquatic and terrestrial systems by an order of magnitude (Reich and Oleksyn, 2004; Borer et al., 2013). Importantly, the C:nutrient ratios of detritus often far exceed those of decomposers (Sterner and Elser, 2002; Martinson et al., 2008), particularly for terrestrial leaf litter, which re-translocates ca. 50% of its nutrients prior to senescence (Kobe et al., 2005). Thus, during decomposition, nutrients limit use of C in detritus by decomposers, so detritus (along with its microbial decomposers) generally exhibits an initial period of nutrient uptake or "immobilization" during decomposition, followed by a period of nutrient release in proportion to mass loss (Staaf and Berg, 1981; Parton et al., 2007; Hobbie, 2008). The quantity of nutrients immobilized is influenced by the initial nutrient concentration: for litter with low initial nutrient concentrations, nutrients limit decomposition and litter exhibits more nutrient immobilization, whereas for litter with high initial nutrient concentrations, C limits decomposition, and litter exhibits nutrient release (Parton et al., 2007; Hobbie, 2008).

After the initial immobilization period, detritus generally releases nutrients in proportion to mass loss (Gessner et al., 1999; Hobbie, 2000; Parton et al., 2007; Hobbie, 2008). Thus, the rate of nutrient release from detritus is influenced by all of the various factors influencing decomposition rate, such as temperature and moisture, as well as bottom-up and top-down factors (i.e., the chemistry of primary producer biomass and decomposer characteristics, respectively). For instance, decomposition rates are slowed by high concentrations of complex, low energy-yielding molecules such as lignin (Melillo et al., 1982; Cornwell et al., 2008). Similarly, decomposition is slowed by low concentrations of nutrients in detritus. Most work has focused on concentrations of N or P (Enriquez et al., 1993; Cornwell et al., 2008), but other elements are particularly important when the dominant decomposers have unique nutritional requirements (Kaspari and Yanoviak, 2009), such as Ca in the case of lumbricid earthworms (Hobbie et al., 2006), sodium (Na) in the case of termites (Kaspari et al., 2009), or manganese (Mn) in the case of lignin-degrading fungi (Berg et al., 2010). Primary producers may also influence decomposition by releasing labile organic substrates that can "prime" decomposition of more recalcitrant organic matter, stimulating microbial activity in both aquatic (Danger et al., 2013) and terrestrial (Phillips et al., 2011) ecosystems.

As a research community, we are far from being able to predict the nature of bottom-up control of decomposition; in particular, whether C quality (i.e., the complexity of C molecules) or nutrient concentration (and which nutrient) will limit decomposition in a particular ecosystem. Studies suggest that patterns of microbial decomposer nutrient limitation only partially follow those of nutrient limitation of NPP: when supplied as fertilizer, N does not always limit decomposition where N limits NPP, but P more consistently limits decomposition where P limits NPP, such as in tropical forests on highly weathered soils (Hobbie and Vitousek, 2000; Wieder et al., 2009). Further, the role of micronutrients, which are seldom measured, is becoming increasingly apparent in influencing decomposition rates (Kaspari et al., 2008; Berg et al., 2010).

Top-down forces in decomposer food webs

In decomposer food webs, top-down factors that influence rates and patterns of nutrient release from decomposing organic matter include the energy and nutritional requirements of decomposers. These requirements in turn are influenced by the nutrient stoichiometry of decomposer biomass and decomposer carbon use efficiency (i.e., the proportion of C consumed that contributes to growth as opposed to respiration). This has long been recognized and demonstrated in aquatic ecosystems (Redfield, 1958; Hall et al., 2011), where it is clear that the stoichiometry of bacteria can influence the stoichiometry of nutrients in the water column. In terrestrial ecosystems, direct demonstration of linkages between microbial characteristics and decomposing litter nutrient dynamics is hindered by the challenges associated with measuring and manipulating decomposer physiology and stoichiometry in soils. Nevertheless, theory suggests that variation in terrestrial decomposer stoichiometry and carbon use efficiency should affect patterns of nutrient immobilization and release from litter (Manzoni et al., 2010).

Indeed, there is strong evidence that microbial stoichiometry can vary in both aquatic and terrestrial ecosystems in ways that should affect nutrient dynamics during decomposition. Bacterial C:P ratios vary by an order of magnitude in lakes (Hall et al., 2011) and marine ecosystems (Vrede et al., 2002). In soils, although average community microbial C:N and C:P ratios are fairly constrained across a wide variety of ecosystems, community microbial C:N ratios can vary by one and microbial C:P ratios by one to two orders of magnitude among different ecosystems (Cleveland and Liptzin, 2007; Manzoni et al., 2010). Such variation should have measurable effects on detrital nutrient dynamics: colonization of detritus by decomposers with narrower C:N or C:P ratios (i.e., with high nutrient requirements) should cause greater immobilization of N or P, respectively, delaying the point at which detritus switches from nutrient immobilization to release during decomposition (Manzoni et al., 2010).

Decomposer carbon use efficiency is another top-down factor that affects patterns of nutrient immobilization. Higher carbon use efficiency should drive greater immobilization: since more C is used for growth, more nutrients are needed to support that growth (Manzoni et al., 2010). Carbon use efficiency is sensitive to environmental conditions, decreasing with warming and increasing with increasing nutrient availability in both aquatic and terrestrial ecosystems (del Giorgio and Cole, 1998; Manzoni et al., 2012). These patterns imply that decomposers may release more nutrients during decomposition of similar substrates under warmer or under more nutrient-poor conditions.

Symbioses: a special case of interacting bottom-up and top-down forces

Nutrient cycling is also influenced by interactions between primary producers and symbionts, which can be conceptualized as interactions between bottom-up (primary producer) and top-down (symbiont) forces. An obvious example is the association of primary producers with bacteria that are able to acquire atmospheric N_2 through the process of N-fixation. In terrestrial ecosystems, these symbiotic associations occur primarily between plants in the pea family and bacteria called Rhizobia; between plants in other families and the bacterium *Frankia*; and between cyanobacteria or other N-fixing bacteria and a variety of hosts, including bryophytes, liverworts, hornworts, and cycads (Adams and Duggan, 2008). Examples in aquatic ecosystems include cyanobacteria that form symbiotic associations with various algae and the water fern *Azolla*, and other N-fixing bacteria that associate closely with roots of seagrasses (Welsh, 2000). These symbioses not only enhance the N nutrition of the primary producers involved in the symbioses, but also of neighboring organisms, as the N-rich tissues of N-fixing plants die, decompose, and release N into the environment (Lee et al., 2003). Indeed, this process underlies the use of "green manures" in both terrestrial ecosystems (e.g., use of plants in the pea family as cover crops) and aquatic ecosystems (e.g., use of *Azolla* in rice agriculture).

Terrestrial plants also influence nutrient cycling through their associations with mycorrhizal fungi; mycorrhizae generally increase the ability of plants to acquire nutrients from soils by increasing the volume of soil exploited.

In addition, different types of mycorrhizae are able to access (and influence the cycling of) different nutrients. For example, ectomycorrhizae and ericoid mycorrhizae – unlike arbuscular mycorrhizae or many non-mycorrhizal roots – have the ability to decompose organic matter and thus can access organic N pools, likely enhancing ecosystem N retention (Read, 1991; Lambers et al., 2008; Phillips et al., 2013).

Finally, some primary producers can enhance the supply and cycling of P through the production of specialized structures and enzymes. For example, some plants produce "cluster roots" that exude organic acids that increase the solubility of soil P (Lambers et al., 2008). These plants and others also secrete root phosphatase enzymes that cleave organically bound phosphate (Treseder and Vitousek, 2001; Lambers et al., 2008); these plants thus need not rely solely on decomposers to mineralize P for their uptake. Similarly, several phytoplankton taxa can excrete alkaline phosphatase and can thus enhance P-cycling in P-limited marine ecosystems (Hoppe, 2003).

Herbivores and nutrient cycling

A number of reviews have been published on the effects of herbivores on nutrient cycling that organize thinking around this issue in different ways (e.g., Hunter, 2001; Wardle, 2002; Bardgett and Wardle, 2003; Hartley and Jones, 2004; and many others). Although a review of this literature is beyond the scope of this chapter, we have chosen to highlight some of the mechanisms by which herbivores can both *accelerate* and *decelerate* rates of nutrient cycling in terrestrial and aquatic ecosystems.

Acceleration of nutrient cycling by herbivores

One of the most widely recognized effects of herbivores on nutrient cycling is the acceleration that can occur when herbivores excrete waste products that are more nutrient-rich than the primary producers that they consume, a process regulated by an interaction between bottom-up factors (the C:nutrient ratios of primary producers) and top-down factors (the production efficiencies and nutrient requirements of herbivores) (Hobbs, 1996; Bardgett and Wardle, 2003). If not consumed by herbivores, primary production enters the decomposer food web. In the case of leaf litter, this occurs after plants re-translocate some fraction of foliar nutrients, further reducing the nutritional content of plant litter. However, through their consumption of leaves prior to senescence, and their subsequent digestion, respiration, and excretion and egestion of wastes, some herbivores effectively partially break down organic material and reduce its C:nutrient ratio before it enters the decomposer pathway, thereby potentially accelerating microbial nutrient cycling (Hobbs, 1996; Lovett et al., 2002).

In aquatic ecosystems, various classes of herbivores increase nutrient cycling by releasing nutrients at a higher C:N but lower N:P stoichiometric ratio than in the ambient water (Vanni, 2002; Sereda and Hudson, 2011; but see

Atkinson et al., 2013). For instance, nutrient excretion by benthic invertebrates (e.g., insect larvae, annelids, mussels, and crustaceans) influences primary producer nutrient limitation and dynamics in freshwater and marine systems (Haertel-Borer et al., 2004; Conroy and Edwards, 2005; Alves et al., 2010; Atkinson et al., 2013). Similarly, nutrient recycling by zooplankton (e.g., copepods and cladocerans) can contribute significantly to the nutrient demand by phytoplankton (e.g., Attayde and Hansson, 1999; Pérez-Aragón et al., 2011). Vertebrate herbivores such as fish tend to excrete nutrients at lower rates than invertebrates but with similar N:P ratios (Attayde and Hansson, 1999; Sereda and Hudson 2011). They hence also significantly affect nutrient mineralization and primary productivity (Schaus and Vanni, 2000; Flecker et al., 2002; Burkepile et al., 2013).

In terrestrial ecosystems, N-rich frass (insect excreta) can stimulate N mineralization and leaching (Hunter, 2001; Frost and Hunter, 2007; but see Lovett et al., 2002). Furthermore, dead insects themselves may be more nutrient rich than the leaves they consume (and the litter that would otherwise fall to the forest floor), providing a nutrient-rich energy source for decomposers (Hunter, 2001). Accordingly, forest insect outbreaks have been linked in some instances to increased nitrate export in streams, although other systems strongly retain N transferred to the forest floor via frass and insect cadavers (Lovett et al., 2002; Hartley and Jones, 2004).

Herbivores also can accelerate decomposition processes and potentially nutrient cycling by concentrating, harboring, or cultivating decomposers in their guts, mounds, or gardens that are capable of breaking down compounds such as lignin or cellulose that the herbivores themselves cannot. Arguably the most well known terrestrial example of this is the ruminant ungulates, although termites also harbor intestinal protists and bacteria that can digest cellulose (Tokuda and Watanabe, 2007; Jouquet et al., 2011). Ruminant animals provide an example of the reduction in C:nutrient ratio that occurs with herbivory, described above. They are relatively inefficient at extracting N from their food, particularly from N-rich food, but are relatively efficient at extracting energy, because of their gut symbionts that aid in the digestion of cellulose and hemicellulose (Hobbs, 1996; Jarvis, 2000; Reece, 2013). Thus, the urine and dung produced by ruminants can be nutrient rich, particularly where herbivores are consuming relatively N-rich plant species (Ruess and McNaughton, 1988; Bardgett and Wardle, 2003). Consequently, grazing ungulates have been shown to increase rates of N cycling, with the best known examples in Serengeti National Park, Tanzania, Africa and Yellowstone National Park, Wyoming, USA. Ungulates increased N mineralization and gaseous N losses in both Serengeti (Ruess and McNaughton, 1988; McNaughton et al., 1997) and Yellowstone, where grazers also increased denitrification, but not N leaching (Frank and Groffman, 1998a; 1998b; Frank et al., 2000). In domesticated livestock systems, ruminants promote N losses via multiple pathways, including leaching, denitrification, and ammonia volatilization, and P losses via erosion (Jarvis, 2000). Many aquatic herbivorous vertebrates (e.g., fish, turtles, and dugongs) also have specialized gut morphologies (e.g., hindgut) and microbiota (e.g., cellulose-decomposing bacteria) that help them to process aquatic plants and likely influence nutrient cycling (Stevens and Hume, 1998; Mountfort et al., 2002; Wu et al., 2012).

Other examples of enhanced nutrient cycling associated with herbivore–symbiont systems can be found among insects that cultivate fungi (Mueller et al., 2005). For example, fungus-growing termites (Macrotermitinae) collect plant detritus in their mounds and tunnels, and use it to cultivate white-rot fungi in a microclimate that is highly conducive to decomposition compared to the surrounding arid landscapes where they occur (Jouquet et al., 2011; Schuurman, 2012). The fungi specialize in lignin degradation and produce fungal nodules that are eaten by the termites. This greatly increases the digestibility of the nutrients contained in the litter gathered by the termites, improving its C quality and reducing its C:N ratio, sometimes by orders of magnitude (Mueller et al., 2005; Schuurman, 2012). Termite–fungal mutualisms can substantially increase overall decomposition rates in arid ecosystems, particularly for wood (Cornwell et al., 2009), with termites mediating 10–20% of the C mineralization in these ecosystems (Schuurman, 2012). The nutrients that flow via litter into macrotermitine mounds ultimately may be transferred to higher trophic levels, as termites constitute a significant source of nutrition for a variety of predators that are able to capture foraging termites or dispersing alates (winged termites), or break open termite mounds (Schuurman, 2012). Similarly, leaf-cutting ants in the Neotropics carry leaves (and other plant biomass) into their subterranean nests, where they are degraded by symbiotic fungi that the ants subsequently eat. Thus, fungus-growing termites and leaf-cutting ants similarly concentrate nutrients in their nests, while denuding the surrounding area (Meyer et al., 2013). As with the termites, leaf-cutting ants likely support higher trophic levels (Terborgh et al., 2001).

Physiological responses by primary producers to herbivory may also accelerate nutrient cycling. For example, there is ample evidence that herbivores can alter the chemistry of leaf litter, including its nutrient content, in ways that may accelerate decomposition. An increase in foliar (and litter) nutrient concentration is a common response to herbivory (Hunter, 2001; Bardgett and Wardle, 2003). This occurs because herbivores remove sinks for nutrients, and also because they can disrupt re-translocation and increase the fraction of leaves that fall when they are green, prior to senescence (Lovett et al., 2002; Chapman et al., 2003). Herbivores may also influence nutrient cycling by altering plant allocation of C. For example, in response to removal of aboveground tissues by herbivores, plants may allocate to replace lost tissues, reducing allocation to belowground production, as happens with seagrasses (Vergés et al., 2008). Decreased C availability to microbes in the soil or sediment may reduce nutrient immobilization and increase nutrient availability (Holland and Detling, 1990). On the other hand, herbivory may stimulate exudation of C into the rhizosphere, stimulating microbial activity, which also may promote nutrient cycling (Bardgett and Wardle, 2003).

Although the mechanisms summarized above may increase nutrient cycling rates, whether increased rates of nutrient cycling in turn enhance primary productivity likely depends on several factors. For example, the direct negative impacts of consumption by herbivores on productivity (Milchunas and Lauenroth, 1993) may negate any positive effects of increased nutrient supply. Further, herbivores such as termites and leaf-cutter ants may concentrate nutrient availability spatially and temporally in ways that plants are unable to exploit. Finally, herbivores may promote nutrient losses from the ecosystem via migration, leaching, erosion, or gaseous N losses, leading to long-term decreases in nutrient supply despite short-term increases in nutrient cycling rates (Pastor et al., 2006; Doughty et al., 2013).

Deceleration of nutrient cycling by herbivores

In contrast to the examples cited above, herbivory may reduce rates of nutrient cycling if herbivores sequester nutrients into inaccessible pools, induce defenses, or preferentially feed on palatable species that also have relatively rapidly decomposing tissues, allowing unpalatable species with tissues that are poor quality for decomposers and herbivores alike to increase in abundance. An often-cited example of this mechanism at work is the boreal moose browsing system, for example, on Isle Royale, Michigan, USA (Pastor et al., 1988; 1993; 2006). There, moose prefer the more palatable deciduous species and fir (*Abies*), and avoid the less palatable spruce (*Picea*). Areas with long-term exclusion of moose exhibit reduced spruce abundance and higher rates of N cycling. But even without species turnover, herbivores may induce chemical changes in plants that reduce the decomposition rates of their tissues and slow rates of nutrient cycling. For example, terrestrial and aquatic primary producers can respond to herbivory by producing chemical defenses (Cronin and Hay, 1996; Karban and Baldwin, 2007; also, see Chapter 8). Given the close correspondence between traits of primary producers that deter herbivores and those that slow rates of decomposition (Grime et al., 1996), these induced defenses might be expected to decrease decomposition and nutrient cycling rates as well (Choudhury, 1988), although comparisons of decomposition rates among plants or algae with and without induced defenses are scarce.

An extreme example of selective feeding that depresses nutrient cycling occurs when herbivores preferentially feed on N-fixers and reduce their abundance in ways that measurably reduce N-fixation rates and, ultimately, soil and plant N pools (Ritchie and Tilman, 1995; Knops et al., 2000). However, chronic herbivory does not always lead to shifts in plant community composition toward dominance by less palatable species, for reasons that are not entirely clear (reviewed in Hunter, 2001). For example, in a prairie ecosystem, grasshoppers preferentially fed on species with lower tissue N concentrations, thereby increasing abundance of species with high N tissue and accelerating N cycling (Belovsky and Slade, 2000).

Herbivores may also sequester nutrients in their bodies, potentially decreasing the amount of nutrients available for primary producers. This nutrient sequestration is especially significant for long-lived organisms and for vertebrates that have a P-rich skeleton (Vanni et al., 2013). For instance, in an oligotrophic, tropical stream, armored catfish (fam. Loricariidae) had a high body P content (because of their bony plates made of calcium phosphate) but fed on P-poor periphyton, which limited their growth (Hood et al., 2005). In turn, because of these biological constraints, armored catfish had very low P excretion rates and thus they immobilized in their bodies the limited P present in the system.

Top-down and bottom-up effects of predators on nutrient cycling

Predators, from consumers of herbivores and detritivores to top predators, are present in all ecosystems and are represented by a large diversity of organisms (e.g., protozoa, invertebrates, and vertebrates) having diverse functional attributes (e.g., size, anatomy, diet, and hunting strategy). Predation, through its consumptive and non-consumptive effects on lower trophic levels, directly and indirectly controls nutrient cycling (Schmitz et al., 2010). First, we present how predators can have direct top-down effects on nutrient cycling through a trophic cascade along the food chain. Then we show how predators can indirectly control nutrient cycling through top-down, non-consumptive effects. Finally, we explain how predators have a bottom-up effect on nutrient cycling through their release of nutrients.

Top-down control through consumptive effects

Food webs are often simplified as a trophic chain with primary producers, primary consumers, secondary consumers, and top predators (e.g., kelp, herbivorous invertebrates and fishes, carnivorous fishes, and sea otters; Halpern et al., 2006). Predators regulate the density of their prey by consuming individuals, which has cascading effects on the density of the prey of their prey, and ultimately on the density of primary producers and on the amount of nutrients they take up (Leroux and Loreau, 2010). Although most studies have focused on top-down interactions among predators, herbivores, and plants, predators can also have top-down effects on nutrient mineralization through cascading trophic effects on detritivores and bacteria (Dunham, 2008; Schmitz, 2010; Crotty et al., 2012).

Real food webs are often more complex than the classic trophic chain because many species are omnivorous, some carnivores feed on both herbivores and secondary consumers, and because interaction strengths vary across predator–prey pairs (DeRuiter et al., 2005; Thompson et al., 2012). Therefore, the effect of a predator species on primary productivity depends on its trophic position and on the structure of the food web (Bruno and O'Connor, 2005; Dunham, 2008; Moksnes et al., 2008). Consequently, the overall top-down control of predation on nutrient cycling is determined by both the vertical and horizontal structure of the food web (Schindler et al., 1997; Duffy et al., 2007).

Top-down control through non-consumptive effects

Predators exert a density-mediated effect on their prey by killing individuals, but at the same time they can also have non-consumptive effects (i.e., trait-mediated effects; Schmitz and Suttle, 2001; Schmitz et al., 2004). Indeed, many organisms are able to detect predation cues and then change their physiology, behavior, and/or diet when facing a high predation risk. Changes in foraging activity and food selectivity under high predation risk have been reported for various herbivores: aquatic and terrestrial invertebrates (Schmitz and Suttle, 2001; Byrnes et al., 2006), ungulates (Frank, 2008), and marine vertebrates (Burkholder et al., 2013); also, see Chapters 4 and 5 for detailed examples in terrestrial ecosystems. Such changes in prey biology could be explained by a shift from N-limitation to C-limitation of metabolism and growth with increasing predation risk (Hawlena and Schmitz, 2010; Leroux et al., 2012). These changes in herbivore diet and activity can modify plant community composition (Schmitz, 2006), and thus primary productivity and nutrient cycling (Schmitz, 2008; Reynolds and Sotka, 2011). In addition, trait-mediated effects of predation on the nutrient content of prey can impact nutrient cycling by changing soil composition, which ultimately decreases litter decomposition (Hawlena et al., 2012). The change in prey traits can also directly affect mineralization by changing the contribution of prey to biophysical processes. For instance, in the presence of predatory fish, some benthic invertebrates spend less time feeding at the sediment surface and burrow deeper in the sediment, thus enhancing bioturbation, which results in higher mineralization rates (Stief and Hölker, 2006). Ultimately, the trait-mediated effects of predation can drive evolutionary processes (see Chapter 13).

Top-down control on nutrient fluxes between ecosystems

Trophic cascades most often affect nutrient cycling locally as predators and prey generally occur in the same ecosystem. However, many prey move between habitats and contribute to the spatial flux of nutrients (see Chapters 1 and 6). Predators of these mobile prey can thus initiate trophic cascades across ecosystem boundaries (e.g., from fish to riparian plants through aquatic predatory insects and terrestrial pollinator insects; Knight et al., 2005) or alternatively reduce the flux of nutrients from other ecosystems by targeting the vectors of allochthonous inputs (e.g., foxes preventing birds nesting on islands, Maron et al., 2006). The role of predators as vectors of nutrients across ecosystem boundaries is detailed in the next section of this chapter.

Bottom-up control through nutrient recycling

Predators, as all animals, release dissolved metabolic wastes, such as ammonium, urea and phosphates, which are directly available, and often limiting, for primary producer growth. Hence, predators participate in the bottom-up control of food webs (Clarholm, 1985; Vanni, 2002; Schmitz et al., 2010). Although the microbial loop has long been seen as the main driver of mineralization, evidence has been accumulating during the last decade that animals can contribute to a significant portion of nutrient cycling, in both aquatic (Vanni et al., 2006; Roman and McCarthy, 2010; Allgeier et al., 2013; Layman et al., 2013) and terrestrial ecosystems (Clarholm, 1985; Beard et al., 2002). For example, grazing of soil bacteria by protists increases rates of N mineralization and N acquisition by plants (Clarholm, 1985; Schimel and Bennett, 2004). Rates and stoichiometry of nutrient excretion by animals are highly variable both within and among species (Vanni et al., 2002; Pilati and Vanni, 2007; Villéger et al., 2012a; 2012b). Indeed, excretion is the result of the balance between the nutrients assimilated and the nutrient demand for growth and maintenance. Therefore, diet quality, ingestion rate, assimilation efficiency, growth rates, and body nutrient content interact to determine the nutrient budget of an animal. Body mass, through its effect on metabolic rate, is the main factor influencing mass-specific excretion rates (Vanni et al., 2002; Hall et al., 2007; Sereda et al., 2008a), but nutritional and physiological variables also have a significant effect (Anderson et al., 2005). For instance, excretion rates of N and P tend to be higher for carnivorous fishes than for herbivorous species because of the lower N and P content of primary producers (Sereda and Hudson, 2011; Burkepile et al., 2013). Nutrient recycling by predators has an important effect on primary productivity (Vanni et al., 2006; McIntyre et al., 2008; Burkepile et al., 2013), especially when animals aggregate in local patches where they can create biogeochemical hotspots (Meyer and Schultz, 1985; Boulêtreau et al., 2011; Capps and Flecker, 2013), or when they speed up the incorporation of allochthonous nutrients (e.g., fish-consuming terrestrial insects; Wurtsbaugh, 2007; Small et al., 2011).

In contrast to excretion of metabolic waste, egestion does not directly contribute to release of nutrients, as the organic matter contained in feces (i.e., the portion of food not assimilated in the gut) must still be mineralized by decomposers. Rates and stoichiometry generally differ between excretion and egestion products, with, for example, feces having lower N:P ratios than dissolved wastes in a carnivorous fish species (Meyer and Schultz, 1985).

Besides being a source of nutrients, predators can also be nutrient sinks because of the large amount of nutrients they store in their bodies (e.g., in

N-rich proteins and nucleic acids, and P-rich fat and bones) and that are thus not available for other organisms (Sereda et al., 2008b; Vanni et al., 2013). This is particularly true for the P trapped in the bony structures of predators that are not easily decomposed and can thus be buried in the sediment before being mineralized (Vanni et al., 2013). On the other hand, nutrient storage in the body can thwart the export of nutrients to other habitats. For example, in oligotrophic lakes, the addition of fish preying upon zooplankton and storing consumed nutrients during winter prevents the sedimentation of P to the bottom of the lake, and hence sustains the amount of nutrients available for primary production during summer (Schindler et al., 2001).

Organisms mediate spatial transfer of nutrients across habitats and ecosystem boundaries

Ecological systems are not isolated units, as they export and receive matter and/or organisms from adjacent and even remote ecosystems. The concept of meta-ecosystems, extending the concept of metacommunities proposed for community ecology to account for source–sink dynamics in local species abundances (Leibold et al., 2004), has been recently proposed to emphasize the importance of these reciprocal fluxes of nutrients and organisms for ecosystem structure and functioning (Gravel et al., 2010; Massol et al., 2011; also, see Chapter 1). Indeed, most of the boundaries between spatially distinct units are permeable to the flows of nutrients, detritus, and living organisms, provided that physical vectors or mobile species allow their transport across these boundaries (Polis et al., 1997). These exchanges are often bidirectional, but their ecological consequences are noticeable, and hence studied, only when the allochthonous inputs (those from outside the ecosystem) correspond to a significant proportion of the autochthonous pool of nutrients (those from within the ecosystem). Since the seminal paper by Polis et al. (1997), which called for an integration of landscape and food web ecology, many assessments of these "trophic subsidies" have been done on various systems, from local to global scales. These studies have demonstrated that allochthonous inputs of nutrients and organisms significantly affect bottom-up and top-down control of nutrient cycling within the recipient ecosystem (Nakano and Murakami, 2001; Leroux and Loreau, 2010; Bartels et al., 2012; also, see Chapter 6).

The fluxes between systems are diverse in terms of what is moving (i.e., nutrients, detritus, plants, or animals), how it is moving (i.e., passively or actively), and which types of source and sink ecosystems are involved (Fig. 9.2). Here, we first present how primary producers can incorporate allochthonous nutrients in primary production. Then we present how primary producers and animals can incorporate the allochthonous organic matter passively transported by abiotic vectors. Finally, we show the unique role of mobile animals that can actively translocate nutrients across ecosystem boundaries from small to large scales.

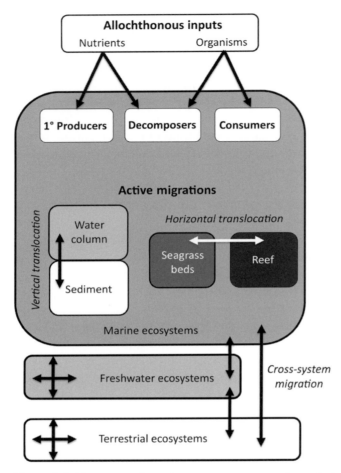

Figure 9.2 Movement of nutrients across ecosystem boundaries can occur among marine, freshwater, and terrestrial ecosystems because of allochthonous inputs of dead organisms and nutrients and because of migration. Similarly, within any of these broad ecosystem types, migration or movement of nutrients or dead organisms can occur between different habitat types (e.g., in marine ecosystems, migration can happen vertically between the water column and the sediment, or horizontally between reef and seagrass habitats).

Allochthonous flux of nutrients affects bottom-up control by primary producers

Global hydrological and atmospheric circulation strongly structures the flow of nutrients among ecosystems and can contribute to the bottom-up control of food webs. Atmospheric circulation continuously transports large amounts of dust that can fertilize terrestrial and aquatic ecosystems from local to inter-continental scales. For example, dust from Australian deserts not only affects adjacent forests and lakes, but also contributes to sustaining the productivity of the western Pacific Ocean through iron input (McTainsh and Strong, 2007). This physical transfer of nutrients can even connect ecosystems on different continents. The most striking example is the flow of 10^6 metric tons of dust per year from the Sahara Desert to the Amazon basin. This dust, carried over the Atlantic Ocean, is estimated to bring more than $1\,kg$ of $P\,ha^{-1}\,y^{-1}$ and thus to support the high primary productivity of the Amazonian rainforest, which is growing on nutrient-poor soils (Swap et al., 1992).

Nutrients transported in precipitation can also subsidize ecosystems signifi-cantly (Kuhn, 2001; Chuyong et al., 2004). However, the main role of water as a vector of nutrients and organic matter is within and among aquatic ecosystems. Dissolved nutrients and particulate organic matter are transferred from soils to rivers, then to estuaries, and finally to marine ecosystems. Water flow hence subsidizes coastal marine ecosystems in nutrients and organic matter originat-ing from freshwater ecosystems. For instance, some marine benthic food webs rely strongly on terrestrial particulate organic matter (Darnaude, 2005; Gillson, 2011). In marine ecosystems, large-scale currents also continuously transport nutrients horizontally and vertically (see Chapter 2). For instance, the powerful Humboldt Current along the west coast of South America creates an upwelling of nutrients that sustains high primary and secondary productivity in this region (Thiel et al., 2007).

Passive flux of plants and animals

Besides transporting nutrients and dissolved and particulate organic matter, air and water can also transport plants and animals across ecosystem boundaries. Organisms that are passively transported by these physical vectors are either dead (e.g., leaves in a forest, fragments of macroalgae in marine ecosystems, and animal carrion) or small animals that cannot escape the physical flow (e.g., insects and zooplankton). These organisms and their waste products are incorporated at different trophic levels in recipient food webs by detritivores, herbivores, and predators, and can represent a significant proportion of their resources. For instance, inputs of plant material (e.g., litterfall, fruit, seeds, and wood) and insects to freshwater ecosystems are an important source of organic matter for aquatic animals (Schindler and Scheuerell, 2002; Lamberti et al., 2010; also, see Chapter 6). Reciprocally, aquatic ecosystems can export organic matter to adjacent terrestrial ecosystems (Bastow et al., 2002; Crawley et al., 2009). For example, tides and waves leave large amounts of detached marine macrophytes on shores (Crawley et al., 2009), as well as animal carrion (e.g., fish and birds; Rose and Polis 1998). This aquatic organic matter significantly subsidizes terrestrial herbivores (e.g., arthropods) and predators (e.g., lizards, birds, and coyotes) in habitats with low productivity, such as remote islands (Barrett et al., 2005) or desert coastal areas (Rose and Polis, 1998). Marine allochthonous resources thus have direct positive effects on terrestrial consumers (Dugan et al., 2003), as well as indirect effects on terrestrial plants through the nutrient recycling done by consumers (Spiller et al., 2010). Water flow can also carry subsidies among distant marine ecosystems, such as particulate organic matter from the open ocean to fringing coral reefs (Wyatt et al., 2010), or fragments of kelp from reefs to seagrass meadows (Hyndes et al., 2012). In the latter case, macroalgae fragments subsidize both grazers and decomposers living in seagrass habitats and subsequently contribute indirectly to the nutrients assimilated by seagrass.

Animal migrations between ecosystems

Besides passive fluxes of nutrients and organisms across ecosystem boundaries, active dispersion of organisms also contributes to the transfer of nutrients among ecosystems (Polis et al., 1997). Indeed, mobile animals have the unique ability to move nutrients across ecosystem boundaries through migrations on short (day/night) or long (season/year) temporal scales. More importantly, these active movements can be in the opposite direction to passive fluxes due to physical vectors (e.g., gravity and water flow). One of the most famous examples of an allochthonous subsidy supported by an animal migration is the anadromous salmon species (genus *Onchorhynchus*) in Western North America (Helfield and Naiman, 2001; Janetski et al., 2009). These species spawn in rivers in the USA and Canada, juvenile fish move to the Pacific Ocean, and after several years of growth in the marine ecosystem, mature adults migrate up their native river to reproduce and then die. Anadromous salmon thus transfer large amounts of marine-derived nutrients in their bodies, and their carcasses fertilize the recipient freshwater ecosystems (Moore et al., 2007). Indeed, carcasses are a resource for bacteria and aquatic macroinvertebrates (Wipfli et al., 1998; Winder et al., 2005), and they thus indirectly benefit invertivorous, juvenile anadromous salmonids and even non-migratory species (Wipfli et al., 2003). In addition, spawning salmonids are a prey for top predators (mainly bear), which transfer the marine nutrients contained in fish to riparian ecosystems through defecation. This allochthonous input of marine nutrients mediated by two animal species significantly influences the productivity and biodiversity of several components of riparian ecosystems, such as trees (Helfield and Naiman, 2001) and insects (Hocking and Reimchen, 2002; Hocking and Reynolds, 2011). Freshwater ecosystems also export nutrients to riparian ecosystems through the dispersion of flying insects; larvae grow on benthic subsidies before adults emerge and fly to surrounding habitat to reproduce and die. These allochthonous subsides can represent a large flux of nutrients around lakes and rivers (Nakano and Murakami, 2001; Gratton and Vander Zanden, 2009; also, see Chapter 6).

Animal migrations between habitats

Besides these massive and temporally limited animal migrations, animals also contribute to moving nutrients between neighboring habitats on a daily basis. Indeed, many species migrate between areas favorable for foraging and areas favorable for resting (e.g., that are less risky in term of predation). These daily movements, coupled with the inertia of the digestive process, contribute to subsidizing the resting habitats with nutrients coming from foraging areas that are often more productive. These allochthonous fluxes thus occur among habitats within aquatic or terrestrial ecosystems. For instance, herbivores such as roe deer graze in meadows, but rest in forest patches where they urinate and egest feces rich in N and P, thus subsidizing forests with nutrients derived from cultivated fields (Abbas et al., 2012). Birds can also contribute significantly to the allochthonous input of nutrients in forests (Fujita and Koike, 2009). In aquatic ecosystems, herbivorous and carnivorous vertebrates also transfer nutrients between habitats. For instance, haemulid fish feeding on invertebrates in seagrass patches during the night and resting near coral reefs during the day significantly increase coral growth through the excretion and egestion of nutrients coming from seagrass patches (Meyer and Schultz, 1985). Fish can also bring nutrients into enclosed habitats, such as small caves in rocky reefs, which are poorly connected with the water column (e.g., zooplanktivorous damselfish; Bray et al., 1981). Some fish species can even deliberately move far outside their feeding area for egesting their feces (Brown et al., 1996; Krone et al., 2008).

The potential of a species to transfer nutrients from its foraging habitat to its resting place is constrained by the lag between food ingestion and waste production. Therefore, most species transfer nutrients short distances (< 1 km) within aquatic or terrestrial ecosystems. However, some amphibious animals transfer nutrients across ecosystem boundaries. For example, oceanic birds feeding on marine prey (mainly fish) subsidize the terrestrial habitats where they rest or reproduce. The allochthonous nutrients released by birds (i.e., guano) shape the composition of plant communities and their productivity (Sanchez-Pinero and Polis, 2000; Fukami et al., 2006; Young et al., 2010). Large amphibious herbivores, such as *Hippopotamus*, also have the potential to move nutrients across the aquatic/terrestrial boundary (Naiman and Rogers, 1997).

Animal vertical migrations in stratified ecosystems

In addition to moving nutrients horizontally, animals also move nutrients between the vertical strata of ecosystems. For instance, in savanna, ground-dwelling predators feed on both ground and arboreal prey, and thus increase the coupling between the two food webs (Pringle and Fox-Dobbs, 2008). In freshwater and coastal marine ecosystems, many fish species feed on benthic prey and then swim in the water column (pelagic zone), where they release nutrients initially contained in their benthic prey. This vertical and horizontal translocation of benthic nutrients can significantly increase planktonic productivity (Schaus and Vanni, 2000). In the ocean, whales foraging at depth move nutrients from deep habitats to the euphotic zone, which is often nutrient limited (Lavery et al., 2010; Roman and McCarthy, 2010). This "whale pump" thus counterbalances the sedimentation process of dead plants and animals that sink to the aphotic zone and helps sustain the productivity of the ocean.

The connection of nutrient cycles among and within ecosystems is mediated by primary producers and consumers through the transport and/or the incorporation of nutrients into food webs. Trophic subsidies vary in terms of the nature, intensity, and stability of the incoming flow of nutrients, the type of vector, and the characteristics of the local food web (Likens and Bormann, 1974; Polis et al., 1997; Lamberti et al., 2010). Comparisons among systems (e.g., aquatic versus terrestrial) are thus required for a global understanding of the interactive effects of species traits (e.g., diet, mobility, and metabolism) and food web characteristics (e.g., number of compartments and connectivity among them) on the impact of allochthonous fluxes on bottom-up control of food webs (Baxter et al., 2005; Gravel et al., 2010; Massol et al., 2011; Marcarelli et al., 2011).

Conclusion

Human activities are increasingly affecting nutrient cycles through the addition of anthropogenic N and P, as well as by modifying the composition and abundance of primary producer, decomposer, herbivore, and predator species in aquatic and terrestrial ecosystems. Therefore, future investigations are needed to quantify how these global changes affect the interactions between the bottom-up and top-down effects of organisms on nutrient cycles (see Chapter 14).

For instance, the diversity and abundance of predators are severely affected by human activities in many terrestrial and aquatic ecosystems, through exploitation but also through introduction of non-native species (Cucherousset and Olden, 2011; Estes et al., 2011). Developing an integrated approach that accounts for both their top-down and bottom-up effects on nutrient cycling, including indirect effects, is thus an urgent need. Toward this objective, functional traits (e.g., size, morphology, metabolic, and excretion rates) could offer an operational framework (Schmitz et al., 2010; Massol et al., 2011; also, see Chapter 8).

Human activities also impact allochthonous fluxes between ecosystems, directly by removing or adding organisms able to transport nutrients (e.g., through fishing and introduction of non-native species) and by changing ecosystem productivity (e.g., through pollution), and indirectly by altering the connectivity between ecosystems (e.g., through construction and removal of dams). Therefore, future developments of the meta-ecosystem concept (Gravel et al., 2010; Massol et al., 2011; also, see Chapter 1) should help to integrate spatial and functional ecology.

References

Abbas, F., Merlet, J., Morellet, N., et al. (2012). Roe deer may markedly alter forest nitrogen and phosphorus budgets across Europe. *Oikos*, **121**, 1271–1278.

Adams, D. G. and Duggan, P. S. (2008). Cyanobacteria-bryophyte symbioses. *Journal of Experimental Botany*, **59**, 1047–1058.

Allgeier, J. E., Yeager, L. A. and Layman, C. A. (2013). Consumers regulate nutrient limitation regimes and primary production in seagrass ecosystems. *Ecology*, **94**, 521–529.

Alves, J., Caliman, A., Guariento, R. D., et al. (2010). Stoichiometry of benthic invertebrate nutrient recycling: interspecific variation and the role of body mass. *Aquatic Ecology*, **44**, 421–430.

Anderson, T. R., Hessen, D. O., Elser, J. J. and Urabe, J. (2005). Metabolic stoichiometry and the fate of excess carbon and nutrients in consumers. *The American Naturalist*, **165**, 1–15.

Atkinson, C. L., Vaughn, C. C., Forshay, K. J. and Cooper, J. T. (2013). Aggregated filter-feeding consumers alter nutrient limitation: consequences for ecosystem and community dynamics. *Ecology*, **94**, 1359–1369.

Attayde, J. L. and Hansson, L. A. (1999). Effects of nutrient recycling by zooplankton and fish on phytoplankton communities. *Oecologia*, **121**, 47–54.

Bardgett, R. D. and Wardle, D. A. (2003). Herbivore-mediated linkages between aboveground and belowground communities. *Ecology*, **84**, 2258–2268.

Barrett, K., Anderson, W. B., Wait, D. A., et al. (2005). Marine subsidies alter the diet and abundance of insular and coastal lizard populations. *Oikos*, **109**, 145–153.

Bartels, P., Cucherousset, J., Steger, K., et al. (2012). Reciprocal subsidies between freshwater and terrestrial ecosystems structure consumer resource dynamics. *Ecology*, **93**, 1173–1182.

Bastow, J., Sabo, J., Finlay, J. and Power, M. (2002). A basal aquatic-terrestrial trophic link in rivers: algal subsidies via shore-dwelling grasshoppers. *Oecologia*, **131**, 261–268.

Baxter, C. V., Fausch, K. D. and Saunders, W. C. (2005). Tangled webs: reciprocal flows of invertebrate prey link streams and riparian zones. *Freshwater Biology*, **50**, 201–220.

Beard, K. H., Vogt, K. A. and Kulmatiski, A. (2002). Top-down effects of a terrestrial frog on forest nutrient dynamics. *Oecologia*, **133**, 583–593.

Belovsky, G. E. and Slade, J. B. (2000). Insect herbivory accerlates nutrient cycling and increase plant production. *Proceedings of the National Academy of Sciences of the USA*, **97**, 14412–14417.

Berg, B., Davey, M. P., De Marco, A., et al. (2010). Factors influencing limit values for pine needle litter decomposition: a synthesis for boreal and temperate pine forest systems. *Biogeochemistry*, **100**, 57–73.

Borer, E. T., Bracken, M. E., Seabloom, E. W., et al. (2013). Global biogeography of autotroph chemistry: is insolation a driving force? *Oikos*, **122**, 1121–1130.

Boulêtreau, S., Cucherousset, J., Villéger, S., Masson, R. and Santoul, F. (2011). Colossal aggregations of giant alien freshwater fish as a potential biogeochemical hotspot. *PLoS One*, **6**, e25732.

Bray, R. N., Miller, A. C. and Geesey, G. G. (1981). The fish connection: a trophic link between planktonic and rocky reef communities? *Science*, **214**, 204–205.

Brown, G. E., Chivers, D. P. and Smith, R. J. F. (1996). Effects of diet on localized defecation by Northern pike, *Esox lucius*. *Journal of Chemical Ecology*, **22**, 467–475.

Bruno, J. F. and O'Connor, M. I. (2005). Cascading effects of predator diversity and omnivory in a marine food web. *Ecology Letters*, **8**, 1048–1056.

Burkepile, D. E., Allgeier, J. E., Shantz, A. A., et al. (2013). Nutrient supply from fishes facilitates macroalgae and suppresses corals in a Caribbean coral reef ecosystem. *Scientific Reports*, **3**, 1493.

Burkholder, D. A., Heithaus, M. R., Fourqurean, J. W., Wirsing, A. and Dill, L. M. (2013). Patterns of top-down control in a seagrass ecosystem: could a roving apex predator induce a behaviour-mediated trophic cascade? *Journal of Animal Ecology*, **82**, 1192–1202.

Byrnes, J., Stachowicz, J. J., Hultgren, K. M., et al. (2006). Predator diversity strengthens trophic cascades in kelp forests by modifying herbivore behaviour. *Ecology Letters*, **9**, 61–71.

Capps, K. A. and Flecker, A. S. (2013). Invasive fishes generate biogeochemical hotspots in a nutrient-limited system. *PLoS One*, **8**, e54093.

Cebrian, J. and Lartigue, J. (2004). Patterns of herbivory and decomposition in aquatic and terrestrial ecosystems. *Ecological Monographs*, **74**, 237–259.

Chapman, S. K., Hart, S. C., Cobb, N. S., Whitham, T. G. and Koch, G. W. (2003). Insect herbivory increases litter quality and decomposition: an extension of the acceleration hypothesis. *Ecology*, **84**, 2867–2876.

Choudhury, D. (1988). Herbivore induced changes in leaf-litter resource quality: a neglected aspect of herbivory in ecosystem nutrient dynamics. *Oikos*, **51**, 389–393.

Chuyong, G. B., Newbery, D. M. and Songwe, N. C. (2004). Rainfall input, throughfall and stemflow of nutrients in a central African rain forest dominated by ectomycorrhizal trees. *Biogeochemistry*, **67**, 73–91.

Clarholm, M. (1985). Interactions of bacteria, protozoa and plants leading to mineralization of soil nitrogen. *Soil Biology and Biochemistry*, **17**, 181–187.

Cleveland, C. C. and Liptzin, D. (2007). C: N: P stoichiometry in soil: is there a "Redfield ratio" for the microbial biomass? *Biogeochemistry*, **85**, 235–252.

Conroy, J. D. and Edwards, W. J. (2005). Soluble nitrogen and phosphorus excretion of exotic freshwater mussels (*Dreissena* spp.): potential impacts for nutrient

remineralisation in western Lake Erie. *Freshwater Biology*, **50**, 1146–1162.

Cornwell, W. K., Cornelissen, J. H. C., Amatangelo, K., et al. (2008). Plant species traits are the predominant control on litter decomposition rates within biomes worldwide. *Ecology Letters*, **11**, 1065–1071.

Cornwell, W. K., Cornelissen, J. H., Allison, S. D., et al. (2009). Plant traits and wood fates across the globe: rotted, burned, or consumed? *Global Change Biology*, **15**, 2431–2449.

Crawley, K. R., Hyndes, G. A., Vanderklift, M. A., Revill, A. T. and Nichols, P. D. (2009). Allochthonous brown algae are the primary food source for consumers in a temperate, coastal environment. *Marine Ecology Progress Series*, **376**, 33–44.

Cronin, G. and Hay, M. E. (1996). Induction of seaweed chemical defenses by amphipod grazing. *Ecology*, **77**, 2287–2301.

Crotty, F. V, Adl, S. M., Blackshaw, R. P. and Murray, P. J. (2012). Protozoan pulses unveil their pivotal position within the soil food web. *Microbial Ecology*, **63**, 905–918.

Cucherousset, J. and Olden, J. D. (2011). Ecological impacts of nonnative freshwater fishes. *Fisheries*, **36**, 215–230.

Cyr, H. and Pace, M. L. (1993). Magnitude and patterns of herbivory in aquatic and terrestrial ecosystems. *Nature*, **361**, 148–150.

Danger, M., Cornut, J., Chauvet, E., Chavez, P., Elger, A. and Lecerf, A. (2013). Benthic algae stimulate leaf litter decomposition in detritus-based headwater streams: a case of aquatic priming effect? *Ecology*, **94**(7), 1604–1613.

Darnaude, A. M. (2005). Fish ecology and terrestrial carbon use in coastal areas: implications for marine fish production. *Journal of Animal Ecology*, **74**, 864–876.

del Giorgio, P. A. and Cole, J. J. (1998). Bacterial growth efficiency in natural aquatic systems. *Annual Review of Ecology and Systematics*, **29**, 503–541.

DeRuiter, P. C., Wolters, V., Moore, J. C. and Winemiller, K. O. (2005). Food web ecology.

Playing Jenga and beyond. *Science*, **309**, 68–70.

Doughty, C. E., Wolf, A. and Malhi, Y. T. (2013). The legacy of the Pleistocene megafauna extinctions on nutrient availability in Amazonia. *Nature Geoscience*, **6**, 761–764.

Duffy, J. E., Cardinale, B. J., France, K. E., et al. (2007). The functional role of biodiversity in ecosystems: incorporating trophic complexity. *Ecology Letters*, **10**, 522–538.

Dugan, J. E., Hubbard, D. M., McCrary, M. D. and Pierson, M. O. (2003). The response of macrofauna communities and shorebirds to macrophyte wrack subsidies on exposed sandy beaches of southern California. *Estuarine, Coastal and Shelf Science*, **58**, 25–40.

Dunham, A. E. (2008). Above and below ground impacts of terrestrial mammals and birds in a tropical forest. *Oikos*, **117**, 571–579.

Enriquez, S., Duarte, C. M. and Sand-Jensen, K. (1993). Patterns in decomposition rates among photosynthetic organisms: the importance of detritus C:N:P content. *Oecologia*, **94**, 457–471.

Estes, J. A., Terborgh, J., Brashares, J. S., et al. (2011). Trophic downgrading of planet Earth. *Science*, **333**, 301–306.

Flecker, A. S., Taylor, B. W. and Bernhardt, E. S. (2002). Interactions between herbivorous fishes and limiting nutrients in a tropical stream ecosystem. *Ecology*, **83**, 1831–1844.

Frank, D. A. (2008). Evidence for top predator control of a grazing ecosystem. *Oikos*, **117**, 1718–1724.

Frank, D. A. and Groffman, P. M. (1998a). Denitrification in a semi-arid grazing ecosystem. *Oecologia*, **117**, 564–569.

Frank, D. A. and Groffman, P. M. (1998b). Ungulate vs. landscape control of soil C and N processes in grasslands of Yellowstone National Park. *Ecology*, **79**, 2229–2241.

Frank, D. A., Groffman, P. M., Evans, R. D. and Tracy, B. F. (2000). Ungulate stimulation of nitrogen cycling and retention in Yellowstone Park grasslands. *Oecologia*, **123**, 116–121.

Frost, C. J. and Hunter, M. D. (2007). Recycling of nitrogen in herbivore feces: plant recovery, herbivore assimilation, soil retention, and leaching losses. *Oecologia*, **151**, 42–53.

Fujita, M. and Koike, F. (2009). Landscape effects on ecosystems: birds as active vectors of nutrient transport to fragmented urban forests versus forest-dominated landscapes. *Ecosystems*, **12**, 391–400.

Fukami, T., Wardle, D. A., Bellingham, P. J., et al. (2006). Above- and below-ground impacts of introduced predators in seabird-dominated island ecosystems. *Ecology Letters*, **9**, 1299–1307.

Gessner, M. O., Chauvet, E. and Dobson, M. (1999). A perspective on leaf litter breakdown in streams. *Oikos*, **85**, 377–384.

Gillson, J. (2011). Freshwater flow and fisheries production in estuarine and coastal systems: where a drop of rain is not lost. *Reviews in Fisheries Science*, **19**, 168–186.

Gratton, C. and Vander Zanden, M. J. (2009). Flux of aquatic insect productivity to land: comparison of lentic and lotic ecosystems. *Ecology*, **90**, 2689–2699.

Gravel, D., Guichard, F., Loreau, M. and Mouquet, N. (2010). Source and sink dynamics in meta-ecosystems. *Ecology*, **91**, 2172–2184.

Grime, J. P., Cornelissen, J. H. C., Thompson, K. and Hodgson, J. G. (1996). Evidence of a causal connection between anti-herbivore defence and the decomposition rate of leaves.

Haertel-Borer, S. S., Allen, D. M. and Dame, R. F. (2004). Fishes and shrimps are significant sources of dissolved inorganic nutrients in intertidal salt marsh creeks. *Journal of Experimental Marine Biology and Ecology*, **311**, 79–99.

Hall, E., Maixner, F., Franklin, O., et al. (2011). Linking microbial and ecosystem ecology using ecological stoichiometry: a synthesis of conceptual and empirical approaches. *Ecosystems*, **14**, 261–273.

Hall, R. O., Jr., Koch, B. J. and Marshall, M. C. (2007). How body size mediates the role of animals in nutrient cycling in aquatic ecosystems. In *Body size: The Structure and Function of Aquatic Ecosystems*, ed. A. G. Hildrew, D. G. Raffaelli and R. Edmonds-Brown. Cambridge, UK: Cambridge University Press, pp. 286–305.

Halpern, B. S., Cottenie, K. and Broitman, B. R. (2006). Strong top-down control in southern California kelp forest ecosystems. *Science*, **312**, 1230–1232.

Hartley, S. E. and Jones, T. H. (2004). Insect herbivores, nutrient cycling, and plant productivity. In *Insects and Ecosystem Function*, ed. W. W. Weisser and E. Siemann. Berlin: Springer-Verlag, pp. 27–52.

Hawlena, D. and Schmitz, O. J. (2010). Herbivore physiological response to predation risk and implications for ecosystem nutrient dynamics. *Proceedings of the National Academy of Sciences of the USA*, **107**, 15503–15507.

Hawlena, D., Strickland, M. S., Bradford, M. A. and Schmitz, O. J. (2012). Fear of predation slows plant-litter decomposition. *Science*, **336**, 1434–1438.

Helfield, J. M. and Naiman, R. J. (2001). Effects of salmon-derived nitrogen on riparian forest growth and implications for stream productivity. *Ecology*, **82**, 2403–2409.

Hobbie, S. E. (2000). Interactions between lignin and nutrient availability during decomposition in Hawaiian montane forest. *Ecosystems*, **3**, 484–494.

Hobbie, S. E. (2008). Nitrogen effects on litter decomposition: a five-year experiment in eight temperate grassland and forest sites. *Ecology*, **89**, 2633–2644.

Hobbie, S. E. and Vitousek, P. M. (2000). Nutrient regulation of decomposition in Hawaiian montane forests: do the same nutrients limit production and decomposition? *Ecology*, **81**, 1867–1877.

Hobbie, S. E., Reich, P. B., Oleksyn, J., et al. (2006). Tree species effects on decomposition and forest floor dynamics in a common garden. *Ecology*, **87**, 2288–2297.

Hobbs, N. T. (1996). Modification of ecosystems by ungulates. *The Journal of Wildlife Management*, **60**, 695–713.

Hocking, M. D. and Reimchen, T. E. (2002). Salmon-derived nitrogen in terrestrial invertebrates from coniferous forests of the Pacific Northwest. *BMC Ecology*, **14**, 1–14.

Hocking, M. D. and Reynolds, J. D. (2011). Impacts of salmon on riparian plant diversity. *Science*, **331**, 1609–1612.

Holland, E. A. and Detling, J. K. (1990). Plant response to herbivory and belowground nitrogen cycling. *Ecology*, **71**, 1040–1049.

Hood, J. M., Vanni, M. J. and Flecker, A. S. (2005). Nutrient recycling by two phosphorus-rich grazing catfish: the potential for phosphorus-limitation of fish growth. *Oecologia*, **146**, 247–57.

Hoppe, H. (2003). Phosphatase activity in the sea. *Hydrobiologia*, **493**, 187–200.

Hunter, M. D. (2001). Insect population dynamics meets ecosystem ecology: effects of herbivory on soil nutrient dynamics. *Agricultural and Forest Entomology*, **3**, 77–84.

Hyndes, G., Lavery, P. and Doropoulos, C. (2012). Dual processes for cross-boundary subsidies: incorporation of nutrients from reef-derived kelp into a seagrass ecosystem. *Marine Ecology Progress Series*, **445**, 97–107.

Janetski, D. J., Chaloner, D. T., Tiegs, S. D. and Lamberti, G. A. (2009). Pacific salmon effects on stream ecosystems: a quantitative synthesis. *Oecologia*, **159**, 583–595.

Jarvis, S. C. (2000). Soil-plant-animal interactions and impact on nitrogen and phosphorus cycling and recycling in grazed pastures. In *Grassland Ecophysiology and Grazing Ecology*, ed. G. Lemaire, J. Hodgson, A. de Moraes, C. Nabinger and P. C. de F. Carvalho. New York: CAB International, pp. 191–207.

Jouquet, P., Traoré, S., Choosai, C., Hartmann, C. and Bignell, D. (2011). Influence of termites on ecosystem functioning: ecosystem services provided by termites. *European Journal of Soil Biology*, **47**, 215–222.

Karban, R. and Baldwin, I. T. (2007). *Induced Responses to Herbivory*. Chicago: University of Chicago Press.

Kaspari, M. and Yanoviak, S. P. (2009). Biogeochemistry and the structure of tropical brown food webs. *Ecology*, **90**, 3342–3351.

Kaspari, M., Garcia, M. N., Harms, K. E., et al. (2008). Multiple nutrients limit litterfall and decomposition in a tropical forest. *Ecology Letters*, **11**, 35–43.

Kaspari, M., Yanoviak, S. P., Dudley, R., Yuan, M. and Clay, N. A. (2009). Sodium shortage as a constraint on the carbon cycle in an inland tropical rainforest. *Proceedings of the National Academy of Sciences of the USA*, **106**, 19405–19409.

Knight, T. M., McCoy, M. W., Chase, J. M., McCoy, K. A. and Holt, R. D. (2005). Trophic cascades across ecosystems. *Nature*, **437**, 880–883.

Knops, J. M., Ritchie, M. and Tilman, D. (2000). Selective herbivory on a nitrogen fixing legume (*Lathyrus venosus*) influences productivity and ecosystem nitrogen pools in an oak savanna. *Ecoscience*, **7**(2), 166–174.

Kobe, R. K., Lepczyk, C. A. and Iyer, M. (2005). Resorption efficiency decreases with increasing green leaf nutrients in a global data set. *Ecology*, **86**(10), 2780–2792.

Krone, R., Bshary, R., Paster, M., et al. (2008). Defecation behaviour of the lined bristletooth surgeonfish *Ctenochaetus striatus* (Acanthuridae). *Coral Reefs*, **27**, 619–622.

Kuhn, M. (2001). The nutrient cycle through snow and ice, a review. *Aquatic Sciences*, **63**, 150–167.

Lambers, H., Raven, J. A., Shaver, G. R. and Smith, S. E. (2008). Plant nutrient-acquisition strategies change with soil age. *Trends in Ecology and Evolution*, **23**, 95–103.

Lamberti, G. A., Chaloner, D. T. and Hershey, A. E. (2010). Linkages among aquatic ecosystems. *Journal of the North American Benthological Society*, **29**, 245–263.

Lavery, T. J., Roudnew, B., Gill, P., et al. (2010). Iron defecation by sperm whales stimulates carbon export in the Southern Ocean. *Proceedings of the Royal Society B: Biological Sciences*, **277**, 3527–3531.

Layman, C. A., Allgeier, J. E., Yeager, L. A. and Stoner, E. W. (2013). Thresholds of

ecosystem response to nutrient enrichment from fish aggregations. *Ecology*, **94**, 530–536.

Lee, T. D., Reich, P. B. and Tjoelker, M. G. (2003). Legume presence increases photosynthetic and N concentration of co-occurring non-fixers but does not modulate their response to carbon dioxide enrichment. *Oecologia*, **137**, 22–31.

Leibold, M. A., Holyoak, M., Mouquet, N., et al. (2004). The metacommunity concept: a framework for multi-scale community ecology. *Ecology Letters*, **7**, 601–613.

Leroux, S. J. and Loreau, M. (2010). Consumer-mediated recycling and cascading trophic interactions. *Ecology*, **91**, 2162–2171.

Leroux, S. J., Hawlena, D. and Schmitz, O. J. (2012). Predation risk, stoichiometric plasticity and ecosystem elemental cycling. *Proceedings of the Royal Society of London Series B*, **279**, 4183–4191.

Likens, G. E. and Bormann, F. H. (1974). Linkages between terrestrial and aquatic ecosystems. *BioScience*, **24**, 447–456.

Lindeman, R. L. (1942). The trophic-dynamic aspect of ecology. *Ecology*, **23**, 399–418.

Lovett, G. M., Christenson, L. M., Groffman, P. M., et al. (2002). Insect defoliation and nitrogen cycling in forests. *BioScience*, **52**, 335–341.

Manzoni, S., Trofymow, J. A., Jackson, R. B. and Porporato, A. (2010). Stoichiometric controls on carbon, nitrogen, and phosphorus dynamics in decomposing litter. *Ecological Monographs*, **80**, 89–106.

Manzoni, S., Taylor, P., Richter, A., Porporato, A. and Ågren, G. (2012). Environmental and stoichiometric controls on microbial carbon-use efficiency in soils. *New Phytologist*, **196**, 79–91.

Marcarelli, A. M., Baxter, C. V., Mineau, M. M. and Hall, R. O. (2011). Quantity and quality: unifying food web and ecosystem perspectives on the role of resource subsidies in freshwaters. *Ecology*, **92**, 1215–1225.

Maron, J. L., Estes, J. A., Croll, D. A., et al. (2006). An introduced predator alters Aleutian Island plant communities by thwarting nutrient subsidies. *Ecological Monographs*, **76**, 3–24.

Martinson, H. M., Schneider, K., Gilbert, J., et al. (2008). Detritivory: stoichiometry of a neglected trophic level. *Ecological Research*, **23**, 487–491.

Massol, F., Gravel, D., Mouquet, N., et al. (2011). Linking community and ecosystem dynamics through spatial ecology. *Ecology Letters*, **14**, 313–323.

McIntyre, P. B., Flecker, A. S., Vanni, M. J., et al. (2008). Fish distributions and nutrient cycling in streams: can fish create biogeochemical hotspots? *Ecology*, **89**, 2335–2346.

McNaughton, S., Banyikwa, F. and McNaughton, M. (1997). Promotion of the cycling of diet-enhancing nutrients by African grazers. *Science*, **278**, 1798–1800.

McTainsh, G. and Strong, C. (2007). The role of aeolian dust in ecosystems. *Geomorphology*, **89**, 39–54.

Melillo, J. M., Aber, J. D. and Muratore, J. F. (1982). Nitrogen and lignin control of hardwood leaf litter decomposition dynamics. *Ecology*, **63**, 621–626.

Meyer, J. L. and Schultz, E. T. (1985). Migrating Haemulid fishes as a source of nutrients and organic matter on coral reefs. *Limnology and Oceanography*, **30**, 146–156.

Meyer, S. T., Neubauer, M., Sayer, E. J., et al. (2013). Leaf-cutting ants as ecosystem engineers: topsoil and litter perturbations around *Atta cephalotes* nests reduce nutrient availability. *Ecological Entomology*, **38**, 497–504.

Milchunas, D. G. and Lauenroth, W. K. (1993). Quantitative effects of grazing on vegetation and soils over a global range of environments. *Ecological Monographs*, **63**, 327–366.

Moksnes, P., Gullstro, M., Tryman, K. and Baden, S. (2008). Trophic cascades in a temperate seagrass community. *Oikos*, **117**, 763–777.

Moore, J. W., Schindler, D. E., Carter, J. L., et al. (2007). Biotic control of stream fluxes: spawning salmon drive nutrient and matter export. *Ecology*, **88**, 1278–1291.

Mountfort, D. O., Campbell, J. and Clements, K. D. (2002). Hindgut fermentation in three species of marine herbivorous fish. *Applied and Environmental Microbiology*, **68**, 1374–1380.

Mueller, U. G., Gerardo, N. M., Aanen, D. K., Six, D. L. and Schultz, T. R. (2005). The evolution of agriculture in insects. *Annual Review of Ecology, Evolution, and Systematics*, **36**, 563–595.

Naiman, R. J. and Rogers, K. H. (1997). Large animals and system-level characteristics in river corridors. *BioScience*, **47**, 521–529.

Nakano, S. and Murakami, M. (2001). Reciprocal subsidies: dynamic interdependence. *Proceedings of the National Academy of Sciences of the USA*, **98**, 166–170.

Parton, W. A., Silver, W. L., Burke, I. C., et al. (2007). Global-scale similarities in nitrogen release patterns during long-term decomposition. *Science*, **315**, 361–364.

Pastor, J., Naimen, R. J., Dewey, B. and McInnes, P. (1988). Moose, microbes, and the boreal forest. *BioScience*, **38**, 770–777.

Pastor, J., Dewey, B., Naiman, R. J., MiInnes, P. F. and Cohen, Y. (1993). Moose browsing and soil fertility in the boreal forests of Isle Royale National Park. *Ecology*, **74**, 467–480.

Pastor, J., Cohen, Y. and Hobbs, N. T. (2006). The roles of large herbivores in ecosystem nutrient cycles. In *Large Mammalian Herbivores, Ecosystem Dynamics, and Conservation*, ed. K. Danell, K. Bergström, P. Duncan, J. Pastor and H. Olff. Cambridge: Cambridge University Press, pp. 289–325.

Pérez-Aragón, M., Fernandez, C. and Escribano, R. (2011). Nitrogen excretion by mesozooplankton in a coastal upwelling area: seasonal trends and implications for biological production. *Journal of Experimental Marine Biology and Ecology*, **406**, 116–124.

Phillips, R. P., Finzi, A. C. and Bernhardt, E. S. (2011). Enhanced root exudation induces microbial feedbacks to N cycling in a pine forest under long-term CO_2 fumigation. *Ecology Letters*, **14**, 187–194.

Phillips, R. P., Brzostek, E. and Midgley, M. G. (2013). The mycorrhizal-associated nutrient economy: a new framework for predicting carbon-nutrient couplings in temperate forests. *New Phytologist*, **199**, 41–51.

Pilati, A. and Vanni, M. J. (2007). Ontogeny, diet shifts, and nutrient stoichiometry in fish. *Oikos*, **116**, 1663–1674.

Polis, G. A., Anderson, W. B. and Holt, R. D. (1997). Toward an integration of landscape and food web ecology: the dynamics of spatially subsidized food webs. *Annual Review of Ecology and Systematics*, **28**, 289–316.

Pringle, R. M. and Fox-Dobbs, K. (2008). Coupling of canopy and understory food webs by ground-dwelling predators. *Ecology Letters*, **11**, 1328–1337.

Read, D. J. (1991). Mycorrhizas in ecosystems. *Experientia*, **47**, 376–391.

Redfield, A. C. (1958). The biological control of chemical factors in the environment. *American Scientist*, **46**, 205–221.

Reece, W. O. (2013). *Functional Anatomy and Physiology of Domestic Animals*. Singapore: Wiley-Blackwell.

Reich, P. B. and Oleksyn, J. (2004). Global patterns of plant leaf N and P in relation to temperature and latitude. *Proceedings of the National Academy of Sciences of the USA*, **101**, 11001–11006.

Reynolds, P. L. and Sotka, E. E. (2011). Non-consumptive predator effects indirectly influence marine plant biomass and palatability. *Journal of Ecology*, **99**, 1272–1281.

Ritchie, M. E. and Tilman, D. (1995). Responses of legumes to herbivores and nutrients during succession on a nitrogen-poor soil. *Ecology*, **76**, 2648–2655.

Roman, J. and McCarthy, J. J. (2010). The whale pump: marine mammals enhance primary productivity in a coastal basin. *PLoS One*, **5**, e13255.

Rose, M. D. and Polis, G. A. (1998). The distribution and abundance of coyotes: the effects of allochthonous food subsidies from the sea. *Ecology*, **79**, 998–1007.

Ruess, R. W. and McNaughton, S. J. (1988). Grazing and the dynamics of nutrient and energy regulated microbial processes in the Serengeti grasslands. *Oikos*, **49**, 101–110.

Sanchez-Pinero, F. and Polis, G. A. (2000). Bottom-up dynamics of allochthonous input: direct and indirect effects of seabirds on islands. *Ecology*, **81**, 3117–3132.

Schaus, M. H. and Vanni, M. J. (2000). Effects of gizzard shad on phytoplankton and nutrient dynamics: role of sediment feeding and fish size. *Ecology*, **81**, 1701–1719.

Schimel, J. P. and Bennett, J. (2004). Nitrogen mineralization: challenges of a changing paradigm. *Ecology*, **85**, 591–602.

Schindler, D. E. and Scheuerell, M. D. (2002). Habitat coupling in lake ecosystems. *Oikos*, **98**, 177–189.

Schindler, D. E., Carpenter, S. R., Cole, J. J., Kitchell, J. F. and Pace, M. L. (1997). Influence of food web structure on carbon exchange between lakes and the atmosphere. *Science*, **277**, 248–251.

Schindler, D. E., Knapp, R. A. and Leavitt, P. R. (2001). Alteration of nutrient cycles and algal production resulting from fish introductions into mountain lakes. *Ecosystems*, **4**, 308–321.

Schmitz, O. J. (2006). Predators have large effects on ecosystem properties by changing plant diversity, not plant biomass. *Ecology*, **87**, 1432–1437.

Schmitz, O. J. (2008). Effects of predator hunting mode on grassland ecosystem function. *Science*, **319**, 952–954.

Schmitz, O. J. (2010). *The Green World and the Brown Chain in Resolving Ecosystem Complexity*. Princeton, NJ: Princeton University Press.

Schmitz, O. J. and Suttle, K. B. (2001). Effects of top predator species on direct and indirect interactions in a food web. *Ecology*, **82**, 2072–2081.

Schmitz, O. J., Krivan, V. and Ovadia, O. (2004). Trophic cascades: the primacy of trait-mediated indirect interactions. *Ecology Letters*, **7**, 153–163.

Schmitz, O. J., Hawlena, D. and Trussell, G. C. (2010). Predator control of ecosystem nutrient dynamics. *Ecology Letters*, **13**, 1199–1209.

Schuurman, G. W. (2012). Ecosystem influences of fungus-growing termites in the dry paleotropics. In *Soil Ecology and Ecosystem Services*, ed. D. Wall, R. D. Bardgett, V. Behan-Pelletier, et al. Oxford: Oxford University Press, pp. 173–188.

Sereda, J. M. and Hudson, J. J. (2011). Empirical models for predicting the excretion of nutrients (N and P) by aquatic metazoans: taxonomic differences in rates and element ratios. *Freshwater Biology*, **56**, 250–263.

Sereda, J. M., Hudson, J. J. and Mcloughlin, P. D. (2008a). General empirical models for predicting the release of nutrients by fish, with a comparison between detritivores and non-detritivores. *Freshwater Biology*, **53**, 2133–2144.

Sereda, J. M., Hudson, J. J., Taylor, W. D. and Demers, E. (2008b). Fish as sources and sinks of nutrients in lakes: direct estimates, comparison with plankton and stoichiometry. *Freshwater Biology*, **53**, 278–289.

Shurin, J. B., Gruner, D. S. and Hillebrand, H. (2006). All wet or dried up? Real differences between aquatic and terrestrial food webs. *Proceedings of the Royal Society B: Biological Sciences*, **273**, 1–9.

Small, G. E., Pringle, C. M., Pyron, M. and Duff, J. H. (2011). Role of the fish *Astyanax aeneus* (Characidae) as a keystone nutrient recycler in low-nutrient neotropical streams. *Ecology*, **92**, 386–397.

Spiller, D. A., Piovia-Scorr, J., Wright, A. N., et al. (2010). Marine subsidies have multiple effects on coastal food webs. *Ecology*, **91**, 1424–1434.

Staaf, H. and Berg, B. (1981). Accumulation and release of plant nutrients in decomposing Scots pine needle litter. Long-term decomposition in a Scots pine forest II. *Canadian Journal of Botany*, **60**, 1561–1568.

Sterner, R. W. and Elser, J. J. (2002). *Ecological Stoichiometry: The Biology of Elements from Molecules to the Biosphere*, Princeton, NJ: Princeton University Press.

Stevens, C. E. and Hume, I. D. (1998). Contributions of microbes in vertebrate gastrointestinal tract to production and conservation of nutrients. *Physiological Reviews*, **78**, 393–427.

Stief, P. and Hölker, F. (2006). Trait-mediated indirect effects of predatory fish on microbial mineralization in aquatic sediments. *Ecology*, **87**, 3152–3159.

Swap, R., Garstang, M. and Greco, S. (1992). Saharan dust in the Amazon Basin. *Tellus*, **44**, 133–149.

Terborgh, J., Lopez, L., Nunez, P., et al. (2001). Ecological meltdown in predator-free forest fragments. *Science*, **294**, 1923–1926.

Thiel, M., Macaya, E. C., Acuna, E., et al. (2007). The Humboldt current system of Northern and Central Chile. *Oceanography and Marine Biology: An Annual Review*, **45**, 195–344.

Thompson, R. M., Brose, U., Dunne, J., et al. (2012). Food webs: reconciling the structure and function of biodiversity. *Trends in Ecology and Evolution*, **27**, 689–697.

Tokuda, G. and Watanabe, H. (2007). Hidden cellulases in termites: revision of an old hypothesis. *Biology Letters*, **3**, 336–339.

Treseder, K. K. and Vitousek, P. M. (2001). Effects of soil nutrient availability on investment in acquisition of N and P in Hawaiian rain forests. *Ecology*, **82**, 946–954.

Vanni, M. J. (2002). Nutrient cycling by animals in freshwater ecosystems. *Annual Review of Ecology and Systematics*, **33**, 341–370.

Vanni, M. M. J., Flecker, A. A. S., Hood, J. M. and Headworth, J. L. (2002). Stoichiometry of nutrient recycling by vertebrates in a tropical stream: linking species identity and ecosystem processes. *Ecology Letters*, **5**, 285–293.

Vanni, M. J., Bowling, A. M., Dickman, E. M., et al. (2006). Nutrient cycling by fish supports relatively more primary production as lake productivity increases. *Ecology*, **87**, 1696–1709.

Vanni, M. J., Boros, G. and McIntyre, P. B. (2013). When are fish sources versus sinks of nutrients in lake ecosystems? *Ecology*, **94**, 2195–2206.

Vergés, A., Pérez, M., Alcoverro, T. and Romero, J. (2008). Compensation and resistance to herbivory in seagrasses: induced responses to simulated consumption by fish. *Oecologia*, **155**, 751–760.

Villéger, S., Ferraton, F., Mouillot, D. and de Wit, R. (2012a). Nutrient recycling by coastal macrofauna: intra- versus interspecific differences. *Marine Ecology Progress Series*, **452**, 297–303.

Villéger, S., Grenouillet, G., Suc, V. and Brosse, S. (2012b). Intra- and interspecific differences in nutrient recycling by European freshwater fish. *Freshwater Biology*, **57**, 2330–2341.

Vrede, K., Heldal, M., Norland, S. and Bratbak, G. (2002). Elemental composition (C, N, P) and cell volume of exponentially growing and nutrient-limited bacterioplankton. *Applied and Environmental Microbiology*, **68**, 2965–2971.

Wardle, D. A. (2002). *Communities and Ecosystems: Linking the Aboveground and Belowground Components*. Princeton, NJ: Princeton University Press.

Welsh, D. T. (2000). Nitrogen fixation in seagrass meadows: regulation, plant-bacteria interactions and significance to primary productivity. *Ecology Letters*, **3**, 58–71.

Wieder, W. R., Cleveland, C. C. and Townsend, A. R. (2009). Controls over leaf litter decomposition in wet tropical forests. *Ecology*, **90**, 3333–3341.

Winder, M., Schindler, D. E., Moore, J. W., Johnson, S. P. and Palen, W. J. (2005). Do bears facilitate transfer of salmon resources to aquatic macroinvertebrates? *Canadian Journal of Fisheries and Aquatic Sciences*, **2293**, 2285–2293.

Wipfli, M. S., Hudson, J. and Caouette, J. (1998). Influence of salmon carcasses on stream productivity: response of biofilm and benthic macroinvertebrates in southeastern Alaska, USA. *Canadian Journal of Fisheries and Aquatic Sciences*, **1511**, 1503–1511.

Wipfli, M., Hudson, J. P. and Caouette, J. P. (2003). Marine subsidies in freshwater ecosystems: salmon carcasses increase the growth rates of stream-resident salmonids. *Transactions of the American Fisheries Society*, **132**, 371–381.

Wu, S., Wang, G., Angert, E. R., et al. (2012). Composition, diversity, and origin of the bacterial community in grass carp intestine. *PLoS One*, **7**, e30440.

Wurtsbaugh, W. A. (2007). Nutrient cycling and transport by fish and terrestrial insect nutrient subsidies to lakes. *Limnology and Oceanography*, **52**, 2715–2718.

Wyatt, A., Lowe, R., Humphries, S. and Waite, A. (2010). Particulate nutrient fluxes over a fringing coral reef: relevant scales of phytoplankton production and mechanisms of supply. *Marine Ecology Progress Series*, **405**, 113–130.

Young, H. S., McCauley, D. J., Dunbar, R. B. and Dirzo, R. (2010). Plants cause ecosystem nutrient depletion via the interruption of bird-derived spatial subsidies. *Proceedings of the National Academy of Sciences of the USA*, **107**, 2072–2077.

CHAPTER TEN

The role of bottom-up and top-down interactions in determining microbial and fungal diversity and function

THOMAS W. CROWTHER

Yale University, New Haven, CT, USA

and

HANS-PETER GROSSART

Leiibniz Institute of Freshwater Ecology and Inland Fisheries, Stechlin, Germany; and Potsdam University, Germany

Background

Microbes (Bacteria, Archaea, and Fungi) are the most diverse and numerous organisms on earth. They are essential components of every ecosystem and occupy almost every trophic position and niche space (i.e., as heterotrophs, photoheterotrophs, chemoautotrophs, and photoautotrophs; see DeLong et al., 2014). Saprotrophic or heterotrophic species are unique in their capacity to break down and mineralize organic matter (OM) that forms the basis of the carbon, nutrient, and energy cycles. As such, they are the biochemical engineers responsible for the mobilization and mineralization of non-living OM in terrestrial and aquatic ecosystems, and they contribute extensively to the respiratory exchanges of carbon (mainly CO_2 and CH_4) between the biosphere and the atmosphere. The overwhelming dominance of microbes in OM decomposition has led to extensive exploration into the top-down and bottom-up processes regulating their biomass and diversity, and their consequences for biogeochemical cycling. These heterotrophic microbes form the basis of the decomposer food web, supporting a wide diversity of detritivorous invertebrates and protists, and in turn, a wide range of vertebrate and invertebrate predators.

Unlike autotrophic food webs, where living plants are the primary trophic level, detrital food webs are based on the production of non-living organic material that cannot grow in response to reduced consumption rates, but can accumulate when introduced from adjacent ecosystems. In this sense, the detritus-based food web (often referred to as the "brown" food web) is a "donor-controlled"

Trophic Ecology: Bottom-Up and Top-Down Interactions across Aquatic and Terrestrial Systems, eds T. C. Hanley and K. J. La Pierre. Published by Cambridge University Press. © Cambridge University Press 2015.

system, relying on allochthonous inputs from the autotrophic (plant-based) sub-system. As such, these systems can never be exclusively top-down controlled, because microbial predation cannot directly affect the production of OM. Instead, top-down and bottom-up processes are on a continuum of relative importance, and ultimately, the effects of predation propagate as increased or decreased rates of OM decomposition, and thus OM accumulation. In essence, as microbial activity increases, decomposition rates (i.e., the initial loss of OM) generally increase, but when invertebrate predators (henceforth referred to as "grazers") exert top-down control of microbial communities, their ability to decompose OM can be altered. The direction and extent of these grazer effects vary depending on the relative importance of top-down and bottom-up processes within the system (Zheng et al., 1999; Lenoir et al., 2007). In this chapter, we consider how the relative importance of regulatory (top-down and bottom-up) forces varies between aquatic and terrestrial ecosystems, depending on (i) environmental features, (ii) the nature of the organic inputs, (iii) the microbial and fungal groups, and (iv) the grazer community.

Characteristics of the soil environment

In terrestrial ecosystems, approximately 98% of plant material escapes herbivory and enters the detritus-based nutrient pool (Bardgett, 2005). Although the magnitude of this pool varies substantially between environments, determined predominantly by the nature of the dominant plant community, the vast majority of the plant OM remains for several years after senescence, before eventually entering the soil organic matter (SOM) pool – prompting the question "why is the ground brown" (Allison, 2006)? As with the analogous aboveground question (i.e., "why is the world green?"), the top-down view is that animal "grazers" limit the activity of microbes, preventing them from completely decomposing all organic material (Allison, 2006). The alternative view is that structural complexity and physical protection of OM in soil restricts degradation by microbes, limiting the availability of nutrients both to them and to their predators.

Given the huge abundance of detritus entering the belowground sub-system, it is not the quantity, but the accessibility of nutrients that limits microbial growth (Schmidt et al., 2011). Analogous to autotrophic systems, the relative importance of top-down and bottom-up control on the intermediate trophic level (in this case, microbes, which are intermediate between detritus and grazers) is determined by whether organisms are experiencing absolute (resource-limited) or relative (time-limited) limitation (see Chapter 8). Consequently, ecosystems with a consistent supply of readily accessible (i.e., biochemically simple) nutrients are strongly controlled by top-down processes, while those receiving recalcitrant nutrient inputs are generally controlled by bottom-up processes (Wardle and Yeates, 1993; Moore et al., 2003). This promotes the importance of top-down control, as the extent to which grazers consume microbes determines the rates

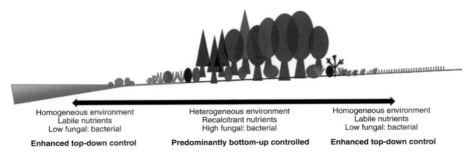

Figure 10.1 Diagram illustrating how the heterogeneity of environments and the recalcitrance of nutrients available to microbes varies among ecosystems. These attributes determine fungal:bacterial ratios and ultimately, the relative importance of top-down and bottom-up control in the microbial food web.

of nutrient mineralization; consequently, microbivorous grazers can control the proportion of nutrients assimilated. Therefore, the structural complexity of plant nutrients entering the soil determines the extent to which microbial communities are controlled predominantly by top-down or bottom-up processes.

Another consequence of the huge abundance of nutrients entering the soil is that the detritus-based food web is often highly complex, with potentially thousands of microbial and animal species interacting within a square meter of soil (Currie et al., 2004). This complexity is expected to dampen the effects of top-down control on any particular microbial group. The highly interconnected network is more resistant to disturbance, limiting the effects of trophic cascades on microbial community composition and diversity in soil, in comparison to aboveground communities (Polis and Strong, 1996; Dyer and Letourneau, 2003). Furthermore, at the scale of individual microbes, the soil environment is extremely heterogeneous, with dramatic gradients in moisture, temperature, pH, and nutrient availability. These vary within and between individual soil aggregates, further contributing to food web complexity, and limiting the potential for grazers to access their microbial prey. However, as the accessibility of nutrients and heterogeneity of soil can vary substantially across landscapes, depending on both the nature of the plant inputs and the geological history of the area, the effects of top-down control can vary substantially among ecosystems (Fig. 10.1).

Bottom-up control in soils

Sources of OM inputs in soils

There are two dominant mechanisms through which plant nutrients enter the soil: root exudates and senesced litter (wood and leaf material). The chemical complexity of these nutrient inputs determines the accessibility of nutrients to microbes, and consequently the relative importance of top-down and bottom-up processes. Root exudates typically consist of simple, low molecular weight compounds (e.g., sugars, amino acids, and organic acids) that are readily accessible for microbial uptake and immobilization. Plant litter is more chemically complex, and the recalcitrance of this material varies dramatically within and between individual plants. It is widely accepted that plant traits ultimately control the accessibility of nutrients within litter. A suite of biochemical traits (e.g., carbon:nitrogen (C:N) ratios, lignin concentrations, and secondary compounds (e.g., phenolics)) have been shown to correlate negatively with rates of plant litter decomposition (Wardle et al., 1998; Bardgett, 2005), and recent empirical evidence suggests that physical properties (e.g., leaf toughness and woodiness) may be equally, if not more, important in limiting the accessibility of nutrients to microbes (e.g., Cornwell et al., 2008). The palatability of living plant material to herbivores serves as a good predictor of decomposition rates of that material following senescence (Grime et al., 1996), suggesting that nutrient acquisition is constrained by a similar suite of mechanisms in microbes and animals.

It is possible to predict soil microbial functioning across landscapes, thereby linking microbial traits to plant biogeography. There is a consensus view among community ecologists that plant species adapted to a given habitat will share sets of ecophysiological traits (e.g., McGill et al., 2006; Smith et al., 2013). These "response traits" are often correlated with "effect traits" that determine how species influence ecosystem processes (e.g., decomposition rates; McGill et al., 2006; Allison, 2012). For example, a number of studies have shown how plant traits, including leaf tissue strength (Cornelissen and Thompson, 1997), longevity (Wardle et al., 1998), and nutrient use efficiency (Aerts and Berendse, 1988) − all of which enable species to tolerate a certain set of environmental conditions − can influence the accessibility of nutrients to microbes, and ultimately, decomposition rates. Plant species also vary predictably in their relative contribution of C and nutrients to above- and belowground components, so the relative importance of litter and root exudates varies between environments (Bradford and Crowther, 2013; Street et al., 2013). These plant traits, along with the aforementioned chemical and physical traits, are associated with distinct geographical distributions, and so the accessibility of nutrients to microbes will vary predictably across landscapes (Fierer et al., 2009), potentially selecting for decomposer food webs with certain basic attributes (Wardle et al., 2004).

The most widely documented example of plant-mediated decomposer community structuring is the distinction between fungal- and bacterial-dominated ecosystems (Wardle et al., 2004; Strickland and Rousk, 2010). Fast-growing plant species, which dominate fertile or early succession soils (e.g., grasslands), allocate the majority of their nutrients to rapid leaf and root growth. Consequently, these plant communities produce leaf litter of high nutrient quality, and allocate a considerable fraction of their C belowground in the form of labile root exudates, promoting bacterial-dominated food webs associated with the "fast" cycling of C and nutrients. In contrast, slow-growing species that proliferate in nutrient-poor or late-succession soils (e.g., forests) are more conservative of C and nutrients. These species produce nutrient-poor litter and allocate a relatively large proportion of their energy to the production of recalcitrant compounds such as lignin and phenolics, which select for fungal-dominated food webs, and 'slow' cycling of nutrients (Coleman et al., 1983; Wardle et al., 2004; Strickland and Rousk, 2010). The basis for this distinction is that bacteria are capable of rapidly accessing labile nutrients, but high metabolic rates and short generation times lead to the fast release of mineralized nutrients, which in turn benefit the development of fast-growing plant species adapted to fertile soils. In contrast, a powerful cocktail of oxidative and hydrolytic enzymes allows fungal communities to access the nutrients locked up in recalcitrant litter, but they are also more conservative of their nutrients, and their slow mineralization and release of nutrients back into the soil favors slow-growing plants that are adapted to infertile soils (Bardgett, 2005). Although they are not discrete ecosystem types (almost all soils contain representatives of both fungal and bacterial communities, and their respective predators), there is a clear trend that in moving toward more fertile soils with fast-growing species, the relative importance of the bacterial energy channel increases in relation to the fungal energy channel, based on the increased accessibility of organic nutrients (Moore et al., 2003). This relationship between plant and microbial communities, itself an example of bottom-up control, provides a useful framework to explore the relative importance of top-down and bottom-up processes in soil (Fig. 10.1).

Fungal-based food web

The recalcitrant nature of plant inputs that characterize fungal-dominated ecosystems suggests that fungal-based food webs are likely to be limited by nutrient accessibility, rather than time (Crowther et al., 2013). The abundance and chemical complexity of low-quality litter entering these systems also support a highly diverse/complex food web (Setälä et al., 2005). Theoretical models, therefore, predict that fungal biomass is predominantly, if not exclusively, controlled by bottom-up forces (Wardle and Yeates, 1993; Moore et al., 2003). There is extensive, and unequivocal evidence from field- (e.g., Chatterjee et al., 2008; Rinnan et al., 2013; Crowther et al., 2014) and laboratory- (e.g., Harold et al., 2005) based studies showing that equilibrium abundances of fungi and their mycophagous grazers are consistently positively correlated with OM inputs to the soil. Increased primary productivity generally leads to increased detrital inputs, which fuel increased fungal growth, and ultimately facilitate increased abundance of the grazer community (Shurin et al., 2006). Such increases in biomass have been observed across natural gradients in primary productivity, both in response to increased herbivore activity (as herbivores excrete waste products that are more nutrient-rich than the primary producers that they consume; see Chapter 9), and following human-induced alteration of the landscape (via land-use change and nutrient deposition), showing consistently that the abundance of fungi is tightly linked with the abundance and accessibility of their nutrients.

Although the complexity of the decomposer food web has historically obscured efforts to identify the factors affecting fungal community composition, this very complexity is thought to favor bottom-up as opposed to top-down control (Strong, 1992; Moore et al., 2003). Early experimental evidence suggested that nutrient availability could favor specific fungal groups (Donnelly and Boddy, 1998), with the potential to influence diversity, but it was not until the advent of molecular tools that we could comprehend the full extent of fungal community restructuring in response to changes in OM. A suite of studies have since reported effects of nutrient availability on fungal diversity in tropical (Lodge, 1997; McGuire et al., 2012), boreal (Lindahl et al., 2007), and temperate (Blackwood et al., 2007; U'Ren et al., 2010) regions, but effects have been found to vary dramatically, with positive, negative, and neutral effects of increasing OM concentrations being recorded in different ecosystems (Kerekes et al., 2013). The relationship between fungal diversity and nutrient availability has been proposed to fit the classic productivity–diversity distribution (*sensu* Tilman, 1982), with a unimodal curve along a productivity gradient (Kerekes et al., 2013). The relative importance of neutral and niche processes in structuring fungal communities across landscapes is still under extensive debate (Feinstein and Blackwood, 2013), but consistent differences in fungal community composition between vegetation types provides strong support for the potential of bottom-up forces in structuring fungal communities, with direct consequences for higher trophic levels. For example, the community composition (Crowther et al., 2014) and abundance (A'Bear et al., 2013) of highly abundant mesofauna (springtails and mites) are very responsive to changes in the relative abundance of basidiomycete and ascomycete fungi in woodland soil.

Bacterial-based food web

As with fungal biomass, correlations between rates of primary productivity and bacterial biomass (e.g., Clarholm and Rosswall, 1980; Moore-Kucera and Dick, 2008; Crowther et al., 2014), particularly in ecosystems characterized by plant species producing nutritious litter with a low C:N ratio (Fierer et al., 2007), provide strong support for the role of bottom-up forces in limiting bacterial biomass. Increased production of labile plant nutrients generally leads to increased bacterial biomass, eventually supporting a greater abundance of nematode grazers and their predators (Shurin et al., 2006). Unlike the fungal system, however, the increase of populations within different trophic levels is not always proportional; initial increases in bacterial biomass following nutrient inputs are generally reversed following increased microfaunal activity, suggesting that top-down control is likely to have a greater role in bacterial-dominated ecosystems (Clarholm and Rosswall, 1980). The nature of this bottom-up control also varies between ecosystems; although N amendment often leads to increased bacterial growth and respiration, effects can be negligible in warm temperate and tropical ecosystems, which are commonly limited by P availability (Gallardo and Schlesinger, 1994).

The rapid development of phospholipid fatty acid (PLFA) analysis and next generation DNA sequencing techniques over the past decade has revealed considerable evidence for the dominant role of plant nutrient inputs in structuring bacterial communities and diversity. Plant biomass and species identity have consistently been found to be dominant drivers of bacterial community composition at both small spatial scales (e.g., Zak et al., 2003; Bakker et al., 2012) and across landscapes (Fierer and Jackson, 2006; Fierer et al., 2009). These studies highlight that it is the nature of root architecture and exudates in the rhizosphere that predominantly drive antagonistic and mutualistic interactions among bacteria, ultimately structuring their community compositions (Bakker et al., 2012). Plant nutrient effects are highly specific; even within-species genotypic variation has been shown to exert selective pressures on bacterial communities (Crutsinger et al., 2006). As with fungi, the consequences of this bottom-up control of bacterial community composition for grazers and their associated predators remain unclear, but plant-induced shifts in the soil fungal:bacterial ratio has been shown to influence the abundance of microbivorous nematodes in soil (Viketoft et al., 2005). The consequences of this bottom-up restructuring of bacterial and grazer communities can have direct consequences for the cycling and sequestration of C and nutrients in soil (e.g., Briones et al., 2004), although the directional effects of soil biodiversity can vary substantially between environments.

Top-down control in soils

Grazers

The relative importance of top-down control on microbial communities varies depending on plant resource quality and soil fertility (Lenoir et al., 2007), but the extent to which this control modifies microbial communities can vary depending on the invertebrate community composition. Soil fauna contribute extensively to the functional and species diversity within soil ecosystems, and they influence microbial communities indirectly via the comminution of litter (mixing soil and producing fecal pellets), and directly via grazing interactions. Numerically, and in terms of ecosystem processes, the most important soil fauna include Protozoa, Nematoda, Oligochaeta (earthworms and enchytraeid worms), Mollusca (slugs and snails), and Arthropoda (e.g., woodlice, millipedes, springtails, termites, ants, and oribatid mites) – the vast majority of which feed directly or indirectly on soil microbes. Soil fauna classification is frequently based on their ecotype (surface-dwelling, epigeic; burrowing, endogeic; or deep burrowing, anecic) or trophic group (predatory, herbivorous, detritivorous, or microbivorous), but body size ($< 100\,\mu m$, micro-; $100\,\mu m$–$2\,mm$, meso-; or $> 2\,mm$, macrofauna) has provided the most useful functional classification because size correlates with metabolic rate, generation time, population density, and prey size (Bardgett, 2005).

As well as regulating the activity of microbes, microbivory is an essential process in the mineralization of nutrients that are immobilized by microbial biomass. This process is thought to accentuate the positive effects of the fungal and bacterial energy channels on their respective vegetation types (Moore et al., 2003). Bacteria have low C:N ratios and low concentrations of defense chemicals compared to fungi, making them readily susceptible to the grazing effects of microfauna, including nematodes and protozoa. This can stimulate the release of nutrients to promote soil fertility, while the more recalcitrant nature of fungal tissues (with a relatively high C:N ratio and high concentrations of chitin and toxic secondary metabolites) makes them more resistant to grazers, promoting the slow cycling of nutrients in late-successional, undisturbed soils (Wardle and Yeates, 1993; Moore et al., 2003). Arthropods are the dominant grazers within the fungal-based food web, and the majority of research on this mycophagy (fungus-feeding) has focused on highly abundant mesofauna (collembola and mites), eusocial insects (termites and ants), and the large macrofauna (woodlice and millipedes), which have the greatest capacity to decimate fungal colonies (Crowther et al., 2012a).

Fungal-based food web

The most prominent examples of top-down control of fungal communities involve the eusocial insects, which are capable of maintaining their fungal food sources in a monoculture. Through the use of physical grooming, antibiotics, and grazing, "fungal-farming" termite and ant species are directly responsible for maintaining biomass and diversity of fungi within their colonies (Crowther et al., 2012a). In most soil systems, however, the prominent role of bottom-up forces in regulating fungal biomass has led to the assumption that fungal-based communities are exclusively controlled by the accessibility of nutrients. The capacity of fungi to immobilize and retain nutrients partially explains their positive effect on the growth and dominance of slow-growing plant species in relatively infertile soil.

A large number of studies have shown that mycophagous grazers have the capacity to limit fungal growth (e.g., Hanlon and Anderson, 1979; Crowther et al., 2011a; 2011b) and thereby increase rates of nutrient release (Bardgett and Chan, 1999) from fungi within controlled microcosm experiments. These effects vary substantially between systems, depending on grazer identity (predominantly controlled by size and feeding preference; Crowther et al., 2012a) and density (with positive effects on fungal biomass at low density and negative effects at high density; Crowther et al., 2012a). Trophic cascades have also been identified, with predatory vertebrates (Wyman, 1998) and invertebrates (Hedlund and Ohrn, 2000; Lenoir et al., 2007) influencing the capacity of grazers to regulate fungal biomass and decomposition. However, these studies have commonly been conducted in artificial (often homogeneous) laboratory conditions, where fungi are limited in their capacity to avoid grazers; consequently, evidence for top-down control in natural, heterogeneous ecosystems is rare.

Field studies across gradients of primary productivity also lend little support to top-down control in fungal-based systems. For example, the 'Green World' (or HSS) hypothesis (Hairston et al., 1960) argues that in a three-level food chain (in this case, OM, fungi, and grazers), strong top-down control would mean that, in response to a nutrient influx, any initial increase in biomass of the intermediate level would be consumed efficiently by the top consumer, and increased grazer biomass would ultimately decrease fungal biomass. However, in most fungal-based communities, increased OM availability results in increased biomass of all trophic levels, suggesting that the food web is predominantly controlled by bottom-up forces (Ponsard et al., 2000). It is expected that fungi are generally limited by the accessibility of nutrients. However, the high enzymatic capabilities of certain groups can prevent this nutrient-limitation. For example, lignocellulytic, basidiomycete fungi are uniquely capable of readily unlocking the nutrients in even the most complex structural polymers within recalcitrant leaf litter. However, in the absence of nutrient-limitation, biomass of these fungi can be substantially reduced by high-intensity invertebrate grazing (Crowther et al., 2013).

As well as biomass, the potential for trophic cascades to alter fungal community composition and diversity in natural ecosystems is also under debate. Most mycophagous invertebrates are generalists, capable of feeding on a wide variety of food sources (Setälä et al., 2005). This omnivorous nature dramatically increases the number of trophic links, dampening the effects of any particular grazing interaction (Polis and Strong, 1996). Given the hyper-diverse nature of most soil fungal communities, the indiscriminate feeding of individuals is not expected to exert strong selective pressures on fungal communities. Although preferential feeding on individual fungi has been shown to increase the relative abundances of unpalatable or resistant fungi in two-species microcosm settings (Newell, 1984; Klironomos et al., 1992; Crowther et al., 2011c), this process appears highly dependent on the environmental conditions (i.e., predominantly temperature and moisture; Crowther et al., 2012b). Furthermore, most studies conducted in complex, semi-natural soil environments have revealed no effects of invertebrate grazers on fungal diversity or community composition

(Parkinson et al., 1979; McLean et al., 1996; Kaneko et al., 1998). However, as with fungal biomass, top-down control can have important regulatory effects on fungal diversity given a certain set of ecosystem characteristics. In woodland soils, cord-forming basidiomycete fungi dominate the microbial community near areas of coarse woody debris. By selectively ingesting these large macro-fungi, macro-invertebrate grazers can prevent the competitive exclusion of smaller litter fungi, increasing diversity within highly complex soil communities (Crowther et al., 2013). This keystone process requires that a community is dominated by a particular group of fungi, and the intensity of selective grazing is high enough to prevent the competitive exclusion of other, less competitive fungal groups.

Bacterial-based food web

In contrast to fungal-dominated communities, the importance of top-down forces within the bacterial-based food web is well-acknowledged, and considered an essential process in the rapid cycling of nutrients to facilitate plant growth (Wardle and Yeates, 1993; Moore et al., 2003; Bardgett, 2005). The reliance of bacterial communities on labile nutrients means that they are less limited by nutrient accessibility, and their small size and limited structural and biochemical defenses make bacteria relatively vulnerable to the effects of nematode and protozoa grazers. Laboratory- and field-based studies provide strong support for this top-down control, with increased bacterial biomass in response to reduced grazer (Caron, 1987; Rønn et al., 2002) and bacteriophage (Allen et al., 2010) abundance. The capacity of predatory nematodes and mites to control the abundances of microbivorous grazers means that trophic cascades are common within the bacterial energy channel, with direct consequences for the mineralization of organic nutrients (Santos et al., 1981). Although nutrient addition was found to enhance the growth of bacteria and their nematode predators in a complex, multi-trophic system, the inclusion of predators limited the equilibrium biomass of bacteria, leading Mikola and Setälä (1998) to conclude that bacteria are controlled simultaneously by top-down and bottom-up forces – the relative importance of which varies depending on the fertility of the soil (Fig. 10.2).

Top-down control also has the potential to exert selective pressures on bacterial communities, although mechanistic understanding of this process is limited. However, with minimal expression of secondary metabolites and anti-feedants, grazer preferences are unlikely to be based on their capacity to withstand specific toxins. Instead, selection is based on bacterial size, with larger protozoa consuming larger bacterial resources and decreasing the average size of their bacterial prey (Rønn et al., 2002). Structural differences between bacteria can also lead to selective grazing, with protozoa preferentially consuming gram-negative, as opposed to gram-positive, species (Rønn et al., 2002). Although there is limited evidence for top-down control of bacterial diversity in soil, selective

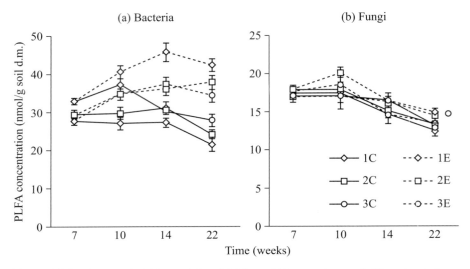

Figure 10.2 Changes in (a) bacterial and (b) fungal biomass (mean ± SE; estimated using phospholipid fatty acid analysis (PFLA)) over time within soil microcosm environments exposed to different nutrient and invertebrate regimes. Numbers indicate the number of trophic levels present in the food web; 'E' indicates energy added (5 mg glucose added in 30 µl sterilized water); and 'C' indicates a control for energy addition (30 µl sterilized water added alone). The presence of microbivorous nematodes significantly reduced bacterial biomass in all treatments, although the magnitude of this effect was greatest when energy limitation was alleviated by glucose addition. Fungal biomass was not significantly affected by microbivory, but the addition of energy had a positive effect, irrespective of trophic length. Figure modified from Mikola and Setälä (1998).

grazing is predicted to be an important control on the diversity and functioning of bacteria across a wide range of terrestrial and aquatic environments (Torsvik et al., 2002).

Characteristics of aquatic systems

In contrast to terrestrial systems, the major organic carbon (OC) pool in aquatic ecosystems is comprised of dissolved organic carbon (DOC; Cole et al., 2007). Aquatic DOC consists of a heterogeneous mixture of various C compounds of different chemical quality, including both autochthonous DOC (e.g., C released by phytoplankton and macrophytes) and allochthonous DOC (e.g., C originating from terrestrial areas surrounding aquatic ecosystems). Allochthonous C inputs to aquatic ecosystems, in particular in the dissolved form, can greatly vary in time and quality (see Chapter 6). For example, substantial amounts of leaves can fall into small, shallow lakes and dissolve in the water during leaf fall (e.g., Vander Zanden and Gratton, 2011), whereas during the growing season, most DOC originates from autochthonous sources. Both quantity and quality of autochthonous and allochthonous C sources may alter the internal C-cycling of the recipient aquatic ecosystem. Terrestrial organic C input into freshwater systems reaches *ca.* 2.9 petagrams C per year worldwide (Tranvik et al., 2009), and contributions of terrestrial DOC to aquatic systems have further increased in the past decades (Hansson et al., 2013).

Recent analytical techniques reveal that the OM pool in aquatic systems forms a continuum of gel-like and particulate OM ranging from the nanometer-scale up to the meter-scale (e.g., Chin et al., 1998). This leads to a patchy distribution of OM in aquatic systems, with large spatio-temporal fluctuations (e.g., Simon et al., 2002). Thus, in addition to OM variability, the architecture of the aquatic microenvironment can greatly vary among systems. For example, the higher ionic strength in marine water commonly favors aggregation of smaller particles into larger ones (Simon et al., 2002). A high potential for aggregate formation is partly attributed to coagulation of transparent exopolysaccharide particles (TEP; Logan et al., 1994), depending on physico-chemical conditions of the surrounding water and retention time in the water column. On the other hand, food web structure (e.g., the absence of important herbivores, such as daphniids (water fleas) in marine habitats, and salps (barrel-shaped, planktonic tunicates), pteropods (specialized free-swimming pelagic sea snails), and sea slugs (marine, opisthobranch gastropods) in freshwater habitats) may lead to pronounced differences in OM composition in different water bodies. The high complexity and spatio-temporal variability of OM in aquatic systems facilitates the evolution of a broad repertoire of physiological adaptations of bacteria to their environment (Grossart, 2010), similar to that in terrestrial systems (e.g., Janssen, 2006).

Bottom-up control

Sources of OM inputs in aquatic systems

Although terrestrial ecosystems are supplied to a large extent by particulate organic matter (POM), the importance of dissolved organic matter (DOM) is estimated to be ten times greater in aquatic systems. However, the majority of the autochthonous POM pool (e.g., total amino acids and carbohydrates of diatom aggregates) is often readily available for aquatic microorganisms (Grossart, 2010). Consequently, in most aquatic environments, autochthonous OM, of both algal and macrophytic origin, is an important source of both POM and DOM (Bertilsson and Jones, 2003). DOM production from aquatic algae and macrophytes occurs via various bottom-up and top-down mechanisms, including extracellular release (i.e., a significant portion of the net primary production is released during active growth of both algae and macrophytes), viral, phytoplankton, and bacterial cell lysis, and protozoan grazing (see below). Thus, the assessment of functional links between heterotrophic microorganisms and aquatic primary producers, and their coupling, is crucial for understanding the ecological role of autochthonously produced OM.

In addition to autochthonous OM sources, allochthonous OM can substantially contribute to total OM in aquatic systems (see Chapter 6). DOM discharges from terrestrial to aquatic systems often lead to increased DOM concentrations in aquatic systems (e.g., Freeman et al., 2001). Aside from fossil, humic matter-rich DOM, fresh allochthonous DOM (e.g., leachates; Gasith and Hasler, 1976) adds substantial amounts of nutrients, as well as organic C, to the water column, which is readily available for heterotrophic microorganisms (Attermeyer et al., 2013). However, due to its structural and chemical features, a large fraction of this allochthonous C pool (e.g., in the form of leaves and wood) is not readily available (Webster and Meyer, 1997).

The importance of structural (e.g., molar volume, chain length, molecular connectivity, etc.) and physico-chemical properties (e.g., molecular weight) for bioavailability and reactivity of chemical compounds has been well demonstrated (e.g., Stumm, 1990). Similar to terrestrial ecosystems, the structure of OM defines the traits of microorganisms (e.g., free-living versus particle-associated lifestyles), which are crucial for OM mineralization. Thereby, bacterial traits and community composition may be linked to certain environmental niches (Sjöstedt et al., 2014), including higher organisms such as phytoplankton and zooplankton (e.g., Grossart et al., 2005; Sher et al., 2011). The same is true for bacteria that specifically depend on the structural and chemical quality of the OM present, whereby OM bioavailability is greatly determined by the source (e.g., allochthonous versus autochthonous). Although studies of biogeographical patterns of aquatic microorgansims are just in their infancy, aquatic microbes seem to be characterized by high capacities for both dispersal and gene flow (Walsh et al., 2013). Therefore, biogeographical patterns of aquatic microorganisms (mainly bacteria and single cell fungi) are less obvious than for terrestrial microorganisms, and may greatly depend on the spatial and temporal scale taken into consideration (see Chapter 11). In particular, for marine systems, water currents can disperse bacteria over huge distances; thus hydrography is a main driver of bacterial community composition in the ocean (Galand et al., 2010).

Fungal-based food web
While bacteria dominate the oceans and deep lakes, fungi are accepted as the dominant decomposers of plant matter in shallow water and lotic systems, especially those rich in recalcitrant, lignocellulosic fibers (Fig. 10.3). Thereby, fungi introduce allochthonous leaf litter compounds into the food web of streams and render it available for bacteria and higher organisms such as

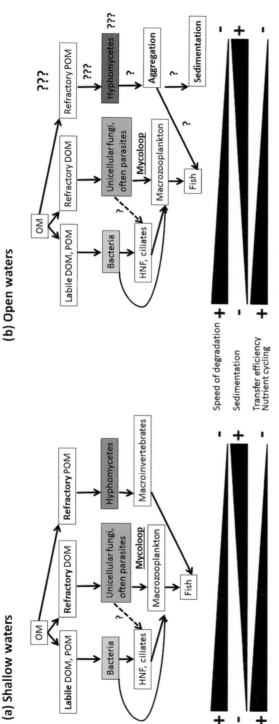

Figure 10.3 Schematic overview of the ecological role of bacteria and fungi in (a) shallow and (b) open waters. In open waters, in particular, there are many unknown trophic interactions and processes due to a lack of detailed studies on the role of fungi, especially of hyphomycetes (i.e., filamentous fungi that require surfaces like organic aggregates to grow on). The bars on the bottom of each figure represent the general trends in three important biogeochemical processes when moving from labile dissolved (DOM) and particulate (POM) organic matter to refractory DOM and POM. In general, aquatic fungi seem to be important for the degradation of more refractory organic matter, but the role of aggregates as a matrix for microbial degradation still needs to be examined in more depth. HNF refers to heterotrophic nanoflagellates – potential grazers of bacteria and single cell fungi. Solid lines represent known and well-established pathways, whereas dashed lines represent potential pathways that require further study.

macroinvertebrates (Gessner and Van Ryckegem, 2003). In aquatic ecosystems, "true fungi" are osmo-organotrophs that absorb nutrients across their cell walls, and most of them have a filamentous growth stage during their life cycle. This cell morphology allows them to successfully invade deep into substrates and to directly digest POM for growth and reproduction. Fungal filaments vary in length from several micrometers for the "rhizoids" of Chytridiomycetes to several millimeters or meters for hyphae or hyphal networks (Wurzbacher et al., 2011). Due to these traits, filamentous aquatic fungi are of great importance for leaf litter and wood degradation in small rivers and streams, as well as in the littoral zone of small and shallow lakes characterized by an extensive OM input from the surrounding land (reviewed in Wurzbacher et al., 2010). Importantly, current knowledge of fungal ecology in lakes is very rudimentary, and only a few studies have examined the effects of fungi on allochthonous leaf litter in lakes (reviewed by Wurzbacher et al., 2010). Depending largely on lake size and the surrounding vegetation, allochthonous leaf litter may contribute up to 10% of production in small lakes (Gasith and Hasler, 1976). In general, the role of fungi in aquatic systems is less well studied than in terrestrial ecosystems and thus merits further investigation, particularly with respect to shifting bottom-up and top-down control.

Running waters provide a highly diverse habitat for aquatic fungi. For example, lowland streams consist of wider channels in which litter inputs from riparian plants are low in relation to autochthonous inputs (Vannote et al., 1980). The submerged litter in these streams is dominated by hyphomycetes, the dominant order of aquatic fungi (Baldy et al., 2002), but the fungal community composition greatly differs from upland streams. In lowland streams, biomass accumulation is often limited by competition with other microorganisms, substrate burial, and lower oxygen availability (Bärlocher, 1992). In addition, fungal diversity in the littoral zone can be extremely high; for example, over 600 species of fungi have been recorded from the litter of *Phragmites australis* alone (Gessner and Van Ryckegem, 2003).

This diverse array of fungi (particularly saprophytic fungi) facilitates transfer of OM directly to higher trophic levels of aquatic food webs. For example, the foregut content of various aquatic insects collected on submerged wood revealed that 66% of the insects ingested fungi as food (Pereira et al., 1982), and evidence of aquatic fungi was also found in fish feces (Sridhar and Sudheep, 2011). Attermeyer and coauthors (2013) also showed that in the benthos of shallow lakes, fungi form the basis of the "benthic shortcut," which channels terrestrial OM and energy efficiently to higher trophic levels.

Despite the dominant role of fungi in the decomposition within lotic ecosystems, aquatic fungi have to compete with many other heterotrophs, especially prokaryotes. Both Bacteria and Archaea occur at high cell densities

(10^5–10^7 cells ml^{-1} in open water and 10^9–10^{11} cells g^{-1} in sediments; Moss, 2010), and both rapidly take up labile DOM as soon as it becomes available. Consequently, cell numbers and biomasses of aquatic fungi in the open water and sediments are substantially lower, with the greatest abundances occurring on particles of living and non-living litter (Wurzbacher et al., 2010). There is no evidence that free-floating fungal colonies occur in the water column, suggesting that eukaryotic fungi have a much lower impact than bacteria in aquatic systems. However, there are cases in which fungi may play a more important role in aquatic systems: (i) if the habitat is relatively constant over time (e.g., in stagnant waters, on biofilms (Sampaio et al., 2007), on algae (Kagami et al., 2007), or in aggregates and extracellular polysaccharides (Masters, 1971)); (ii) if the nutrient source is highly recalcitrant and/or requires specific enzymes (e.g., polyaromatic hydrocarbons (Augustin et al., 2006), humic matter (Claus and Filip, 1998), chitin (Reisert and Fuller, 1962), and lignocellulose (Fischer et al., 2009)); and (iii) if fungi are involved in specialized interspecific interactions (e.g., mutualism (Gimmler, 2001), commensalism (Lichtwardt, 2004), parasitism (Whisler et al., 1975; Ibeling et al., 2004), and predation (Barron, 1992)).

Fungal parasites are of particular significance due to their role in shaping aquatic food web structure by facilitating energy transfer and controlling disease. For example, parasitic chytrids can control the development of various phytoplankton blooms via active cell lysis (Kagami et al., 2007), and thus potentially play an important role in shaping and stabilizing aquatic systems (Fig. 10.3; Kagami et al., 2014). Further, zoospores, a free-living stage of chytrids, serve as excellent food for zooplankton in terms of size, shape, and nutritional quality, constituting a new and potentially important pathway in aquatic food webs termed the "mycoloop" (Kagami et al., 2007).

Bacterial-based food web

The "microbial loop" concept (Azam, 1998) refers to bacteria as basically the sole organism remineralizing DOM and thereby releasing inorganic nutrients required for the growth of photoautotrophs. The most important ecological function of pelagic bacteria is the production of biomass, which then becomes available for higher trophic levels. In marine systems, the abundance of DOM increases in relation to POM moving from the shores and littoral zones into the deep, open waters (Fig. 10.3; Hedges, 1992). The change in structure and composition of OM is essential since, in quantitative terms, bacteria are important for DOM uptake and remineralization. A synthesis of 70 published studies from marine and freshwater environments revealed that, on average, bacterial production equals 30% of total primary production at a given location, and bacterial production was consistently related to chlorophyll concentrations (Cole et al., 1988). Interestingly, the ratio between bacterial and phytoplankton biomass can be highly variable, and usually decreases with higher phytoplankton abundance (Cho and Azam, 1990). Overall, in low productivity regions (e.g., the open ocean or oligotrophic lakes), bacterial stocks seem to be limited by resource availability, whereas in high productivity regions (e.g., coastal and littoral systems), top-down forces are more important.

A recent study by Attermeyer et al. (2014) highlights the importance of DOM quantity for bacterial community composition. Increasing DOC concentrations led to higher bacterial biomass. In addition, higher DOC resulted in relatively high bacterial growth efficiencies due to enhanced utilization of the humic matter fraction when nutrients were not limiting. In both aquatic and terrestrial systems, physical properties and chemical quality of the OM pool have a great impact on OM turnover relative to bacterial community composition (Grossart, 2010).

Similar to terrestrial ecosystems, bacteria are an active component of aquatic food webs. As mentioned above, bacteria on particles can differ substantially from those in the surrounding water. Conversion of polymeric material to monomers by extracellular enzymatic activities is the rate-limiting step for OM utilization of most aquatic bacteria; thus, differences in enzyme production (e.g., Martinez et al., 1996) select for certain bacterial species, determining the outcome of competition. Further, physiological differences, such as uptake systems of monomeric substrates, have a strong impact on bacterial community composition in aquatic ecosystems. The effects of differences in physiology of free-living and attached bacteria (Walsh et al., 2013) on bacterial community structure have been recently studied intensively using modern molecular tools such as metatranscriptomics (e.g. Smith et al., 2013) and whole genome sequencing (e.g. Garcia et al., 2013). In particular, relationships and interactions between bacteria and phytoplankton (e.g., Grossart et al., 2005), as well as bacteria and zooplankton (reviewed in Tang et al., 2010), seem to be of great importance for structuring bacterial communities, indicating a clear role of bottom-up and top-down processes in determining bacterial community composition. In addition, other important biological processes structuring bacterial communities include grazing by macro- and micro-zooplankton and viral lysis (Riemann, 2002). For example, protozoan predation and viral lysis of free-living bacteria affect OM transformation not only in the surrounding water, but also on living and/or non-living particles.

Notably, particle-associated bacteria in freshwater and saltwater are generally phylogenetically distinct from free-living bacteria (DeLong et al., 1993; Allgaier and Grossart, 2006), but several lineages have the capability to switch between lifestyles depending on the given environmental conditions (Polz et al., 2006). In addition to rapidly adapting to changing environmental conditions, bacterial colonizers need to efficiently use available resources, while also successfully competing with specialist bacteria.

Top-down control

Grazers

Predation is an important factor altering microbial community structure and composition through various mechanisms, including direct effects of decreases in prey abundance, as well as indirect effects of interfering with and altering competitive interactions among prey species (Leibold, 1996). Protists have been shown to obey optimal foraging rules, selectively consuming the most common and the most metabolically active cells (Leibold, 1996). Thus, protists can serve as keystone predators, affecting bacterial morphology, activity, and community composition.

In lakes, protist grazing on bacteria can act as an ecological switch, alternatively recycling dissolved nutrients and organic C back to the environment, or shuttling bacterial biomass to larger organisms via food web interactions with zooplankton (Yannarell and Kent, 2009). Interestingly, community assembly mechanisms may shift in the presence of a predator. Whereas generalist predators may increase stochastic processes, such as neutral or priority effects, specialist predators may increase deterministic processes, like environmental filtering and species interactions (Ryberg et al., 2012). Macrozooplankton, like filter-feeding cladocerans, have the potential to alter the structure and functioning of the whole aquatic community and food web (Jürgens, 1994). For example, *Daphnia magna*, a large-bodied zooplankton, ingests organisms of a wide size range and simultaneously exerts predation effects at multiple trophic levels – namely, bacteria, heterotrophic nanoflagellates, ciliates, and phytoplankton (Sanders and Wickham, 1993). In particular, its utilization of zoospores, as well as other single-celled fungi, renders these grazers important components of the "mycoloop" (Fig. 10.3; reviewed by Kagami et al., 2014). *Daphnia magna* exerts direct, negative effects on bacterial and fungal communities via consumption, as well as indirect, positive effects by suppressing heterotrophic nanoflagellates and other competitors, and by releasing nutrients and OM due to "sloppy feeding" (Lampert, 1978).

Fungal-based food web

For streams and small lakes, it is well known that fungi colonizing leaf litter and other recalcitrant OM such as wood can serve as a valuable food source for macroinvertebrate grazers (Hanlon and Anderson, 1979; Suberkropp and Klug, 1980; Dray et al., 2014). Mechanical fragmentation of POM greatly depends on the extent of fungal colonization (Gulis et al., 2009), which leads to maceration of deciduous leaf litter (Dray et al., 2014). In addition, increases in N and protein content, and partial digestion of refractory POM, such as plant polymers, due to fungal colonization and enzymatic activities render this refractory OM a more nutritious and palatable food resource for invertebrate consumers (Dray et al.,

2014). These studies also show that grazing can greatly stimulate microbial activity by balancing fungal and bacterial abundance, with direct consequences for microbial community structure.

In contrast to the well-studied fungal food webs in streams and the littoral zone of shallow lakes (e.g., Attermeyer et al., 2014), little is known about fungal food webs in the open waters (i.e., pelagic zone) of freshwater and marine systems (Fig. 10.3). Except for the recently proposed "mycoloop" concept (Kagami et al., 2007; 2014), the top-down control of aquatic fungi – aside from those that can be directly ingested by macrozooplankton (e.g., chytridiomycota, cryptomycota, and yeasts) – has been relatively unexplored in pelagic environments. However, in littoral and coastal zones – which are similar to terrestrial systems in that refractory, polymeric DOM and POM are highly abundant – fungi may serve as subsidies for micro- and macrozooplankton grazers, but they seldom form the basis of aquatic food webs. However, as parasites, fungi can be of exceptional importance in pelagic systems by controlling and structuring interactions between the different food web compounds (e.g., Lepère et al., 2008; Gleason et al., 2008; Fig. 10.3). Despite the importance of fungi for biotic interactions of higher organisms, at this point there is little evidence for top-down control of fungal community composition in these ecosystems.

Bacterial-based food web

Of particular importance in aquatic, bacterial-based food webs is predation by heterotrophic nanoflagellates, which has the potential to affect the abundance, size structure, community composition, and diversity of bacterial communities (Langenheder and Jürgens, 2001; Pernthaler, 2005). Top-down control by specific grazers, such as protozoans, has the potential to increase prey species diversity because the ability to withstand predation and the ability to capture resources constitute a fundamental tradeoff (e.g., Kneitel and Chase, 2004). Consequently, predation by heterotrophic flagellates and ciliates can alter the dominance of a few competitive species, driving less-efficient species to extinction. However, defense mechanisms are less effective against larger macrozooplankton grazers (e.g., daphniids; Pernthaler, 2005). Daphniids can be regarded as a biotic disturbance that strongly affects microbial food web structure in a number of ways (e.g., Jürgens, 1994). First, the entire microbial community, including bacteria, can be suppressed when other *Daphnia* resources, such as heterotrophic nanoflagellates and phytoplankton, are limited. Second, when alternative resources are abundant, bacteria will be released from grazing pressure. Thus, macrozooplankton like daphniids have both direct, negative grazing effects and indirect, positive cascading effects on bacteria. Consequently, daphniids may have strong effects on bacterial communities at the local scale, decreasing bacterial biomass (e.g., Pernthaler, 2005) and leading to changes in bacterial community composition (e.g., Compte et al., 2009).

Conclusions

Bacteria and fungi are among the dominant agents of biogeochemical cycling in terrestrial and aquatic systems. Understanding the top-down and bottom-up processes governing patterns of microbial biomass and community composition across landscapes is, therefore, essential, especially in predicting the responses of ecosystems to environmental change. Despite fundamental differences in the physical structure of terrestrial and aquatic systems, we highlight several similarities across these systems. Ecosystems dominated by large, late-succession, or biochemically complex plants (e.g., forests, coastal seas, shallow lakes, and streams) generally give rise to highly heterogeneous environments, fueled by highly recalcitrant, detrital inputs. These ecosystems favor the dominance of fungi and fungal-based arthropod communities. Fungi are large and biochemically complex in relation to bacteria, making them more resistant to consumers; increased environmental heterogeneity selects for a highly complex fungal-based food web. Under these conditions, the microbial community is structured predominantly by bottom-up control, and microbial growth and activity are limited by the accessibility of organic nutrients (e.g., from plants). As ecosystems become more homogeneous and the abundance of POM generally decreases in relation to DOM (e.g., grasslands, deep lakes, and open oceans), this selects for a metabolically active, bacterial-dominated community. Small, and biochemically simple, these communities are more vulnerable to the effects of consumers, and the relative importance of top-down control increases in relation to bottom-up control. A simple tradeoff between predator tolerance and nutrient acquisition governs these food web differences, leading to clear differences in total ecosystem C and nutrient dynamics.

Despite these generalizations, a variety of biotic and abiotic factors can alter the strength and interaction of top-down and bottom-up forces, with both processes being apparent at different locations within most ecosystems. Furthermore, the strength of these generalizations remains unclear due to the relative dearth of studies looking at the importance of biotic interactions at broad spatial scales. The effects of nutrient inputs (i.e., bottom-up forces) in structuring microbial communities across landscapes are well acknowledged (Fierer et al., 2009; Lauber et al., 2009), but microbivory is assumed to be a local-scale process, and so the importance of top-down processes at the landscape scale has not been extensively tested. However, the context-dependent effects of microbivorous grazers on microbial biomass and community compositions suggests that the relative importance of top-down processes is predominantly determined by the nature of nutrient inputs into the detrital sub-system – a clear indication of the importance of bottom-up and top-down interactions in shaping microbial communities and processes across aquatic and terrestrial ecosystems.

Acknowledgments

We thank Dr. Hefin Jones for helpful discussion and comments on the manuscript. This work was funded through a Yale Climate and Energy Institute Fellowship.

References

A'Bear, A. D., Boddy, L. and Jones, T. H. (2013). Bottom-up determination of soil collembola diversity and population dynamics in response to interactive climatic factors. *Oecologia*, **3**, 1083–1087.

Aerts, R. and Berendse, F. (1988). The effect of increased nutrient availability on vegetation dynamics in wet heathlands. *Vegetatio*, **76**, 63–69.

Allen, B., Willner, D., Oechel, W. C. and Lipson, D. (2010). Top-down control of microbial activity and biomass in an Arctic soil ecosystem. *Environmental Microbiology*, **12**, 642–648.

Allgaier, M. and Grossart H. P. (2006). Seasonal dynamics and phylogenetic diversity of free-living and particle-associated bacterial communities in four lakes in northeastern Germany. *Aquatic Microbial Ecology*, **45**, 115–128.

Allison, S. D. (2006). Brown ground: a soil carbon analogue for the green world hypothesis? *The American Naturalist*, **167**, 619–627.

Allison, S. D. (2012). A trait-based approach for modelling microbial litter decomposition. *Ecology Letters*, **15**, 1058–1070.

Attermeyer, K., Premke, K., Hornick, T., Hilt, S. and Grossart, H. P. (2013). Ecosystem-level studies of terrestrial carbon reveal contrasting bacterial metabolism in different aquatic habitats. *Ecology*, **94**, 2754–2766.

Attermeyer, K., Hornick, T., Kayler, Z. E., et al. (2014). Increasing addition of autochthonous to allochthonous carbon in nutrient-rich aquatic systems stimulates carbon consumption but does not alter bacterial community composition. *Biogeosciences Discussions*, **10**, 14261–14300.

Augustin, T., Schlosser, D., Baumbach, R., et al. (2006). Biotransformation of 1-naphthol by a strictly aquatic fungus. *Current Microbiology*, **52**, 216–220.

Azam, F. (1998). Microbial control of oceanic carbon flux: the plot thickens. *Science*, **280**, 694–696.

Bakker, M. G., Bradeen, J. M. and Kinkel, L. L. (2012). Effects of plant host species and plant community richness on streptomycete community structure. *FEMS Microbiology Ecology*, **83**, 596–606.

Baldy, V., Chauvet, E., Charcosset, J. Y. and Gessner, M. O. (2002). Microbial dynamics associated with leaves decomposing in the mainstem and floodplain pond of a large river. *Aquatic Microbial Ecology*, **28**, 25–36.

Bardgett, R. D. (2005). *The Biology of Soil*. Oxford, UK: Oxford University Press.

Bardgett, R. D. and Chan, K. F. (1999). Experimental evidence that soil fauna enhance nutrient mineralization and plant nutrient in montane grassland ecosystems. *Soil Biology Biochemistry*, **31**, 23–33.

Bärlocher, F. (1992). *The Ecology of Aquatic Hyphomycetes*. Berlin: Springer.

Barron, G. (1992). Lignolytic and cellulolytic fungi as predators and parasites. In *The Fungal Community: Its Organization and Role in the Ecosystem*, Vol. 9, 2nd edn, ed. G Carroll and D. Wicklow. New York: Marcel Dekker, pp. 311–326.

Bertilsson, S. and Jones Jr., J. B. (2003). Supply of dissolved organic matter to aquatic ecosystems: autochthonous sources. Aquatic ecosystems. In: *Interactivity of Dissolved Organic Matter*. A volume in Aquatic Ecology, ed. S. E. G. Findlay and R. L. Sinsabaugh. Cambridge, MA: Elsevier Inc.

Blackwood, C. B., Waldrop, M. P., Zak, D. R. and Sinsabaugh, R. L. (2007). Molecular analysis of fungal communities and laccase genes in decomposing litter reveals differences among forest types but no impact of nitrogen deposition. *Environmental Microbiology*, **9**, 1306–1316.

Bradford, M. A. and Crowther, T. W. (2013). Carbon use efficiency and storage in terrestrial ecosystems. *New Phytologist*, **199**, 7–9.

Briones, M. J. I., Poskitt, J. and Ostle, N. (2004). Influence of warming and enchytraeid activities on soil CO_2 and CH_4 fluxes. *Soil Biology and Biochemistry*, **3**, 1851–1859.

Caron, D. A. (1987). Grazing of attached bacteria by heterotrophic microflagellates. *Microbial Ecology*, **13**, 203–218.

Chatterjee, A., Vance, G. F., Pendall, E. and Stahl, P. D. (2008). Timber harvesting alters soil carbon mineralization and microbial community structure in coniferous forests. *Soil Biology Biochemistry*, **40**, 1901–1907.

Chin, W., Orellana, M. V. and Verdugo, P. (1998). Formation of microgels by spontaneous assembly of dissolved marine biopolymers. *Nature*, **391**, 568–572.

Cho, B. C. and Azam, F. (1990). Biogeochemical significance of bacterial biomass in the ocean's euphotic zone. *Marine Ecological Progress Series*, **63**, 253–2259.

Clarholm, M. and Rosswall, T. (1980). Biomass and turnover of bacteria in a forest soil and a peat. *Soil Biology and Biochemistry*, **12**, 49–57.

Claus, H. and Filip, Z. (1998). Degradation and transformation of aquatic humic substances by laccase-producing fungi *Cladosporium cladosporioides* and *Polyporus versicolor*. *Acta Hydrochimica et Hydrobiologica*, **26**, 180–185.

Cole, J. J., Findlay, S. and Pace, M. L. (1988). Bacterial production in fresh and saltwater ecosystems: a cross-system overview. *Marine Ecological Progress Series*, **43**, 1–10.

Cole, J. J., Prairie, Y. T., Caraço, N. F., et al. (2007). Plumbing the global carbon cycle: integrating inland waters into the terrestrial carbon budget. *Ecosystems*, **10**, 171–184.

Coleman, D. C., Reid, C. P. P. and Cole, C. V. (1983). Biological strategies of nutrient cycling in soil systems. *Advances in Ecological Research*, **13**, 1–55.

Compte, J., Brucet, S., Gascón, S., et al. (2009). Impact of different developmental stages of *Daphnia magna* (Straus) on the plankton community under different trophic conditions. *Hydrobiologia*, **635**, 45–56.

Cornelissen, J. H. C. and Thompson, K. (1997). Functional leaf attributes predict litter decomposition rate in herbaceous plants. *New Phytologist*, **135**, 109–114.

Cornwell, W. K., Cornelissen, J. H., Amatangelo, K., et al. (2008). Plant species traits are the predominant control on litter decomposition rates within biomes worldwide. *Ecology Letters*, **11**, 1065–1071.

Crowther, T. W., Boddy, L. and Jones, T. H. (2011a). Species-specific effects of soil fauna on fungal foraging and decomposition. *Oecologia*, **167**, 535–545.

Crowther, T. W., Jones, T. H. and Boddy, L. (2011b). Species-specific effects of grazing invertebrates on mycelial emergence and growth from woody resources into soil. *Fungal Ecology*, **4**, 333–341.

Crowther, T. W., Boddy, L. and Jones, T. H. (2011c). Outcomes of fungal interaction are determined by soil invertebrate grazers. *Ecology Letters*, **14**, 1134–1142.

Crowther, T. W., Boddy, L. and Jones, T. H. (2012a). Functional and ecological consequences of saprotrophic fungus-grazer interactions. *The ISME Journal*, **6**(11), 1992–2001.

Crowther, T. W., Littleboy, A., Jones, T. H. and Boddy, L. (2012b). Interactive effects of warming and invertebrate grazing on the outcomes of competitive fungal interactions. *FEMS Microbiology Ecology*, **81**, 419–426.

Crowther, T. W., Stanton, D., Thomas, S., et al. (2013). Top-down control of soil fungal community composition by a globally distributed keystone species. *Ecology*, **94**, 2518–2528.

Crowther, T. W., Maynard, D. S., Leff, J. W., et al. (2014). Explaining the vulnerability of soil biodiversity to deforestation: a cross-biome study. *Global Change Biology*. **20**, 2983–2994.

Crutsinger, G. M., Collins, M. D., Fordyce, J. A., et al. (2006). Plant genotypic diversity predicts community structure and governs an ecosystem process. *Science*, **313**, 966–968.

Currie, D. J., Mittelbach, G. G., Cornell, H. V., et al. (2004). A critical review of species-energy theory. *Ecology Letters*, **7**, 1121–1134.

DeLong, E .F., Franks, D. G. and Alldredge, A. L. (1993). Phylogenetic diversity of aggregate-attached vs. free-living marine bacterial assemblages. *Limnology and Oceanography*, **38**, 924–934.

DeLong, E. F., Rosenberg, E., Lory, S., Stackebrandt, E. and Thompson, F., eds. (2014). *The Prokaryotes*, 4th edition, 11 volumes. New York, NY: Springer-Science-Business-Media.

Donnelly, D. P. and Boddy, L. (1998). Developmental and morphological responses of mycelial systems *of Stropharia caerulea* and *Phanerochaete velutina* to soil nutrient enrichment. *New Phytologist*, **138**, 519–531.

Dray, M. W., Crowther, T. W., Thomas, S. M., et al. (2014). Effects of elevated CO_2 on litter chemistry and subsequent invertebrate detritivore feeding responses. *PloS One*, **9**(1), e86246.

Dyer, L. A. and Letourneau, D. (2003). Top-down and bottom-up diversity cascades in detrital vs. living food webs. *Ecology Letters*, **6**, 60–68.

Feinstein, L. M. and Blackwood, C. B. (2013). The spatial scaling of saprotrophic fungal beta diversity in decomposing leaves. *Molecular Ecology*, **22**, 1171–1184.

Fierer, N. and Jackson, R. B. (2006). The diversity and biogeography of soil bacterial communities. *Proceedings of the National Academy of Sciences of the USA*, **103**, 626–631.

Fierer, N., Bradford, M. A. and Jackson, R. B. (2007). Toward an ecological classification of soil bacteria. *Ecology*, **88**, 1354–1364.

Fierer, N., Strickland, M. S., Liptzin, D., Bradford, M. A. and Cleveland, C. C. (2009). Global patterns in belowground communities. *Ecology Letters*, **12**, 1238–1249.

Fischer, H., Bergfur, J., Goedkoop, W. and Tranvik, L. (2009). Microbial leaf degraders in boreal streams: bringing together stochastic and deterministic regulators of community composition. *Freshwater Biology*, **54**, 2276–2289.

Freeman, C., Evans, C. D., Monteith, D. T., Reynolds, B. and Fenner, B. (2001). Export of organic carbon from peat soils. *Nature*, **412**, 785–785.

Galand, P. E., Potvin, M., Casamayor, E. O. and Lovejoy, C. (2010). Hydrography shapes bacterial biogeography of the deep Arctic Ocean. *The ISME Journal*, **4**, 564–576.

Gallardo, A. and Schlesinger, W. H. (1994). Factors limiting microbial biomass in the mineral soil and forest floor of a warm-temperate forest. *Soil Biology and Biochemistry*, **26**, 1409–1415.

Garcia, S. L., McMahon, K. D., Martinez-Garcia, M., et al. (2013). Metabolic potential of a single cell belonging to one of the most abundant lineages in freshwater bacterioplankton. *The ISME Journal*, **1**, 137–147.

Gasith, A. and Hasler, A. D. (1976). Airborne litterfall as a source of organic matter in lakes. *Limnology and Oceanography*, **21**, 253–258.

Gessner, M. O. and Van Ryckegem, G. (2003). Water fungi as decomposers in freshwater ecosystems. In *Encyclopaedia of Environmental Microbiology*, ed. G. Bitton. New York, NY: Wiley, pp. 1–38.

Gimmler, H. (2001). Mutalistic relationships between algae and fungi (excluding lichens). In *Progress in Botany*, Vol. 62. ed. K. Esser, U. Lüttge, J. W. Kadereit and W. Beyschlag. Berlin: Springer, pp. 194–214.

Gleason, F. H., Kagami, M., Lefèvre, E. and Sime-Ngando, T. (2008). The ecology of chytrids in aquatic ecosystems: roles in food web dynamics. *Fungal Biology Reviews*, **22**, 17–25.

Grime, J. P., Cornelissen, J. H. C., Thompson, K. and Hodgson, J. G. (1996). Evidence of a causal connection between anti-herbivore defence and the decomposition rate of leaves. *Oikos*, 489–494.

Grossart, H. P. (2010). Ecological consequences of bacterioplankton lifestyles: changes in concepts are needed. *Environmental Microbiology*, **2**, 706–714.

Grossart, H. P., Levold, F., Allgaier, M., Simon, M. and Brinkhoff, T. (2005). Marine diatom species harbour distinct bacterial communities. *Environmental Microbiology*, **7**, 860–873.

Gulis, V., Kuehn, K. A. and Suberkropp, K. (2009). Fungi. In *Encyclopedia of Inland Waters*, ed. Gene E Likens. Cambridge, MA: Elsevier Inc., pp. 233–243.

Hairston, N. G., Smith, F. E. and Slobodkin, L. B. (1960). Community structure, population control and competition. *American Naturalist*, **94**, 357–425.

Hanlon, R. D. G. and Anderson, J. M. (1979). Effects of collembola grazing on microbial activity in decomposing leaf litter. *Oecologia*, **38**, 93–99.

Hansson, L.-A., Nicolle, A., Granéli, W., et al. (2013). Food-chain length alters community responses to global change in aquatic systems. *Nature Climate Change*, **3**, 228–233.

Harold, S., Tordoff, G. M., Jones, T. H. and Boddy, L. (2005). Mycelial responses of *Hypholoma fasciculare* to collembola grazing: effect of inoculum age, nutrient status and resource quality. *Mycological Research*, **109**, 927–935.

Hedges, J. I. (1992). Global biogeochemical cycles: progress and problems. *Marine Chemistry*, **39**, 67–93.

Hedlund, K. and Ohrn, M. S. (2000). Tritrophic interactions in a soil community enhance decomposition rates. *Oikos*, **88**, 585–591.

Ibelings, B. W., Bruin, A. D., Kagami, M., et al. (2004). Host parasite interactions between freshwater phytoplankton and chytrid

fungi (Chytridiomycota). *Journal of Phycology*, **40**, 437–453.

Janssen, P. H. (2006). Identifying the dominant soil bacterial taxa in libraries of 16SrRNA and 16SrRNA genes. *Applied Environmental Microbiology*, **72**, 1719–1728.

Jürgens, K. (1994). Impact of *Daphnia* on planktonic microbial food webs: a review. *Marine Microbiology Food Webs*, **8**, 295–324.

Kagami, M., de Bruin, A., Ibelings, B. W. and Van Donk, E. (2007). Parasitic chytrids: their effects on phytoplankton communities and food-web dynamics. *Hydrobiologia*, **578**, 113–29.

Kagami, M., Miki, T. and Takimoto, G. (2014). Mycoloop: chytrids in aquatic food webs. *Frontiers in Microbiology*, **5**, 166.

Kaneko, N., McLean, M. A. and Parkinson, D. (1998). Do mites and Collembola affect pine litter fungal biomass and microbial respiration? *Applied Soil Ecology*, **9**, 209–213.

Kerekes, J., Kaspari, M., Stevenson, B., et al. (2013). Nutrient enrichment increased species richness of leaf litter fungal assemblages in a tropical forest. *Molecular Ecology*, **22**, 2827–2838.

Klironomos, J. N., Widden, P. and Deslandes, I. (1992). Feeding preferences of the collembolan, *Folsomia candida*, in relation to microfungal successions on decaying litter. *Soil Biology and Biochemistry*, **24**, 685–692.

Kneitel, J. M. and Chase, J. M. (2004). Trade-offs in community ecology: linking spatial scales and species coexistence. *Ecology Letters*, **7**, 69–80.

Lampert, W. (1978). Release of dissolved organic carbon by grazing zooplankton. *Limnology and Oceanography*, **23**, 831–834.

Langenheder, S. and Jürgens, K. (2001). Regulation of bacterial biomass and community structure by metazoan and protozoan predation. *Limnology and Oceanography*, **46**, 121–134.

Lauber, C. L., Hamady, M., Knight, R. and Fierer, N. (2009). Pyrosequencing-based assessment of soil pH as a predictor of soil bacterial community structure at the continental scale. *Applied and Environmental Microbiology*, **75**, 5111–5120.

Leibold, M. A. (1996). A graphical model of keystone predation: effects of productivity on abundance, incidence and ecological diversity in communities. *American Naturalist*, **147**, 784–812.

Lenoir, L., Persson, T., Bengtsson, J., Wallander, H. and Wiren, A. (2007). Bottom-up or top-down control in forest soil microcosms? Effects of soil fauna on fungal biomass and C/N mineralisation. *Biology and Fertility of Soils*, **43**, 281–294.

Lepère, C., Domaizon, I. and Debroas, D. (2008). Unexpected importance of potential parasites in the composition of the freshwater small-eukaryote community. *Applied and Environmental Microbiology*, **74**, 2940–2949.

Lichtwardt, R. (2004). Trichomycetes: fungi in relationship with insects and other arthropods. In *Symbiosis Mechanisms and Model Systems, Cellular Origin, Life in Extreme Habitats and Astrobiology*, Vol. 4, ed. J. Seckbach. Dordrecht: Kluwer Academic Publishers, pp. 575–588.

Lindahl, B. D., Ihrmark K. and Boberg, J. (2007). Spatial separation of litter decomposition and mycorrhizal nitrogen uptake in a boreal forest. *New Phytologist*, **173**, 611–620.

Lodge, D. (1997). Factors related to diversity of decomposer fungi in tropical forests. *Biodiversity and Conservation*, **688**, 681–688.

Logan, B. E., Grossart, H. P. and Simon, M. (1994). Direct observation of phytoplankton, TEP and aggregates on polycarbonate filters using brightfield microscopy. *Journal of Plankton Research*, **16**, 1811–1815.

Martinez, J., Smith, D. C., Steward, G. F. and Azam, F. (1996). Variability in ectohydrolytic enzyme activities of pelagic marine bacteria and its significance for substrate processing in the sea. *Aquatic Microbiology Ecology*, **10**, 223–230.

Masters, M. (1971). The occurrence of *Chytridium marylandicum* on *Botryococcus braunii* in School Bay of the Delta Marsh. *Canadian Journal of Botany*, **49**, 1479–1485.

McGill, B. J., Enquist, B. J., Weiher, E. and Westoby, M. (2006). Rebuilding community ecology from functional traits. *Trends in Ecology and Evolution*, **21**, 178–185.

McGuire, K. L., Fierer, N., Bateman, C., Treseder, K. K. and Turner, B. L. (2012). Fungal community composition in neotropical rain forests: the influence of tree diversity and precipitation. *Microbial Ecology*, **63**, 804–812.

McLean, M. A., Kaneko, N. and Parkinson, D. (1996). Does selective grazing by mites and collembola affect litter fungal community structure? *Pedobiologia*, **40**, 97–105.

Mikola, J. and Setälä, H. (1998). Productivity and trophic-level biomasses in a microbial-based soil food web. *Oikos*, **82**, 158–168.

Moore, J. C., Mccann, K., Setälä, H. and De Ruiter, P. C. (2003). Top-down is bottom-up: does predation in the rhizosphere regulate aboveground dynamics? *Ecology*, **84**, 846–857.

Moore-Kucera, J. and Dick, R. P. (2008). PLFA profiling of microbial community structure and seasonal shifts in soils of a Douglas-fir chronosequence. *Microbial Ecology*, **55**, 500–511.

Moss, B. (2010). *Ecology of Freshwaters : A View for the Twenty-first Century*. 4th edition. New York, NY: Wiley-Blackwell.

Newell, K. (1984). Interaction between two decomposer Basidiomycetes and a collembolan under Sitka spruce: grazing and its potential effects on fungal distribution and litter decomposition. *Soil Biology and Biochemistry*, **16**, 235–239.

Parkinson, D., Visser, S. and Whittaker, J. B. (1979). Effects of collembolan grazing on fungal colonization of leaf litter. *Soil Biology and Biochemistry*, **11**, 529–535.

Pereira, C. R. D., Anderson, N. H. and Dudley, T. (1982). Gut content analysis of aquatic insects from wood substrates. *Melanderia*, **39**, 23–33.

Pernthaler, J. (2005). Predation on prokaryotes in the water column and its ecological implications. *National Reviews of Microbiology*, **3**, 537–546.

Polis, G. A. and Strong, D. R. (1996). Food web complexity and community dynamics. *American Naturalist*, **147**, 813–846.

Polz, M. F., Hunt, D. E., Preheim, S. P., Weinreich, D. M. and Weinreich, D. M. (2006). Patterns and mechanisms of genetic and phenotypic differentiation in marine microbes. *Philosophical Transactions of the Royal Society of London B*, **361**, 2009–2021.

Ponsard, S., Arditi, R. and Jost, C. (2000). Assessing top-down and bottom-up control in a litter-based soil macroinvertebrate food chain. *Oikos*, **89**, 524–540.

Reisert, P. and Fuller, M. (1962). Decomposition of chitin by chytriomyces species. *Mycologia*, **54**, 647–657.

Riemann, L. (2002). Population dynamics of bacterioplankton: regulation and importance. PhD thesis, University of Copenhagen, Denmark.

Rinnan, R., Michelsen, A. and Bååth, E. (2013). Fungi benefit from two decades of increased nutrient availability in tundra heath soil. *PloS One*, **8**(2), e56532.

Rønn, R., McCaig, A. E., Griffiths, B. S. and Prosser, J. I. (2002). Impact of protozoan grazing on bacterial community structure in soil microcosms. *Applied and Environmental Microbiology*, **68**, 6094–6105.

Ryberg, W. A., Smith, K. G. and Chase, J. M. (2012). Predators alter the scaling of diversity in prey metacommunities. *Oikos*, **121**, 1995–2000.

Sampaio, A., Sampaio, J. P. and Leão, C. (2007). Dynamics of yeast populations recovered from decaying leaves in a nonpolluted stream: a 2-year study on the effects of leaf litter type and decomposition time. *FEM Yeast Research*, **7**, 595–603.

Sanders, R. W. and Wickham, S. A. (1993). Planktonic protozoa and metazoa: predation, food quality and population control. *Marine Microbiology Food Webs*, **7**, 197–223.

Santos, P. F., Phillips, J. and Whitford, W. G. (1981). The role of mites and nematodes in early stages of buried litter decomposition in a desert. *Ecology*, **62**, 664–669.

Schmidt, M. W. I., Torn, M. S., Abiven, S., et al. (2011). Persistence of soil organic matter as an ecosystem property. *Nature*, **478**, 49–56.

Setälä, H., Berg P. M. and Jones, T. H. (2005). Trophic structure and functional redundancy in soil communities. In *Biological Diversity and Function in Soils*, ed. R. D. Bardgett, M. B. Usher and D. W Hopkins. Cambridge, UK: Cambridge University Press, pp. 236–249.

Sher, D., Thompson, J. W., Kashtan, N., Croal, L. and Chisholm, S. W. (2011). Response of *Prochlorococcus* ecotypes to co-culture with diverse marine bacteria. *The ISME Journal*, **5**, 1125–1132.

Simon, M., Grossart, H. P., Schweitzer, B. and Ploug, H. (2002). Microbial ecology of organic aggregates in aquatic ecosystems. *Aquatic Microbial Ecology*, **28**, 175–211.

Shurin, J. B., Gruner, D. S. and Hillebrand, H. (2006). All wet or dried up? Real differences between aquatic and terrestrial food webs. *Proceedings of the Royal Society B*, **273**, 1–9.

Sjöstedt, J., Martiny, J. B. H., Munk, P. and Riemann, L. (2014). Abundance of broad bacterial taxa explained by environmental conditions but not water mass in the Sargasso Sea. *Applied and Environmental Microbiology*, **80**, 2786–2795.

Smith, A. B., Sandel, B., Kraft, N. J. B. and Carey, S. (2013a). Characterizing scale-dependent community assembly using the functional-diversity–area relationship. *Ecology*, **94**, 2392–2402.

Smith, M. W., Allen, A. E., Herfort, L. and Simon, H. M. (2013b). Contrasting genomic properties of free-living and particle-attached microbial assemblages within a coastal ecosystem. *Frontiers in Microbiology*, **4**, 1–20.

Sridhar, K. R. and Sudheep, N. M. (2011). Do the tropical freshwater fishes feed on aquatic fungi? *Frontiers of Agriculture in China*, **5**(1), 77–86. DOI: 10.1007/s11703-011-1055-9.

Street L. E., Subke J.-A., Sommerkorn, M., et al. (2013). The role of mosses in carbon uptake and partitioning in arctic vegetation. *New Phytologist*, **199**, 163–175.

Strickland, M. S. and Rousk, J. (2010). Considering fungal: bacterial dominance in soils – methods, controls and ecosystem implications. *Soil Biology and Biochemistry*, **42**, 1385–1395.

Strong, D. R. (1992). Are trophic cascades all wet? Differentiation and donor-control in speciose ecosystems. *Ecology*, **73** 747–754.

Stumm, W. (1990). *Aquatic Chemical Kinetics. Reaction Rates of Processes in Natural Waters.* New York, NY: Wiley-Interscience Publication, John Wiley & Sons.

Suberkropp, K. and Klug, M. J. (1980). The maceration of deciduous leaf litter by aquatic hyphomycetes. *Canadian Journal of Botany*, **58**, 1025–1031.

Tang, K. W., Turk, V. and Grossart, H. P. (2010). Zooplankton makes a difference to the microbial world. *Aquatic Microbial Ecology*, **61**, 261–277.

Tilman, D. (1982). *Resource Competition and Community Structure.* Princeton, NJ: Princeton University Press.

Torsvik, V., Ovreas, L. and Thingstad, T. F. (2002). Prokaryotic diversity – magnitude, dynamics and controlling factors. *Science Signaling*, **296**, 1064.

Tranvik, L. J., Downing, J. A., Cotner, J. B., et al. (2009). Lakes and reservoirs as regulators of carbon cycling and climate. *Limnology and Oceanography*, **54**, 2298–2314.

U'Ren, J. M., Lutzoni, F., Miadlikowska, J. and Arnold, A. E. (2010). Community analysis reveals close affinities between endophytic and endolichenic fungi in mosses and lichens. *Microbial Ecology*, **60**, 340–353.

Vander Zanden, M. J. and Gratton, C. (2011). Blowin' in the wind: reciprocal airborne carbon fluxes between lakes and land.

Canadian Journal of Fisheries and Aquatic Sciences, **68**, 170–182.

Vannote, R. L., Minshall, G. W., Cummins, K. W., Sedell, J. R. and Cushing, C. E. (1980). The river continuum concept. *Canadian Journal of Aquatic Science*, **37**, 130–137.

Viketoft, M., Palmborg, C., Sohlenius, B., Huss-Danell, K. and Bengtsson, J. (2005). Plant species effects on soil nematode communities in experimental grasslands. *Applied Soil Ecology*, **30**, 90–103.

Walsh, D. A., Lafontaine, J. and Grossart, H. P. (2013). On the eco-evolutionary relationships of fresh and salt water bacteria and the role of gene transfer in their adaptation. In *Lateral Gene Transfer in Evolution*, ed. U. Gophna. New York, NJ: Springer Science and Business Media, pp. 55–77.

Wardle, D. A. and Yeates, G. W. (1993). The dual importance of competition and predation as regulatory forces in terrestrial ecosystems: evidence from decomposer food-webs. *Oecologia*, **93**, 303–306.

Wardle, D. A., Barker, G. M., Bonner, K. I. and Nicholson, K. S. (1998). Can comparative approaches based on plant ecophysiological traits predict the nature of biotic interactions and individual plant species effects in ecosystems? *Journal of Ecology*, **86**, 405–420.

Wardle, D. A., Bardgett, R. D., Klironomos, J. N., et al. (2004). Ecological linkages between aboveground and belowground biota. *Science*, **304**, 1629–1633.

Webster, J. R. and Meyer, J. L. (1997). Organic matter budgets for streams: a synthesis. *Journal of the North American Benthological Society*, **16**, 141–161.

Whisler, H. C., Zebold, S. L. and Shemanchuk, J. A. (1975). Life history of *Coelomomyces psorophorae. Proceedings of the National Academy of Sciences of the USA*, **72**, 693–696.

Wurzbacher, C., Bärlocher, F. and Grossart, H. P. (2010). Review: aquatic fungi in lake ecosystems. *Aquatic Microbial Ecology*, **59**, 125–149.

Wurzbacher, C., Kerr J. and Grossart H.-P. (2011). Aquatic fungi. In *The Dynamical Processes of Biodiversity – Case Studies of Evolution and Spatial Distribution*, ed. O. Grillo and G. Venora. InTech, pp. 227–258.

Wyman, R. L. (1998). Experimental assessment of salamanders as predators of detrital food webs: effects on invertebrates, decomposition and the carbon cycle. *Biodiversity and Conservation*, **7**, 641–650.

Yannarell, A. C. and Kent, A. D. (2009). Bacteria, distribution and community structure.

In *Encyclopedia of Inland Waters*, Vol. 3, ed. G. E. Likens. Cambridge, MA; Elsevier Inc., pp. 201–210.

Zak, D. R., Holmes, W. E., White, D. C., Peacock, A. D. and Tilman, D. (2003). Plant diversity, soil microbial communities, and ecosystem function: are there any links? *Ecology*, **84**, 2042–2050.

Zheng, D. W., Ågren, G. I. and Bengtsson, J. (1999). How do soil organisms affect total organic nitrogen storage in soils and substrate nitrogen to carbon ratio? A theoretical analysis. *Oikos*, **86**, 430–442.

CHAPTER ELEVEN

The question of scale in trophic ecology

LEE A. DYER

University of Nevada, Reno, NV, USA

TARA J. MASSAD

University of Sao Paulo, Brazil

and

MATTHEW L. FORISTER

University of Nevada, Reno, NV, USA

Introduction

The thousands of studies on determinants and effects of top-down and bottom-up trophic forces (hereafter, TDBU) in aquatic and terrestrial communities include approaches that usually focus on specific scales; however, as a combined body of work, these studies span temporal scales from minutes to centuries and spatial scales from bench-top microcosms (cm^3) to entire forests, lakes, or oceans (km^3). How do empiricists synthesize this extensive literature and how do theorists calculate meaningful model parameters given these massive scale disparities? Levin (1992) argues that issues of scale are among the most important in ecology and perhaps all of the sciences, and there are both practical and theoretical reasons for using a variety of spatial and temporal scales in ecological studies. However, spatial and temporal scales are often ignored in syntheses of literature on trophic interactions. For example, a $4\,m^2$ plot and a $100\,cm^3$ microcosm are the most appropriate scales for units of replication in experiments examining how arthropod foraging affects alpha diversity of primary producers in long leaf pine understories and ephemeral pool communities respectively, but these studies cannot in either case demonstrate population-level effects and should not be included in syntheses that examine top-down effects on beta diversity for entire forests or streams. In this chapter, we examine scaling issues associated with empirical studies of trophic interactions (for theoretical considerations, see Chapter 1, this volume).

We are interested in the extent to which processes and mechanisms observed at particular spatial and temporal scales are relevant for processes and patterns

Trophic Ecology: Bottom-Up and Top-Down Interactions across Aquatic and Terrestrial Systems, eds T. C. Hanley and K. J. La Pierre. Published by Cambridge University Press. © Cambridge University Press 2015.

at other scales. To that end, we present examples of a concerted research effort with a model terrestrial system (ant-plants and associated rain forest communities), as well as multiple examples from freshwater and marine systems that have utilized a mix of experimental, observational, and modeling approaches across a continuum of spatial and temporal scales. In addition to a detailed discussion of scale and trophic ecology in these focal systems, we investigate the distribution of studies from the empirical literature with respect to spatial and temporal scale, and specifically ask whether effect sizes vary across scales using a meta-analysis of TDBU studies in terrestrial and aquatic systems. Finally, we review practical and theoretical reasons for using specific scales, and end with future directions for TDBU research focusing on issues of evolution and global change. Throughout, we emphasize the complexity of studying trophic biology across spatial and temporal scales, but we also find that the situation might not be as intractable as one might expect. Specifically, results from the meta-analysis suggest that insight into TDBU forces can be found across a surprising range of scales. Throughout the chapter, we focus on spatial and temporal *grain*, which refers to the sampling unit utilized in empirical studies, but our discussion extends to an examination of the ecological effects that can be inferred from the different research approaches.

The literature on TDBU is vast (reviewed by chapters in Tscharntke and Hawkins, 2001; Terborgh and Estes, 2010) as is the literature on scaling in ecology (Chave, 2013). All hypotheses involving trophic interactions are plagued with unique scaling issues as well as the common problem of how to infer larger scale patterns from small-scale experiments or work with smaller organisms. In this chapter, we focus on the most extensive subset of this literature – empirical studies that examine direct and indirect interactions between primary producers, herbivores, and predators and parasites at multiple spatial and temporal scales. The "top-down" forces examined here are effects of consumers (herbivores, predators, parasites, parasitoids) on other consumers and on primary producers, while "bottom-up" are effects of resources (light and mineral nutrients), producers, and consumers on higher trophic levels. The parameters measured or manipulated for different trophic levels can be densities, diversities, or traits, such as behavior.

We also focus on a commonly studied and important indirect effect – the trophic cascade (Hairston et al., 1960) – as it has been a cornerstone of TDBU research for decades. In fact, Strong (1992) suggested that documentation of four-level trophic cascades in aquatic systems was among the most important ecological discoveries of the preceding decade. For this chapter, we define a trophic cascade as measurable change in a trophic level parameter due to indirect effects from non-adjacent trophic level parameters (Dyer and Letourneau, 2013). The top-down trophic cascade and other indirect trophic effects are important topics in the discussion of scaling issues because they typically require a

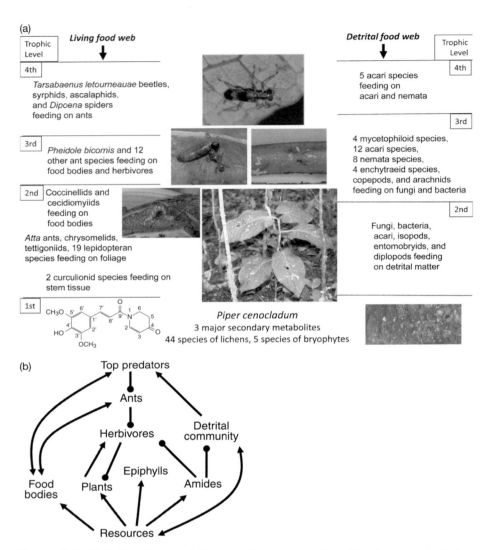

Figure 11.1 (a) A slice of a typical *Piper* ant-plant community, with a focus on the living and detrital communities associated with *P. cenocladum* shrubs, which were used to examine TDBU effects on living and detrital communities. The living and detrital communities are summarized by listing the number of different taxa and their resources at different trophic levels. Organisms depicted in the images are associated with the living food web, *clockwise from bottom right:* lichens and mosses found on *Piper* leaves; an experimental *P. cenocladum* shrub; structure of a defensive amide; *Eois nympha* (Geometridae – specializes on *P. cenocladum*); *Pheidole bicornis* (mutualistic ant) worker attacking specialist caterpillar, *Quadrus cerealis* (Hesperiidae); *Tarsobaenus letourneauae* (Cleridae), specialist beetle that lives in petioles and preys on resident ants; colony of mutualistic ants in dissected petiole – queen, workers, brood, and food bodies (smaller white spheres) are visible. (b) A summary of some of the documented TDBU relationships

detailed understanding of the natural history of the study system plus a combination of experiments and observations ranging from the smallest temporal and spatial scales to decade-long experiments over entire landscapes. For example, Carson and Root (2000) used insecticides over 10 years to reduce insect herbivore abundance in large replicated plots in a goldenrod-dominated old-field in central New York, for which the natural history was well known. It was not until an outbreak of a native chrysomelid beetle occurred in year 8 of the experiment that it became clear that these beetles could devastate goldenrod communities and accelerate succession by allowing the invasion of trees. A decade of censusing beetle boom and bust cycles throughout central New York provided data verifying that the outbreaks were common enough, and at a large enough spatial scale to regulate goldenrod communities in the region and likely elsewhere (Root and Cappuccino, 1992; Carson and Root, 2000). In the next section, we turn to another example of long-term research that has explored a variety of TDBU forces at multiple scales.

Tests of TDBU hypotheses across multiple scales in a model terrestrial system

Letourneau and colleagues investigated TDBU influences on insect and plant densities in a four-trophic level system in Central America at multiple spatial and temporal scales (reviewed by Letourneau, 2004; Letourneau and Dyer, 2005; Dyer and Letourneau, 2013). The focal community in this system is associated with a common understory ant-plant in Costa Rica, *Piper cenocladum* (Piperaceae), numerous specialist and generalist herbivores, resident mutualistic ants (*Pheidole bicornis*), and a specialist predatory beetle, *Tarsobaenus letourneauae* (Cleridae; Letourneau, 1983; 1990; Fig. 11.1). Work also focused on arthropod communities associated with other species of *Piper* throughout the Neotropics. The initial research used the focal ant-plant to test four prominent models of herbivore regulation: (1) top-down trophic cascades in which predators regulate their prey (Hairston et al., 1960); (2) the thermodynamic model of bottom-up trophic cascades in which herbivores are regulated by plant biomass (Slobodkin, 1960); (3) the green desert model of bottom-up forces in which herbivores are regulated through plant nutritional quality or secondary compounds (Abe

---

(connecting lines) between plant resources, plant chemistry, *Piper* species, other understory plant species, and associated communities, with arrows indicating the flow of energy or other positive effects and filled circles indicating consumption or other negative effects. *Resources* included light and soil nutrients; *amides* were secondary metabolites produced by *Piper* plants; *food bodies* were nutritive cells produced in *Piper* petioles; *plants*, *herbivores*, *ants*, and *top predators* all included densities and diversity measures for taxa in these trophic levels; the *detrital community* and *epiphylls* included all of the species summarized in Fig. 11.1a.

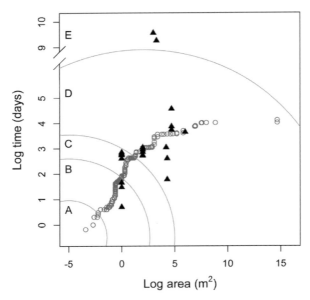

Figure 11.2 Spatial and temporal scales of 24 studies focused on the trophic interactions involving *Piper* (triangles) and of 287 other studies from a meta-analysis of aquatic and terrestrial TDBU studies (circles). Letters correspond to the ideal scales for conducting TDBU research that focuses on: (A) mechanisms mediating consumer–resource interactions, such as chemical defense against enemies or handling time of prey; (B) studies uncovering density or trait effects on one trophic level in response to a manipulation at another trophic level, whose measured effects can be used to parameterize models or to guide larger observational studies; (C) natural history observations; (D) mathematical models or observational studies to test hypotheses about large-scale effects based on information yielded from smaller studies; and (E) comparisons of emergent properties across landscapes or phylogeny generation across multiple trophic levels.

and Higashi, 1991; see Chapter 8, this volume); and (4) a non-trophic structure model focusing on omnivory and incorporating competition, natural enemies, and food quality factors (Polis and Strong, 1996). Work with this focal system was then expanded to the wider community to examine top-down and bottom-up influences on the diversity of arthropods and understory plants. Good tests of these four models and of tropical diversity hypotheses require long-term experiments and observational data at the largest scales possible, but it was clear from this work that smaller scale studies also provide important insights into key mechanisms. Although there are distinct theoretical considerations for processes at different spatial and temporal scales, the temporal and spatial scales of empirical studies are usually correlated, and this was usually the case for work with the *Piper* community (see Fig. 11.2 for a summary of spatial and temporal scales across which this system has been studied).

Study system

Piper cenocladum occurs throughout the Atlantic lowlands of Costa Rica and is one of many species of *Piper* understory shrubs that are ant-plants (Letourneau and Dyer, 1998a). The ant colonies reside in hollow petioles that produce opalescent food bodies, rich in lipids and proteins; these food bodies are produced when petioles are occupied by mutualistic ants or predatory clerid beetles. There are many other predatory species of ants, beetles, flies, and spiders associated with *Piper* shrubs, but none of these induce food body production. *Piper* ant-plants contain defensive amides and other defensive compounds in high concentrations, as is the case for most species of *Piper* (Dyer and Palmer, 2004). Generalist insect herbivores can cause extensive damage to the plant, but the most important herbivores are specialist lepidopteran larvae in the genus *Eois* (Geometridae; Dyer et al., 2010), which reduce the fitness of *Piper* clones (Dyer et al., 2004c). Most *Eois* caterpillars feed exclusively on *Piper* and each *Eois* species feeds on an average of two *Piper* species (Connahs et al., 2009; Rodríguez-Castañeda et al., 2010). For multiple *Piper* shrubs, mutualisms with predatory ants have evolved in response to TDBU forces, with ants providing a bottom-up, nutrient procurement role in regions with low nutrients and high parasitism, and ants providing a top-down protective role in forests with high levels of herbivory (Tepe et al., 2009; Rodríguez-Castañeda et al., 2011).

The ant-plant study system (Fig. 11.1) was used to directly compare the suppression of herbivores by predators versus chemical defenses and to assess the effects of plant resource availability on upper trophic levels via secondary metabolites. Experimental manipulations at multiple scales included additions of nutrients to soils, modifications of light availability, additions of top predators (4th trophic level), removals of ants (3rd trophic level), herbivore removals, manipulations of plant densities, and manipulations of plant chemistry. Observations included measures of densities and diversity at multiple trophic levels at sites from Mexico to Brazil and with data relevant to time periods from a few years to decades (including syntheses of all data collected). Overall, this work demonstrated that the presence of the predatory beetles in *P. cenocladum* plants cascades down through the ants and herbivores, resulting in smaller plants (Letourneau and Dyer, 1998a; 1998b) and affecting the diversity of associated arthropods and surrounding understory plants. In this same system, strong direct bottom-up effects of enhanced light and nutrients on plant biomass and chemistry were found (Dyer and Letourneau, 1999a; 1999b; Dyer et al., 2001; 2003; 2004a; 2004b). The effects on biomass did not cascade up to other trophic levels, but the effects on chemistry did, which is more consistent with a green desert model than a thermodynamic bottom-up model. Specialist herbivores sequestered secondary compounds, which made them more susceptible to parasitism, and generalist herbivores avoided plants with high levels of defensive chemistry. Were these general experimental results consistent across all temporal and spatial scales (Fig. 11.2)?

Multiple spatial and temporal scales

At the smallest spatial scales, experiments and observations utilized individual potted plants in shadehouse structures, individual plants in the forest, and experimental plots that varied in size from 6 m^2 plots to 100 m^2 plots. Intermediate scales examined responses of plants to manipulations and natural variation within an entire forest (at least 1600 ha in size), and large scales covered the entire range of the *Piper* species examined. The smallest temporal scales were represented by laboratory experiments examining effects of plant chemistry on herbivorous caterpillars over several months, while most field and shadehouse experiments were conducted at intermediate time scales of 18 months to 5 years in length. The longest studies compiled observational and experimental data collected over 10–30 years. These studies were accompanied by modeling and phylogenetic work, which address questions at the largest spatial and temporal scales (Fig. 11.2). To examine how these different scales were utilized with this system, we discuss three research goals below: (1) documenting community-wide trophic cascades; (2) understanding the role of plant chemistry in TDBU interactions; and (3) uncovering evolutionary responses to TDBU interactions.

Trophic cascades

The first tests of the trophic cascade hypotheses within this model system focused on both the largest and smallest spatial scales. Observational data from entire forests across Costa Rica uncovered patterns of alternating high and low densities of beetles, ants, herbivores, and plants (Letourneau and Dyer, 1998a) and these patterns were consistent with the hypothesis that high predatory beetle densities within a forest would cascade down through the ant and herbivore trophic levels to decrease plant densities at the whole forest scale (1600 ha and larger). This work was coupled with 18-month experiments in 100 m^2 plots (Letourneau and Dyer, 1998b), which uncovered mechanisms by which the predatory beetles lower ant densities, resulting in higher herbivory and lower plant biomass. Although these experiments indicated that bottom-up factors were swamped by the top-down cascade, 15 years of subsequent experiments conducted mostly at small spatial and temporal scales revealed many interesting details of the original large-scale observation, including the important bottom-up influences of plant chemistry. For example, ant predators had the strongest negative effects on specialist caterpillars in the system, while plant chemical defenses had the strongest effects on generalists. Both of these scales are appropriate, as documenting cascades requires large-scale observational data, while associated experiments at small spatial scales allow for tests of mechanistic hypotheses about factors that mediate cascades.

Trophic cascade hypotheses traditionally focus on entire ecosystem parameters, such as total primary productivity, the loss of entire trophic levels, or ubiquitous outbreaks of herbivores (reviewed by Dyer and Letourneau, 2013); thus, there is another scaling issue associated with these results: all density effects in this model system were focused on *Piper* shrubs. Even for the large-scale observations across forests (Letourneau and Dyer, 1998b), the focus was on a small component of the ecosystems studied, and it is not clear if these effects scale up to ecosystem parameters. In response to this problem, years of follow up experiments examined effects of these chemically mediated trophic interactions on an important community parameter, taxonomic diversity. Results at multiple scales support the concept of a *community-wide* top-down trophic cascade and provide evidence for the idea that one keystone species can affect the diversity of all plants and animals in a community (Dyer and Letourneau, 2003). At the smallest scale, individual plants were the replicates, and the response variables were diversity measures of the entire arthropod community on the plant surface and living within the plant tissues. The addition of top predators increased the diversity of consumers supported by *P. cenocladum*, whereas balanced plant resources (light and nutrients) increased the diversity of primary through tertiary consumers in a species-rich detrital food web (Dyer and Letourneau, 2003). Furthermore, increases in chemical defenses indirectly increased lichen diversity by decreasing herbivory (Dyer and Letourneau, 2007). At larger scales, within several forests that were at least 1600 ha in area, the effects of top predators on *P. cenocladum* cascaded out to change understory plant diversity, and this effect was modified by soil quality.

Two additional approaches were utilized to test the trophic cascade hypotheses at the largest temporal and spatial scales to examine population regulation and geographical variation. First, to explore hypotheses at temporal scales longer than decades, the empirical results just described were expanded to examine population dynamics and regulation with a modeling approach (Matlock and Dyer, unpublished model). Second, the generality of some trophic interaction results was examined for different *Piper* species (Fincher et al., 2008), different ecosystems (Connahs et al., 2009; Rodríguez-Castañeda et al., 2010; 2011), and across large elevational gradients (Rodríguez-Castañeda et al., 2010; 2011).

Long-term data on multi-trophic interactions over many decades are rare; studies typically do not measure the regulation of populations, but rather short-term or small-scale density effects of one trophic level on another (Hunter, 2001). This includes the numerous empirical studies on trophic cascades in terrestrial and aquatic systems (e.g., empirical studies described in Terborgh and Estes (2010) and studies described below in the aquatic section) – these studies document density effects of top trophic levels on herbivores and primary producers that may not reflect population dynamics over time (Hunter, 2001). Nevertheless, understanding the regulation of herbivore and plant populations is one of the most fundamental questions in ecology. Matlock and Dyer (unpublished data) utilized the multi-scale experimental and observational data from the *P. cenocladum* system (Fig. 11.1) to parameterize an analytical model that illustrates long-term dynamics of predator populations that limit herbivore populations. The four trophic levels were modeled with a system of eight ordinary differential equations. Population densities of free-living adults of both the predatory beetles and the mutualistic ants are very difficult to census, thus the SIR (susceptible, infected, removed) formulation from epidemic models was used to model the densities of plants with resident ants, with beetles, and without symbionts. This analytical model allowed for a realistic investigation of regulatory mechanisms that operate at long time scales (e.g., hundreds or thousands of generations). When initial *Eois* caterpillar densities were set to 0, the model predicted that the predatory beetle would be extant and the predatory ant would be extinct, which is consistent with observations across four forests where either the beetles or ants dominate (Letourneau and Dyer, 1998a). Because herbivory reduces the clerid beetle's capacity for population growth and because ant-plants are defended from *Eois* herbivory by the predatory ants, the addition of *Eois* herbivores destabilizes the monomorphic beetle equilibrium, allowing ants to invade. By tuning key parameters, the model can readily be made to switch between ant and beetle dominated equilibria, as documented with empirical data at multiple scales. Modeling work such as this enhances the experimental and observational approaches to understanding strong TDBU forces at the largest ecological scales.

The second approach to broadening the scale of trophic cascades results documented for the *Piper* system involved compiling data collected across large geographical areas to examine interactions at larger scales and to identify factors that are potential ecological and evolutionary consequences of the small-scale *Piper*–*Eois*–enemy relationships. Correlations between levels of herbivore specialization on *Piper*, levels of parasitism, plant diversity, natural enemy diversity, and *Piper* arthropod diversity were documented from Panama to Ecuador across multiple elevational and resource availability gradients. For example, based on over 25 000 rearings of caterpillars, from Mexico to Argentina, it is clear that *Piper* species often experience high levels (over 10%) of folivory from *Eois* caterpillars, which in turn experience high levels of parasitism (over 10%), and both herbivory and parasitism vary predictably with climate and topography (Fincher et al., 2008; Connahs et al., 2009; Dyer, unpublished data). *Eois* species also form distinct communities on *Piper* host plants across elevational gradients in Ecuador and Costa Rica, with high alpha diversity (at the local forest scale = 1000 ha) and high beta diversity (across 2000 m in elevation) (Brehm and Fiedler, 2003; Rodríguez-Castañeda et al., 2010; Brehm et al., 2011). Along these large elevational gradients, the bottom-up effect of *Piper* diversity only partly explains *Eois* diversity, with degree of specialization and species packing on host plant species responsible for over 50% of the variance in *Eois* diversity in Ecuador and

Costa Rica (Rodríguez-Castañeda et al., 2010). In sum, *Eois* parasitism rates, *Eois* diet breadth, *Piper* density and diversity, and *Eois* density and diversity are all correlated across broad geographical gradients, corroborating diversity cascade predictions based on small-scale experimental studies (Dyer and Letourneau, 2013). As described in the aquatic examples below, classic examples of trophic cascades in freshwater and marine systems (e.g., Estes and Palmisano, 1974; Power, 1990) were similarly preceded and followed by decades of study at different scales, including careful natural history observations, experiments, and data collection across large gradients.

Chemical ecology of TDBU factors

Just as large-scale observational data in aquatic and terrestrial systems are important for documenting patterns predicted by trophic cascades, biogeographical data on phytochemistry are necessary for understanding the roles of plant chemistry in mediating TDBU forces in both terrestrial and aquatic systems (e.g., Coley et al., 1985; Hay, 1996). However, for terrestrial and aquatic systems, most studies of the chemical ecology of trophic relationships are short term and occur in the laboratory, greenhouse, mesocosms, or small field plots (see Chapter 8, this volume). Studies at this scale reveal a great deal about potential mechanisms by which plant chemistry can affect large TDBU processes, such as trophic cascades and the evolution of herbivore specificity (see Chapter 13, this volume). For the *Piper* model system, a combination of small-scale field studies demonstrated that amide concentrations in leaf tissues varied significantly based on manipulations of plant resources and also due to the presence or absence of ants or beetles in all experiments, and these results were corroborated by the larger scale observational studies. Starting from the smallest spatial and temporal scales, greenhouse experiments on individual *P. cenocladum* fragments revealed defensive tradeoffs that occur within a few hours; when mutualistic ants are excluded from plants, allocation to ant rewards decreases, and levels of defensive amides increase 3–5 times in concentration (Dyer et al., 2001). These tradeoffs scale up to larger spatial and temporal scales, since interspecies and intra-species comparisons throughout the Neotropics reveal that different *Piper* species and individuals invest more in one type of defense than the other (Fincher et al., 2008; Rodríguez-Castañeda et al., 2011). Similarly, small-scale manipulations of individual plants over 6–36 months show that levels of amides are higher in balanced versus unbalanced resource (light and nutrients) conditions, and these bottom-up effects scale up to entire forests (Dyer et al., 2004a; 2004b: unpublished data).

How do these changes in chemistry affect upper trophic levels? The most notable tri-trophic effects of variation in *Piper* plant chemistry are that specialist caterpillars are rendered more susceptible to parasitism via sequestration of toxic amides, which disrupt their parasitoid encapsulation processes. The three amides in *P. cenocladum* also have a variety of synergistic negative effects on specialist and generalist herbivores, predators, and parasitoids in laboratory experiments (Dyer et al., 2003; 2004a; 2004b). While these are all fascinating aspects of chemically mediated TDBU interactions, it is not clear how such influences of chemistry scale up to affect populations of any of the trophic levels examined, but the modeling and large-scale sampling approaches described above would be the best methods for exploring these possibilities. Similarly, the long-term evolutionary consequences of these chemically mediated trophic interactions are best studied using a phylogenetic approach, as described in the next section.

The longest temporal and spatial scales: evolutionary consequences of TDBU forces

The consequences of well-studied trophic cascades are often applied to conservation issues, but the evolutionary implications are also very interesting (see Chapter 13, this volume) and not always considered. The concept of the "phylogenetic cascade" combines trophic cascade dynamics with predictions of coevolution and posits that evolutionary histories may be shared across more than two trophic levels (Forister and Feldman, 2011). Such evolutionary consequences of trophic interactions associated with *Piper* were examined utilizing a phylogeny of *Eois* species that are important specialist herbivores on *Piper* host plants. This approach was utilized to address evolutionary questions about how top-down (parasitoids) and bottom-up (host plant chemistry) forces can shape the evolution of herbivore host affiliation. Multiple genetic markers were used from *Eois*, *Piper*, and *Parapanteles* (braconid parasitoid wasps that attack *Eois*) species to examine phylogenetic patterns and evolutionary relationships between the three trophic levels (Wilson et al., 2012). The phylogenies included species across much of the range of *Eois* in the Neotropics. Again, this approach included analyses of potential evolutionary consequences of TDBU forces that are persistent across all temporal and spatial scales. Host–parasite associations were mapped onto phylogenies, and non-random patterns of host use within both the moth and wasp phylogenies were uncovered. Most notably, the *Eois–Piper* associations are characterized by Pleistocene radiations of caterpillars associated with the ant-plant, *P. cenocladum* and close relatives; similar radiations were uncovered on other individual *Piper* species within distinct ecosystems (Wilson et al., 2012). Classic chemically mediated trophic interactions could be part of these diversifications. For example, the sites where *Eois* diversified on *P. cenocladum* are characterized by high variance in nutrient and light availability, and the small-scale experiments described in the "chemical ecology" section above demonstrate that chemical defenses of *P. cenocladum* vary with resource availability; in turn, these changes in chemistry have fitness consequences for the *Eois* herbivores and

Parapanteles parasitoids (Dyer et al., 2004a; 2004b). The high levels of specializa-tion and geographical variation in parasitism and chemical defense documented for *Piper* therefore suggest higher trophic levels can develop isolated gene pools on host plants of the same species growing under different TDBU conditions. Fur-thermore, such interactions might have long-term consequences, as two small radiations of parasitic wasps were discovered specializing on *Eois* (Wilson et al., 2012), and this could be one of the first examples of a bottom-up trophic cascade being associated with a phylogenetic cascade, given that there is an indirect effect of diversification at one trophic level (*Piper*) on a non-adjacent trophic level (parasitic wasps; Forister and Feldman, 2011).

Insights about scale in terrestrial systems

In sum, the research approaches applied to uncover large-scale, long-term con-sequences of TDBU forces in the ant-plant system relied on a variety of exper-imental, observational, and modeling studies at multiple spatial and tempo-ral scales (Fig. 11.2). To test longstanding hypotheses relevant to trophic cas-cades and chemical ecology, observations over three decades (Letourneau, 1983; Letourneau, 1990; Dyer and Letourneau, 2013) and across ecosystems from Mexico to Brazil provided the insight necessary to guide all of the smaller scale experimental work, as well as the modeling and phylogenies. The experimental work, in turn, provided key details about mechanisms by which TDBU forces function in this system. While it is tempting to argue that all terrestrial studies of TDBU should focus on large spatial and temporal scales (e.g., Terborgh and Estes, 2010), large-scale studies do not necessarily provide details that allow for greater generalizations and for a clear understanding of processes that could underlie many of the large patterns. Thus, decreases in productivity or plant diversity may be correlated with the absence of top predators (e.g., clerid beetles or jaguars), but this pattern could be driven by subtle changes in the mesopredator (e.g., ants or birds) community, changes in parasite loads, mesopredator behavioral changes, or many other potential mechanisms. Of course, not all small-scale mechanisms, such as sequestration of secondary metabolites by *Eois* caterpillars, which are a part of large-scale cascades or other TDBU patterns, are relevant to larger organisms; however, the approach of combining large-scale work with smaller scale, focused studies still applies. For example, examining the effects of tree chemistry on mammals in the laboratory or in small-scale field trials is relevant to TDBU relationships of mammals in an ecosystem context (Bryant et al., 1991). Furthermore, small-bodied animals and small, confined ecosys-tems (such as endophytic communities or islands) are the empirical equivalent of mathematical models – examining the reasons why these smaller systems function the way they do provides insight into larger systems and can provide relevant hypotheses to test in the larger systems. Of course, mechanisms and processes at one scale may also be of inherent biological interest even if they do not have consequences at smaller or larger scales, but it is always important to ask to what extent factors have consistent or scalable effects across spatial and temporal levels.

Tests of TDBU across multiple scales in freshwater and marine systems

From the trophic cascade in lakes to smaller field studies and mesocosms

Menge and Olson (1990) argue that predictive theoretical frameworks should be hierarchical, with models of local community processes nested within ecosystem-wide, landscape, and, ultimately, global models. For TDBU patterns in freshwater systems, the ecosystem scale is the best studied, and producing good predictions requires large spatial and temporal empirical data. Among the most ambitious, illustrative studies of trophic cascades are the large-scale, long-term studies from Carpenter and colleagues (e.g., Carpenter et al., 2001; also reviewed by Carpenter et al., 2010). Carpenter et al. (2001) manipulated nutrient availability and predator densities in four lakes over 7 years and effectively created ecosystems with three versus four trophic levels. This work demonstrated clear density-mediated cascades across all trophic levels, as well as more nuanced changes (as reviewed in several other chapters in this volume). In addition to demonstrating strong density-mediated effects, Carpenter et al. (2001) also demonstrated that planktivory affects zooplankton size structure more than total biomass. These non-density, or "trait-mediated," effects on intermediate trophic levels are a ubiquitous component of terrestrial and aquatic ecosystems and are usually investigated at smaller spatial scales, such as experimental plots or mesocosms, and shorter time periods, such as the course of a single field season (reviewed by Schmitz et al., 2004).

The scale of the experiment by Carpenter and colleagues, which encompassed large, stratified lakes, was sufficient in size to demonstrate that top-down trophic cascades operate at an ecosystem scale even in heterogeneous environments. This is relevant to the assertion that cascades are rare in terrestrial ecosystems because they are too complex and reticulate for trophic cascades to structure populations (Polis and Strong, 1996). Carpenter et al. (2010) point out that in spite of such predictions, there is evidence for cascades in open marine systems as well as the large, heterogeneous lakes they studied, and they suggest that large-scale, long-term studies are necessary for detecting cascades. However, without knowing more explicit details of factors that mediate consumption and productivity, it is hard to predict how other ecosystems, with different species assemblages and different abiotic stresses, are likely to respond to changes in nutrient availability or predation pressure. Chemical defenses and other trait-mediated effects, as described above, can drastically modify trophic cascades and the outcomes of TDBU interactions. Schmitz et al. (2004) review some of the scaling issues associated with detecting trait-mediated trophic cascades and suggest that studies at all spatial and temporal scales are important for understanding the mechanisms by which indirect trophic effects function in terrestrial and aquatic systems.

In contrast to the entire lake experimental approach, mesocosm studies are frequently utilized to examine trophic interactions in aquatic systems. As with small-scale terrestrial studies, the limited complexity inherent to these studies allows for clear mechanistic interpretations but does not necessarily scale up to complete ecosystems (Bell et al., 2003). Low complementarity among predators, modified behaviors of predators and prey, limited diversity of resources, and strong effects of subtle abiotic variations are among the most obvious limitations of the mesocosm approach. A recent example comes from a study conducted in large tanks (183 cm diameter) and run for 46 days (Wesner, 2010). Single species of predatory fish were added to the tanks that were inoculated with aquatic insect larvae from a local stream. Strong top-down effects of the fish were measured, reducing the biomass of aquatic insect larvae by 55%. Similar to the compositional changes measured in the lake experiments conducted by Carpenter and colleagues, this top-down effect in the mesocosms was species specific, and only predatory dragonfly larvae were significantly reduced by fish predation (Wesner, 2010). One strength of this mesocosm approach is that the species most susceptible to fish predation was identified, and this type of information can generate further hypotheses regarding life history traits important for facilitating trophic cascades at larger scales.

More can be learned from both whole-lake and mesocosm studies when they are complemented with additional research at smaller and larger scales respectively. For example, an especially intriguing aspect of the mesocosm results reported by Wesner (2010) is that the reduction in dragonfly numbers by fish predation could have cross-ecosystem effects by altering terrestrial top-down and bottom-up interactions, as dragonflies are both important predators and prey in riparian zones (Wesner, 2010). Aquatic insects have, in fact, been studied in intermediate-sized field studies along streams as links from aquatic to terrestrial systems (see Chapter 6, this volume; Knight et al., 2005). In the first of such studies, an invasive trout was added to 27.5 m reaches of a natural stream in Japan, and its negative effect on the emergence of aquatic insects was almost equal to the effect of physically covering the stream with screen. This top-down reduction in aquatic insect biomass then cascaded up to reduce the numbers of spiders (Tetragnathidae) that rely on emerging insects as prey (Baxter et al., 2004). Similarly, Allen and colleagues (2012) combined both mesocosm and medium-scale field studies to examine the effects of mussel diversity on algal biomass and the emergence of aquatic insects. In 100 L mesocosms, mussel richness had a positive effect on insect emergence, which cascaded up to enhance

tetragnathid spider reproduction. Mussel richness did not affect algal accumulation, but isotope analyses showed mussel-derived nitrogen was important
for algal growth. The companion field study was conducted in replicated 100 m
reaches in two streams, and insects were sampled in traps over the course
of 1 week. Despite the short time frame of the data collection, the research
uncovered the same pattern of increased insect emergence with higher mussel
richness (Allen et al., 2012). Once again, the combination of studies at different
spatial scales is a more nuanced approach to understanding how diversity and
other factors are affected by TDBU interactions, and multi-scale work provides
evidence of ecological mechanisms suggested by less spatially integrated studies.

TDBU and aquatic biodiversity

Some of the best studies of diversity are an expansion of trophic cascade research
to diversity cascades work (Dyer and Letourneau, 2013) and are conducted
at multiple, complementary scales, because local diversity parameters affect
community-wide properties across space and time. Thus, the best approach
to diversity cascades studies is to examine the scaling hierarchy in its entirety.
Interactions between local diversity and primary productivity across a landscape
may be an important part of the multivariate productivity–diversity hypothesis
(MPD), which connects older ideas suggesting that enhanced resources cause
greater diversity with more recent thoughts that higher diversity promotes
biomass accumulation (Cardinale et al., 2009). The MPD model specifies that
the local (small-scale) richness of competitors is a better predictor of biomass
than the total richness of a pool of colonists. This hypothesis was tested in
20 streams in the eastern Sierra Nevada in California and the authors found
support for their hypothesis that bottom-up forces directly affected primary
producer biomass, as well as the number of potential colonists that successfully
co-established. Nutrient additions altered the composition of algal communities
across streams, despite the fact that species' identities remained unique between
the streams (Cardinale et al., 2009). Patches with higher species richness also
had higher biomass, and the direct positive effect of richness on biomass was
stronger than the direct effect of nutrients on biomass (Cardinale et al., 2009).
This integration of localized and community-wide parameters is similar to metacommunity research, where examination of small-scale processes clarifies larger
scale patterns of diversity.

In fact, metacommunity studies provide a perfect example of integrating multiple scales in TDBU research for most ecosystems, and aquatic systems that are
more confined are especially well suited for such an approach. Howeth and
Leibold (2010) utilized a metacommunity–mesocosm experiment to examine
the effects of predator identity and prey dispersal between communities on
trophic interactions, thus linking small-scale processes (mesocosms) with entire

communities. Theory suggests that regional dispersal of prey caused by differences in predation pressure can result in temporal variation in the outcome of species sorting and large-scale population stability in spite of changes at local scales. Howeth and Leibold (2010) tested for the effects of dispersal and predator identity on community-wide diversity using two aquatic insect predators and controlling the dispersal rate of plankton prey between mesocosms. Metacommunities with high dispersal rates were more homogeneous and less diverse than communities with low dispersal rates, a result which increased in strength over the 6 weeks of the experiment. Low regional diversity with high dispersal was primarily caused by effective dispersal abilities of a select number of prey species as opposed to post-dispersal interactions between species. Predator identity did not interact with dispersal to affect the diversity of the regional prey communities, but the identity of the predators did matter for the size structure of the prey community, a fairly common result in aquatic studies (Pace et al., 1998; Carpenter et al., 2001; Clemente et al., 2011).

In addition to examining top-down effects of diversity at multiple scales, this type of work is relevant to some of the big scaling questions about diversity and neutrality, which are typically addressed in terrestrial systems. Hubbell's neutral view (2001) is that local processes in a metacommunity are irrelevant. According to neutral theory, competitive interactions along with predation, herbivory, resource limitation, and other top-down and bottom-up effects at local scales are irrelevant due to dispersal limitations and the makeup of the overall species pool. The aquatic mesocosm experiments conducted by Howeth and Leibold (2010), however, demonstrated that predation can alter the composition of prey at local scales, even while this effect is mediated by regional dispersal. Together, these studies demonstrate that different insights emerge from examining connections between biodiversity and TDBU at different scales. These results also emphasize the value of regional replication for establishing the ubiquity of large-scale patterns, as well as the necessity of controlled mesocosms for teasing apart the functioning of such patterns.

Marine ecosystems: MPAs and variable TDBU forces

Marine systems have provided some of the strongest examples of prevalent TDBU effects and trophic cascades, and these effects are usually examined at the largest scales, allowing for the best landscape or global-sized predictions (see Chapter 2, this volume). One common large-scale approach is to contrast relatively intact ecosystems in marine-protected areas (MPAs) with heavily fished sites. Classic examples come from the study of urchins, which are well known for altering benthic communities when released from predation (Estes and Palmisano, 1974). The settlement of the urchin, *Diadema* aff. *antillarum*, is also affected by temperature. This type of biotic variation with abiotic variables illustrates the importance of integrating studies at different scales to fully understand trophic relationships. Urchin densities in three MPAs were paired with sites open to fishing that encompassed a temperature gradient across the Canary Islands and were monitored to examine top-down effects of fish predation. The MPAs were at least 9 years old and at least 775 ha in size. Urchin behavior, as well as abundance, changed with protected status; in general, urchins preferred cryptic habitats until they reached a larger size in MPAs, where fish predation was more intense. A tethering experiment was also conducted to measure predation pressure in and outside of MPAs, and over 70% of tethered urchins were preyed on in the MPAs, but only about 40% of urchins were consumed in fished sites. Assemblages of fish predators changed across the temperature gradient and this variation in predator composition was associated with a change in the size of urchins commonly preyed on (Clemente et al., 2011). While marine studies allow for addressing questions at the largest spatial scales over decades, there is still much to be learned from smaller scale experiments, such as the prey-tethering approach.

Temperature is just one of many locally varying environmental conditions that can alter the importance of TDBU forces at multiple scales – nutrient availability, depth, wave action, and water clarity are all local processes that can determine outcomes at larger scales. Another study with urchins found that the indirect positive effects of predatory fish on algal growth are weaker in nutrient-rich upwelling zones, where kelp grow vigorously (Shears et al., 2008). In yet another example, effects of MPAs were studied in six sites around New Zealand, and negative effects of predators on urchins and positive effects on algal growth were only detected in two locations. In the remaining sites, reductions in urchin numbers were likely due to sediments or wave exposure, illustrating context-dependent trophic effects (Shears et al., 2008) that can be elucidated with smaller scale experiments.

Longer temporal scales for ecological studies, comparable in scale to the decades-long studies with *Piper* or goldenrods (Carson and Root, 2000) in terrestrial ecosystems, have also been examined to understand persistence of TDBU forces in MPAs. Another study in the Canary Islands focused on a single MPA and investigated potential restoration of trophic cascades over time. Data on predatory fish, urchins, and seaweed were collected 3 and 7 years after reserve establishment in no-take areas and in an area with semi-restricted fishing. Seaweed recovered quickly in the no-take zone, and urchins decreased and seaweed cover increased markedly over the second 4 years of the study (Sangil et al., 2012). However, results from a meta-analysis combining data from MPAs of different ages indicate that studies of trophic recovery in MPAs must span even longer periods of time to capture the life cycles of large top predators (Micheli et al., 2004). Importantly, community composition was found to continue changing over decades, emphasizing the need for long-term studies to accompany changes in patterns of TDBU interactions. The meta-analysis also revealed high variability between fish recovery in MPAs of similar age but in different parts of the globe, corroborating the result that local environmental conditions modify TDBU interactions and trophic recovery (Micheli et al., 2004). Similarly, a time series study including temperate and tropical sites from around the globe found that direct effects of MPAs (the recovery of fished populations) appeared after roughly 5 years, whereas indirect trophic effects resulting from trophic cascades only became apparent after 13 years (Babcock et al., 2010). It seems likely that long-term TDBU studies are essential for freshwater and terrestrial systems as well, and that large temporal scales are necessary for designing a TDBU predictive hierarchy, but there are few datasets available for drawing such conclusions.

Insights about scale in aquatic systems

Environmental change has created a demand for predictions of ecological functioning to be made at global scales. Thus, ecologists have addressed important top-down questions examining the consequences of large-scale predator extinctions or range shifts of parasites and bottom-up questions focused on forest-wide or lake-wide resource enrichment or increases in invasive primary producers. This is especially true for aquatic and marine ecosystems, where whole-lake studies and large-scale investigations in MPAs have uncovered important patterns of trophic interactions and have documented strong trophic cascades. Nevertheless, our conclusions here are the same as the insights provided from years of careful multi-scale study with a model, multi-trophic, terrestrial system. As demonstrated in this brief review of aquatic studies at multiple scales, basic ecological theory still demands investigation at all scales, including the smallest scale studies, especially for clarification of mechanisms and for understanding links between different levels of organization. Returning to Menge and Olson's (1990) argument that predictive theoretical frameworks should be hierarchical, it is clear that this is a goal that has not yet been achieved in freshwater, marine, or terrestrial ecosystems.

Considering aquatic, terrestrial, large and small scales – do effect sizes differ?

In addition to detailed qualitative comparisons of the aquatic and terrestrial examples, such as those provided above, it is useful to quantitatively compare results from empirical studies conducted at different scales. We generated a sample of terrestrial and aquatic TDBU studies and used a meta-analysis to examine differences in effect size for studies conducted at different spatial and temporal scales (Fig. 11.2). Specifically, we wanted to know if effects were more easily observed at certain spatial and temporal scales, which would suggest that there might be optimal scales for studying TDBU processes. To allow for an

Table 11.1 *Results from a meta-analysis on the strength of trophic relationships in aquatic (A) and terrestrial (T) studies with top-down (T), bottom-up (B), direct (D), or cascading (C) interactions*

Comparison between effect sizes	dsi	CI	N[1]
$Q_B = 0.03$, n = 8, P > 0.1	ABD = 0.64	0.80	16
	ABC = 0.32	0.81	16
	ATD = 0.96	**0.37**	**73**
	ATC = 0.74	**0.49**	**42**
	TBD = 0.36	0.60	26
	TBC = 0.51	**0.51**	**35**
	TTD = 0.85	**0.45**	**47**
	TTC = 0.36	0.55	30

[1] N = the number of studies for each combination of ecosystem and interaction. Effect sizes (dsi) in bold are significant (95% confidence interval, CI, does not cross zero).

adequate sampling of longer and larger scale studies, we searched the trophic cascades literature, as tests for trophic cascades require larger and longer-term studies (Terborgh and Estes, 2010). Detailed methods for the meta-analysis are provided in the Appendix at the end of this chapter.

One clear result from the sample of studies included in the meta-analysis is that spatial and temporal scales of empirical studies are usually correlated, and this was also the case for the model *Piper* system discussed above (Fig. 11.2). The other clear pattern that emerged was that many TDBU effects are similar across different ecosystems and different scales. The strength of trophic interactions (effect sizes) did not differ depending on the environment (aquatic/terrestrial), direction (top-down/bottom-up), or proximity (direct/cascade) of the interactions tested (Table 11.1). Significant effects of trophic interactions were measured for aquatic top-down studies with both three and four trophic levels, meaning that predators effectively limit prey and enable primary production through control of herbivores (Table 11.1). In terrestrial systems, there were significant bottom-up cascades, indicating that resources enhanced higher trophic levels. Predators and herbivores in terrestrial studies also had significant negative effects on their hosts (Table 11.1).

Significant effect sizes were found for experiments of all sizes and temporal scales (Fig. 11.2), with few studies exhibiting differences in effect sizes (Fig. 11.3). The fact that for a random sample of studies, a broad range of spatial scales (10^{-4} to 10^{14} m^2) and temporal scales (10^0 to 10^4 days) yielded significant effect sizes could be a consequence of publication bias and researcher hesitancy to

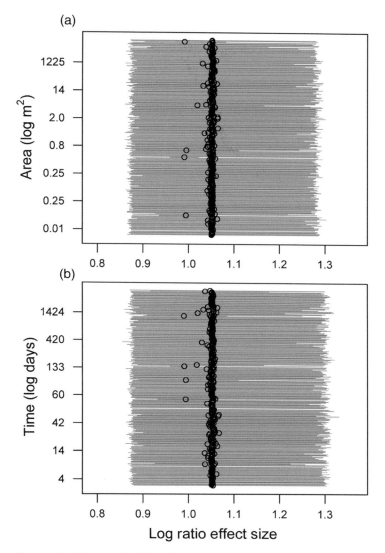

Figure 11.3 Log ratio effect sizes with 95% credibility intervals for 287 empirical studies. Credibility intervals were estimated using a hierarchical Bayesian analysis. Intervals greater than zero are considered to be significant effect sizes and intervals that overlap are not different. Studies are arranged according to: (a) spatial scale (natural log m^2) and (b) temporal scale (natural log days).

publish negative results, but there are convincing arguments that such biases are oversold (Koricheva, 2003). On the other hand, these large effect sizes across scales could also indicate that TDBU forces are pervasive at all scales, and the fact that effect sizes that were reported are not significantly different among different study sizes supports this hypothesis. As we outlined in the empirical examples from terrestrial and aquatic systems, there are clearly differences in the details that are revealed when research is conducted at different scales, and there are effects that cannot be uncovered without long-term or large-scale approaches, such as the effects of outbreaking insects (Dyer et al., 2012). However, if the directions and magnitudes of effect sizes are the same, then there are positive consequences for planning research and trying to scale results up or down. This is especially true for large-scale or long-term studies, where there is concern that low sample sizes undermine inferential statistics – if TDBU effect sizes are large regardless of scale, small sample sizes are less of a concern, because statistical power is likely to be high. This is evident in the comparisons of fished versus MPAs, where there is a striking contrast between intact and predator-free ecosystems (the largest spatial scales depicted in Figs. 11.2–11.3). Similarly, the longest studies with strong effect sizes were from studies of MPAs where predator pressure differed dramatically between treatment levels (the largest temporal scales depicted in Figs. 11.2–11.3). When such a strong difference between control and experimental conditions is not present, longer-term (and larger scale) studies may require larger sample sizes, which may not be possible in many instances. Greater abiotic and biotic variation is a natural component of these larger grained studies, and though this may dampen effect sizes, TDBU research designs should include studies that incorporate natural variation to help expand the scale of inference. Small-scale, short-term work has contributed fundamentally to our knowledge of the role of predators in limiting prey and promoting primary production, as well as the effects of resources in fortifying higher trophic levels (Winter and Rostás, 2010; Castejón-Silvo et al., 2012; Coverdale et al., 2012; Verreydt et al., 2012), but a greater effort should be made to measure trophic interactions in all their complexity – especially if the effect sizes are large enough to warrant smaller sample sizes for these studies.

Future directions for work at large scales: evolution and global change

We close by raising issues of scale in two distinct areas where future work on TDBU forces could profitably be focused: evolutionary processes and global change. How is our understanding of TDBU forces affected by a consideration of the time scales over which evolutionary processes act? How does the time frame relevant to anthropogenic alteration of the biosphere affect our expectations for the action of TDBU forces? How do the temporal scales for both of these issues interact with the spatial scales discussed above?

With respect to evolutionary scales, it is interesting to note that the situation is inherently more complicated than it was just a generation ago. In the not too distant past, it was reasonable to contrast evolutionary and ecological time scales (Hutchinson, 1965), but that dichotomy has become less useful as evidence has steadily accumulated for evolution on time scales that can be observed by a single researcher. The coeval nature of evolutionary and ecological processes has been reviewed in a number of places (Reznick and Ghalambor, 2001; Carroll et al., 2007; Schoener, 2011). Hairston et al. (2005) suggest that rapid evolution be defined as change that is sufficiently rapid that evolutionary shifts in ecologically relevant characters are large enough to impact measured variation in ecological processes (e.g., in response to experimental perturbations). Examples of recent and rapid evolution include not only traits of single species (e.g., flowering time in response to extended drought, Franks et al., 2007), but also complex community effects that are relevant to TDBU processes (Miller and Travis, 1996; Post and Palkovacs, 2009). Consider an evolutionary perspective of top-down effects on the planktonic crustacean *Daphnia*, which is a focal taxon in the classic trophic cascades work described earlier by Carpenter et al. (2001). In a top-down evolutionary cascade, the presence of alewife (a zooplanktivorous fish) affected the life history evolution of *Daphnia*, which in turn affected primary phytoplankton production (Walsh et al., 2012).

While it is now clear that evolution can be rapid and associated ecological effects can be pronounced, it is still important to distinguish between evolutionary processes at different temporal scales. Specifically, dispersal and drift change allele frequencies at the most immediate scales (within a single generation or from one generation to the next); natural selection produces phenotypic change by acting on standing genetic variation typically at intermediate scales (few to many generations), while evolution can proceed from novel mutation at much longer time scales. Issues of temporal scale in evolution are of course an issue unto themselves (Barton and Keightley, 2002; Salamin et al., 2010), but here we highlight a number of questions focused on evolution at different time scales in response to TDBU forces.

At the shortest time scales, more studies of natural selection interacting with multi-trophic processes would clearly be welcome (Schmitz et al., 2008). Considering longer time scales, will major changes in TDBU forces be more likely to result from *in situ* evolution of top predators, herbivores, or primary producers, or from the immigration of such players? It is clear that certain pairwise interactions (e.g., predator–prey) can evolve in a predictable direction over enormous time scales (Vermeij, 1994; but see Hunt, 2007); can we expect associated indirect effects to evolve with the same consistency? These are some of the more interesting issues for understanding TDBU forces at evolutionary scales; Abdala-Roberts and Mooney (Chapter 13, this volume) present a related discussion, particularly with respect to the evolution of herbivore and plant traits. We might also ask how evolutionary scale interacts specifically with the issues of spatial scale that we have highlighted in the rest of the chapter. For studies of natural selection on both small spatial and temporal scales, to what extent can results be extrapolated to other scales? Recent progress comes from a meta-analysis by Siepielski et al. (2013), who investigated spatially replicated studies of natural selection, and found that selection within any particular system tends to vary spatially in magnitude but not in direction. Thus, as with our meta-analysis of studies at different ecological scales, we could predict that the short-term selection studies focused on consequences of trophic interactions might indeed provide insight into longer-term evolutionary change.

The consideration of interactions between evolutionary time scales and spatial scales of TDBU processes is perhaps less urgent than understanding how TDBU forces might be affected by global change, which presents challenges of both temporal and spatial scale. Based on the meta-analysis we have presented, we suggest that studies conducted at smaller spatial and temporal scales have relevance for predicting biotic responses to anthropogenic change over the coming decades and centuries across scales, but that they should be linked to decades-long studies with landscape-level spatial grain. As an example, Free Air CO_2 Enrichment (FACE) studies suggest that herbivory might decrease with higher CO_2, but measurements spanning multiple years demonstrate that this is not a straightforward response. The FACE project in North Carolina showed herbivory decreased with high CO_2 for multiple tree species in the first year of measurements, but there were no detectable changes in herbivory in the following 2 years. A potential bottom-up explanation, based on years of short physiological ecology studies, is that foliar C:N ratios were generally higher with excess CO_2 in the year with reduced herbivory (Knepp et al., 2005). In other analyses from the same FACE project, Lepidoptera and Coleoptera showed a negative response to increased CO_2, while predatory spiders and Hymenoptera increased (Hamilton et al., 2012). Four years of work in another FACE experiment in broadleaf temperate forest showed CO_2 and O_3 have contrasting and independent effects on insect communities. O_3 in particular reduced parasitoid abundance, while CO_2 increased two parasitoid families and increased ant abundance (Hillstrom and Lindroth, 2008). Top-down and bottom-up changes can therefore be expected to accompany increasing greenhouse gases, which will likely have consequences for trophic interactions and ecosystem functioning that are, as of yet, difficult to predict (see Chapter 14, this volume, for further discussion).

Aquatic systems are particularly vulnerable to bottom-up global changes, and similarly complex issues of scaling can be expected. Eutrophication due to P pollution and high N inputs from agricultural runoff both affect the balance of aquatic systems, which motivated a large-scale experiment encompassing 30 streams in two mountain ranges in California (Nelson et al., 2013). Algal primary productivity was measured in response to N and P augmentation, and GPP, NPP, and respiration all increased with nutrients. However, no variable responded to N and P in the same way across all streams, and primary producer biomass increased in one ecosystem but not the other, demonstrating again that large-scale or spatially replicated studies are necessary before generalizations can be drawn regarding ecosystem responses to global change. Diversity, however, consistently declined with nutrient additions, a response common to many ecosystems (Isbell et al., 2013). The authors from the stream study concluded by suggesting that in spite of spatial variability, NPP and GPP are the most reliable, cross-ecoregion indicators of a response to nutrient additions (Nelson et al., 2013).

In conclusion, we suggest that whether trophic interactions are occurring on land or in the water, model systems examined across spatial and temporal scales will continue to provide the necessary natural history information and a clear picture of the nuts and bolts of interactions. With these study systems, long-term, large-scale work (and patience) will be required to document current changes in top-down and bottom-up interactions and to predict the ecological and evolutionary consequences of these changes. Much progress in both aquatic and terrestrial systems has been made, and it is noteworthy that our small sample from the published literature demonstrates considerable coverage of spatial and temporal scales (i.e., published studies are not clustered into one or a few frequently studied combinations of scales; Fig. 11.2). Furthermore, we sound the hopeful note that TDBU processes can be studied effectively across all of those scales, as revealed by our meta-analysis finding consistent effect sizes across very different studies (Fig. 11.3). However, much remains to be learned, particularly at larger temporal and spatial scales where the coverage from the literature is less dense (Fig. 11.2), particularly as more complex variables, such as diversity and trait-mediated interactions, are better incorporated into trophic ecology research.

Acknowledgments

We thank W. Carson and D. Letourneau for contributing ideas and insight for this manuscript, A. Smilanich and J. Harrison for helpful comments, and T. Hanley and K. La Pierre for providing excellent editorial advice and inviting us to submit this chapter. LAD and MLF were supported by NSF grant DEB 1145609 and DEB 1145609 while working on this chapter.

Appendix: meta-analysis methods

Using the Web of Science search engine, all articles published between 2008 and July 2013 fitting the search term "trophic cascade*," and excluding agricultural and invasive species studies, were collected. Data were gathered from 122 papers, producing 287 effect sizes that included 16 studies of aquatic, bottom-up, direct interactions; 16 studies of aquatic, bottom-up, cascading interactions; 73 studies of aquatic, top-down, direct interactions; 42 studies of aquatic, top-down, cascading interactions; 26 studies of terrestrial, bottom-up, direct interactions; 35 studies of terrestrial, bottom-up, cascading interactions; 48 studies of terrestrial, top-down, direct interactions; and 30 studies of terrestrial, top-down, cascading interactions. Means, sample sizes, and standard errors or standard deviations were gleaned from all these studies.

To calculate effect sizes, means were assigned to two levels, one (X1) that is hypothesized to yield higher responses and the other (X2) that is hypothesized to yield unchanged or lower responses. For example, if the independent variable was predator removal, and the dependent variable was herbivory, then X1 would be from the predator removal treatment (because the expectation is for higher herbivory when predators are absent). The data were organized in this way to maintain consistent directions of the treatment effects, regardless of the treatments and responses. The expectation that responses of X1 would be larger than responses to X2 was not always met by the data, however. For example, there were instances when herbivory was greater when predators were present. Effect sizes were calculated as Hedges' d, which is the treatment mean relative to the control divided by the combined variance of the treatment and control and corrected for sample size (Gurevitch and Hedges, 2001). We also calculated log ratio effect sizes and used these in a hierarchical Bayesian analysis (SAS 9.3; SAS Institute, Cary, NC); this analysis was conducted to allow for stronger inference when there were no differences between categories compared. This analysis also yields 95% credibility intervals for all studies included in the analysis. Hedges' d effect sizes were considered significant if their 95% confidence intervals did not cross zero, while log ratio effect sizes were considered significantly different from each other if their 95% credibility intervals did not intersect.

For Hedges' d, differences between effect sizes in different categories were evaluated with the between-class heterogeneity statistic Q_B (Gurevitch and Hedges, 2001). In this way, differences between all the treatment combinations (e.g., aquatic*top-down*direct vs. terrestrial*top-down*direct) were determined. In addition, differences due to the size and the duration of the studies were explored. Study sizes were treated as continuous variables – with days as the unit for time and m^2 (surface area in aquatic studies) as the unit for spatial scale. When individuals or plots were nested within larger experimental areas, the area subjected to the treatment was used to define the area (e.g., a 1 m^2 quadrat in a large marine reserve was classified at the scale of the reserve and not the quadrat).

References

Abe, T. and Higashi, M. (1991). Cellulose centered perspective on terrestrial community structure. *Oikos*, **60**, 127–133.

Allen, D. C., Vaughn, C. C., Kelly, J. F., Cooper, J. T. and Engel, M. H. (2012). Bottom-up biodiversity effects increase resource subsidy flux between ecosystems. *Ecology*, **93**, 2165–2174.

Babcock, R. C., Shears, N. T., Alcala, A. C., et al. (2010). Decadal trends in marine reserves reveal differential rates of change in direct and indirect effects. *Proceedings of the National Academy of Sciences of the USA*, **107**, 18256–18261.

Barton, N. H. and Keightley, P. D. (2002). Understanding quantitative genetic variation. *Nature Reviews Genetics*, **3**, 11–21.

Baxter, C. V., Fausch, K. D., Murakami, M. and Chapman, P. L. (2004). Fish invasion restructures stream and forest food webs by interrupting reciprocal prey subsidies. *Ecology*, **85**, 2656–2663.

Bell, T., Neill, W. E. and Schluter, D. (2003). The effect of temporal scale on the outcome of trophic cascade experiments. *Oecologia*, **134**, 578–586.

Brehm, G. and Fiedler, K. (2003). Faunal composition of geometrid moths changes with altitude in an Andean montane rain forest. *Journal of Biogeography*, **30**, 431–440.

Brehm, G., Bodner, F., Strutzenberger, P., Hunefeld, F. and Fiedler, K. (2011). Neotropical *Eois* (Lepidoptera: Geometridae): checklist, biogeography, diversity, and description patterns. *Annals of the Entomological Society of America*, **104**, 1091–1107.

Bryant, J. P., Provenza, F. D., Pastor, J. T., et al. (1991). Interactions between woody-plants and browsing mammals mediated by secondary metabolites. *Annual Review of Ecology and Systematics*, **22**, 431–446.

Cardinale, B. J., Bennett, D. M., Nelson, C. E. and Gross, K. (2009). Does productivity drive diversity or vice versa? A test of the multivariate productivity–diversity hypothesis in streams. *Ecology*, **90**, 1227–1241.

Carpenter, S. R., Cole, J. J., Hodgson, J. R., et al. (2001). Trophic cascades, nutrients, and lake productivity: whole-lake experiments. *Ecological Monographs*, **71**, 163–186.

Carpenter, S. R., Cole, J. J., Kitchell, J. F. and Pace, M. L. (2010). Trophic cascades in lakes: lessons and prospects. In *Trophic Cascades*, ed. J. Terborgh and J. Estes. Washington, DC: Island Press, pp. 55–69.

Carroll, S. P., Hendry, A. P., Reznick, D. N. and Fox, C. W. (2007). Evolution on ecological time-scales. *Functional Ecology*, **21**, 387–393.

Carson, W. P. and Root, R. B. (2000). Herbivory and plant species coexistence: Community regulation by an outbreaking phytophagous insect. *Ecological Monographs*, **70**, 73–99.

Castejón-Silvo, I., Terrados, J., Domínguez, M. and Morales-Nin, B. (2012). Epiphyte response to *in situ* manipulation of nutrient availability and fish presence in a *Posidonia oceanica* (L.) Delile meadow. *Hydrobiologia*, **696**, 159–170.

Chave, J. (2013). The problem of pattern and scale in ecology: what have we learned in 20 years? *Ecology Letters*, **16**, 4–16.

Clemente, S., Hernandez, J. C. and Brito, A. (2011). Context-dependent effects of marine protected areas on predatory interactions. *Marine Ecology Progress Series*, **437**, 119–133.

Coley, P. D., Bryant, J. P. and Chapin III, F. S. (1985). Resource availability and plant antiherbivore defense. *Science*, **230**, 895–899.

Connahs, H., Rodríguez-Castañeda, G., Walters, T., Walla, T. R. and Dyer, L. A. (2009). Geographical variation in host-specificity and parasitoid pressure of an herbivore (Geometridae) associated with the tropical genus *Piper*. *Journal of Insect Science*, **9**(28), 1–11; available online: insectscience. org/9.28.

Coverdale, T. C., Altieri, A. H. and Bertness, M. D. (2012). Belowground herbivory increases vulnerability of New England salt marshes to die-off. *Ecology*, **93**, 2085–2094.

Dyer, L. A. and Letourneau, D. K. (1999a). Trophic cascades in a complex, terrestrial community. *Proceedings of the National Academy of Sciences of the USA*, **96**, 5072–5076.

Dyer, L. A. and Letourneau, D. K. (1999b). Relative strengths of top-down and bottom-up forces in a tropical forest community. *Oecologia*, **119**, 265–274.

Dyer, L. A. and Letourneau, D. K. (2003). Top-down and bottom-up diversity cascades in detrital versus living food webs. *Ecology Letters*, **6**, 60–68.

Dyer, L. A. and Letourneau, D. K. (2007). Determinants of lichen diversity in a rainforest understory. *Biotropica*, **39**, 525–539.

Dyer, L. A. and Letourneau, D. K. (2013). Can climate change trigger massive diversity cascades in terrestrial ecosystems? *Diversity*, **5**, 1–35.

Dyer, L. A. and Palmer, A. N. (2004). Piper. *A Model Genus for Studies of Evolution, Chemical Ecology, and Trophic Interactions*. Boston: Kluwer Academic Publishers.

Dyer, L. A., Dodson, C. D., Beihoffer, J. and Letourneau, D. K. (2001). Trade offs in anti-herbivore defenses in *Piper cenocladum*: ant mutualists versus plant secondary metabolites. *Journal of Chemical Ecology*, **27**, 581–592.

Dyer, L. A., Dodson, C. D., Stireman III, J. O., et al. (2003). Synergistic effects of three *Piper* amides on generalist and specialist herbivores. *Journal of Chemical Ecology*, **29**, 2499–2514.

Dyer, L. A., Dodson, C. D., Letourneau, D. K., et al. (2004a). Ecological causes and consequences of variation in defensive chemistry of a neotropical shrub. *Ecology*, **85**, 2795–2803.

Dyer, L. A., Dodson, C. D. and Richards, J. (2004b). Isolation, synthesis, and evolutionary ecology of *Piper amides*. In Piper. *A Model Genus for Studies of Evolution, Chemical Ecology, and Trophic Interactions*, ed. L. A. Dyer and A. N. Palmer. Boston: Kluwer Academic Publishers, pp. 117–139.

Dyer, L. A., Gentry, G. and Tobler, M. (2004c). Fitness consequences of herbivory: impacts on asexual reproduction of tropical rainforest understory plants. *Biotropica*, **36**, 68–73.

Dyer, L. A., Letourneau, D. K., Vega Chavarria, G. and Salazar Amoretti, D. (2010). Herbivores on a dominant understory shrub increase local plant diversity in rain forest communities. *Ecology*, **91**, 3707–3718.

Dyer, L. A., Carson, W. P. and Leigh, E. G. (2012). Insect outbreaks in tropical forests: patterns, mechanisms, and consequences. In *Insect Outbreaks Revisited*, ed. P. Barbosa, D. K. Letourneau and A. A. Agrawal. New Jersey: Wiley-Blackwell, pp. 219–245.

Estes, J. A. and Palmisano, J. F. (1974). Sea otters: their role in structuring nearshore communities. *Science*, **185**, 1058–1060.

Fincher, R. M., Dyer, L. A., Dodson, C. D., et al. (2008). Inter- and intraspecific comparisons of antiherbivore defenses in three species of rainforest understory shrubs. *Journal of Chemical Ecology*, **34**, 558–574.

Forister, M. L. and Feldman, C. R. (2011). Phylogenetic cascades and the origins of tropical diversity. *Biotropica*, **43**, 270–278.

Franks, S. J., Sim, S. and Weis, A. E. (2007). Rapid evolution of flowering time by an annual plant in response to a climate fluctuation. *Proceedings of the National Academy of Sciences of the USA*, **104**, 1278–1282.

Gurevitch, J. and Hedges, L. V. (2001). Meta analysis: combining the results of independent experiments. In *Design and Analysis of Ecological Experiments*, ed. S. M. Scheiner and J. Gurevitch. New York: Oxford University Press, pp. 347–369.

Hairston, N. G., Smith, F. E. and Slobodkin, L. B. (1960). Community structure, population control, and competition. *American Naturalist*, **94**, 421–424.

Hairston, N. G., Ellner, S. P., Geber, M. A., Yoshida, T. and Fox, J. A. (2005). Rapid evolution and the convergence of ecological and evolutionary time. *Ecology Letters*, **8**, 1114–1127.

Hamilton, J., Zangerl, A. R., Berenbaum, M. R., et al. (2012). Elevated atmospheric CO_2 alters the arthropod community in a forest understory. *Acta Oecologica*, **43**, 80–85.

Hay, M. E. (1996). Marine chemical ecology: what's known and what's next? *Journal of Experimental Marine Biology and Ecology*, **200**, 103–134.

Hillstrom, M. L. and Lindroth, R. L. (2008). Elevated atmospheric carbon dioxide and ozone alter forest insect abundance and community composition: carbon dioxide and ozone alter forest insect communities. *Insect Conservation and Diversity*, **1**, 233–241.

Howeth, J. G. and Leibold, M. A. (2010). Prey dispersal rate affects prey species composition and trait diversity in response to multiple predators in metacommunities: multiple predators in prey metacommunities. *Journal of Animal Ecology*, **79**, 1000–1011.

Hubbell, S. P. (2001). *The Unified Neutral Theory of Biodiversity and Biogeography*. Princeton: Princeton University Press.

Hunt, G. (2007). The relative importance of directional change, random walks, and stasis in the evolution of fossil lineages. *Proceedings of the National Academy of Sciences of the USA*, **104**, 18404–18408.

Hunter, M. D. (2001). Multiple approaches to estimating the relative importance of top-down and bottom-up forces on insect populations: experiments, life tables, and time-series analysis. *Basic and Applied Ecology*, **2**, 295–309.

Hutchinson, G. E. (1965). *The Ecological Theater and the Evolutionary Play*. New Haven: Yale University Press.

Isbell, F., Reich, P. B., Tilman, D., et al. (2013). Nutrient enrichment, biodiversity loss, and consequent declines in ecosystem productivity. *Proceedings of the National Academy of Sciences of the USA*, **110**, 11911–11916.

Knepp, R. G., Hamilton, J. G., Mohan, J. E., et al. (2005). Elevated CO_2 reduces leaf damage by insect herbivores in a forest community. *New Phytologist*, **167**, 207–218.

Knight, T. M., McCoy, M. W., Chase, J. M., McCoy, K. A. and Holt, R. D. (2005). Trophic cascades across ecosystems. *Nature*, **437**, 880–883.

Koricheva, J. (2003). Non-significant results in ecology: a burden or a blessing in disguise? *Oikos*, **102**, 397–401.

Letourneau, D. K. (1983). Passive aggression: an alternative hypothesis for the *Piper-Pheidole* Association. *Oecologia*, **60**, 122–126.

Letourneau, D. K. (1990). Code of ant-plant mutualism broken by parasite. *Science*, **248**, 215–217.

Letourneau, D. K. (2004). Mutualism, antiherbivore defense, and trophic cascades: ant-plants as a mesocosm for experimentation. In Piper. *A Model Genus for Studies Of Evolution, Chemical Ecology, and Trophic Interactions*, ed. L. A. Dyer and A. N. Palmer. Boston, MA: Kluwer Academic Publishers, pp. 5–32.

Letourneau, D. K. and Dyer, L. A. (1998a). Density patterns of *Piper* ant-plants and associated arthropods: top predator cascades in a terrestrial system? *Biotropica*, **30**, 162–169.

Letourneau, D. K. and Dyer, L. A. (1998b). Experimental manipulations in lowland tropical forest demonstrate top-down cascades through four trophic levels. *Ecology*, **79**, 1678–1687.

Letourneau, D. K. and Dyer, L. A. (2005). Multi-trophic interactions and biodiversity: beetles, ants, caterpillars, and plants. In *Biotic Interactions in the Tropics: Their Role in the Maintenance of Species Diversity*, ed. D. F. R. P. Burslem, M. A. Pinard and S. E. Hartley. Cambridge, UK: Cambridge University Press, pp. 366–385.

Levin, S. A. (1992). The problem of pattern and scale in ecology. *Ecology*, **73**, 1943–1967.

Menge, B. A. and Olson, A. M. (1990). Role of scale and environmental-factors in regulation of community structure. *Trends in Ecology and Evolution*, **5**, 52–57.

Micheli, F., Halpern, B. S., Botsford, L. W. and Warner, R. R. (2004). Trajectories and correlates of community change in no-take marine reserves. *Ecological Applications*, **14**, 1709–1723.

Miller, T. E. and Travis, J. (1996). The evolutionary role of indirect effects in communities. *Ecology*, **5**, 1329–1335.

Nelson, C. E., Bennett, D. M. and Cardinale, B. J. (2013). Consistency and sensitivity of stream periphyton community structural and functional responses to nutrient enrichment. *Ecological Applications*, **23**, 159–173.

Pace, M. L., Cole, J. J. and Carpenter, S. R. (1998). Trophic cascades and compensation: differential responses of microzooplankton in whole-lake experiments. *Ecology*, **79**, 138–152.

Polis, G. A. and Strong, D. R. (1996). Food web complexity and commuity dynamics. *American Naturalist*, **147**, 813–846.

Post, D. M. and Palkovacs, E. P. (2009). Eco-evolutionary feedbacks in community and ecosystem ecology: interactions between the ecological theatre and the evolutionary play. *Philosophical Transactions of the Royal Society of London B: Biological Sciences*, **364**, 1629–1640.

Power, M. E. (1990). Effects of fish in river food webs. *Science*, **250**, 811–814.

Reznick, D. N. and Ghalambor, C. K. (2001). The population ecology of contemporary adaptations: what empirical studies reveal about the conditions that promote adaptive evolution. *Genetica*, **112**, 183–198.

Rodríguez-Castañeda, G., Dyer, L. A., Brehm, G., et al. (2010). Tropical forests are not flat: how mountains affect herbivore diversity. *Ecology Letters*, **13**, 1348–1357.

Rodríguez-Castañeda, G., Forkner, R. E., Tepe, E. J., Gentry, G. L. and Dyer, L. A. (2011). Weighing defensive and nutritive roles of ant mutualists across a tropical altitudinal gradient. *Biotropica*, **43**, 343–350.

Root, R. B. and Cappuccino, N. (1992). Patterns in population change and the organization of an insect community associated with goldenrod, *Solidago altissima*. *Ecological Monographs*, **62**, 393–420.

Salamin, N., Wüest, R. O., Lavergne, S., Thuiller, W. and Pearman, P. B. (2010). Assessing rapid evolution in a changing environment. *Trends in Ecology and Evolution*, **25**, 692–698.

Sangil, C., Clemente, S., Martín-García, L. and Hernández, J. C. (2012). No-take areas as an effective tool to restore urchin barrens on subtropical rocky reefs. *Estuarine, Coastal and Shelf Science*, **112**, 207–215.

Schmitz, O. J., Krivan, V. and Ovadia, O. (2004). Trophic cascades: the primacy of trait-mediated indirect interactions. *Ecology Letters*, **7**, 153–163.

Schmitz, O. J., Grabowski, J. H., Peckarsky, B. L., et al. (2008). From individuals to ecosystem function: toward an integration of evolutionary and ecosystem ecology. *Ecology*, **89**, 2436–2445.

Schoener, T. W. (2011). The newest synthesis: understanding the interplay of evolutionary and ecological dynamics. *Science*, **331**, 426–429.

Shears, N. T., Babcock, R. C. and Salomon, A. K. (2008). Context-dependent effects of fishing: variation in trophic cascades across environmental gradients. *Ecological Applications*, **18**, 1860–1873.

Siepielski, A. M., Gotanda, K. M., Morrissey, M. B., et al. (2013). The spatial patterns of directional phenotypic selection. *Ecology Letters*, **16**, 1382–1392.

Slobodkin, L. B. (1960). Ecological energy relationships at the population level. *American Naturalist*, **94**, 213–236.

Strong, D. R. (1992). Are trophic cascades all wet? Differentiation and donor-control in speciose ecosystems. *Ecology*, **73**, 747–754.

Tepe, E. J., Kelley, W. A., Rodríguez-Castañeda, G. and Dyer, L. A. (2009). Characterizing the cauline domatia of two newly discovered Ecuadorian ant-plants in *Piper*: An example of convergent evolution. *Journal of Insect Science*, **9**, available online: insectscience. org/9.27.

Terborgh, J. and Estes, J. A. (2010). *Trophic Cascades: Predators, Prey, and the Changing Dynamics of Nature*. Washington: Island Press.

Tscharntke, T. and Hawkins, B. A. (2001). *Multitrophic Level Interactions*. Cambridge, UK: Cambridge University Press.

Vermeij, G. J. (1994). The evolutionary interaction among species: selection, escalation, and coevolution. *Annual Review of Ecology and Systematics*, **25**, 219–236.

Verreydt, D., De Meester, L., Decaestecker, E., et al. (2012). Dispersal-mediated trophic interactions can generate apparent patterns of dispersal limitation in aquatic metacommunities: dispersal-mediated metacommunity responses. *Ecology Letters*, **15**, 218–226.

Walsh, M. R., DeLong, J. P., Hanley, T. C. and Post, D. M. (2012). A cascade of evolutionary change alters consumer-resource dynamics and ecosystem function. *Proceedings of the Royal Society B: Biological Sciences*, **279**, 3184–3192.

Wesner, J. S. (2010). Aquatic predation alters a terrestrial prey subsidy. *Ecology*, **91**, 1435–1444.

Wilson, J. S., Forister, M. L., Dyer, L. A., et al. (2012). Host conservatism, host shifts and diversification across three trophic levels in two Neotropical forests. *Journal of Evolutionary Biology*, **25**, 532–546.

Winter, T. R. and Rostás, M. (2010). Nitrogen deficiency affects bottom-up cascade without disrupting indirect plant defense. *Journal of Chemical Ecology*, **36**, 642–651.

CHAPTER TWELVE

The role of species diversity in bottom-up and top-down interactions

JEROME J. WEIS

Yale University, New Haven, CT, USA

Introduction

The flow of energy, carbon, and nutrients through food webs is constrained by the abilities of constituent organisms to consume, assimilate, and excrete resources. There is a growing recognition that the range of these abilities is positively correlated with community diversity and thus, variation in diversity can directly impact the functioning of food webs and ecosystems (Loreau et al., 2001; Naeem, 2002; Hooper et al., 2005). For instance, the strong influence of trophic diversity, or the number of trophic levels in a food web, is recognized as an important predictor of the standing stock and productivity of primary producers and herbivores (Hairston et al., 1960; Carpenter et al., 1985). In addition to trophic diversity, the diversity of competitors, from genotypic diversity within a species (Whitham et al., 2006; Hughes et al., 2008), to species diversity (Loreau et al., 2001; Naeem, 2002; Hooper et al., 2005), to broad phylogenetic diversity (Cadotte et al., 2008), can strongly influence ecosystem functions, including the uptake and assimilation of limiting nutrients and the stability of food webs through time.

In this chapter, I review the current understanding of the direct influences of competitor species richness on the transfer of energy and nutrients through food webs; a topic that has received a huge amount of attention over the past two decades (Naeem, 2008; Hooper et al., 2012). In part, the motivation to understand the influence of species richness on ecosystem functions has been prompted by a need to understand the ecosystem consequences of global species losses (Loreau et al., 2001; Naeem, 2002, Hooper et al., 2005; 2012; also, see Chapter 14). However, diversity varies across landscapes for a number of deterministic and stochastic reasons (MacArthur, 1967; Rosenzweig, 1995), and understanding the causes and consequences of this variation has resulted in a more broadly informative integration of community and ecosystem ecology with relevance beyond the implications of global species loss (Loreau et al., 2001).

Trophic Ecology: Bottom-Up and Top-Down Interactions across Aquatic and Terrestrial Systems, eds T. C. Hanley and K. J. La Pierre. Published by Cambridge University Press. © Cambridge University Press 2015.

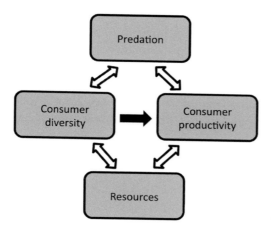

Figure 12.1 The influence of competitor diversity in food webs. Bottom-up and top-down environmental factors, such as resource supply and predation rates, interact with mid-level consumer trophic levels to influence the biomass, productivity, and diversity of consumers (white arrows). In this chapter, I focus on the direct influences of consumer diversity on the standing biomass and productivity of the consumer trophic level (black arrow).

In addressing the influences of species richness, my focus here is somewhat different from the focus of other food web or trophic studies. Many of these studies, as well as many of the topics covered in previous chapters of this book, focus on the direct effects of bottom-up environmental factors, such as the supply of inorganic nutrients, and top-down drivers, such as predation, on the productivity and standing stock biomass of trophic levels in ecosystems (Fig. 12.1). Here, I review studies that ask how changes or variations in species richness within a trophic level directly influence the flow of energy and nutrients through that trophic level (e.g., consumer diversity; Fig. 12.1).

Generalizing the effects of species richness across trophic levels

The large and growing number of theoretical and empirical studies that address the role of competitor species on ecosystem and food web functions has allowed for useful comparisons across ecosystem types and trophic levels (Balvanera et al., 2006; Cardinale et al., 2006; Duffy et al., 2007). With a few exceptions, empirical evidence of a mechanistic link between species richness and the consumption of resources comes from experimental manipulations of species richness or manipulations of the abundance of organisms at higher herbivore and predator trophic levels. The relative rarity of comparable observational studies reflects the difficulty in partitioning the direct effects of variation in species richness on resource consumption from variation in other environmental factors that may simultaneously influence species richness and resource consumption rates (Fig. 12.1; Loreau et al., 2001; Hooper et al., 2005). Many of the early experimental and theoretical studies in this area of research addressed the interactions of primary producer "consumers" and inorganic nutrient "resources" (Tilman et al., 1996; 1997; Hector et al., 1999). However, a more general two-trophic level consumer–resource framework has proven useful in understanding the direct effects of species richness at a variety of trophic levels (Hooper et al., 2005;

Cardinale et al., 2006). Throughout this chapter, I compare results across a number of consumer–resource food webs, including primary producer–inorganic resource, herbivore–primary producer, predator–prey, and detritivore–detritus food webs. Later in the chapter, I consider the effects of species richness in three-trophic level food webs, and compare studies in a conceptual predator–consumer–resource food web (Fig. 12.1). This information comes from a variety of predator–herbivore–primary producer and herbivore–primary producer–inorganic resource food webs.

Comparing the results of empirical studies across ecosystems and trophic levels often requires generalizing results from studies that have measured different response variables, such as the standing biomass of trophic levels or the rate of resource consumption (Duffy et al., 2007). In comparing consumer–resource food webs, it is conceptually useful to ask how species richness influences a single response variable, the rate of resource consumption. However, it is often empirically difficult to measure this rate, generally as a flux of carbon or nutrients from the resource to consumer trophic level. Instead, studies frequently quantify the influences of consumer or resource species richness by comparing the standing stock of consumer or resource trophic levels across treatments. Here, I assume that the rate of resource consumption is positively correlated with the standing stock or biomass of the consumer trophic level. For instance, I make the general assumption that grassland plant biomass is positively correlated with plant primary productivity and the consumption of soil nitrogen. I also assume that the rate of resource consumption is negatively correlated with the standing stock of resources. For instance, in predator–prey food webs, I assume standing stocks of prey are negatively correlated with predation rates among treatments. These assumptions are generally informative across a variety of trophic levels and experimental designs. However, it is important to consider (i) which response variables are measured and (ii) the relative time scales of experiments, where consumer or resource trophic levels may or may not have time to grow and reproduce (Duffy et al., 2007; also, see Chapter 11).

Competitor species richness within a trophic level

In this section, I review the role of competitor species richness on the rate of resource consumption from a two-trophic level consumer–resource food web perspective (Fig. 12.2a,b). The majority of theoretical and experimental studies have focused on the effects of competitor species richness at a single trophic level; some focus on the top-down effect of consumer species richness on the consumption of resources (Fig. 12.2a), and others on the bottom-up effect of resource species richness on the consumption of resources (Fig. 12.2b). The prevalence of similarly designed studies has allowed ecologists to conduct more systematic reviews and meta-analyses, and begin to generalize the influences of consumer or resource species richness within ecosystem types (Cardinale et al.,

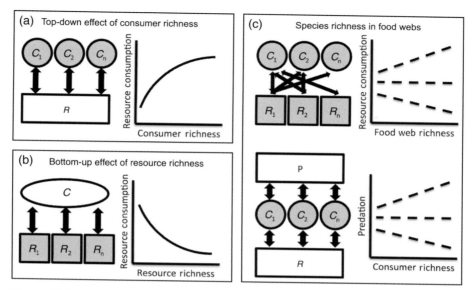

Figure 12.2 Competitor species richness and resource consumption. In a two-trophic level, consumer–resource (C_i–R_i) food web: (a) consumer species richness is expected to have a positive but saturating influence on the rate of resource consumption; and (b) resource species richness is expected to have a negative influence on the rate of resource consumption. (c) In more complex food webs, the influence of competitor species richness depends on food web characteristics. Competitor richness at adjacent trophic levels may have a positive, negative, or neutral influence on resource consumption. In a three-trophic level, predator–consumer–resource (P–C_i–R) food web, consumer species richness may have a positive, negative, or neutral influence on the predation of consumers and predator biomass.

2007; Stachowicz et al., 2007; Srivastava et al., 2009; Cardinale et al., 2011), and across ecosystems and trophic levels (Hooper et al., 2005; Balvanera et al., 2006; Cardinale et al., 2006; Duffy et al., 2007; Griffin et al., 2013). In this section, I rely on the results of these meta-analyses and reviews to highlight general trends in the strength and direction of the relationship between competitor species richness and resource consumption. I also include examples from individual studies, but these examples do not necessarily represent a systematic analysis of all available studies.

Top-down effects of consumer richness

Theory based on local species sorting or niche partitioning generally predicts a positive and saturating influence of consumer species richness on the consumption of resources (Fig. 12.2a; Tilman et al., 1997; Loreau, 1998). This positive effect of consumer species richness is understood to result from two non-exclusive phenomena: selection and complementarity effects of diversity (Huston, 1997; Tilman et al., 1997; Loreau, 1998; Loreau and Hector, 2001). Positive selection effects are understood to arise in environments that favor the dominance of a few species that consume or assimilate resources at high rates relative to other species. Species-rich assemblages are expected to include these highly productive species more frequently, resulting in higher resource consumption and productivity, on average, in species-rich assemblages compared to species-poor assemblages. Positive complementarity effects are understood to arise in environments where species coexist by exploiting distinct niches in the environment, or through facilitative interactions with competitors. In such environments, species-rich assemblages are expected to exploit a higher percentage of available resources than species-poor assemblages.

Empirical tests of the top-down effects of consumer richness confirm that consumer richness is most frequently positively correlated with consumer biomass and resource consumption across a range of ecosystems and trophic levels, but that negative and non-significant relationships are also common (Hooper et al., 2005; Balvanera et al., 2006; Cardinale et al., 2006; Stachowicz et al., 2007). A common experimental approach involves directly manipulating consumer species richness in replicated experimental units, such as terrestrial-field plots or aquatic mesocosms, and then comparing either the rate of resource consumption or standing stocks of resources or consumers among treatments varying in species richness (Loreau et al., 2001; Hooper et al., 2005). Several of the first and most well-known experiments manipulated the species richness of grassland plants in replicated field plots and found that higher plant richness led to greater plant cover (Tilman et al., 1996), greater total plant biomass (Naeem et al., 1996; Hector et al., 1999), and greater uptake of soil nitrate (Tilman et al., 1996). Other experiments have tested the top-down influence of consumer species richness on consumer biomass or resource consumption across a range of ecosystems and trophic levels, including freshwater microalgae (Fox, 2004), marine invertebrate herbivores (Duffy et al., 2001), intertidal marine predators (Bruno and O'Connor, 2005), terrestrial invertebrate predators (Snyder et al., 2006), detritivorous fungi (Bärlocher and Corkum, 2003), and heterotrophic bacteria (Bell et al., 2005). Meta-analyses of these manipulative studies have confirmed that, on average, species-rich assemblages of consumers tend to produce more biomass and consume more resources than the average of any single consumer species grown alone in monoculture (Balvanera et al., 2006; Cardinale et al., 2006; 2011; Griffin et al., 2013).

Positive relationships between consumer richness and consumer productivity can often be attributed to positive selection and complementarity effects in experiments. In many cases, consumer productivity is best explained by a positive selection effect of consumer richness, where one or a few highly influential species grow to dominate consumer biomass or have strong individual influences on resource biomass in species-rich assemblages (Srivastava and Vellend, 2005; Cardinale et al., 2006; Stachowicz et al., 2007). For instance, in an aquatic microcosm experiment, Weis and others (2008) reported a positive influence of phytoplankton species richness on phytoplankton biomass that could be explained by the dominance and high yield of a single species in species-rich polycultures. In a manipulation of invertebrate predators and aphid prey in an agricultural ecosystem, Snyder and others (2006) reported a positive influence of predator species richness on the reduction of two aphid species. However, two of the predator species grown alone reduced aphids to similar densities to the high richness predator treatments. The strong influence of these two predator species suggests a positive selection effect of predator species richness on the consumption of aphids in this experimental setting.

There is also increasing experimental evidence of positive complementarity in a range of ecosystems and trophic levels. For instance, in a meta-analysis of grassland plant experiments, Cardinale and others (2007) found evidence of positive complementarity effects, measured as a higher average relative biomass of all species in species-rich polycultures compared to monocultures (Loreau and Hector, 2001), which increased over the course of many growing seasons. A more mechanistic analysis of one of these grassland experiments suggested that complementarity in that experiment was the result of increased levels of plant nitrogen and more efficient nitrogen use at higher levels of plant species richness (Fargione et al., 2007). In general, it is expected that a quantifiable outcome of niche complementarity is that species-rich assemblages of consumers should yield more biomass than any single consumer species grown alone (Srivastava and Vellend, 2005; Cardinale et al., 2006), a phenomenon known as transgressive overyielding (Vandermeer, 1989). Meta-analyses have shown that transgressive overyielding is actually relatively rare across experiments in a range of ecosystems and trophic levels (Cardinale et al., 2006; Stachowicz et al., 2007; Griffin et al., 2013). However, manipulations of rocky intertidal seaweed species (Stachowicz et al., 2008) and grassland plant communities (Cardinale et al., 2007) have shown that transgressive overyielding tends to increase with time. Many authors point out that a general lack of transgressive overyielding and niche complementarity should be common in experiments, where short duration times combined with the simplified environments of plots and microcosms are likely to preclude niche partitioning and favor the dominance of one or a few species (Loreau et al., 2001; 2003; Symstad et al., 2003; Hooper et al., 2005). More recent studies that have explicitly included forms of resource heterogeneity or environmental variability have confirmed high yields of multiple competing species, leading to transgressive overyielding (Griffin et al., 2009; Venail et al., 2010; Cardinale, 2011; Godbold et al., 2011). For instance, Cardinale (2011) manipulated the richness of benthic microalgae in experimental streams and found that species-rich communities produced more biomass and took up more dissolved nitrate than any species grown alone in monoculture, but only in experimental units that included heterogeneity in stream flow and disturbance. In contrast, in experimental streams with no variation in flow or disturbance, Cardinale (2011) observed a positive but saturating relationship between algal species richness and biomass, driven by a positive selection effect of a single highly productive species that grew to dominance in this relatively homogeneous environment. In a heterotrophic microbial example, Venail and others (2010) addressed the role of heterotrophic bacterial richness, using strains that had evolved in a laboratory setting on different carbohydrate resources. They manipulated the richness of bacteria in regional metacommunities consisting of patches that contained different carbohydrate resources. They did find a positive relationship between bacterial richness and bacterial abundance, but only at the metacommunity scale consisting of multiple patches, where high richness assemblages could partition resources.

A few recent observational studies have addressed the role of variation in competitor species richness in more realistic, less-controlled environments. To date, most of these studies have focused on phytoplankton communities, asking how variation in phytoplankton species richness correlates with stocks of phytoplankton biomass and dissolved inorganic resources (Ptacnik et al., 2008; Cardinale et al., 2009; Korhonen et al., 2011; Zimmerman and Cardinale, 2013). Ptacnik and others (2008) address the relationship between phytoplankton genus richness and the uptake of inorganic phosphorus following a conceptual framework that articulates the distinction between the direct influence of resource abundance on consumer abundance, and the influence of consumer species richness on resource use (Loreau et al., 2001; Schmid, 2002; Gross and Cardinale, 2007). They showed that among Scandinavian lake and Baltic Sea phytoplankton communities, phytoplankton genus richness was positively correlated with resource use efficiency (the ratio of phytoplankton biomass to total inorganic phosphorus). Cardinale and others (2009) used a structural equation modeling approach to look at phytoplankton species richness in Norwegian lakes; they found support for the multivariate hypothesis that phytoplankton species richness and phytoplankton biomass were significantly correlated with total abundance and balance of limiting inorganic resources. In addition, phytoplankton species richness had an independent, positive correlation with phytoplankton biomass. Overall, the conclusions of this growing group of observational studies are consistent with theoretical predictions of a positive influence of consumer richness on resource consumption, but it is clear that more observational studies in a wider variety of ecosystems and trophic levels are needed.

While meta-analyses and reviews of experimental manipulations of consumer richness confirm a positive influence of consumer richness on consumer biomass and resource uptake in most cases, these analyses also confirm a range of outcomes including negative and non-significant influences of consumer species richness (Hooper et al., 2005; Srivastava and Vellend, 2005; Cardinale et al., 2006; 2011). In some environments, dominant competitor species in polyculture are relatively unproductive, resulting in what can be thought of as a negative selection effect of diversity (Loreau, 1998; Loreau and Hector, 2001). For instance, several manipulations of freshwater phytoplankton have observed a negative relationship between phytoplankton richness and biomass, due largely to the dominance of a low yielding species in polyculture (Weis et al., 2007; Schmidtke et al., 2010). In general, antagonistic interactions among consumer species or defensive responses of resources to particular consumer species can also lead to lower consumption rates and consumer productivity at higher levels of consumer species richness (Sih et al., 1998).

While a single conceptual framework is useful for understanding the primary roles of consumer species richness across a range of ecosystems and trophic levels, individual studies often highlight distinctions within trophic levels or ecosystems. Many of these distinctions involve the basic differences between animals and plants (reviewed in Duffy et al., 2007). In particular, predators may demonstrate intraguild predation and omnivory, and these more complicated trophic interactions can change the role of consumer species richness on resource uptake in a variety of ways (Polis and Holt, 1992; Ives et al., 2005; Casula et al., 2006). Animal behaviors, including antagonistic responses to competitors and defensive or allusive responses to predators, are also known to influence the relationship between consumer species richness and the consumption of resources at higher trophic levels (Sih et al., 1998; Schmitz, 2009). In contrast, similar plastic responses to competition and consumption by plants and microbes (Agrawal, 2001) are not as frequently recognized as important factors in studies that address the direct effects of consumer species richness. Animal behaviors may have uniquely strong implications for the relationship between consumer species richness and the consumption of resources, but several factors related mostly to experimental logistics may also explain the recognized influences of animal behavior in experiments. First, plastic responses may be generally important across trophic levels, but animal behaviors may be easier to quantify than other plastic responses, such as physiological changes in cell or colony size in microbes or changes in plant chemical or physical defenses. Second, the importance of plastic responses in general may be linked to competitor density, which is often more closely scrutinized in experiments looking at longer-lived and larger animals, and thus often cannot be run for long enough time periods to allow for population changes (Byrnes and Stachowicz, 2009). In contrast, the long-term results of experiments that address species richness in microbes, plants, and some small parthenogenic animals are run for a long enough time period for populations to grow and change in response to inter- and intra-specific competition and environmental conditions, and are therefore less sensitive to initial densities (Weis et al., 2007; Hanley et al., unpublished data).

Overall, there is growing evidence that the influence of consumer species richness on resource consumption is mechanistically similar across ecosystems and trophic levels. In experiments that manipulate consumer species richness, meta-analyses have shown that on average there is a positive influence of consumer species richness on consumer biomass and resource uptake (Balvanera et al., 2006; Cardinale et al., 2006; 2007; Srivastava et al., 2009; Cardinale et al., 2011; Griffin et al., 2013). Concurrently, meta-analyses recognize that negative or non-significant relationships between consumer diversity and resource consumption are also common across this same range of trophic levels and ecosystems.

Bottom-up effects of resource richness

For roughly the same deterministic reasons that a positive top-down influence of consumer richness on resource consumption is expected, a negative bottom-up influence of resource richness on resource consumption is expected (Fig. 12.2b; Duffy, 2002; Duffy et al., 2007). Common hypotheses predict that as the species richness of resources increases, the likelihood that a given assemblage of consumers can efficiently exploit the overall resource base decreases; either because of the limited consumptive breadth of specialist consumers (Root, 1973; Ostfeld and LoGiudice, 2003), or because of the increased chance of including defended or inedible resource species at higher levels of resource species richness (Leibold, 1989; Duffy, 2002).

Empirical support for a negative relationship between resource species richness and the total uptake of resources has come largely from comparative meta-analyses of consumer removals across environments varying in resource species richness (Duffy et al., 2007; Srivastava et al., 2009; Cardinale et al., 2011). Hillebrand and Cardinale (2004) analyzed field and microcosm experiments that tested the effect of herbivore removal on the biomass of periphytic microalgae and found that, among studies, herbivores reduced the algal biomass by smaller proportions at higher levels of algal species richness. In a similar analysis of predator removals in rocky intertidal communities, Edwards and others (2010) showed that the reduction of sessile prey abundance in the presence of predators was negatively correlated with prey species richness. Jactel and Brockerhoff (2007) addressed herbivory by insects in individual tree species located in mixed or monoculture forest stands and found that the consumption by herbivores was generally reduced in mixed stands consisting of many tree species.

Experiments that have directly manipulated resource species richness have shown positive and negative relationships between resource species richness and resource consumption (Duffy et al., 2007; Narwani and Mazumder, 2010; Cardinale et al., 2011). Duffy and others (2007) outline two studies that report a negative relationship between resource species richness and the consumption of resource biomass in a freshwater zooplankton–phytoplankton food web (Steiner, 2001) and a marine crustacean predator–herbivore food web (Duffy et al., 2005). In a recent example from an agricultural invertebrate predator–herbivore food web, Wilby and Orwin (2013) manipulated the species richness of aphids in the presence of aphid predators and found that the suppression of aphids by predators decreased with increased aphid species richness. However, not all studies have reported consistently negative influences of resource species on resource consumption, and some studies have reported a variety of influences within the same experiment. For instance, Narwani and Mazumder (2010) manipulated phytoplankton species richness in the presence of a number of herbivorous crustacean zooplankton species and found that herbivory rates tended to decrease with increased phytoplankton species richness in the majority of cases. However, they note that in a subset of cases, phytoplankton species richness increased consumption rates by a generalist herbivore species.

To date, there are relatively few meta-analyses of manipulations of resource richness like the examples outlined above for consumer richness. Srivastava and others (2009) compared the strength of top-down and bottom-up effects of species richness in brown food webs consisting of dead plants as detrital resources and detritivorous consumers including bacteria, fungi, and animals. They found examples of positive and negative bottom-up influences of detrital species richness on rates of resource depletion and report no general positive or negative trend. Cardinale and others (2011) also found examples of positive and negative relationships between primary producer species richness and the effect of herbivory on total primary producer biomass. They explicitly state that they did not find general support for the hypothesis that primary producer richness reduces the impact of herbivory on primary producer biomass, but were careful to note that their analysis only included 13 data points, making it difficult to make any broad generalizations.

While I have thus far highlighted niche-based hypotheses that resource species richness should have a negative influence on resource consumption, there are at least two relatively unrelated hypotheses that predict the positive influence of resource richness observed in the studies detailed above (reviewed in Duffy et al., 2007). One hypothesis, which Duffy and others (2007) call the "balanced diet hypothesis," suggests that species-rich assemblages of resources may directly benefit consumer growth rates, ultimately resulting in increased consumption (DeMott, 1998; Pfisterer et al., 2003). For instance, DeMott (1998) compared the growth of several species of herbivorous crustacean zooplankton in monocultures and mixtures of two phytoplankton species: (i) a green algal species that was easily consumed by zooplankton, but had a higher cellular carbon to phosphorus (C:P) ratio than zooplankton, and (ii) a cyanobacterial species that was more difficult for zooplankton to consume, but had a lower C:P ratio and thus was more optimal for zooplankton growth. DeMott (1998) showed that zooplankton growth was highest in mixtures of both phytoplankton species. He attributed this increased zooplankton growth to the more balanced consumption of limiting nutrients (C and P) in mixed phytoplankton assemblages. Another hypothesis suggests that increased interspecific competition at high levels of resource species richness forces competing resource species to make behavioral or physiological changes that make them more vulnerable to consumption (Narwani and Mazumder, 2010; Toscano et al., 2010). For instance, in a marine predator–prey food web, Toscano and others (2010) showed that in mixed assemblages of fish or crab prey species, dominant prey species displaced submissive prey species from a seagrass refuge, increasing the overall predation rate on prey assemblages when multiple prey species were present.

Overall, there does not appear to be strong evidence of systematic differences in the bottom-up influence of resource species richness on resource consumption across ecosystems and trophic levels. Comparative analyses of consumer removals in freshwater and terrestrial herbivore–primary producer food webs (Andow, 1991; Hillebrand and Cardinale, 2004; Jactel and Brockerhoff, 2007), and marine predator–prey food webs (Edwards et al., 2010) report negative correlations between resource species richness and resource consumption. Results from studies that systematically manipulate resource species richness have been less conclusive, showing a variety of positive and negative relationships between resource species richness and resource consumption in herbivore–primary producer (Cardinale et al., 2011) and detritivore–detritus food webs (Srivastava et al., 2009). In part, the lack of a general trend in the strength and direction of the influences of resource species richness is likely the result of the relative lack of studies that manipulate resource species richness directly (Cardinale et al., 2011). However, understanding the variety of influences of resource species richness may require a more integrated approach to understanding competitor richness in more realistic food webs. In many of the studies that manipulate resource species richness outlined above, the focal "resource" trophic levels are also consuming and competing for limiting resources at lower trophic levels, resulting in a more complicated interplay of bottom-up and top-down influences on competitor richness (Fig. 12.2c). In the next section, I review a more integrated, multiple trophic level approach to understanding the influences of competitor species richness in food webs.

Competitor richness in food webs

There is growing evidence that across trophic levels and ecosystems the effect of consumer and resource species richness often depends on two basic principles: (i) consumers vary in their ability to consume available resources; and (ii) resources vary in their vulnerability to consumption. Theory predicts a positive top-down effect of consumer species richness on resource consumption and a negative bottom-up effect of resource species richness on resource consumption (Fig. 12.2a,b). However, in realistic food webs, the flow of energy and nutrients through food webs depends on variation in species richness at multiple trophic levels (Fig. 12.2c; Worm and Duffy, 2003; Duffy et al., 2007). At many trophic levels, species compete for limiting resources and are themselves resources to higher trophic levels. In general, the potentially opposing positive top-down and negative bottom-up effects of species richness in the same food web require that the overall influence of species richness in food webs depends on the balance of top-down and bottom-up interactions (Duffy, 2002; Thebault and Loreau, 2003). Incorporating trophic complexity also increases the number of trophic levels at which competitor or trophic diversity can be manipulated, and thus increases the variety of specific questions these studies address. Here, I review two related approaches to address more complex and realistic variation in species diversity in food webs (Fig. 12.2c). I first focus on the interactive effects of simultaneous variation in species richness of adjacent resource and consumer trophic levels. I then consider the role of consumer richness in a three-trophic level, predator–consumer–resource food web.

Species richness at adjacent trophic levels

Variation in species richness at adjacent trophic levels may have positive or negative influences on the flow of energy and nutrients between those two trophic levels, depending on factors related to food web structure (Thebault and Loreau, 2003). In general, experiments that factorially manipulate species richness at two trophic levels are logistically difficult because of the number of treatments required, but a few have been reported in the past decade (Fox, 2004; Aquilino et al., 2005; Gamfeldt et al., 2005; Douglass et al., 2008). Fox (2004) observed a positive but transient top-down effect of phytoplankton species richness on phytoplankton biomass in the absence of ciliate herbivores. However, Fox (2004) did not observe significant effects of phytoplankton species richness on herbivore biomass, of herbivore species richness on herbivore biomass, or of herbivore species richness on phytoplankton biomass. In a similarly designed manipulation in marine microbial food webs, Gamfeldt and others (2005) observed no significant effect of phytoplankton species richness on phytoplankton biomass, but observed a positive effect of herbivore species richness on herbivore biomass and a negative effect of herbivore species richness on phytoplankton biomass. In another marine example, Douglass and others (2008) factorially manipulated predator and herbivore species richness in a predator–herbivore–primary producer food web. They found that the species richness of herbivores did not have a strong influence on herbivore density, nor on primary producer biomass. However, predator species richness had a negative influence on average herbivore biomass.

To date, the limited number of experiments that have manipulated species richness at adjacent trophic levels has not allowed for any systematic comparative analyses. One potentially useful approach to summarizing studies may be to compare bottom-up and top-down influences of competitor diversity across trophic levels. For instance, in Fox's (2004) study, the strongest effect of species richness appears to be a positive but transient top-down effect of phytoplankton species richness on the production of phytoplankton biomass. In the study by Gamfeldt and others (2005), the strongest effect appears to be a positive top-down effect of herbivore species richness on herbivore biomass and phytoplankton consumption. Similarly, in the study by Douglass and others (2008), the strongest effect appears to be a positive top-down effect of predator species richness on crustacean prey density. In all of these similarly designed studies, a positive top-down effect of consumer species richness is stronger than any bottom-up effects, but the trophic level at which this strong top-down effect occurred varied across studies. If top-down effects are generally stronger than bottom-up effects of competitor species richness in food webs, then high species richness at any trophic level may lead to greater food web productivity and uptake of inorganic resources. However, a comparison of just three studies is certainly not sufficient to make any general conclusions about the strength of top-down and bottom-up effects of species richness in food webs. As one counter example from a study with a subtly different experimental design, Aquilino and others (2005) manipulated predator and primary producer species richness in a predator–herbivore–primary producer agricultural food web and found a positive top-down effect of predator species richness on the predation of herbivores and a roughly equal negative effect of primary producer species richness on the predation of herbivores. In this case, predator capture efficiency of the single herbivore species varied depending on the plant species that predators hunted on, resulting in lower overall capture rates in species-rich plant communities. While species richness was not manipulated at adjacent trophic levels, the negative effect of plant species richness on predation could be interpreted as a strong negative bottom-up effect of resource richness. There is a need for more studies that simultaneously manipulate consumer and resource diversity to make more systematic and quantitative comparisons of the bottom-up and top-down influences of competitor species richness in these experiments.

Consumer richness in a three-trophic level food web

A growing group of studies consider the effects of competitor species richness in a three-trophic level food web, where species at a mid-level consumer trophic level compete for limiting resources and are consumed by a higher predator trophic level (Fig. 12.3a). Theoretical studies that have addressed the role of consumer species richness in three-trophic level predator–consumer–resource

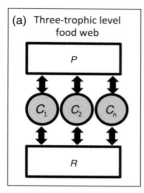 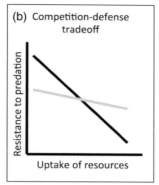 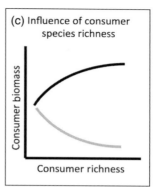

Figure 12.3 Consumer species richness in three-trophic level food webs. (a). Consumer species richness influences the standing biomass of consumers in three-trophic level, predator–consumer–resource food webs. (b) Theory generally assumes that consumer species coexist in the presence of predators because of a "competition–defense" tradeoff between the uptake of limiting resources and resistance to predation. This tradeoff may be strong (black line) or weak (gray line). (c) Theory predicts that the relationship between consumer species richness and consumer biomass depends on the strength of the competition–defense tradeoff. If the tradeoff is strong (black line), a positive relationship between consumer species richness and consumer biomass is predicted. If the tradeoff is weak (gray line), a negative relationship is predicted.

food webs generally make similar assumptions to the top-down and bottom-up studies outlined above: (i) consumers vary in their ability to consume limiting resources; and (ii) consumers vary in their vulnerability to predation (Holt and Loreau, 2002; Fox, 2003; Thebault and Loreau, 2003; Weis and Vasseur, 2014). These studies have often focused on parameter values that allow competing consumer species to coexist in the presence of a predator population through a tradeoff between uptake of limiting resources and resistance to predation (Holt et al., 1994; Leibold, 1996), referred to as a competition-defense tradeoff (Viola et al., 2010; Lind et al., 2013; Fig. 12.3b).

Given this basic framework, it is informative to compare the effect of consumer species richness on resource consumption in the presence and absence of predators (Holt and Loreau, 2002; Weis and Vasseur, 2014). In the absence of predators, consumer community composition is dominated by species that are better able to consume limiting resources (i.e., superior uptake-competitors) (Holt et al., 1994; Leibold, 1996), resulting in a positive selection effect of consumer species richness on consumer biomass and the consumption of resources (Fig. 12.2a; Tilman et al., 1997; Holt and Loreau, 2002; Weis and Vasseur, 2014). In the presence of predators, the strength of the competition–defense tradeoff influences consumer community composition and the relationship between consumer species richness and consumer biomass (Fig. 12.3c; Fox, 2003; Weis and Vasseur, 2014). If the competition–defense tradeoff is weak, consumer community composition is dominated by superior uptake-competitors in the presence and absence of predators. However, in the presence of predators, superior uptake-competitors drive a negative selection effect in species-rich assemblages because they are more vulnerable to predators and are consumed at higher rates. In other words, consumer species richness can actually lead to a negative relationship between consumer species richness and consumer biomass in the presence of predators (Fig. 12.3c). Alternatively, if the competition–defense tradeoff is strong, consumer community composition is dominated by consumer species that are better able to resist predation (defense-competitors) in the presence of predators. This results in a positive selection effect driven by defense-competitors at high levels of consumer species richness and a positive relationship between consumer species richness and consumer biomass (Fig. 12.3c).

The hypothesis that consumer species richness may positively influence consumer biomass in the absence of predators and negatively influence consumer biomass in the presence of predators has been confirmed in a number of individual empirical studies (reviewed in Duffy et al., 2007), but hasn't been quantitatively tested across studies. Mulder and others (1999) addressed the influence of grassland plant species richness on plant biomass in a herbivore–plant–inorganic resource food web and found that plant species richness had a stronger positive influence on plant standing biomass in treatments where invertebrate herbivore abundance had been reduced. Similarly, in a freshwater predator–snail–algae food web, Wojdak (2005) showed that standing biomass of algae was lower at high snail species richness in the absence of predators (a positive relationship between consumer (snail) species richness and resource (phytoplankton) uptake). However, in the presence of the predator, algal biomass was higher at high snail species richness (a negative relationship between consumer species richness and resource uptake). Both of these studies suggest that the presence of an upper "predator" trophic level decreases the influence of consumer species richness on consumer biomass. In contrast, a wide variety of results have been reported in studies that have manipulated the species richness of phytoplankton in the presence and absence of ciliate or crustacean herbivores (Fox, 2004; Gamfeldt et al., 2005; Behl et al., 2012). Fox (2004) found a positive but transient relationship between phytoplankton species richness and phytoplankton biomass in the absence but not in the presence of herbivores. Gamfeldt and others (2005) found no significant effect of phytoplankton species richness on phytoplankton biomass in the absence or presence of herbivores. In contrast, Behl and others (2012) found positive relationships between phytoplankton species richness and phytoplankton biomass in the absence and presence of herbivores.

Overall, several studies have shown that the presence of a predator trophic level may reduce the generally positive relationship between consumer species richness and consumer biomass in predator–consumer–resource food webs

(Mulder et al., 1999; Fox, 2004; Wojdak, 2005). This empirical result is consistent with theoretical predictions that the presence of predators can result in null or negative relationships between consumer species richness and consumer biomass if competition–defense tradeoffs are weak (Fox, 2003; Weis and Vasseur, 2014). However, it is difficult to make any general conclusions from this limited number of studies. It is still unclear how frequently the top-down effects of predation or herbivory can influence positive or negative relationships between consumer species richness and consumer biomass at lower trophic levels. It is also unclear if the direction of this relationship is related to the strength of competition–defense tradeoffs. However, if the presence of a predator trophic level frequently shifts the influence of consumer species richness on consumer biomass from positive to negative, this is an important caveat to the often-cited generalization that consumer species richness has a positive influence on consumer biomass (Mulder et al., 1999; Wojdak, 2005). As the number of experiments that manipulate species richness at intermediate consumer trophic levels increases, it will be possible to quantify the general influences of consumer species richness in more complex food webs.

In addition to the influence of consumer species richness on consumer biomass (Fig. 12.3), a related and important question in predator–consumer–resource food webs asks how consumer species richness influences overall predation rates on consumers and the stock of predator biomass in three-trophic level food webs (Fig. 12.2c). If consumers vary in their abilities to take up limiting resources and resist predation, the effect of consumer richness on predator biomass will depend on the balance of a positive top-down influence of consumer species richness on resource uptake and a negative bottom-up influence of consumer species richness on predator consumption. A few experiments provide some insight into the influence of consumer species richness on predator biomass. In particular, three-trophic level herbivore–phytoplankton–nutrient microcosm studies are often run for long enough time periods to observe changes in phytoplankton and herbivore populations in response to phytoplankton richness treatments. Fox (2004) showed no significant influence of phytoplankton species richness on ciliate herbivore biomass. In contrast, Gamfeldt and others (2005) and Behl and others (2012) found that herbivore biomass was positively correlated with phytoplankton species richness. Like other questions relating to the influences of species richness in more realistic food webs, more theory and experiments are needed to understand more generally how consumer species richness influences the flow of energy and nutrients to higher trophic levels.

Conclusions

Competitor species richness can influence the uptake of limiting nutrients and the flow of energy and nutrients through food webs (Hooper et al., 2005;

Cardinale et al., 2006; Duffy et al., 2007). Some have even argued that changes in competitor species richness, at least under experimental conditions, can have as large an impact on these ecosystem functions as changes in nutrients or additions/removals of trophic levels (Hooper et al., 2012; Tilman et al., 2012). In general, effects of competitor species richness on food webs are frequently realized through the competitive interactions that result in niche partitioning and local species sorting (Tilman et al., 1997; Loreau and Hector, 2001), highlighting the potentially strong influence that competitive interactions can have on other food web interactions. In this chapter, I have outlined the increasing theoretical and empirical evidence that variation in competitor species richness can directly influence the rates at which competitor communities consume resources, produce biomass, and are consumed by higher predator and herbivore trophic levels. Strikingly, meta-analyses of experiments to date suggest that species richness has relatively similar influences in aquatic, terrestrial, and marine environments (Balvanera et al., 2006; Cardinale et al., 2006). However, differences in the effects of species richness among ecosystems associated with differences in the strength of trophic cascades (Shurin et al., 2002; Borer et al., 2005) may become apparent as experiments incorporate more realistic food web structure.

Similar frameworks have been used to understand the effects of other measures of biological diversity in ecosystems and food webs. A growing number of studies have shown that phenotypic variation among genotypes within a species can also have strong influences on a number of ecosystem functions, including the production of biomass and the consumption of resources (Crutsinger et al., 2006; Whitham et al., 2006; Hughes et al., 2008). Given the assumption that genotypes within a species compete and coexist under similar mechanisms to species, the effects of genotypic richness on the flow of energy and nutrients through food webs are expected to be mechanistically similar to the effects of species richness (Hughes et al., 2008). Finally, many authors have argued that addressing variables other than species richness can more accurately inform the influences of competitive interactions and individual trophic levels on the flow of energy and nutrients through ecosystems. Some recognize that in many ecosystems, functions such as biomass production and invasion resistance are most strongly influenced by one or a few highly influential or dominant species, and understanding the functioning of these particular species in more detail can be more informative than focusing on species richness per se (Smith and Knapp, 2003; Smith et al., 2004; Hillebrand et al., 2008). Other authors argue that assessing the diversity of particular functional traits, rather than species richness, allows for more accurate assessments of ecosystem functioning (Díaz and Cabido, 2001). All of these approaches have been useful in informing the roles that diversity and competitive interactions can play in ecosystems. While in many cases metrics other than species richness may be useful, species richness is an important and accessible concept to biologists and non-biologists alike.

Understanding the consequences of species richness in food webs and ecosystems has proven an important step in understanding how energy and nutrients flow through food webs.

Acknowledgments

I am grateful for financial support from a National Science Foundation Graduate Research Fellowship and from the Yale University Graduate School of Arts and Sciences. I thank Torrance Hanley and Kimberly La Pierre for helpful comments that improved this chapter.

References

Agrawal, A. A. (2001). Phenotypic plasticity in the interactions and evolution of species. *Science*, **294**, 321–326.

Andow, D. (1991). Vegetational diversity and arthropod population response. *Annual Review of Entomology*, **36**, 561–586.

Aquilino, K. M., Cardinale, B. J. and Ives, A. R. (2005). Reciprocal effects of host plant and natural enemy diversity on herbivore suppression: an empirical study of a model tritrophic system. *Oikos*, **108**, 275–282.

Balvanera, P., Pfisterer, A. B., Buchmann, N., et al. (2006). Quantifying the evidence for biodiversity effects on ecosystem functioning and services. *Ecology Letters*, **9**, 1146–1156.

Bärlocher, F. and Corkum, M. (2003). Nutrient enrichment overwhelms diversity effects in leaf decomposition by stream fungi. *Oikos*, **101**, 247–252.

Behl, S., de Schryver, V., Diehl, S. and Stibor, H. (2012). Trophic transfer of biodiversity effects: functional equivalence of prey diversity and enrichment? *Ecology and Evolution*, **2**, 3110–3122.

Bell, T., Newman, J. A., Silverman, B. W., Turner, S. L. and Lilley, A. K. (2005). The contribution of species richness and composition to bacterial services. *Nature*, **436**, 1157–1160.

Borer, E. T., Seabloom, E., Shurin, J. B., et al. (2005). What determines the strength of a trophic cascade? *Ecology*, **86**, 528–537.

Bruno, J. F. and O'Connor, M. I. (2005). Cascading effects of predator diversity and omnivory in a marine food web. *Ecology Letters*, **8**, 1048–1056.

Byrnes, J. E. and Stachowicz, J. J. (2009). The consequences of consumer diversity loss: different answers from different experimental designs. *Ecology*, **90**, 2879–2888.

Cadotte, M. W., Cardinale, B. J. and Oakley, T. H. (2008). Evolutionary history and the effect of biodiversity on plant productivity. *Proceedings of the National Academy of Sciences of the USA*, **105**, 17012–17017.

Cardinale, B. J. (2011). Biodiversity improves water quality through niche partitioning. *Nature*, **472**, 86–U113.

Cardinale, B. J., Srivastava, D. S., Duffy, J. E., et al. (2006). Effects of biodiversity on the functioning of trophic groups and ecosystems. *Nature*, **443**, 989–992.

Cardinale, B. J., Wright, J. P., Cadotte, M. W., et al. (2007). Impacts of plant diversity on biomass production increase through time because of species complementarity. *Proceedings of the National Academy of Sciences of the USA*, **104**, 18123–18128.

Cardinale, B. J., Hillebrand, H., Harpole, W. S., Gross, K. and Ptacnik, R. (2009). Separating the influence of resource "availability" from resource "imbalance" on productivity-diversity relationships. *Ecology Letters*, **12**, 475–487.

Cardinale, B. J., Matulich, K. L., Hooper, D. U., et al. (2011). The functional role of producer diversity in ecosystems. *American Journal of Botany*, **98**, 572–592.

Carpenter, S. R., Kitchell, J. F. and Hodgson, J. R. (1985). Cascading trophic interactions and lake productivity. *BioScience*, **35**, 634–639.

Casula, P., Wilby, A. and Thomas, M. B. (2006). Understanding biodiversity effects on prey in multi-enemy systems. *Ecology Letters*, **9**, 995–1004.

Crutsinger, G. M., Collins, M. D., Fordyce, J. A., et al. (2006). Plant genotypic diversity predicts community structure and governs an ecosystem process. *Science*, **313**, 966–968.

DeMott, W. R. (1998). Utilization of a cyanobacterium and a phosphorus-deficient green alga as complementary resources by daphnids. *Ecology*, **79**, 2463–2481.

Díaz, S. and Cabido, M. (2001). Vive la difference: plant functional diversity matters to ecosystem processes. *Trends in Ecology & Evolution*, **16**, 646–655.

Douglass, J. G., Duffy, J. E. and Bruno, J. F. (2008). Herbivore and predator diversity interactively affect ecosystem properties in an experimental marine community. *Ecology Letters*, **11**, 598–608.

Duffy, J. E. (2002). Biodiversity and ecosystem function: the consumer connection. *Oikos*, **99**, 201–219.

Duffy, J. E., Macdonald, K. S., Rhode, J. M. and Parker, J. D. (2001). Grazer diversity, functional redundancy, and productivity in seagrass beds: an experimental test. *Ecology*, **82**, 2417–2434.

Duffy, J. E., Richardson, J. P. and France, K. E. (2005). Ecosystem consequences of diversity depend on food chain length in estuarine vegetation. *Ecology Letters*, **8**, 301–309.

Duffy, J. E., Cardinale, B. J., France, K. E., et al. (2007). The functional role of biodiversity in ecosystems: incorporating trophic complexity. *Ecology Letters*, **10**, 522–538.

Edwards, K. F., Aquilino, K. M., Best, R. J., Sellheim, K. L. and Stachowicz, J. J. (2010). Prey diversity is associated with weaker consumer effects in a meta-analysis of benthic marine experiments. *Ecology Letters*, **13**, 194–201.

Fargione, J., Tilman, G. D., Dybzinski, R., et al. (2007). From selection to complementarity: shifts in the causes of biodiversity-productivity relationships in a long-term biodiversity experiment. *Proceedings of the Royal Society B: Biological Sciences*, **274**, 871–876.

Fox, J. W. (2003). The long-term relationship between plant diversity and total plant biomass depends on the mechanism maintaining diversity. *Oikos*, **102**, 630–640.

Fox, J. W. (2004). Effects of algal and herbivore diversity on the partitioning of biomass within and among trophic levels. *Ecology*, **85**, 549–559.

Gamfeldt, L., Hillebrand, H. and Jonsson, P. R. (2005). Species richness changes across two trophic levels simultaneously affect prey and consumer biomass. *Ecology Letters*, **8**, 696–703.

Godbold, J. A., Bulling, M. T. and Solan, M. (2011). Habitat structure mediates biodiversity effects on ecosystem properties. *Proceedings of the Royal Society B: Biological Sciences*, **287** (1717), 1–10.

Griffin, J. N., Jenkins, S. R., Gamfeldt, L., et al. (2009). Spatial heterogeneity increases the importance of species richness for an ecosystem process. *Oikos*, **118**, 1335–1342.

Griffin, J. N., Byrnes, J. E. and Cardinale, B. J. (2013). Effects of predator richness on prey suppression: a meta-analysis. *Ecology*, **94**, 2180–2187.

Gross, K. and Cardinale, B. J. (2007). Does species richness drive community production or vice versa? Reconciling historical and contemporary paradigms in competitive communities. *The American Naturalist*, **170**, 207–220.

Hairston, N. G., Smith, F. E. and Slobodkin, L. B. (1960). Community structure, population control, and competition. *The American Naturalist*, **94**, 421–425.

Hector, A., Schmid, B., Beierkuhnlein, C., et al. (1999). Plant diversity and productivity experiments in European grasslands. *Science*, **286**, 1123–1127.

Hillebrand, H. and Cardinale, B. J. (2004). Consumer effects decline with prey diversity. *Ecology Letters*, **7**, 192–201.

Hillebrand, H., Bennett, D. M. and Cadotte, M. W. (2008). Consequences of dominance: a review of evenness effects on local and regional ecosystem processes. *Ecology*, **89**, 1510–1520.

Holt, R. D. and Loreau, M. (2002). Biodiversity and ecosystem functioning: the role of trophic interactions and the importance of system openness. In *The Functional Consequences of Biodiversity: Empirical Progress and Theoretical Extensions*, ed. A. P. Kinzig, S. W. Pacala and D. Tilman. Princeton: Princeton University Press, pp. 246–262.

Holt, R. D., Grover, J. and Tilman, G. D. (1994). Simple rules for interspecific dominance in systems with exploitative and apparent competition. *The American Naturalist*, **144**, 741–771.

Hooper, D. U., Chapin, F., Ewel, J., et al. (2005). Effects of biodiversity on ecosystem functioning: a consensus of current knowledge. *Ecological Monographs*, **75**, 3–35.

Hooper, D. U., Adair, E. C., Cardinale, B. J., et al. (2012). A global synthesis reveals biodiversity loss as a major driver of ecosystem change. *Nature*, **486**, 105–108.

Hughes, A. R., Inouye, B. D., Johnson, M. T. J., Underwood, N. and Vellend, M. (2008). Ecological consequences of genetic diversity. *Ecology Letters*, **11**, 609–623.

Huston, M. (1997). Hidden treatments in ecological experiments: re-evaluating the ecosystem function of biodiversity. *Oecologia*, **110**, 449–460.

Ives, A. R., Cardinale, B. J. and Snyder, W. E. (2005). A synthesis of subdisciplines: predator-prey interactions, and biodiversity and ecosystem functioning. *Ecology Letters*, **8**, 102–116.

Jactel, H. and Brockerhoff, E. G. (2007). Tree diversity reduces herbivory by forest insects. *Ecology Letters*, **10**, 835–848.

Korhonen, J. J., Wang, J. and Soininen, J. (2011). Productivity-diversity relationships in lake plankton communities. *PLoS One*, **6**, e22041.

Leibold, M. A. (1989). Resource edibility and the effects of predators and productivity on the outcome of trophic interactions. *American Naturalist*, **134** (6), 922–949.

Leibold, M. A. (1996). A graphical model of keystone predators in food webs: Trophic regulation of abundance, incidence, and diversity patterns in communities. *The American Naturalist*, **147**, 784–812.

Lind, E. M., Borer, E. T., Seabloom, E., et al. (2013). Life-history constraints in grassland plant species: a growth-defence trade-off is the norm. *Ecology Letters*, **16**, 513–521.

Loreau, M. (1998). Biodiversity and ecosystem functioning: a mechanistic model. *Proceedings of The National Academy of Sciences of the USA*, **95**, 5632–5636.

Loreau, M. and Hector, A. (2001). Partitioning selection and complementarity in biodiversity experiments. *Nature*, **412**, 72–76.

Loreau, M., Naeem, S., Inchausti, P., et al. (2001). Ecology – biodiversity and ecosystem functioning: current knowledge and future challenges. *Science*, **294**, 804–808.

Loreau, M., Mouquet, N. and Gonzalez, A. (2003). Biodiversity as spatial insurance in heterogeneous landscapes. *Proceedings of The National Academy of Sciences of the USA*, **100**, 12765–12770.

MacArthur, R. H. (1967). *The Theory of Island Biogeography*. Princeton: Princeton University Press.

Mulder, C., Koricheva, J., Huss-Danell, K., Hogberg, P. and Joshi, J. (1999). Insects affect relationships between plant species richness and ecosystem processes. *Ecology Letters*, **2**, 237–246.

Naeem, S. (2002). Ecosystem consequences of biodiversity loss: the evolution of a paradigm. *Ecology*, **83**, 1537–1552.

Naeem, S. (2008). Advancing realism in biodiversity research. *Trends in Ecology & Evolution*, **23**, 414–416.

Naeem, S., Hakansson, K., Lawton, J. H., Crawley, M. J. and Thompson, L. J. (1996). Biodiversity and plant productivity in a model assemblage of plant species. *Oikos*, **76**, 259–264.

Narwani, A. and Mazumder, A. (2010). Community composition and consumer identity determine the effect of resource species diversity on rates of consumption. *Ecology*, **91**, 3441–3447.

Ostfeld, R. S. and LoGiudice, K. (2003). Community disassembly, biodiversity loss, and the erosion of an ecosystem service. *Ecology*, **84**, 1421–1427.

Pfisterer, A. B., Diemer, M. and Schmid, B. (2003). Dietary shift and lowered biomass gain of a generalist herbivore in species-poor experimental plant communities. *Oecologia*, **135**, 234–241.

Polis, G. and Holt, R. D. (1992). Intraguild predation – the dynamics of complex trophic interactions. *Trends in Ecology & Evolution*, **7**, 151–154.

Ptacnik, R., Solimini, A. G., Andersen, T., et al. (2008). Diversity predicts stability and resource use efficiency in natural phytoplankton communities. *Proceedings of the National Academy of Sciences of the USA*, **105**, 5134–5138.

Root, R. B. (1973). Organization of a plant-arthropod association in simple and diverse habitats: the fauna of collards (*Brassica oleracea*). *Ecological Monographs*, **43**, 95–124.

Rosenzweig, M. L., (1995). *Species Diversity in Space and Time*. Cambridge: Cambridge University Press.

Schmid, B. (2002). The species richness-productivity controversy. *Trends in Ecology and Evolution*, **17**, 113–114.

Schmidtke, A., Gaedke, U. and Weithoff, G. (2010). A mechanistic basis for underyielding in phytoplankton communities. *Ecology*, **91**, 212–221.

Schmitz, O. J. (2009). Effects of predator functional diversity on grassland ecosystem function. *Ecology*, **90**, 2339–2345.

Shurin, J. B., Borer, E. T., Seabloom, E., et al. (2002). A cross-ecosystem comparison of the strength of trophic cascades. *Ecology Letters*, **5**, 785–791.

Sih, A., Englund, G. and Wooster, D. (1998). Emergent impacts of multiple predators on prey. *Trends in Ecology and Evolution*, **13**, 350–355.

Smith, M. D. and Knapp, A. (2003). Dominant species maintain ecosystem function with non-random species loss. *Ecology Letters*, **6**, 509–517.

Smith, M. D., Wilcox, J., Kelly, T. and Knapp, A. (2004). Dominance not richness determines invasibility of tallgrass prairie. *Oikos*, **106**, 253–262.

Snyder, W. E., Snyder, G. B., Finke, D. L. and Straub, C. S. (2006). Predator biodiversity strengthens herbivore suppression. *Ecology Letters*, **9**, 789–796.

Srivastava, D. S. and Velland, M. (2005). Biodiversity-ecosystem function research: is it relevant to conservation? *Annual Review of Ecology, Evolution, and Systematics*, **36**, 267–294.

Srivastava, D. S., Cardinale, B. J., Downing, A. L., et al. (2009). Diversity has stronger top-down than bottom-up effects on decomposition. *Ecology*, **90**, 1073–1083.

Stachowicz, J. J., Bruno, J. F. and Duffy, J. E. (2007). Understanding the effects of marine biodiversity on communities and ecosystems. *Annual Review of Ecology, Evolution, and Systematics*, **38**, 739–766.

Stachowicz, J. J., Best, R. J., Bracken, M. E. S. and Graham, M. H. (2008). Complementarity in marine biodiversity manipulations: reconciling divergent evidence from field and mesocosm experiments. *Proceedings of the National Academy of Sciences of the USA*, **105**, 18842–18847.

Steiner, C. F. (2001). The effects of prey heterogeneity and consumer identity on the limitation of trophic-level biomass. *Ecology*, **82**, 2495–2506.

Symstad, A., Chapin, F., Wall, D., et al. (2003). Long-term and large-scale perspectives on the relationship between biodiversity and ecosystem functioning. *BioScience*, **53**, 89.

Thebault, E. and Loreau, M. (2003). Food-web constraints on biodiversity-ecosystem functioning relationships. *Proceedings of the National Academy of Sciences of the USA*, **100**, 14949–14954.

Tilman, G. D., Wedin, D. and Knops, J. M. H. (1996). Productivity and sustainability influenced by biodiversity in grassland ecosystems. *Nature*, **379**, 718–720.

Tilman, G. D., Lehman, C. and Thomson, K. (1997). Plant diversity and ecosystem productivity: theoretical considerations. *Proceedings of the National Academy of Sciences of the USA*, **94**, 1857–1861.

Tilman, G. D., Reich, P. B. and Isbell, F. (2012). Biodiversity impacts ecosystem productivity as much as resources, disturbance, or herbivory. *Proceedings of the National Academy of Sciences of the USA*, **109**, 10394–10397.

Toscano, B. J., Fodrie, F. J., Madsen, S. L. and Powers, S. P. (2010). Multiple prey effects: agonistic behaviors between prey species enhances consumption by their shared predator. *Journal of Experimental Marine Biology and Ecology*, **385**, 59–65.

Vandermeer, J. H. (1989). *The Ecology of Intercropping*. Cambridge: Cambridge University Press.

Venail, P. A., Maclean, R. C., Meynard, C. N. and Mouquet, N. (2010). Dispersal scales up the biodiversity-productivity relationship in an experimental source-sink metacommunity. *Proceedings of the Royal Society B: Biological Sciences*, **277**, 2339–2345.

Viola, D. V., Mordecai, E. A., Jaramillo, A. G., et al. (2010). Competition-defense tradeoffs and the maintenance of plant diversity.

Proceedings of the National Academy of Sciences of the USA, **107**, 17217–17222.

Weis, J. J. and Vasseur, D. A. (2014). Differential predation drives overyielding of prey species in a patchy environment. *Oikos*, **123**, 79–88.

Weis, J. J., Cardinale, B. J., Forshay, K. J. and Ives, A. R. (2007). Effects of species diversity on community biomass production change over the course of succession. *Ecology*, **88**, 929–939.

Weis, J. J., Madrigal, D. S. and Cardinale, B. J. (2008). Effects of algal diversity on the production of biomass in homogeneous and heterogeneous nutrient environments: a microcosm experiment. *PLoS One*, **3**, e2825.

Whitham, T. G., Bailey, J. K., Schweitzer, J. A., et al. (2006). A framework for community and ecosystem genetics: from genes to ecosystems. *Nature Reviews Genetics*, **7**, 510–523.

Wilby, A. and Orwin, K. H. (2013). Herbivore species richness, composition and community structure mediate predator richness effects and top-down control of herbivore biomass. *Oecologia*, **172**, 1167–1177.

Wojdak, J. M. (2005). Relative strength of top-down, bottom-up, and consumer species richness effects on pond ecosystems. *Ecological Monographs*, **75**, 489–504.

Worm, B. and Duffy, J. E. (2003). Biodiversity, productivity and stability in real food webs. *Trends in Ecology and Evolution*, **18**, 628–632.

Zimmerman, E. K. and Cardinale, B. J. (2013). Is the relationship between algal diversity and biomass in North American lakes consistent with biodiversity experiments? *Oikos*, **123**, 267–278.

CHAPTER THIRTEEN

Plant and herbivore evolution within the trophic sandwich

LUIS ABDALA-ROBERTS and KAILEN A. MOONEY

University of California-Irvine, CA, USA

Introduction

Understanding the relative roles of resource availability and natural enemies (i.e., predators, parasites, pathogens) as determinants of species abundance and trait variation has been a research area of fundamental interest to ecologists and evolutionary biologists for decades. Essentially every organism copes with the dual concerns of bottom-up and top-down trophic pressure. Primary producers are positioned between the acquisition of nutrients, water, space, and light versus herbivory and disease; herbivores between plants versus predators, parasitoids, and disease; and predators between prey acquisition versus other predators, parasites, and disease. Consequently, the conceptual framework of bottom-up versus top-down control can be applied uniformly across all trophic levels.

In this chapter, we consider how species evolve in response to pressures imposed by resources and consumers, with a focus on species responses to the so-called "trophic sandwich," where selective pressures are imposed simultaneously from trophic levels both above and below. A consideration of evolutionary processes within the context of trophic dynamics is critical, as it is the evolved traits of the species consuming and being consumed that determine the nature of those interactions (Mooney et al., 2010). We first provide background on the concepts and theories pertaining to the ecological and evolutionary consequences of top-down and bottom-up dynamics. Second, we review work that has addressed (implicitly or explicitly) evolution in the context of bottom-up and top-down trophic dynamics. Third, we explicitly compare aquatic and terrestrial systems. Fourth, we present a framework outlining the mechanisms that determine the combined selective effects of resources and consumers and, based on this framework, consider how common such dynamics might be. Fifth, we present a case study from our research on the interactions between the perennial herb *Ruellia nudiflora* Engelm. and Gray Urban (Acanthaceae), a seed predator

Trophic Ecology: Bottom-Up and Top-Down Interactions across Aquatic and Terrestrial Systems, eds T. C. Hanley and K. J. La Pierre. Published by Cambridge University Press. © Cambridge University Press 2015.

(caterpillars of a noctuid moth), and parasitic wasps attacking the latter. Finally, we outline our perspective on future directions. Because of the long history of studies of plant–herbivore interactions, we center our review of the literature within this setting.

Background

The first context in which top-down and bottom-up effects were explicitly considered was ecological, and traces back over more than half a century (Hunter and Price, 1992; Gripenberg and Roslin, 2007). Hairston, Smith, and Slobodkin (HSS; Hairston et al., 1960), complemented by Oksanen et al. (1981), proposed that trophic levels are alternatively regulated by competition for resources (plants, prey) and by consumers (herbivores, predators), with the number of trophic levels being determined by primary productivity. The HSS framework set up a debate that was polarized by opposing views on the relative strength of primary productivity (bottom-up) and natural enemy (top-down) effects on herbivores (Murdoch, 1966; Lawton and McNeill, 1979). More recently, some agreement has been reached based on cross-ecosystem comparisons of the strength of bottom-up and top-down forcing (Shurin et al., 2002; Borer et al., 2006; Gruner et al., 2008), examinations of the factors underlying these effects (Borer et al., 2005), as well as new insights from integration with other disciplines (Gripenberg and Roslin, 2007). We can now conclude that bottom-up and top-down forcing can and frequently do act in combination (Hunter and Price, 1992; Borer et al., 2006; Gruner et al., 2008).

However, one critical distinction to make is whether such concurrent effects operate additively or non-additively, as distinguishing between the two conditions is necessary to predict the outcomes of bottom-up and top-down control (Wootton, 1994). In the case of additive effects, the combined influence of top-down and bottom-up forcing can be predicted based on the independent effects of each source, i.e., effects by each type of forcing can be understood separately and thus examined individually. However, if effects are non-additive, then combined forcing leads to emergent effects that cannot be predicted on the basis of each independent effect. While empirical studies vary with respect to these dynamics, a meta-analysis of 121 studies that factorially manipulated both predation and nutrient availability concluded that the strength of predator effects is not strongly mediated by nutrients across a range of ecosystem types (Borer et al., 2006).

Research on natural selection from ecological interactions has a parallel, but somewhat separate, history within the paradigm of bottom-up and top-down effects (Feeny, 1976; Lawton and McNeill, 1979; Hunter and Price, 1992; Singer, 2008). From an evolutionary perspective, every plant and herbivore species is recognized to be under selection by both resources and consumers (Coley et al.,

1985; Williams, 1999; Singer and Stireman, 2005). As in the parallel case of ecological dynamics, knowing whether these factors act additively or non-additively
is critical because it provides a mechanistic basis to predict the combined evolutionary influence of resources and consumers (Wootton, 1994). For example, low host plant quality may increase the top-down effects of predators on
insect herbivores by slowing development and prolonging the period in which
herbivores are in relatively vulnerable early life stages (Williams, 1999; Singer
et al., 2012). Moreover, this distinction also has implications for co-evolutionary
dynamics. Under an additive scenario, a species can respond evolutionarily to
multiple selective agents individually, allowing for "pairwise" (co)evolutionary
dynamics. In contrast, non-additive dynamics result in "diffuse" selection and
(co)evolutionary dynamics because the selective impact arising from one source
is mediated by simultaneous influences of other sources of selection (Strauss
et al., 2005; Haloin and Strauss, 2008). Accordingly, the selective pressure
imposed by one source (e.g., consumers) is context-dependent, depending on
the selective pressure imposed by another source (e.g., resources), and may thus
vary dramatically over both space and time (Thompson, 2005).

Effects of bottom-up and top-down forcing on producer evolution

Terrestrial systems

In much of the early work on plant competition (Grime, 1977; Tilman, 1982) and
ecological succession (Connell and Slatyer, 1977; Chapin, 1993), it was implicitly
clear that resources represented a major selective force influencing the evolution of plant life history traits. However, this initial bottom-up view developed
largely independently of evolutionary work on plant–herbivore interactions.
Accordingly, it has been more than half a century since the realization that
plants likely evolved a broad array of chemical defenses (e.g., secondary compounds) targeted at deterring and reducing feeding by herbivores (Fraenkel,
1959; see Chapter 8, this volume, for further discussion of plant defenses). Work
by Marquis (1984) formally showed for the first time that herbivores can act as
selective agents of plants, and subsequent decades witnessed the development of
plant defense theories aimed at predicting the specificity, as well as the ecological and evolutionary benefits and costs of plant defense investment, under
varying environmental conditions (reviewed by Marquis, 1992; Karban and
Baldwin, 1997; Stamp, 2003; Agrawal, 2011). Likewise, other studies have shown
that herbivores are also a major selective force on vegetative (e.g., architecture
and growth rate; Marquis and Whelan, 1996; Turley et al., 2013) and reproductive
traits (e.g., phenology; Elzinga et al., 2007). More recently, phylogenetic studies
on macroevolutionary patterns of plant defense evolution have corroborated
the notion that herbivore top-down forcing has shaped plant trait evolution, as
evidenced by species radiations within certain plant genera where herbivores

appear to have been a major evolutionary driver (e.g., *Bursera*: Becerra et al., 2009; *Asclepias* (milkweeds): Agrawal et al., 2009). Finally, the signature of evolutionary response to consumers can also be seen in plant local adaptation, as shown in studies where native genotypes outperform non-native ones by exhibiting higher resistance and reduced herbivory (e.g., Sork et al., 1993; Cremieux et al., 2008).

Whereas herbivores appear to have played a preponderant role in driving the evolution of plants, particularly defensive traits (relative to soil nutrients), there is also support for the view that bottom-up and top-down forces act in concert to shape plant evolution. This is given mainly by predictions on how soil nutrients and herbivores jointly influence plant growth and defense strategies through evolutionary tradeoffs, assuming that defenses are costly to produce and investment in defenses reduces within-plant resource allocation to growth (Herms and Mattson, 1992; Stamp, 2003). Notably, Coley et al. (1985) proposed the "resource availability hypothesis" stating that soil nutrient availability, mediated by growth–defense tradeoffs, drives relative allocation to growth and defenses, where plants growing in resource-poor environments grow more slowly and invest more in defenses (and vice versa in resource-rich environments). In agreement with this prediction, Mooney et al. (2010) recently documented macroevolutionary patterns of tradeoffs between growth and herbivore defense in milkweeds (genus *Asclepias*; see also Lind et al., 2013). Fine et al. (2004) also demonstrated convergent evolution in multiple plant lineages toward slower growth and increased herbivore defense in resource-poor soils. Interestingly, however, a recent meta-analysis by Endara and Coley (2011) contradicts the idea that herbivores are the major selective force on plant defenses by showing that soil resources are more important than herbivores in shaping the evolution of plant defenses in terrestrial ecosystems. In summary, there is a clear consensus that resources and herbivores frequently work in combination to shape the evolution of terrestrial plants, although resources could play a more important role, as in the case of defensive traits.

Aquatic ecosystems

There is widespread support for the predominant influence of consumers on producer defense evolution in aquatic ecosystems, particularly in marine communities (Gaines and Lubchenco, 1982; Hay, 1991; Hay and Steinberg, 1992). The importance of consumers in shaping the evolution of aquatic producers is reflected in parallel patterns of variation in herbivore pressure and producer defense observed along latitudinal gradients for a number of salt marsh plants (Pennings et al., 2001; 2009) and seaweed species (e.g., Duffy and Hay, 1990; Bolser and Hay, 1996). It is likely that herbivore top-down pressure in aquatic systems has frequently favored the evolution of resistance instead of tolerance traits, as a large extent of biomass and diversity is represented by species of phytoplankton that are reduced in size, and tolerance is of little benefit because predation events result in the consumption of entire organisms. In addition, trophic cascades appear to be stronger in marine and freshwater benthic systems (Shurin et al., 2002), suggesting that predator suppression of herbivore populations may frequently lead to top-down evolutionary indirect effects on producers in aquatic environments (Steinberg et al., 1995).

While there is support for bottom-up control of producer performance and biomass production in aquatic ecosystems (Gruner et al., 2008), there are fewer empirical studies evaluating the evolutionary effects of bottom-up forces (nutrients, temperature, or light) on producers in marine or freshwater ecosystems. Some exceptions are recent studies with phytoplankton in marine and freshwater ecosystems, which have shown that abiotic factors shape life history traits (Thomas et al., 2012), as well as how resource availability influences life history tradeoffs (reviewed by Litchman and Klausmeier, 2008; Edwards et al., in press). Moreover, despite the large body of literature considering how resources and herbivores jointly shape terrestrial plant evolution (see above), such ideas remain relatively unexplored in aquatic systems (Shurin et al., 2006). Nonetheless, there are a few examples of studies showing growth–defense tradeoffs for some groups of organisms, such as seaweeds (Hay and Fenical, 1988) and sponges (Leong and Pawlik, 2010), which provides a substrate for the occurrence of combined effects by resources and herbivores (see also studies in Endara and Coley, 2011). Such studies suggest that hypotheses developed on land for the joint effects of herbivores and resources on plant evolution (e.g., Resource Availability Hypothesis) are applicable to aquatic ecosystems.

Effects of bottom-up and top-down forcing on consumer evolution

Terrestrial systems

Research conducted over the last two decades supports the notion that natural enemies (i.e., predators and parasitoids) are important agents shaping herbivore evolution (Bernays, 1998; Singer, 2008). For example, it has been well documented that predators and parasitoids act as a strong selective force shaping herbivore anti-predator chemical and physical defenses (Abrams, 2000; Dyer, 2008), as well as behaviors associated with predator avoidance (Preisser et al., 2005). Accordingly, the "Enemy-Free Space Hypothesis" poses that much of the evolutionary effects of predators on herbivores can be seen in herbivore feeding behaviors designed to minimize the risk of predation (Bernays, 1998).

Alternatively, if plant defenses are assumed to be ubiquitous, it is likely that herbivores are frequently limited by resources (Murdoch, 1966; Schultz, 1988). From an evolutionary standpoint, much of the early thinking on bi-trophic models of plant–herbivore evolution assumed that herbivore evolution was influenced mainly by plant defensive traits (Dethier, 1954; Fraenkel, 1959; Ehrlich and Raven, 1964; Rhoades and Cates, 1976; Feeny, 1976). More recently, experimental and phylogenetic work has corroborated this view, showing that plant defenses have been a major driver of herbivore evolution, particularly dietary specialization (Singer, 2008). Some of the most convincing examples come from macroevolutionary studies of patterns of host plant use by herbivorous insects. For example, the defensive chemistry of species of the genus *Bursera* plays a central role in determining patterns of host use and physiological mechanisms of detoxification in beetles of the genus *Blepharida* (Becerra, 1997, 2003). Similarly, Berenbaum and Zangerl (1998) showed a high degree of concordance in chemical defense profiles of wild parsnip (*Pastinaca sativa*) and counter-defenses of the parsnip webworm (*Depressaria pastinacella*).

It has also become clear that combined effects of bottom-up and top-down forcing drive key aspects of herbivore evolution in terrestrial systems. Such concurrent effects have led to evolutionary tradeoffs stemming from the benefits (greater access to food resources) and costs (greater apparency and thus risk of predation) of having a broad diet breadth (Lawton and McNeill, 1979; Dyer, 1995; Singer et al., 2004; Singer and Stireman, 2005; Singer, 2008; Mooney et al., 2012). In this sense, the "Slow-Growth/High-Mortality Hypothesis" (reviewed by Williams, 1999) holds that plants mediate herbivore–predator interactions through effects of plant defenses on herbivore developmental time (Moran and Hamilton, 1980). Specifically, slower herbivore growth on more highly defended plants and thus greater time spent in early (more vulnerable) developmental stages should render herbivores at higher risk of predation (see also Mooney et al., 2012). For example, Singer et al. (2012) recently found evidence that the top-down effects of insectivorous birds on caterpillars varied among deciduous tree species as predicted by this hypothesis: caterpillar suppression by birds was highest on species that were of lowest quality to caterpillars. Although widespread support for the Slow-Growth/High-Mortality hypothesis is still lacking (reviewed by Mooney et al., 2012) and the number of studies remains low, these findings suggest that plants and predators can have joint evolutionary effects on herbivores.

Aquatic systems

There has been substantial work examining how top-down forcing shapes herbivore evolution in aquatic systems, particularly in freshwater communities (Berdegue et al., 1996). For example, much of the work on the evolution of anti-predator defenses has developed more or less in parallel in aquatic and terrestrial systems (reviewed by Abrams, 2000), with several examples in freshwater systems of predators shaping the evolution of fish (e.g., Reznick and Endler, 1982; Palkovacs et al., 2011) and crustaceans (e.g., Lampert, 1989; Hays, 2003). Some of the classic examples of anti-predator traits are behavioral traits (diel vertical migration) and morphological traits (helmets, neck teeth) in water fleas

(*Daphnia*) (Dodson, 1989; Loose et al., 1993; Agrawal et al., 1999). Likewise, the importance of predators in shaping herbivore evolution in aquatic systems can be seen in examples of herbivore specialization to predation environments, as shown for several species of fish (e.g., local adaptation; Ingram et al., 2012; Urban, 2013). There are also several ecological studies of predator effects on herbivore diet breadth in aquatic systems (e.g., Stachowicz and Hay, 1999), but it is difficult to draw inferences about evolutionary dynamics from them.

The effects of plant traits and resources in general (i.e., nutrients) on herbivore evolution, mainly life history traits, have also received considerable attention in aquatic systems. For example, variation in algal biomass and quality (measured as C:P or C:N ratio content) has been shown to shape life history traits of planktonic herbivores, such as *Daphnia* and other crustaceans (see studies in Urabe and Sterner, 2001). In contrast to terrestrial systems, however, there are fewer examples of the evolutionary impacts of resources (i.e., plant quality or productivity) on herbivore diet breadth in aquatic systems.

While there are examples of aquatic grazers (e.g., amphipods, gastropods) that selectively feed on highly defended plants to escape predation by sequestering plant defensive compounds, these herbivores, like most aquatic herbivores, are by-and-large dietary generalists (Hay, 1991; Hay and Steinberg, 1992). Thus, the dual effects of predators and plant quality apparently do not select for a narrowing of herbivore diet breadth in aquatic systems (but see Hay and Fenical, 1988; Hay and Steinberg, 1992; Stachowicz and Hay, 1999) as is the case in terrestrial systems (Mooney et al., 2012) where herbivorous insects are composed of a mixture of generalist and specialist species (Strong et al., 1984). Accordingly, there are no aquatic parallels to terrestrial studies testing for plant-mediation of predation effects on herbivore evolution (e.g., Slow-Growth/High-Mortality Hypothesis). Despite such paucity in studies, there is at least one study reporting that predators and resources can jointly influence herbivore evolution in an aquatic system. Walsh and Reznick (2008) showed that the evolutionary effects of predators on killifish were contingent upon the resource environment because predators indirectly influenced killifish life history traits through their effects on resource availability.

Terrestrial versus aquatic: contrasts, comparisons, and the state of knowledge

While the ecological literature has largely embraced the need to consider the joint impacts of resources and consumers (Borer et al., 2006; 2005; Shurin et al., 2006; Gruner et al., 2008), research on the evolutionary implications of such combined effects has lagged behind. The few evolutionary studies conducted to date are almost exclusively terrestrial (see studies reviewed by Williams, 1999; Endara and Coley, 2011). For example, Singer et al. (2012) suggest that herbivore (caterpillar) feeding on high-quality host plants (trees) reduces the strength of predation (by birds), a dynamic that has received some empirical support (Williams, 1999). In addition, the theories developed in terrestrial systems have received little attention in the aquatic literature to date, despite the fact that these predictions need not be limited to land. Moreover, even in terrestrial systems, the majority of evolutionary studies continue to adopt a bi-trophic approach, despite this growing recognition of the importance of a multi-trophic perspective (Williams, 1999; Singer and Stireman, 2005; Singer, 2008; Mooney et al., 2012).

Furthermore, while it is clear that the ecology of terrestrial and aquatic ecosystems differs in several important aspects (e.g., rate of producer growth and nutrient turnover), it is unclear if such differences translate into contrasting evolutionary dynamics. Ecological studies frequently measure changes in biomass or abundance, often of entire trophic guilds, as opposed to changes in the strength of species interactions that are needed to assess effects on fitness and changes in natural selection (Strauss et al., 2005; Benkman, 2013). Moreover, even measuring changes in interaction strength may not necessarily lead to concomitant changes in selection, and to assess the latter would require relating traits to relative fitness (Strauss et al., 2005). A full understanding of the combined effects of two or more selective agents ultimately comes from studies that measure selection from each agent both alone and in combination (e.g., Iwao and Rausher, 1997; Juenger and Bergelson, 1998). We are aware of almost no studies that have taken this approach to evaluate the combined selective effects of resources and consumers in either aquatic or terrestrial systems (but see Walsh and Reznick, 2008).

There are several key ecological differences between terrestrial and aquatic systems relevant to the joint influences of resources and consumers. First, aquatic resources like macrophytes and phytoplankton are composed almost entirely of nutrient-rich photosynthetic foliage, largely lack roots and vascular systems, contain a small fraction of the recalcitrant carbon (C) compounds (e.g., cellulose) found in terrestrial plants, and rely on buoyancy, rather than woody tissue, to rise toward light (Hay and Steinberg, 1992; Choat and Clements, 1998). While seaweeds and terrestrial plants are qualitatively similar in their use of defensive chemistry (Hay and Steinberg, 1992), terrestrial plants have lower nitrogen (N) and phosphorus (P) contents, and higher C:N ratios than most aquatic macrophytes (Hay, 1992; Elser et al., 2000) and phytoplankton (Shurin et al., 2006), likely because of the lack of lignin and comparatively less structural C in aquatic producers (Demment and Van Soest, 1985; Polis and Strong, 1996; Elser et al., 2000). Second, as a result of these trait differences, terrestrial plants are of lower nutritional value and slower growing than their aquatic counterparts (Hay, 1991; Shurin et al., 2006) and suffer lower rates of herbivory (Gruner et al., 2008). Finally, herbivore feeding characteristics differ between ecosystem types, most notably with respect to diet breadths (Strong et al., 1984; Hay, 1991; Shurin et al., 2006). Although terrestrial mammalian herbivores are often dietary generalists, insect herbivory is also important in terrestrial communities (Strong et al., 1984) and these herbivores are relatively specialized in their diets (Bernays, 1998). In contrast, herbivory in marine systems is performed largely by fishes, sea urchins, and some gastropods that commonly feed from 10 or more plant families, and graze non-selectively on turf lawns composed of many species (Hay, 1992; 1996; Choat and Clements, 1998).

Based on the above differences, several predictions can be made for contrasts between the evolutionary forces shaping aquatic and terrestrial producers and herbivores. First, with respect to producers, stronger herbivore pressure in aquatic environments may increase the selective impacts of top-down forces. Second, at the same time, relatively fast-growing aquatic producers may have a greater tolerance to herbivory (Fine et al., 2004; Mooney et al., 2010; Lind et al., 2013), weakening growth–defense tradeoffs (but see Hay and Fenical, 1988; Leong and Pawlik, 2010). Third, with respect to herbivores, the paucity of dietary specialists among aquatic species suggests the tradeoffs maintaining variation in diet breadth among terrestrial species (i.e., enemy-free space and plant quality) may not play a strong role in marine and freshwater habitats. Yet, while ecological dynamics suggest a subordinate importance for non-additive effects in the aquatic domain, empirical tests are sorely needed.

Predicting non-additive effects from resources and consumers

In predicting the occurrence of non-additive (diffuse) selection, the concept of interaction modification is of prime importance. An interaction modification occurs when one species affects the traits of another and, in so doing, alters the pairwise interaction between that species and a third (Wootton, 1994). In an ecological context, interaction modifications lead to trait-mediated indirect effects as, for instance, when predators indirectly reduce herbivory by inducing changes in herbivore behavior (Preisser et al., 2005). In an evolutionary context, these same effects on ecological interactions, in turn, lead to altered selection pressures and, as a result, interaction modifications are recognized as a key ecological mechanism leading to evolutionary changes in selection on species traits (Miller and Travis, 1996; Inouye and Stinchcombe, 2001; Strauss et al., 2005; Mooney and Singer, 2012).

Here we expand upon this general prediction for the importance of interaction modifications with a framework for predicting the circumstances where selection from resources and consumers is additive versus non-additive (Fig. 13.1). Our framework considers selection on a focal species "*B*" that is trophically bracketed by species "*A*" and "*C*" (Fig. 13.1a). We have oriented this framework to consider whether and how *A* might mediate the selective effects of *C* on *B*, thus generating non-additive selection. This perspective is adopted for the sake of simplicity, and this is not to say, for example, that species *C* might not simultaneously mediate selection imposed by *A* on *B*. Species *A, B,* and *C* might represent any three species (e.g., plant, herbivore, and predator, respectively). In addition, the "species" influencing *B* could be envisioned as an abiotic factor (e.g., *A, B,* and *C* might be soil resources, plants and, herbivores, respectively).

Within this framework, selection on *B* is determined by the combined effects of *A* and *C* under any circumstance where *A* alters the interaction strength between *B* and *C*. While ecological interaction strength is often defined on a per capita basis (i.e., the effect of an individual herbivore on an individual plant; Wootton and Emmerson, 2005), the evolutionary consequences of species interactions depend not only on per capita effects, but also on the relative sizes of the interacting populations or species densities (Benkman, 2013). To encompass the total impact at the population level, evolutionary interaction strength must thus account for both per capita effects and population sizes. Accordingly, within our framework, selection on *B* is determined by the non-additive effects of *A* and *C* under any circumstances where *A* alters the relative abundance of interacting individuals of *C* and *B* (i.e., the *C:B* abundance ratio) or the per capita effects of *C* on *B*. As outlined in Fig. 13.1, *A* will alter selection imposed by *C* on *B* based upon one or more of the following criteria: *Criterion 1,* when *A* directly influences the abundance of *C* (i.e., *C:B* ratio change); *Criterion 2,* when *A* directly influences the traits of *B* or *C* that affect the per capita effects of *C* on *B*; *Criterion 3,* when *A* directly influences the traits of *B* or *C* that affect the recruitment response of *C* to *B*; and *Criterion 4,* when the recruitment response of *C* to *B* is non-linear. Figure 13.1 depicts five scenarios (panel sets b–f), one in which none of the above criteria are met, and *A* does not alter selection by *C* on *B* (i.e., the effects of *A* and *C* on *B* act independently or additively; Fig. 13.1b), and subsequent scenarios in which one of the four criteria generating non-additive selection is met (Fig. 13.1c–f).

To help illustrate the above points, consider the specific scenario where *A, B,* and *C* are individual species of plant, herbivore, and predator, respectively, and we ask whether selection on the herbivore (*B*) is driven by the additive or non-additive effects of plant resources (*A*) and predators (*C*). Initially, consider when none of the four criteria is met (Fig. 13.1b). Here, variation in plant resources influences the density of both herbivores and predators from the bottom-up. Herbivore traits are under joint selection from this change in plant resources and the effects of predators, but these two effects operate independently (i.e., additively) because *A* does not alter the selection by *C* on *B* (Fig. 13.1b).

If *A* can directly influence the abundance of *C*, this alters the *C:B* ratio (criterion 1; Fig. 13.1c) and *A* will thus alter the selection by *C* on *B* and any selective effects of *A* and *C* on *B* will occur interactively. With the plant–herbivore–predator example, an increase in plant resources might directly increase predator abundance when, for instance, plants produce domatia, food bodies, or extra-floral nectaries to recruit predators (Rico-Gray and Oliveira, 2007), or when predators

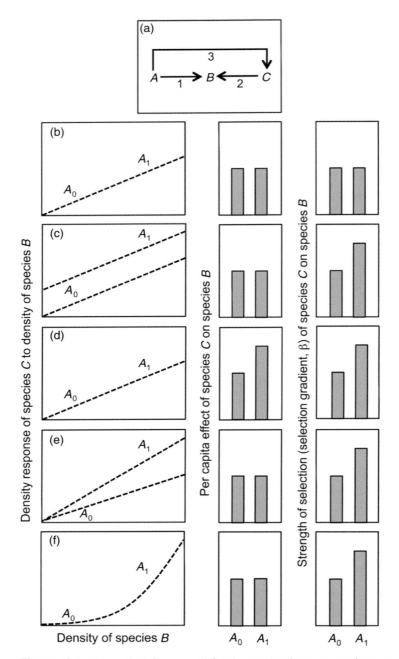

Figure 13.1 Theoretical framework for the mechanisms generating non-additive selection from resources and consumers. (a) Interaction diagram for how species *B* may be simultaneously under selection from species *A* and *C*, and the mechanisms for their

are omnivorous, feeding upon a combination of plant resources and herbivores (Eubanks and Denno, 2000; Rosenheim and Corbett, 2003).

If A can influence traits of C and B that affect the per capita effects of C on B, then any selective effects of A and C on B will occur interactively (criterion 2; Fig. 13.1d) even if the relative density of C and B (the $C{:}B$ ratio) remains unchanged. With the plant–herbivore–predator example, plant structural traits (Marquis and Whelan, 1996) or herbivore-induced plant volatiles (Heil, 2008) may facilitate predator location of prey. Such dynamics are also predicted by the Slow-Growth/High-Mortality Hypothesis, where plant defenses increase herbivore susceptibility to predators by delaying their passage through vulnerable, early life stages (Moran and Hamilton, 1980; Williams, 1999). Similarly, plant provision of carbohydrates to predators can increase protein deficiencies in omnivorous predators, thus increasing their rate of predation (Ness et al., 2009; see also Stenberg et al., 2011).

If A can influence the traits of C and B that affect the recruitment response of C to B, then any selective effects of A and C on B will occur interactively (criterion 3; Fig. 13.1e). With the plant–herbivore–predator example, plant traits may alter the attractiveness of herbivores to predators as, for example, when plants mediate the attractiveness of aphid honeydew (Mooney and Agrawal, 2008; Abdala-Roberts et al., 2012) or when herbivores sequester plant secondary compounds (Hare, 2002). In both cases, this leads to increased or decreased predator recruitment, and thus causes a change in the predator's functional response.

Finally, if the recruitment response of C to B is non-linear, then any selective effects of A and C on B will occur non-additively (criterion 4; Fig. 13.1f). With the

---

potential combined effects. This framework emphasizes how species A might select on B directly (arrow 1), indirectly through C (arrows 3 + 2), and by altering the strength of the B–C interaction through effects on traits of B (arrow 1) and/or traits of C (arrow 3). Not shown (for simplicity) are the potential reciprocal effects of C indirectly through A and by mediating the B–C interaction. (b–f) The ecological and selective effects of C on B under five scenarios for how such interactions might involve species A. The effects of species A are shown by $A0$ and $A1$, which represent presence versus absence of A, two densities of A, or two phenotypes of A. For each scenario, the functional response of C to B is shown in the left column of panels; the per capita effects of C on B are shown in the middle column of panels; and the selective effects of C on B are shown in the right column of panels. Under the first scenario (b) none of the four criteria under which A is predicted to alter selection by C on B are met. Each of the subsequent scenarios (c–f) depict one criterion, including the following: criterion 1, A directly influences the density of C (c); criterion 2, A directly influences the traits of B or C that affect the per capita effects of C on B (d); criterion 3, A directly influences the traits of B or C that affect the functional response of C to B (e); and criterion 4, the functional response of C to B is non-linear (f).

plant–herbivore–predator example, predator recruitment to prey is non-linear, positively density-dependent, and an increase in plant resources may lead to an increase in predation pressure on herbivores. For example, Singer et al. (2012) showed density-dependent recruitment of insectivorous birds to caterpillars across eight tree species, with bird predation rates increasing with host tree quality, because high-quality host trees had higher caterpillar densities.

In each of the four scenarios meeting one of the above criteria, the top-down effects of predation are increased through either a change in the per capita effects of predators (Fig. 13.1d) or the density of predators relative to herbivores (Fig. 13.1c,e,f). In the case of scenario 1 (Fig. 13.1c), such effects occur through a density-mediated indirect interaction (*sensu* Abrams, 1995) where A indirectly affects B through changes in the density (but not traits) of C. In contrast, both scenarios 2 (Fig. 13.1d) and 3 (Fig. 13.1e) can be described as interaction modifications (*sensu* Wootton, 1994), where A mediates the per capita effects of C on B. Where such effects operate through effects on the traits of C, they can also be classified as trait-mediated interactions (*sensu* Abrams, 1995), as A indirectly affects B through changes in the traits (but not density) of C. Finally, scenario four (Fig. 13.1f) occurs in the absence of both indirect effects and interaction modifications.

In each of the above cases, an increase in predator effects may in turn lead to selection for defensive traits in the herbivore, with the strength of such selection being dependent upon the influence of plants on herbivore and/or predator traits and/or densities. While these four scenarios are described separately, they may also act in combination. For instance, as described above, Singer et al. (2012) showed that high-quality trees had higher rates of bird predation on caterpillars due to increased caterpillar density and density-dependent predation (see above; Fig. 13.1f). Yet when caterpillar density was controlled for statistically, bird predation was strongest on low-quality trees, as predicted by the Slow-Growth/High-Mortality Hypothesis (Moran and Hamilton, 1980; Williams, 1999), i.e., slow development rates on poor hosts increased the risk of predation (Fig. 13.1d). Similarly, plant recruitment of predatory ants with extra-floral nectaries can increase predation pressure on herbivores both by increasing ant abundance (Fig. 13.1c) and by changing the per capita effects of those ants by increasing their rate of predation through the provisioning of carbohydrates (Fig. 13.1d; Ness et al., 2009).

Based upon this framework, we can then ask: how common are the criteria that are expected to generate non-additive selection from the joint effects of resources and consumers? First, how often do a species' resources interact directly with that same species' consumers (criterion 1)? For producers – positioned between resources (light, water, nutrients, space, etc.) and herbivores – such linkages are likely weak, except possibly for belowground herbivores in terrestrial systems that might be directly affected by soil characteristics (Erb and Lu, 2013), and any simultaneous influence of water or nutrient flow upon aquatic producers and their herbivores (e.g., effects of leaf litter inputs; Hagen et al., 2012). In contrast, for both aquatic and terrestrial herbivores – positioned between producers and predators – there are numerous mechanisms for direct interactions between producers and predators, including plant volatiles (Heil, 2008), plant architecture (Marquis and Whelan, 1996), and predator rewards (Rico-Gray and Oliveira, 2007). Second, how often are species traits plastic, such that they are likely to influence interactions with resources and consumers (criteria 2 and 3)? Here, evidence from both aquatic and terrestrial systems suggest this is often the case, including herbivore behaviors related to predator avoidance and foraging (Preisser et al., 2005), plant phenotypes related to patterns of growth (Salgado-Luarte and Gianoli, 2012) and defense against herbivores (Karban and Baldwin, 1997), and predator foraging behaviors that respond to herbivore volatiles and herbivore-induced plant volatiles (Heil, 2008). And third, how often is consumer recruitment to resources non-linear, such that functional responses are density-dependent (criterion 4)? Both positive and negative density-dependent effects of consumers are predicted to be common through a variety of mechanisms (reviewed by Abrams and Ginzburg, 2000).

Case study: bottom-up and top-down forcing shapes plant evolution

In this section, we provide an example of a tri-trophic system where elements of the previously described framework can be applied to understand the evolution of species traits under the context of simultaneous and interactive bottom-up and top-down effects. Specifically, the attributes of this plant–herbivore–predator system set up a situation where we can examine the linkages between consumer functional responses, indirect interactions, and changes in patterns of selection on species traits under a multi-trophic setting.

Over the last few years, we have studied the interactions between the perennial herb *Ruellia nudiflora* Engelm. and Gray Urban (Acanthaceae), a seed predator (caterpillars of a noctuid moth), and parasitic wasps attacking the latter. *Ruellia nudiflora* has a mixed mating system and produces chasmogamous flowers (open and visited by pollinators), as well as cleistogamous flowers (have a reduced corolla, do not open, and obligately self-pollinate), with the latter being the most abundant in the first year of growth. Female moths oviposit on recently pollinated flowers and are in turn attacked by up to seven species of parasitic wasps (Hymenoptera) and one fly (Diptera) (Abdala-Roberts et al., 2010). Each seed predator larva matures inside a single developing fruit and, unless parasitized, consumes all the seeds (Abdala-Roberts et al., 2010). Here we summarize the results of work that is described in part by Abdala-Roberts and Mooney (2013) and Abdala-Roberts et al. (2014) to address the ecological and evolutionary consequences of bottom-up and top-down forcing, as well as their interactive effects.

With this system, we sought to address two primary questions. First, how does genetically based variation in plant resources (i.e., fruit number) influence seed predator and parasitoid abundance from the bottom-up? Specifically, by what functional responses do seed predators and parasitoids respond to variation in resource abundance? And second, what are the corresponding selective impacts of seed predators and parasitoids upon fruit number? Specifically, do parasitoids alter the selective impacts of seed predators on the plant, and how do such effects relate to the functional responses describing the bottom-up effects of resource abundance?

We planted a common garden with 14 maternal plant families, performed weekly surveys of fruit number, and collected fruit samples to document seed predator and parasitoid attack, as well as seed production (for details see Abdala-Roberts and Mooney, 2013). Because chasmogamous flowers (< 10% of flowers) were lost to vertebrate herbivory, all measurements were based on cleistogamous flowers. The nature of this system allows for measuring the impacts of seed predators and parasitoids without experimental manipulation. Unparasitized seed predators consume all seeds in a fruit, while parasitized seed predators do not and remaining seeds thus represent the indirect positive effects of parasitoids on plant fitness. The number of seeds in unattacked fruits provides a reliable estimate of the potential seed production in attacked fruits. Based upon this, seed number can be calculated for each plant under a three-trophic level scenario (plant, seed predators, parasitoids; i.e., observed seed number), a two-trophic level scenario (plant, seed predators; i.e., seed number that would occur in the absence of seed rescue by parasitoids), and a one-trophic level scenario (plants alone; i.e., potential seed number in the absence of seed predator attack). Based on seed number, we then calculated relative fitness for each plant under each trophic level scenario, and compared genotypic selection for fruit number (i.e., measure of reproductive display size) among trophic level scenarios with analysis of covariance using family means for fruit number and relative fitness.

With respect to the bottom-up effects of resource abundance on higher trophic levels, we observed contrasting functional responses for seed predators and parasitoids to resource density (Fig. 13.2). The seed predator's recruitment function to fruit abundance was non-linear (i.e., saturating; Fig. 13.2a), such that attack rate showed a negative relationship between plant fruit number and the proportion of fruits attacked (Fig. 13.2a, inset), indicating negative density-dependent attack by the herbivore. In contrast, parasitoid recruitment increased linearly with seed predator abundance (Fig. 13.2b), resulting in density-independent parasitoid attack (Fig. 13.2b, inset).

The top-down selective impacts on fruit number were in turn driven by the contrasting functional responses of seed predators and parasitoids to resources

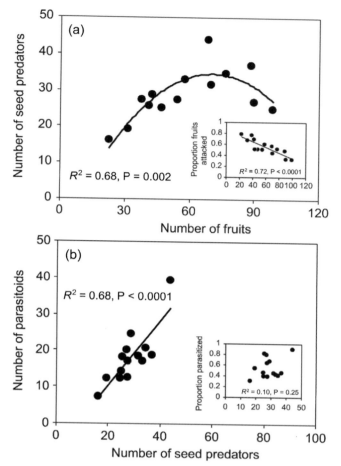

Figure 13.2 Regressions based on *Ruellia nudiflora* maternal family means showing (a) the relationship between fruit number and seed predator abundance ($R^2 = 0.68$, $P = 0.002$), and (b) between seed predator abundance and parasitoid abundance ($R^2 = 0.68$, $P < 0.0001$). Insets have as y-axes the proportion of fruits with seed predator (fruits with seed predator/total fruit number; $R^2 = 0.72$, $P = 0.0001$) and the proportion of parasitized seed predators (parasitized seed predators/total seed predator number; $R^2 = 0.10$, $p = 0.25$), respectively. Modified from Abdala-Roberts and Mooney, © 2013 The Authors.

(Fig. 13.3). There was selection for increased fruit number under the monotrophic scenario (plants alone). Although this result is to be expected (i.e., fruit number is necessarily correlated with fitness, measured as seed number), it is nonetheless relevant as an increase in investment in cleistogamous reproduction in the first year of growth would likely tradeoff with other fitness components, including chasmogamous reproduction during the current year

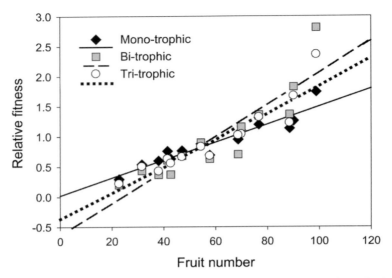

Figure 13.3 Directional selection gradients on cleistogamous fruit display size of *Ruellia nudiflora* under three trophic scenarios: in the absence of seed predator and parasitoid effects on the plant ("mono-trophic"), accounting for seed predator effects but not parasitoid seed rescue ("bi-trophic"), and accounting for both seed predator and parasitoid effects on the plant ("tri-trophic"). Modified from Abdala-Roberts et al. © 2014 The Authors. *Journal of Ecology* © 2014 British Ecological Society). Analysis of covariance showed that the bi-trophic selection gradient was significantly stronger than the mono-trophic gradient ($F_{1,12} = 42.97$, $P < 0.0001$), whereas the tri-trophic gradient was significantly weaker than the bi-trophic gradient ($F_{1,12} = 15.34$, $P = 0.002$) and significantly stronger than the mono-trophic gradient ($F_{1,12} = 34.68$, $P < 0.0001$).

and reproduction by both flower types in subsequent years. Seed predators in turn significantly increased the strength of selection, as the selection gradient was significantly stronger for the bi-trophic relative to the mono-trophic scenario (Fig. 13.3). This increase occurred because of the seed predator's negative density-dependent functional response, with the strongest attack occurring on plant families with low fruit number. Importantly, although parasitoid attack was density-independent, parasitoids dampened (but did not eliminate) seed predator selection (i.e., the tri-trophic selection gradient was weaker than the bi-trophic gradient; Fig. 13.3). The indirect benefit of parasitoids on plant fitness varied among plant families because of the herbivore's non-linear recruitment response. For plant families with small fruit display sizes, herbivore attack rate was high and the benefit of parasitism was large, in terms of a proportional increase in plant fitness. In contrast, for plant families with large display sizes, herbivore attack rate was low and thus the proportional fitness gain from parasitism was comparatively smaller.

These findings have important implications for understanding the evolutionary outcomes of consumer-driven direct and indirect effects (Estes et al., 2013; Abdala-Roberts et al., 2014). Although changes in the function of pairwise species interactions through interaction modifications are an important source of evolutionary change (Miller and Travis, 1996; Inouye and Stinchcombe, 2001; Strauss et al., 2005; Mooney and Singer, 2012), we find that interaction strength may also change if species show non-linear recruitment functions (Mooney and Singer, 2012; Singer et al., 2012). These findings relate to criterion 4 from our conceptual model (Fig. 13.1f), but provide further insight by showing that the occurrence of non-additive selection, in this case represented by parasitoids modifying herbivore selection on plants, may depend upon the specific features of the recruitment functions exhibited by consumers at different trophic levels. These patterns of altered selection regimes from top-down forcing may be common in tri-trophic systems, due to the prevalence of herbivore and predator non-linear functional responses (Mooney and Singer, 2012).

Conclusions

Despite parallel developments in the ecological study of top-down and bottom-up regulation of terrestrial and aquatic ecosystems, within an evolutionary realm the study of aquatic systems has lagged behind. This is particularly true with regards to understanding the interactive evolutionary effects of bottom-up and top-down forcing. Thus, whereas theory stemming from the study of terrestrial plant–herbivore interactions has made some progress toward evaluating the interactive effects of resources and consumers upon plant and herbivore evolution, parallel work is largely undeveloped for aquatic systems, and many hypotheses from the terrestrial literature have not been tested in the aquatic domain (e.g., Resource Availability Hypothesis, Slow-Growth/High-Mortality). Predictions from such theories are in no way limited to terrestrial systems, and more empirical tests are needed to determine whether evolutionary dynamics from bottom-up and top-down forces follow similar patterns on land and in water.

The majority of empirical evolutionary studies to date have focused on the population biology of focal herbivore and plant taxa in response to resources and consumers. Whereas ecological effects on species performance are suggestive of evolutionary dynamics, they do not substitute for studies directly measuring selection and response to selection (Strauss et al., 2005; Abdala-Roberts and Mooney, 2013). Moreover, while there are a few studies considering how species are shaped by the combined influences of multiple selective agents (but see e.g., Iwao and Rausher, 1997; Juenger and Bergelson, 1998; Stinchcombe and Rausher, 2001; Rutter and Rausher, 2004), we are unaware of work that formally tests for multi-trophic non-additive selection from resources and consumers.

Therefore, a push is needed toward more studies measuring the alteration of evolutionary dynamics in response to manipulation of top-down and bottom-up forces. To accurately measure evolutionary effects, as well as to make cross-system comparisons, studies must measure changes in the relationship between-species traits and relative fitness in response to such manipulations (Strauss et al., 2005).

Although this review has focused on joint selection on producers and her-bivores from bracketing trophic levels, there is also increasing evidence for the importance of selection stemming from indirect interactions (from trophic levels higher or lower down). Such a situation is depicted by the case study presented here (Fig. 13.3) and the work of Steinberg et al. (1995) showing indi-rect selection by predators on producers. In addition, Walsh and Reznick (2008) showed that predatory fish had indirect evolutionary effects on killifish through changes in resource availability (stemming indirectly from predator effects on killifish). Nonetheless, except for these studies and two studies on specialized protection mutualisms (Rudgers, 2004; Rutter and Rausher, 2004; reviewed by Kessler and Heil, 2011), the selective indirect effect of predators on non-adjacent lower trophic levels (e.g., plants) has received relatively little attention (Hare, 2002; Mooney and Singer, 2012), but might be a common and underappreci-ated phenomenon (Marquis and Whelan, 1996). While other forms of indirect selection are also likely to be common, such as when plant–herbivore interac-tions are influenced by simultaneous interactions with other competing plants (Hambäck and Beckerman, 2003; Linhart et al., 2005; Terhorst, 2010) and polli-nators (Adler et al., 2001), future work should move beyond examining a focal species and its bracketing trophic levels, to examining the evolutionary conse-quences of top-down and bottom-up indirect effects (Walsh and Reznick, 2008; Estes et al., 2013).

Finally, work conducted over the last decade provides evidence for eco-evolutionary feedbacks (e.g., Post and Palkovacs, 2009), where biotic or abiotic selective agents produce evolutionary change in a species, and such change (pre-sumably occurring over ecological time scales) in turn leads to ecological effects on other members of the community. A particularly appealing context under which to study these dynamics is laid out by community genetics work, which argues that differences in the strength and outcomes of species interactions among plant genotypes may influence selection on heritable plant traits, thus bringing about genetic changes in plant populations which further influence consumers (Whitham et al., 2006; Bailey et al., 2014). The idea of eco-evolutionary feedbacks is thus relevant to the study of bottom-up and top-down forcing, given that resources and consumers can exert interactive controls and thus give rise to these types of dynamics. We therefore call for future work that addresses the eco-logical consequences of rapid evolutionary change stemming from bottom-up and top-down forcing.

Acknowledgments

KAM was supported by NSF grant DEB 1120794. Work by LAR was funded by a GAANN fellowship and a UCMEXUS-CONACyT scholarship.

References

Abdala-Roberts, L. and Mooney, K. A. (2013). Environmental and plant genetic effects on tri-trophic interactions. *Oikos*, **122**, 1157–1166.

Abdala-Roberts, L., Parra-Tabla, V., Salinas-Peba, L., Diaz-Castelazo, C. and Delfin-Gonzalez, H. (2010). Spatial variation in the strength of a trophic cascade involving *Ruellia nudiflora* (Acanthaceae), an insect seed predator and associated parasitoid fauna in Mexico. *Biotropica*, **42**, 180–187.

Abdala-Roberts, L., Agrawal, A. A. and Mooney, K. A. (2012). Ant–aphid interactions on *Asclepias syriaca* are mediated by plant genotype and caterpillar damage. *Oikos*, **121**, 1905–1913.

Abdala-Roberts, L., Parra-Tabla, V., Campbell, D. R. and Mooney, K. A. (2014). Soil fertility and parasitoids shape herbivore selection on plants. *Journal of Ecology*, **102**, 1120–1128.

Abrams, P. A. (1995). Implications of dynamically variable traits for identifying, classifying, and measuring direct and indirect effects in ecological communities. *American Naturalist*, **146**, 112–134.

Abrams, P. A. (2000). The evolution of predator–prey interactions: theory and evidence. *Annual Review of Ecology and Systematics*, **31**, 79–105.

Abrams, P. A. and Ginzburg, L. R. (2000). The nature of predation: prey dependent, ratio dependent or neither? *Trends in Ecology and Evolution*, **15**, 337–341.

Adler, L. S., Karban, R. and Strauss, S. Y. (2001). Direct and indirect effects of alkaloids on plant fitness via herbivory and pollination. *Ecology*, **82**, 2032–2044.

Agrawal, A. A. (2011). Current trends in the evolutionary ecology of plant defence. *Functional Ecology*, **25**, 420–432.

Agrawal, A. A., Laforsch, C. and Tollrian, R. (1999). Transgenerational induction of defences in animals and plants. *Nature*, **401**, 60–63.

Agrawal, A. A., Fishbein, M., Halitschke, R., et al. (2009). Evidence for adaptive radiation from a phylogenetic study of plant defenses. *Proceedings of the National Academy of Sciences of the USA*, **106**, 18067–18072.

Bailey, J. K., Genung, M. A., Ware, I., et al. (2014). Indirect genetic effects: an evolutionary mechanism linking feedbacks, genotypic diversity, and coadaptation in climate change context. *Functional Ecology*, **28**, 87–95.

Becerra, J. X. (1997). Insects on plants: macroevolutionary chemical trends in host use. *Science*, **276**, 253–256.

Becerra, J. X. (2003). Synchronous coadaptation in an ancient case of herbivory. *Proceedings of the National Academy of Sciences of the USA*, **100**, 12804–12807.

Becerra, J. X., Noge, K. and Venable, D. L. (2009). Macroevolutionary chemical escalation in an ancient plant-herbivore arms race. *Proceedings of the National Academy of Sciences of the USA*, **106**, 18062–18066.

Benkman, C. W. (2013). Biotic interaction strength and the intensity of selection. *Ecology Letters*, **16**, 1054–1060.

Berdegue, M., Trumble, J. T., Hare, J. D. and Redak, R. A. (1996). Is it enemy-free space? The evidence for terrestrial insects and freshwater arthropods. *Ecological Entomology*, **21**, 203–217.

Berenbaum, M. R. and Zangerl, A. R. (1998). Chemical phenotype matching between a plant and its insect herbivore. *Proceedings of the National Academy of Sciences of the USA*, **95**, 13743–13748.

Bernays, E. A. (1998). Evolution of feeding behavior in insect herbivores: success seen as different ways to eat without being eaten. *Bioscience*, **48**, 35–44.

Bolser, R. C. and Hay, M. E. (1996). Are tropical plants better defended? Palatability and defenses of temperate vs tropical seaweeds. *Ecology*, **77**, 2269–2286.

Borer, E. T., Seabloom, E. W., Shurin, J. B., et al. (2005). What determines the strength of a trophic cascade? *Ecology*, **86**, 528–537.

Borer, E. T., Halpern, B. S. and Seabloom, E. W. (2006). Asymmetry in community regulation: effects of predators and productivity. *Ecology*, **87**, 2813–2820.

Chapin, F. S. (1993). Physiological controls over plant establishment in primary succession. In *Primary Succession on Land*, ed. J. Miles and D. W. H. Walton. Oxford, Boston: Blackwell Scientific Publications, pp. 161–178.

Choat, J. H. and Clements, K. D. (1998). Vertebrate herbivores in marine and terrestrial environments: a nutritional ecology perspective. *Annual Review of Ecology and Systematics*, **29**, 375–403.

Coley, P. D., Bryant, J. P. and Chapin, F. S. (1985). Resource availability and plant antiherbivore defense. *Science*, **230**, 895–899.

Connell, J. H. and Slatyer, R. O. (1977). Mechanisms of succession in natural communities and their role in community stability and organization. *American Naturalist*, **111**, 1119–1144.

Cremieux, L., Bischoff, A., Smilauerova, M., et al. (2008). Potential contribution of natural enemies to patterns of local adaptation in plants. *New Phytologist*, **180**, 524–533.

Demment, M. W. and Van Soest, P. J. (1985). A nutritional explanation for body-size patterns of ruminant and nonruminant herbivores. *American Naturalist*, **125**, 641–672.

Dethier, V. G. (1954). Evolution of feeding preference in phytophagous insects. *Evolution*, **8**, 33–54.

Dodson, S. I. (1989). The ecological role of chemical stimuli for the zooplankton: predator-induced morphology in Daphnia. *Oecologia*, **78**, 361–367.

Duffy, J. E. and Hay, M. E. (1990). Seaweed adaptations to herbivory: chemical, structural, and morphological defenses are often adjusted to spatial or temporal patterns of attack. *Bioscience*, **40**, 368–375.

Dyer, L. A. (1995). Tasty generalists and nasty specialists: antipredator mechanisms in tropical lepidopteran larvae. *Ecology*, **76**, 1483–1496.

Dyer, L. A. (2008). Tropical tritrophic interactions: nasty hosts and ubiquitous cascades. In *Tropical Forest Community Ecology*, ed. W. P. Carson and S. A. Schnitzer. Oxford: Blackwell Science, pp. 275–293.

Edwards, K. F., Klausmeier, C. A. and Litchman, E. In press. A three-way tradeoff maintains functional diversity under variable resource supply. *American Naturalist*.

Ehrlich, P. R. and Raven, P. H. (1964). Butterflies and plants: a study in coevolution. *Evolution*, **18**, 586–608.

Elser, J. J., Fagan, W. F., Denno, R. F., et al. (2000). Nutritional constraints in terrestrial and freshwater food webs. *Nature*, **408**, 578–580.

Elzinga, J. A., Atlan, A., Biere, A., et al. (2007). Time after time: flowering phenology and biotic interactions. *Trends in Ecology and Evolution*, **22**, 432–439.

Endara, M.-J. and Coley, P. D. (2011). The resource availability hypothesis revisited: a meta-analysis. *Functional Ecology*, **25**, 389–398.

Erb, M. and Lu, J. (2013). Soil abiotic factors influence interactions between belowground herbivores and plant roots. *Journal of Experimental Botany*, **64**, 1295–1303.

Estes, J. A., Brashares, J. S. and Power, M. E. (2013). Predicting and detecting reciprocity between indirect ecological interactions and evolution. *American Naturalist*, **181**, S76–S99.

Eubanks, M. D. and Denno, R. F. (2000). Host plants mediate omnivore-herbivore interactions and influence prey suppression. *Ecology*, **81**, 936–947.

Feeny, P. P. (1976). Plant apparency and chemical defense. *Recent Advances in Phytochemistry*, **10**, 1–40.

Fine, P. V. A., Mesones, I. and Coley, P. D. (2004). Herbivores promote habitat specialization by trees in amazonian forests. *Science*, **305**, 663–665.

Fraenkel, G. S. (1959). Raison d'etre of secondary plant substances. *Science*, **129**, 1466–1470.

Gaines, S. D. and Lubchenco, J. (1982). A unified approach to marine plant-herbivore interactions. 2. Biogeography. *Annual Review of Ecology and Systematics*, **13**, 111–138.

Grime, J. P. (1977). Evidence for existence of 3 primary strategies in plants and its relevance to ecological and evolutionary theory. *American Naturalist*, **111**, 1169–1194.

Gripenberg, S. and Roslin, T. (2007). Up or down in space? Uniting the bottom-up versus top-down paradigm and spatial ecology. *Oikos*, **116**, 181–188.

Gruner, D. S., Smith, J. E., Seabloom, E. W., et al. (2008). A cross-system synthesis of consumer and nutrient resource control on producer biomass. *Ecology Letters*, **11**, 740–755.

Hagen, E. M., Mccluney, K. E., Wyant, K. A., et al. (2012). A meta-analysis of the effects of detritus on primary producers and consumers in marine, freshwater, and terrestrial ecosystems. *Oikos*, **121**, 1507–1515.

Hairston, N. G., Smith, F. E. and Slobodkin, L. G. (1960). Community structure, population control, and competition. *American Naturalist*, **94**, 421–425.

Haloin, J. R. and Strauss, S. Y. (2008). Interplay between ecological communities and evolution review of feedbacks from microevolutionary to macroevolutionary scales. *Year in Evolutionary Biology*, **1133**, 87–125.

Hambäck, P. A. and Beckerman, A. P. (2003). Herbivory and plant resource competition: a review of two interacting interactions. *Oikos*, **101**, 26–37.

Hare, J. D. (2002). Plant genetic variation in tritrophic interactions. In *Multitrophic Level Interactions*, ed. T. Tscharntke and B. A. Hawkins. Cambridge, UK/New York: Cambridge University Press, pp. 278–298.

Hay, M. E. (1991). Marine terrestrial contrasts in the ecology of plant-chemical defenses against herbivores. *Trends in Ecology and Evolution*, **6**, 362–365.

Hay, M. E. (1992). The role of seaweed chemical defenses in the evolution of feeding specialization and in the mediation of complex interactions. In *Ecological Roles of Marine Natural Products*, ed. V. Paul. Ithaca, NY: Cornell University Press.

Hay, M. E. (1996). Marine chemical ecology: what's known and what's next? *Journal of Experimental Marine Biology and Ecology*, **200**, 103–134.

Hay, M. E. and Fenical, W. (1988). Marine plant-herbivore interactions: the ecology of chemical defense. *Annual Review of Ecology and Systematics*, **19**, 111–145.

Hay, M. E. and Steinberg, P. D. (1992). The chemical ecology of plant–herbivore interactions in marine versus terrestrial communities. In *Herbivores: Their Interaction with Secondary Metabolites, Evolutionary and Ecological Processes*, ed. G. A. Rosenthal and M. Berenbaum. San Diego, CA: Academic Press.

Hays, G. C. (2003). A review of the adaptive significance and ecosystem consequences of zooplankton diel vertical migrations. *Hydrobiologia*, **503**, 163–170.

Heil, M. (2008). Indirect defence via tritrophic interactions. *New Phytologist*, **178**, 41–61.

Herms, D. A. and Mattson, W. J. (1992). The dilemma of plants: to grow or defend. *Quarterly Review of Biology*, **67**, 283–335.

Hunter, M. D. and Price, P. W. (1992). Playing chutes and ladders: heterogeneity and the relative roles of bottom-up and top-down forces in natural communities. *Ecology*, **73**, 724–732.

Ingram, T., Svanback, R., Kraft, N. J. B., et al. (2012). Intraguild predation drives evolutionary niche shift in threespine stickleback. *Evolution*, **66**, 1819–1832.

Inouye, B. and Stinchcombe, J. R. (2001). Relationships between ecological interaction modifications and diffuse coevolution: similarities, differences, and causal links. *Oikos*, **95**, 353–360.

Iwao, K. and Rausher, M. D. (1997). Evolution of plant resistance to multiple herbivores: quantifying diffuse coevolution. *American Naturalist*, **149**, 316–335.

Juenger, T. and Bergelson, J. (1998). Pairwise versus diffuse natural selection and the multiple herbivores of scarlet gilia, *Ipomopsis aggregata*. *Evolution*, **52**, 1583–1592.

Karban, R. and Baldwin, I. T. (1997). *Induced Responses to Herbivory*. Chicago, IL: University of Chicago Press.

Kessler, A. and Heil, M. (2011). The multiple faces of indirect defences and their agents of natural selection. *Functional Ecology*, **25**, 348–357.

Lampert, W. (1989). The adaptive significance of diel vertical migration of zooplankton. *Functional Ecology*, **3**, 21–27.

Lawton, J. H. and McNeill, S. (1979). Between the devil and the deep blue sea: on the problems of being an herbivore. In *Population Dynamics. Symposium of the British Ecological Society*, ed. R. M. Anderson, B. D. Turner and L. R. Taylor. Oxford: Blackwell.

Leong, W. and Pawlik, J. R. (2010). Evidence of a resource trade-off between growth and chemical defenses among Caribbean coral reef sponges. *Marine Ecology Progress Series*, **406**, 71–78.

Lind, E. M., Borer, E., Seabloom, E., et al. (2013). Life-history constraints in grassland plant species: a growth-defence trade-off is the norm. *Ecology Letters*, **16**, 513–21.

Linhart, Y. B., Keefover-Ring, K., Mooney, K. A., Breland, B. and Thompson, J. D. (2005). A chemical polymorphism in a multitrophic setting: thyme monoterpene composition and food web structure. *American Naturalist*, **166**, 517–529.

Litchman, E. and Klausmeier, C. A. (2008). Trait-based community ecology of phytoplankton. *Annual Review of Ecology, Evolution, and Systematics*, **39**, 615–639.

Loose, C. J., Vonelert, E. and Dawidowicz, P. (1993). Chemically-induced diel vertical migration in Daphnia: a new bioassay for kairomones exuded by fish. *Archiv Fur Hydrobiologie*, **126**, 329–337.

Marquis, R. J. (1984). Leaf herbivores decrease fitness of a tropical plant. *Science*, **226**, 537–539.

Marquis, R. J. (1992). Selective impact of herbivores. In *Plant Resistance to Herbivores and Pathogens: Ecology, Evolution, and Genetics*, ed. R. S. Fritz and E. L. Simms. Chicago, IL: University of Chicago Press.

Marquis, R. J. and Whelan, C. (1996). Plant morphology, and recruitment of the third trophic level: subtle and little-recognized defenses? *Oikos*, **75**, 330–334.

Miller, T. E. and Travis, J. (1996). The evolutionary role of indirect effects in communities. *Ecology*, **77**, 1329–1335.

Mooney, K. A. and Agrawal, A. A. (2008). Plant genotype shapes ant-aphid interactions: implications for community structure and indirect plant defense. *American Naturalist*, **168**, E195–E205.

Mooney, K. A. and Singer, M. S. (2012). Plant variation in herbivore-enemy interactions in natural systems. In *Trait-Mediated Indirect Interactions: Ecological and Evolutionary Perspectives*, ed. T. Ohgushi, O. J. Schmitz and R. D. Holt. New York, NY: Cambridge University Press.

Mooney, K. A., Halitschke, R., Kessler, A. and Agrawal, A. A. (2010). Evolutionary trade-offs in plants mediate the strength of trophic cascades. *Science*, **327**, 1642–1644.

Mooney, K. A., Pratt, R. T. and Singer, M. S. (2012). The tri-trophic interactions hypothesis: interactive effects of host plant quality, diet breadth and natural enemies on herbivores. *PLoS One*, **7**, e34403.

Moran, N. and Hamilton, W. D. (1980). Low nutritive quality as defense against herbivores. *Journal of Theoretical Biology*, **86**, 247–254.

Murdoch, W. W. (1966). Community structure population control and competition: a critique. *American Naturalist*, **100**, 219–226.

Ness, J. H., Morris, W. F. and Bronstein, J. L. (2009). For ant-protected plants, the best defense is a hungry offense. *Ecology*, **90**, 2823–2831.

Oksanen, L., Fretwell, S. D., Arruda, J. and Niemela, P. (1981). Exploitation ecosystems in gradients of primary productivity. *American Naturalist*, **118**, 240–261.

Palkovacs, E. P., Wasserman, B. A. and Kinnison, M. T. (2011). Eco-evolutionary trophic dynamics: loss of top predators drives trophic evolution and ecology of prey. *PLoS One*, **6**.

Pennings, S. C., Siska, E. L. and Bertness, M. D. (2001). Latitudinal differences in plant palatability in Atlantic coast salt marshes. *Ecology*, **82**, 1344–1359.

Pennings, S. C., Ho, C. K., Salgado, C. S., et al. (2009). Latitudinal variation in herbivore pressure in Atlantic Coast salt marshes. *Ecology*, **90**, 183–195.

Polis, G. A. and Strong, D. R. (1996). Food web complexity and community dynamics. *American Naturalist*, **147**, 813–846.

Post, D. M. and Palkovacs, E. P. (2009). Eco-evolutionary feedbacks in community and ecosystem ecology: interactions between the ecological theatre and the evolutionary play. *Philosophical Transactions of the Royal Society of London B: Biological Sciences*, **364**, 1629–1640.

Preisser, E. L., Bolnick, D. I. and Benard, M. F. (2005). Scared to death? The effects of intimidation and consumption in predator-prey interactions. *Ecology*, **86**, 501–509.

Reznick, D. and Endler, J. A. (1982). The impact of predation on life-history evolution in Trinidadian guppies (Poecilia-Reticulata). *Evolution*, **36**, 160–177.

Rhoades, D. F. and Cates, R. G. (1976). Toward a general theory of plant antiherbivore chemistry. *Recent Advances in Phytochemistry*, **10**, 168–213.

Rico-Gray, V. and Oliveira, P. S. (2007). *The Ecology and Evolution of Ant-Plant Interactions*. Chicago, IL: University of Chicago Press.

Rosenheim, J. A. and Corbett, A. (2003). Omnivory and the indeterminacy of predator function: can a knowledge of foraging behavior help? *Ecology*, **84**, 2538–2548.

Rudgers, J. A. (2004). Enemies of herbivores can shape plant traits: selection in a facultative ant-plant mutualism. *Ecology*, **85**, 192–205.

Rutter, M. T. and Rausher, M. D. (2004). Natural selection on extrafloral nectar production in *Chamaecrista fasciculata*: the costs and benefits of a mutualism trait. *Evolution*, **58**, 2657–2668.

Salgado-Luarte, C. and Gianoli, E. (2012). Herbivores modify selection on plant functional traits in a temperate rainforest understory. *American Naturalist*, **180**, E42–E53.

Schultz, J. C. (1988). Many factors influence the evolution of herbivore diets, but plant chemistry is central. *Ecology*, **69**, 896–897.

Shurin, J. B., Borer, E. T., Seabloom, E. W., et al. (2002). A cross-ecosystem comparison of the strength of trophic cascades. *Ecology Letters*, **5**, 785–791.

Shurin, J. B., Gruner, D. S. and Hillebrand, H. (2006). All wet or dried up? Real differences between aquatic and terrestrial food webs. *Proceedings of the Royal Society B: Biological Sciences*, **273**, 1–9.

Singer, M. S. (2008). Evolutionary ecology of polyphagy. In *Specialization, Speciation, and Radiation: The Evolutionary Biology of Herbivorous Insects*, ed. K. J. Tilmon. Berkeley, CA: University of California Press.

Singer, M. S. and Stireman, J. O. (2005). The tri-trophic niche concept and adaptive radiation of phytophagous insects. *Ecology Letters*, **8**, 1247–1255.

Singer, M. S., Carriere, Y., Theuring, C. and Hartmann, T. (2004). Disentangling food quality from resistance against parasitoids: diet choice by a generalist caterpillar. *American Naturalist*, **164**, 423–429.

Singer, M. S., Farkas, T. E., Skorik, C. M. and Mooney, K. A. (2012). Tri-trophic interactions at a community level: effects of host-plant species quality on bird predation of caterpillars. *American Naturalist*, **179**, 363–374.

Sork, V. L., Stowe, K. A. and Hochwender, C. (1993). Evidence for local adaptation in closely adjacent subpopulations of northern red oak (*Quercus rubra* L) expressed as resistance to leaf herbivores. *American Naturalist*, **142**, 928–936.

Stachowicz, J. J. and Hay, M. E. (1999). Reduced mobility is associated with compensatory feeding and increased diet breadth of marine crabs. *Marine Ecology Progress Series*, **188**, 169–178.

Stamp, N. (2003). Out of the quagmire of plant defense hypotheses. *Quarterly Review of Biology*, **78**, 23–55.

Steinberg, P. D., Estes, J. A. and Winter, F. C. (1995). Evolutionary consequences of food-chain length in kelp forest communities. *Proceedings of the National Academy of Sciences of the USA*, **92**, 8145–8148.

Stenberg, J. A., Lehrman, A. and Bjorkman, C. (2011). Plant defence: feeding your bodyguards can be counter-productive. *Basic and Applied Ecology*, **12**, 629–633.

Stinchcombe, J. R. and Rausher, M. D. (2001). Diffuse selection on resistance to deer herbivory in the ivyleaf morning glory, *Ipomoea hederacea*. *American Naturalist*, **158**, 376–388.

Strauss, S. Y., Sahli, H. and Conner, J. K. (2005). Toward a more trait-centered approach to diffuse (co)evolution. *New Phytologist*, **165**, 81–89.

Strong, D. R., Lawton, J. H. and Southwood, R. (1984). *Insects on Plants: Community Patterns and Mechanisms*. Cambridge, MA: Harvard University Press.

Terhorst, C. P. (2010). Evolution in response to direct and indirect ecological effects in pitcher plant inquiline communities. *American Naturalist*, **176**, 675–685.

Thomas, M. K., Kremer, C. T., Klausmeier, C. A. and Litchman, E. (2012). A global pattern of thermal adaptation in marine phytoplankton. *Science*, **338**, 1085–1088.

Thompson, J. N. (2005). *The Geographic Mosaic of Coevolution*. Chicago, IL: University of Chicago Press.

Tilman, D. (1982). *Resource Competition and Community Structure*. Princeton, NJ: Princeton University Press.

Turley, N. E., Odell, W. C., Schaefer, H., et al. (2013). Contemporary evolution of plant growth rate following experimental removal of herbivores. *American Naturalist*, **181**, S21–S34.

Urabe, J. and Sterner, R. W. (2001). Contrasting effects of different types of resource depletion on life-history traits in Daphnia. *Functional Ecology*, **15**, 165–174.

Urban, M. C. (2013). Evolution mediates the effects of apex predation on aquatic food webs. *Proceedings of the Royal Society B: Biological Sciences*, **280**, 1–9.

Walsh, M. R. and Reznick, D. N. (2008). Interactions between the direct and indirect effects of predators determine life history evolution in a killifish. *Proceedings of the National Academy of Sciences of the USA*, **105**, 594–599.

Whitham, T. G., Bailey, J. K., Schweitzer, J. A., et al. (2006). A framework for community and ecosystem genetics: from genes to ecosystems. *Nature Reviews Genetics*, **7**, 510–523.

Williams, I. S. (1999). Slow-growth, high-mortality: a general hypothesis, or is it? *Ecological Entomology*, **24**, 490–495.

Wootton, J. T. (1994). Putting the pieces together: testing the independence of interactions among organisms. *Ecology*, **75**, 1544–1551.

Wootton, J. T. and Emmerson, M. (2005). Measurement of interaction strength in nature. *Annual Review of Ecology, Evolution, and Systematics*, **36**, 419–444.

CHAPTER FOURTEEN

Bottom-up and top-down interactions across ecosystems in an era of global change

KIMBERLY J. LA PIERRE

University of California, Berkeley, CA, USA

and

TORRANCE C. HANLEY

Northeastern University, Nahant, MA, USA

Introduction

It is undeniable that humans impact the environment in many ways (Vitousek et al., 1997b; Rosa et al., 2004; Ellis, 2011). These impacts have historically been small in comparison to other ecological drivers, such as natural variation in resource availability and disturbance regimes. However, most human impacts are now on par with or even stronger than many ecological drivers, leading to claims that we live on a human-dominated planet (Vitousek et al., 1997b) and have entered the age of the anthropocene (Ellis, 2011). Human-induced global change factors come in many forms, including altered biogeochemical cycles, climate change, and species losses and gains, each of which can be caused or enhanced by habitat modification (Fig. 14.1; Foley et al., 2005). These anthropogenic effects on ecosystems worldwide are expected to increase over time as the human population and levels of affluence continue to grow (Rosa et al., 2004). Therefore, it is imperative that we work to synthesize the broad array of information we have amassed about human impacts on the environment to guide future research and conservation efforts.

A wide variety of trophic interactions are predicted to be strongly affected by global change (Fig. 14.1). As detailed in this chapter, humans have permanently altered many food webs, both by changing the strength of existing abiotic and biotic interactions and by creating new interactions (Emmerson et al., 2004; Strong and Frank, 2010; Blois et al., 2013). These human-induced effects on community and ecosystem structure and function can act from either the bottom-up (i.e., alteration of resource availability; Smith et al., 2009) or top-down (i.e.,

Trophic Ecology: Bottom-Up and Top-Down Interactions across Aquatic and Terrestrial Systems, eds T. C. Hanley and K. J. La Pierre. Published by Cambridge University Press. © Cambridge University Press 2015.

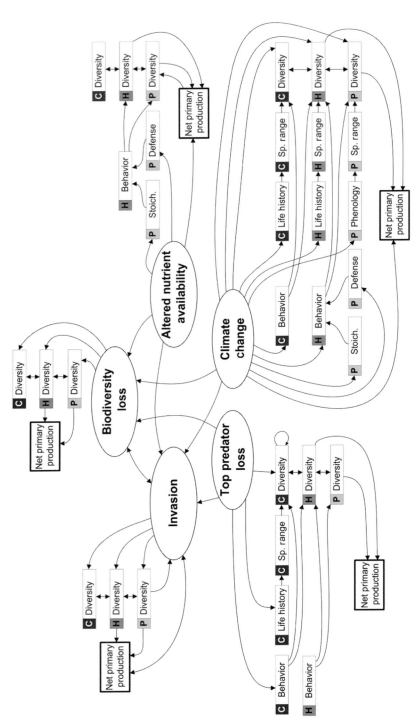

Figure 14.1 Global change factors (ovals) will have direct and/or indirect effects at all trophic levels (boxes; P: primary producer, H: herbivore, C: secondary or tertiary consumer) with consequences for ecosystem function, such as net primary production. Although top predator loss and alteration of nutrient availability primarily act from the top-down and bottom-up, respectively, these global change factors will have indirect consequences for all other trophic levels through alterations in species traits, abundances, and ranges, as well as community-level processes, such as diversity. Climate change, biological invasions, and diversity loss can have direct effects at all trophic levels, as well as indirect effects through cascading trophic interactions. Here, diversity may refer to population (i.e., genetic), species, functional, or phylogenetic diversity (also, see Fig. 14.3).

alteration of consumer pressure; Carpenter et al., 1992; Power, 1992). Perhaps most importantly, global change factors also frequently alter the interactions of bottom-up and top-down forces (Emmerson et al., 2004; Tylianakis et al., 2008; Schmitz, 2013). Explicit study of these interactions is therefore required to understand the consequences of global change across ecosystems.

Human effects on ecosystems are not likely to be consistent around the globe, with variation existing spatially, temporally, and taxonomically (Balmford and Bond, 2005). Further, the effects of global change factors are complex and often manifest at a variety of scales within an ecosystem (see Chapter 11). For example, global change effects can range from changes in individual physiology and population dynamics, to alterations in biotic interactions across species within a community, and finally to changes in ecosystem function (Fig. 14.1; Smith et al., 2009). The determinants of the impacts of global change at these many scales have been extensively studied in aquatic and terrestrial systems (Carpenter et al., 1992; Tylianakis et al., 2008; Field et al., 2010). However, much of this work has been done in isolation, focusing on either aquatic or terrestrial systems, and often solely on specific ecosystem types. Because of the complexity of global change effects on communities and ecosystems, an identification of the commonalities and differences in ecosystem responses to global change factors across a broad array of systems would greatly improve our understanding of global change drivers, and may be essential for predicting the effects of future global change factors on community and ecosystem structure and function.

Here, we examine how several anthropogenically derived global change factors – alterations in nutrient availability, climate change, loss of top predators, changes in biodiversity, and species invasions – affect trophic interactions in a wide variety of aquatic and terrestrial ecosystems, and consider the subsequent consequences for ecosystem structure and function. While many of these factors are traditionally considered to be either bottom-up (e.g., nutrient availability and climate change) or top-down (e.g., loss of top predators) in nature, the examples presented in this chapter illustrate that the bottom-up vs. top-down dichotomy is rarely appropriate, as these factors can have direct and indirect effects at all trophic levels in both aquatic and terrestrial systems. Further, the interaction of these stressors may be additive, synergistic, or antagonistic, which may differ across aquatic and terrestrial ecosystems; consequently, we examine a suite of studies seeking to identify patterns across systems. Finally, we conclude by describing four avenues of research that may prove particularly beneficial in predicting and preventing future global change impacts on ecosystems.

Global change drivers and their effects on community structure and ecosystem function

Alteration of nutrient availability

Humans are currently affecting global nutrient availability through increased atmospheric carbon dioxide (CO_2) concentrations and nitrogen (N) deposition due to fossil fuel combustion, and through N and phosphorus (P) runoff from agricultural and urban settings (Gruber and Galloway, 2008). These changes alter resource ratios within ecosystems, shift which resources are most limiting within a system, and acidify soils and water bodies (Gruber and Galloway, 2008; Sardans et al., 2011). The consequences of these alterations in N, P, and CO_2 availability for abiotic and biotic interactions both within and across trophic levels have been extensively studied in marine, freshwater, and terrestrial habitats.

Changes in resource availability (i.e., bottom-up processes) due to global change can interact with top-down processes to produce alterations in the quantity and quality of primary producers in an ecosystem (see Chapter 9). One body of theory predicts that nutrient enrichment should act to destabilize primary producer–herbivore interactions: increased nutritional quality of primary producers should lead to increased herbivore abundance, resulting in drastic declines in primary producer abundance as a result of increased consumption by these abundant herbivores (Rosenzweig, 1971). Alternatively, consumers may decline in abundance with increased nutrient availability due to higher allocation to primary producer defense (Emmerson et al., 2004). These predictions have been tested in many systems and some commonalities, as well as some key differences, across systems can be identified.

One commonality across aquatic and terrestrial systems is that primary producers tend to be fairly plastic in response to environmental changes in nutrient availability (Sterner and Elser, 2002; Sardans et al., 2011; Sistla and Schimel, 2012). Overall, the quantity (absolute amount) of food available for herbivores generally increases in response to nutrient enrichment, as evidenced by several meta-analyses that have shown increases in primary production in response to nutrient additions in marine, freshwater, and terrestrial ecosystems (Fig. 14.2a; Downing et al., 1999; Elser et al., 2007; Gruner et al., 2008). This shift in primary production with increased nutrient availability may result in altered herbivore resource limitation (e.g., shifts from absolute to relative resource limitation with increasing resource quantity; see Chapter 8).

In contrast to the relatively straightforward effects of nutrient enrichment on changes in primary production, alterations in the quality of food for herbivores are much more complex, ranging from shifts in stoichiometry and defensive characteristics of primary producers to shifts in community composition. Nutrient enrichment, particularly of N and P, tends to decrease the C:nutrient ratios (i.e., increase food quality) in primary producer tissue in all system types. In contrast, elevated atmospheric CO_2 tends to increase C:nutrient ratios (i.e., decrease food quality) in primary producer tissue (Sardans et al., 2011). Based on this tradeoff, concomitant increases in CO_2 and nutrient enrichment may be expected to counteract each other, resulting in no total effect on primary producer tissue chemistry (Throop and Lerdau, 2004). Additionally, while many primary producers exhibit high levels of stoichiometric plasticity, this trait in

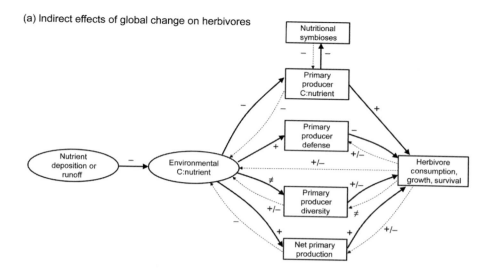

(a) Indirect effects of global change on herbivores

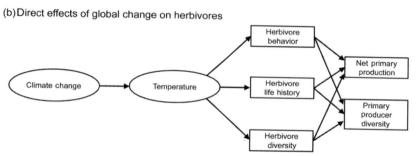

(b) Direct effects of global change on herbivores

Figure 14.2 Global change factors can have both indirect effects on herbivores, resulting from their effects on primary producers, and direct effects on herbivores. (a) For example, nutrient deposition or runoff alters environmental C:nutrient ratios (ovals). This change in environmental C:nutrient ratios has myriad direct effects on primary producers, including alterations in stoichiometry, defense, diversity, and production, which in turn have direct effects on nutritional symbioses and herbivore consumption rates, growth, and survival (solid arrows). These changes could be directional (increases: +, decreases: −) or non-directional (≠). Many of the changes at the primary producer and herbivore levels can feedback (dashed arrows) to further alter environmental nutrient availability. (b) In contrast, increased temperature due to climate change can have direct effects on consumer behavior, life history, and diversity, with cascading effects on net primary production and primary producer diversity.

itself is variable (Sardans et al., 2011). For example, due to differences in physiology, C_4 plant tissue chemistry is generally not affected by elevated atmospheric CO_2, in contrast to C_3 plants (Knops et al., 2007). Similarly, phytoplankton communities in ponds have been shown to vary in their stoichiometric response to altered environmental C:nutrient ratios (Hall et al., 2005).

The above described shifts in primary producer tissue chemistry will likely have major impacts on herbivores, as primary producer nutrient concentration is the best predictor of herbivore performance, particularly for invertebrate herbivores (Throop and Lerdau, 2004; Cebrian et al., 2009; Lindroth, 2010; also, see Chapter 3). Decreased primary producer C:nutrient ratios have been shown to increase herbivore survival, development, growth, and reproduction in both aquatic and terrestrial systems (Fig. 14.2a; Throop and Lerdau, 2004; Lindroth, 2010). Additionally, rates of herbivory may shift with increases in nutrient availability. For example, La Pierre and Smith (unpublished data) observed increased invertebrate herbivore abundances due to increases in primary producer tissue quantity and quality with long-term, chronic nutrient additions in a grassland system. However, per capita rates of herbivory decreased with nutrient enrichment, as individual herbivores were able to consume less of the high-quality primary producer tissue and still gain the nutrients needed for metabolism and growth. As a result of the tradeoff between increased herbivore abundances and decreased per capita rates of herbivory, the overall impact of invertebrate herbivores on primary producers was not altered with nutrient enrichment (La Pierre and Smith, unpublished data). A similar decrease in per capita feeding rates of herbivorous zooplankton with increased food quality has also been observed in freshwater systems (Hillebrand et al., 2009). These decreases in herbivore per capita feeding rates with increased food quality may be why the theorized destabilization of primary producer–herbivore interactions is rarely observed.

In contrast to patterns of increased primary producer quality due to shifts in stoichiometry, many studies have also observed decreases in primary producer quality for herbivores due to increased allocation to defense with nutrient enrichment (see Chapter 8). Increases in primary producer defense with altered nutrient availability may result in decreases in herbivore success and rates of herbivory (Fig. 14.2a; Throop and Lerdau, 2004). In sum, alterations in primary producer quality with changes in nutrient availability may strengthen, weaken, or have no net effect on the interactions between primary producers and herbivores, depending on shifts in primary producer stoichiometry and defense with nutrient alterations. Further research comparing these responses across systems may help elucidate the factors that underlie these varied response types.

Changes in nutrient availability can also impact the community composition of primary producers, with subsequent consequences for consumer communities (Fig. 14.2a). Shifts in nutrient availability play a large role in altering competitive interactions among primary producers in both aquatic and terrestrial systems; recent meta-analyses have shown reductions in species evenness with nutrient enrichment across all systems, and strong declines in species richness in marine and terrestrial systems, but increased richness in freshwater systems (Hillebrand et al., 2007; Johnston and Roberts, 2009; Bobbink et al., 2010). Effects of nutrient enrichment do not apply to all primary producer species equally, and in most systems, species loss due to shifts in nutrient availability primarily affect certain functional groups, with species that are adapted to infertile conditions being more likely to be lost with high levels of anthropogenic nutrient enrichment (Chapin, 2003; Suding et al., 2005). Such shifts in community composition may result in reductions in resource availability for some specialist consumers, thereby resulting in decreased success and abundance of those specialist consumers (Throop and Lerdau, 2004; Haddad et al., 2009). Overall, decreased primary producer diversity may act to decrease food web stability, as has been evidenced in both freshwater and terrestrial systems (Haddad et al., 2011; Hooper et al., 2012). However, the interaction of bottom-up and top-down forces may help to counteract this destabilization: interactive effects of herbivores with nutrient additions have been shown to prevent competitive exclusion of slow-growing producer species through selective consumption of fast-growing species or reduced primary producer densities due to herbivory, thereby actually reducing primary producer species loss with nutrient additions (Worm et al., 2002; Hillebrand et al., 2007; Borer et al., 2014).

Although altered nutrient availability primarily affects higher trophic levels indirectly through changes at the level of primary producers, this global change factor can also have direct effects on higher trophic levels due to pollutant toxicity and acidification. For example, several meta-analyses have shown that acidification in marine systems due to increased atmospheric CO_2 can directly decrease the growth, survival, and consumption rates of higher trophic levels (Dupont et al., 2010; Hendriks et al., 2010; Kroeker et al., 2010). Similar effects have been observed in freshwater systems as well, with consequences for nutrient cycling and primary production (Petrin et al., 2008). Much of the research on this topic has focused on aquatic invertebrates. However, some evidence for direct effects of increased atmospheric CO_2 and nutrient enrichment on belowground consumers in terrestrial systems has also recently been documented (Blankinship et al., 2011).

Mutualisms, particularly nutritional symbioses, can also be affected by anthropogenically derived alterations in nutrient availability (Fig. 14.2a). Global change factors can impact fungal and bacterial mutualists either directly or indirectly through modifications of their symbiotic relationships (Wardle et al., 2004; Johnson et al., 2013). In terrestrial systems, N and P enrichment may increase turnover rates of fungal and rhizobial symbionts, resulting in decreased time associated with their plant hosts and potentially even decreased evolutionary selection for symbiosis (Treseder, 2004; Akçay and Simms, 2011; Johnson et al., 2013). In contrast, increased atmospheric CO_2 may result in an increase in the density of nutritional symbionts and the evolution of more cooperative symbionts (Treseder, 2004; Akçay and Simms, 2011). In marine systems, the symbiosis between endozoic algae and coral is also predicted to decline with increased N and P availability, further magnifying the direct harmful effects of ocean acidification on corals (Hallock, 2001). These altered symbiotic relationships may have large consequences for consumer community structure, primary production, and nutrient cycling (Avolio et al., 2014; Stachowicz, 2001).

Overall, alterations in atmospheric CO_2 concentrations and the deposition and runoff of N and P into aquatic and terrestrial systems have complex consequences for trophic interactions. The direct and indirect effects of altered nutrient availability on multiple trophic levels have important implications for ecosystem functions, such as primary production and nutrient cycling (see Chapter 9), as well as for diversity (see Chapter 12). However, we are still a long way from a complete understanding of the consequences of these C:nutrient alterations for ecosystems. For example, much of our knowledge about the effects of nutrient enrichment on community structure and ecosystem function has been derived from nutrient addition experiments, which are often short in duration and add nutrients at much higher levels than those expected under projected deposition scenarios (Cleland and Harpole, 2010; Knapp et al., 2012). Additionally, the effects of nutrient enrichment are known to be quite variable, depending on a variety of ecosystem characteristics. Responses to nutrient enrichment have been shown to vary geographically, with sensitivity to enrichment depending on both environmental conditions and ecosystem productivity (Matson et al., 2002; Clark et al., 2007), and temporally, with the potential for lagged responses and long-lasting effects even after nutrient inputs have ended (Milchunas and Lauenroth, 1995; Clark et al., 2009; Isbell et al., 2013). Finally, responses to nutrient enrichment can vary within a system due to species trait differences, with enrichment differentially affecting species based on functional type, photosynthetic pathway, and species identity (Xia and Wan, 2008; Lindroth, 2010; Sardans et al., 2011; La Pierre and Smith, 2014). Because of the wide variation in these spatial, temporal, and biological characteristics across ecosystem types, the predictability of system responses to nutrient enrichment remains relatively low (Murphy and Romanuk, 2012). Future research across scales and systems will help elucidate the mechanisms and patterns of altered nutrient availability on trophic interactions.

Climate change

Humans are currently having a large impact on the earth's climate through altered precipitation and temperature patterns. The planet is warming faster than at any time in the past, resulting in shifts in seasonality in both aquatic and terrestrial systems, altered oceanic salinity through glacial melting, and altered air and water currents leading to shifts in upwelling (IPCC, 2013). Additionally, total annual and seasonal rainfall, the intervals between rainfall events, and the intensity of rainfall events are all being impacted around the globe, although the direction and magnitude of these alterations vary spatially and temporally. These changes have resulted in alterations in long-term climatic means, as well as increased frequencies of extreme climatic events (IPCC, 2013). Although ecological study of ecosystem responses to climate change is relatively recent, clear effects of climate change on bottom-up and top-down interactions have been observed in many ecosystems (Walther et al., 2002; van der Putten et al., 2010).

Climate change has been documented to have a large impact on the traits of primary producers, and subsequent indirect effects on herbivores and secondary consumers. As with nutrient enrichment, shifts in temperature and rainfall have been shown to affect herbivores in aquatic and terrestrial systems via altered resource quantity (through changes in primary production; Harrington et al., 1999; Jamieson et al., 2012) and quality (through changes in tissue stoichiometry and defense expression in primary producers; Sardans et al., 2011; Jamieson et al., 2012).

Climate change may also directly impact consumers by altering consumer behavior and life history, resulting in shifts in trophic interactions (Fig. 14.2b). For example, shifts in the canopy position of arthropod consumers have been observed with warming in an old-field system (Barton, 2010). Similarly, shifts in fish behavior, such as altered affinities for littoral vs. benthic zones, have been observed with long-term warming in aquatic systems, particularly in shallow lakes and streams (Jeppesen et al., 2010). In both of these examples, the changes in consumer behavior had cascading consequences for ecosystem function, including shifts in primary production and nutrient cycling.

In addition to changes in behavior, altered temperatures have affected life history traits of many species across aquatic and terrestrial systems, including primary producers, and invertebrate and vertebrate consumers (Winkler et al., 2002; Root et al., 2003; Blenckner et al., 2007; Cleland et al., 2007; Musolin, 2007). Shifts in consumer life history with climate warming may or may not link with shifts in primary producer phenology. Mismatches between changes in the timing of the availability of specific primary producer tissues or species as food for herbivores can have large consequences for food web structure and stability (Throop and Lerdau, 2004; Cleland et al., 2007; HilleRisLambers et al., 2013). For example, earlier stratification of Lake Washington due to long-term climate warming has advanced the spring phytoplankton bloom by about 19 days; some zooplankton species were able to mirror this temporal shift, while others lacked the ability to shift their timing of emergence early enough and therefore declined in abundance (Winder and Schindler, 2004). Similar disruptions in primary producer–herbivore synchrony with warming have been observed in a variety of aquatic and terrestrial systems (Harrington et al., 1999; van Asch and Visser, 2007; Thackeray et al., 2010), although some species have been shown to be able to temporally track the changing climate (Cleland et al., 2012).

Climate change may also be predicted to affect trophic interactions by altering species abundances (Walther et al., 2002; van der Putten et al., 2004; 2010;

Blois et al., 2013). These changes can occur directly through altered environmental selection pressures on existing populations or communities. For example, genetic diversity in forest tree species may be affected by climate change (Theurillat and Guisan, 2001), which has been shown to have important implications for resistance to herbivory and disease (Ayres and Lombardero, 2000). Similar genetic shifts due to climate change have also been documented in primary producers and consumers from a diverse range of ecosystems, with some studies showing decreased genetic diversity, and others showing increased genetic diversity (Bradshaw and Holzapfel, 2006; Hoffmann and Sgro, 2011; Avolio et al., 2013). Additionally, shifts in the species composition of primary producers and consumers in response to climatic alterations have been observed in aquatic and terrestrial systems, which can have important consequences for trophic interactions (Walther et al., 2002).

In addition to changes at the population or community level within a system, many species are also expected to undergo range shifts in response to a changing climate. For example, the ranges of marine taxa across several trophic levels and taxonomic groups have been shown to closely track local climatic shifts (Pinsky et al., 2013), and similar patterns have been observed in other system types as well (Walther et al., 2002). These range shifts can have a variety of effects on trophic interactions, depending on the type and strength of the interaction and the ability of each species involved to shift their ranges. For example, a species may increase in abundance during a range shift as it outruns its competitors or consumers, due to differences in rates of migration among species (Harrington et al., 1999; van der Putten et al., 2010). In contrast, a species may encounter novel biotic interactions as it shifts its range (HilleRisLambers et al., 2013), which may have varying effects on species abundances, depending on the strength and direction of the interactions (e.g., consumption vs. facilitation).

Although it is currently clear that climate change will alter biotic interactions, the direction and magnitude of these changes are likely to vary across space and time (van der Putten et al., 2004; 2010; Blois et al., 2013). One key factor determining ecosystem responses to climate change is the system's existing climate. For example, drought- or warming-induced soil moisture losses in warm, arid terrestrial systems (e.g., sagebrush steppe) have been shown to decrease plant tolerance of herbivory, due to lower resource availability, thereby decreasing nutrient cycling (Sardans et al., 2011; Jamieson et al., 2012). In contrast, in cold, wet terrestrial systems (e.g., arctic tundra), increased temperatures have been shown to increase plant tolerance to herbivory, due to an extended growing season and increased photosynthetic rates, subsequently increasing nutrient cycling (Sardans et al., 2011; Jamieson et al., 2012). Similarly, in aquatic systems, the effects of climate change on trophic interactions may depend on current climate patterns, with warming-induced factors such as ice-off being more important in cold, wet regions, and drought-induced factors, such as reduced stream flow, being more important in arid regions. Overall, complex interactions among organisms and non-linear responses to climate change are expected in ecosystems worldwide (Blois et al., 2013), underscoring the importance of examining trends across aquatic and terrestrial ecosystems and considering bottom-up and top-down interactions in future studies of climate change.

Top predator loss

Humans have changed many food webs permanently, primarily through the loss of top predators (Strong and Frank, 2010). These human impacts on top predators are not a new phenomenon. Early humans are theorized to have caused many late Pleistocene extinctions due to overhunting, and certainly affected food webs through habitat modification, such as burning (Lyons et al., 2004). Current losses of top predators are similarly due to habitat modification, human exploitation of top trophic levels for food, and the inability of humans to coexist with top predators (Ripple et al., 2014). Such human-derived truncation of the food chain is pervasive both on land and in water. Industrialized fisheries have dramatically reduced marine fish populations, especially large predators (Myers and Worm, 2003; FAO, 2012; also, see Chapter 2). Similarly, subsistence and trophy hunting on land and in freshwater are resulting in drastically reduced populations, and even local extinctions, of many top predators (Milner-Gulland and Bennett, 2003; Packer et al., 2011). Additionally, as large-bodied animals often need large areas of contiguous habitat in which to survive, habitat fragmentation is having a disproportionately large impact on top predators (Dobson et al., 2006; Layman et al., 2007). The effects of these human-induced losses of top predators have large and far-reaching consequences for trophic interactions (Larson and Paine, 2007; Travis et al., 2013; Ripple et al., 2014); however, understanding and predicting these consequences has proven to be difficult.

Changes in the abundance of top predators can have cascading effects, strongly impacting ecosystem structure (Fig. 14.1; Larson and Paine, 2007; Estes et al., 2011). Starting at the top of the food chain, loss of apex predators has often been shown to result in increased mesopredator abundances (Strong and Frank, 2010). For example, declines in wolf and lynx populations due to hunting have resulted in increased fox abundances in North America (Elmhagen and Rushton, 2007), and shark and piscivorous fish declines have led to mesopredator increases in the open ocean (Baum and Worm, 2009). These increases in mesopredator abundances have sometimes been shown to result in unexpectedly large decreases in the abundances of their prey (Strong and Frank, 2010).

Herbivores are commonly released from top-down control with the loss of top predators, often with cascading consequences for primary producers (Schmitz et al., 2000; Strong and Frank, 2010; e.g., see Chapters 2–5). A well-studied example of such a trophic cascade occurred with the collapse of sea otter populations, due to hunting for pelts and diet shifts of predatory killer whales (Estes, 1998).

Due to the decline in sea otter abundances, sea urchin (the otter's primary food source) populations increased, resulting in a subsequent decline in kelp abundance (due to increased consumption by the urchins) (Estes, 1998). Similarly, on land, the reintroduction of gray wolves into Yellowstone National Park demonstrates the existence of trophic cascades due to top predator loss; following wolf reintroduction, growth of browse-plant species increased, resulting from both direct predation by wolves on elk and reduced time spent browsing by elk due to the threat of predation (Ripple and Beschta, 2012).

Trophic cascades resulting from the loss of a top predator can lead to the loss of species diversity at lower trophic levels, decreased ecosystem functioning, or even a complete restructuring of the system (Daskalov et al., 2007; Strong and Frank, 2010; Estes et al., 2011; Ripple et al., 2014). For example, the loss of top predators can alter fire regimes due to the build-up of flammable plant litter in terrestrial systems, with potential consequences for carbon sequestration and woody plant abundance (Estes et al., 2011; Ripple et al., 2014; also, see Chapter 4). The loss of top predators has also recently been implicated in increased human exposure to zoonotic diseases, due to the release of carrier populations, primarily rodents, from top-down control (Ostfeld and Holt, 2004; Ripple et al., 2014). In general, ecosystem services provided by top predators are rapidly being lost, and the cascading consequences of top predator loss are resulting in far-reaching changes in ecosystem function.

Historically, there has been much debate regarding the strength of trophic cascades in terrestrial vs. aquatic systems. A large number of independent studies have found strong evidence for trophic cascades in both aquatic and terrestrial habitats (Pace et al., 1999), and even some that cross the boundaries between these ecosystem types (Knight et al., 2005). However, meta-analyses have provided some evidence that terrestrial trophic cascades due to the loss of top predators are weaker than those in aquatic systems (Halaj and Wise, 2001; Shurin et al., 2002). Important differences between aquatic and terrestrial food webs, such as the abundance of detritivores/detritus, the presence/absence of structural tissue in primary producers, the nutritional quality of primary producers, the taxonomy and metabolism of consumers, and the amounts of allochthonous subsidies, may underlie the differences in trophic cascade strength among these systems (Borer et al., 2005; Shurin et al., 2006; Leroux and Loreau, 2008; also, see Chapters 3 and 6).

In addition to differences across aquatic and terrestrial systems, the strength of trophic cascades resulting from top predator losses can vary in space and time within ecosystems. Ecosystem responses to the loss of top predators may be lagged in time and, once they occur, may be long-lasting and/or irreversible (Frank et al., 2011). Thus, the effects of human impacts on top predators may be difficult to detect until the damage has already been done. Additionally, changes due to apex predator loss may not be localized in space. For example, declines of large predators in marine coastal regions have been shown to have consequences throughout the global oceans (Myers and Worm, 2003). Because humans and top predators seem unable to coexist in many regions of the world, either due to exploitation for food or sport, or due to human–wildlife conflict, protected areas are critical for maintaining top predator populations (Strong and Frank, 2010). Thus, continued research examining the effects of top predator losses and the role that protected areas may play in their continued contribution to ecosystem structure and function is needed.

Biodiversity loss

Biodiversity loss is one of the biggest challenges facing the planet, with species rapidly going extinct faster than at any time in previous history (Baillie et al., 2004). Causes of species loss from ecosystems are numerous and varied, ranging from alterations in nutrient availability, climate change, overhunting/fishing, habitat modification, and species invasions (Baillie et al., 2004). In 2002, the Convention on Biological Diversity resulted in a commitment by world leaders to reduce the current rate of biodiversity loss by 2010. However, to date, the rate of loss has not decreased and the anthropogenic pressures known to drive species loss are increasing; this despite increasing policy and management responses to mitigate biodiversity loss, including the expansion of protected and sustainably managed areas and invasive species regulation (Butchart et al., 2010).

The sustained loss of biodiversity despite political and conservation efforts may be due, in part, to time-delayed extinctions, known as the extinction debt (Tilman et al., 1994). This phenomenon is often linked to habitat destruction and fragmentation; species may persist for generations after habitat loss, but the ultimate result is deterministic extinction. The importance of the extinction debt varies with generation time and trophic level: for example, a study of fragmented grasslands in Europe indicated that for long-lived plant species, past landscape characteristics better explained current species richness than current landscape characteristics – evidence for an extinction debt – but for short-lived butterflies, there was no evidence of an extinction debt (Krauss et al., 2010). The future loss of primary producer species as a result of the extinction debt may have cascading effects on arthropod communities, including herbivorous and predatory/parasitoid arthropod species (Haddad et al., 2009). Further, the extinction debt is likely to have a greater impact on bottom-up processes in terrestrial ecosystems, which are dominated by producer species with longer generation times, than in aquatic ecosystems, which are dominated by producer species with shorter generation times, though this merits further investigation.

The global decline in biodiversity has prompted hundreds of studies on biodiversity and ecosystem functioning (BEF) and biodiversity and ecosystem services (BES; Cardinale et al., 2012). While much is now known about the potential effects of biodiversity loss, the magnitude of these effects compared to those of other forms of environmental change remains a critical question. In a recent meta-analysis, Hooper et al. (2012) concluded that the effects of species loss on productivity and decomposition rival those of other global stressors like climate change, acidification, and nutrient pollution, and thus biodiversity loss is "a major driver of ecosystem change." However, future effects of biodiversity loss on ecosystem function depend, in part, on whether extinctions continue at their current rate.

Although dominant species are often thought to account for the majority of ecosystem function (Walker et al., 1999; Loreau et al., 2001), the loss of rare, functionally unique species as a result of declining biodiversity also has important implications for trophic interactions. For example, in a study looking at species loss in a rocky intertidal community, Bracken and Low (2012) found that realistic declines in rare sessile species at the base of the food web (i.e., < 10% of sessile biomass) resulted in a 42–47% decline in consumer biomass, while the removal of equivalent biomass of dominant sessile species had no effect on consumers. Because the abundance of weakly interacting species is positively related to stability, with unique species tending to be weak interactors in marine systems (O'Gorman et al., 2011), the loss of rare species may also lead to decreased stability and reduced resistance/resilience of a community in the long-term. Mouillot et al. (2013) looked at the functional importance of rare species in a variety of systems (i.e., coral reef fishes, alpine plants, and tropical trees) and concluded that even in highly diverse ecosystems, 32–63% of species likely to perform highly vulnerable functions (i.e., exhibit low functional redundancy) were locally rare, and 89–98% were regionally rare. Consequently, even highly diverse ecosystems may not be buffered against declining biodiversity.

Given the generally positive relationship between diversity and stability (Walker, 1992; Ives and Carpenter, 2007), changes in biodiversity resulting from human activities may also impact the stability of food webs. A classic example of the effects of biodiversity on the stability of primary production, an important ecosystem function, comes from a long-term grassland biodiversity manipulation experiment in which inter-annual variation in primary production was less stable through a drought period in plots with lower plant species richness than in plots with higher plant species richness (Tilman, 1996) – an effect which may have been driven by asynchrony in species abundances through time (Hautier et al., 2014). Similarly, in a mesocosm experiment manipulating zooplankton species richness and the presence/absence of immigration in a metacommunity, Downing et al. (2014) found that the communities with immigration were most diverse and most stable under varying environmental conditions. The positive diversity–stability relationship in this study highlights the importance of metacommunity dynamics, but habitat fragmentation makes the stabilizing effects of metacommunities less likely in the face of changing biodiversity (Holt and Loreau, 2001).

Incorporating multiple trophic levels into BEF relationships makes them more complex and non-linear, compared to monotonic changes predicted within trophic levels (Thébault and Loreau, 2005; Duffy et al., 2007). Extinction bias – namely, the greater vulnerability of consumer species (particularly large vertebrates) to anthropogenic activities than producer species (Duffy, 2003) – may shift the relative importance of bottom-up and top-down forces across ecosystems. This "trophic skew" (Duffy, 2003) in the potential effects of biodiversity loss reflects the fact that higher trophic levels are often less diverse and have lower functional redundancy and higher interaction strengths than lower trophic levels; consequently, the loss of a small number of predator species can have comparable effects to a large reduction in primary producer diversity. Functional differences between producers and consumers result in distinct effects on productivity and stability (Thébault and Loreau, 2005), and greater top-down control of aquatic than terrestrial systems (Shurin et al., 2002) may lead to very different effects of biodiversity loss across systems. As a result, it is necessary to consider the trophic position of the species lost (or gained) and the ecosystem type (aquatic or terrestrial) in predicting the effects of changing biodiversity.

To incorporate trophic complexity into predictions of biodiversity loss, it is possible to compare the effects of horizontal diversity (i.e., within trophic levels) and vertical diversity (i.e., across trophic levels) (Fig. 14.3; Duffy et al., 2007). Horizontal and vertical diversity are both measures of functional complexity, with vertical diversity incorporating characteristics of a food web, such as food chain length and degree of omnivory, in contrast to horizontal diversity, which incorporates functional complexity within a trophic level (Duffy et al., 2007). While comparisons of horizontal and vertical diversity often focus on species diversity (see Chapter 12), these measures can also be applied to different levels of diversity (e.g., genetic, functional, or phylogenetic). In a study manipulating the horizontal and vertical diversity of an aquatic insect community within bromeliad plants, Srivastava and Bell (2009) found differing effects of horizontal and vertical diversity on community and ecosystem processes. The loss of horizontal diversity within the detritivore trophic level and the loss of vertical diversity with the elimination of the top trophic level (damselfly larvae) both negatively impacted ciliate communities, resulting in reduced ciliate density and changes in ciliate composition (due in part to secondary extinctions). But each component of diversity also had unique effects on select community and ecosystem processes; for example, only vertical diversity (i.e., damselfly presence) affected detrital processing, while only horizontal diversity (i.e., detritivore diversity) affected rotifer abundance (Srivastava and Bell, 2009). In contrast, in a study manipulating predator and prey diversity of a benthic marine community under strong top-down control, Douglass et al. (2008) found that the effects of species loss on ecosystem function were reliably predicted by diversity within one trophic level (i.e., predator presence and richness), and the incorporation

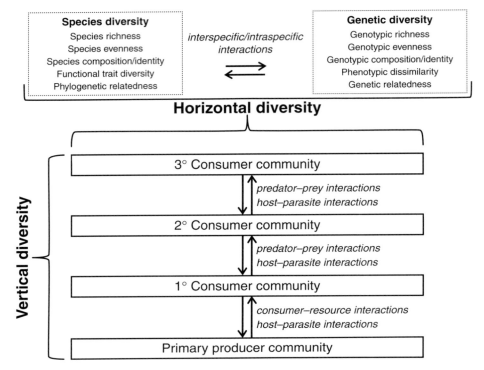

Figure 14.3 Understanding and predicting the effects of changing biodiversity – whether decreasing due to species loss or increasing due to species introductions – requires consideration of both horizontal (i.e., within trophic levels) and vertical (i.e., across trophic levels) diversity. Horizontal diversity includes both inter- and intra-specific diversity (i.e., species and genetic diversity). Richness and evenness are the most common measures of diversity, but composition (i.e., identity) may also play a critical role in horizontal diversity, particularly in the case of dominant or foundation species, or highly successful genotypes. Similarly, measures of trait diversity (e.g., functional trait diversity and phenotypic dissimilarity) have emerged as important components of horizontal diversity. In addition, estimates of relatedness (genetic or phylogenetic) may inform predictions of changing horizontal diversity. Consequently, changing biodiversity may take many forms at the level of horizontal diversity, some of which are more cryptic (e.g., genetic diversity and evolutionary potential). Vertical diversity includes multiple trophic levels, each of which includes these components of horizontal diversity. Changing vertical diversity often affects food chain length, but it may also impact functional complexity via changes in the degree of omnivory. Further, inter- and intra-specific interactions play a key role in determining the effects of species losses and gains (e.g., competitive interactions for horizontal diversity and consumer–resource, predator–prey, and host–parasite interactions for vertical diversity).

of vertical diversity did not increase their predictive power. Overall, predicting the effects of multi-trophic species loss may be difficult, particularly in systems in which bottom-up and top-down processes strongly interact, but incorporating measures of horizontal and vertical diversity into such predictions is one worthwhile strategy for improving our predictive power.

While diversity is often measured in terms of richness (i.e., species number), other measures, such as composition (i.e., species identity), evenness (i.e., relative species abundances), and species interactions (i.e., competition within or between species, mutualism, and trophic interactions such as herbivory, predation, and parasitism) may also impact ecosystem processes and thus merit consideration in predicting the effects of biodiversity loss (Chapin et al., 2000). For example, in a review of studies from across aquatic and terrestrial systems, Worm and Duffy (2003) concluded that both species richness and community composition played a role in many systems, though this does not seem to be a universal pattern (Chapin et al., 2000). Further, because relative species abundances are likely to respond to human-induced changes before richness (Chapin et al., 2000; Smith et al., 2009), species evenness and its relationship to community and ecosystem processes merit further investigation.

The majority of biodiversity research, especially initially, focused on metrics of species diversity, but biodiversity includes multiple levels spanning from genes to populations to species to ecosystems. For example, intra-specific diversity (i.e., genetic diversity) clearly impacts population and community dynamics, ecosystem function, and – perhaps most notably given rapid environmental change – recovery from disturbance (Hughes and Stachowicz, 2004; Hughes et al., 2008). Furthermore, the genotype and its associated phenotype form the fundamental basis for evolutionary change, so knowledge of genetic variation informs predictions of species' adaptation potentials (Schiffers et al., 2012; Pauls et al., 2013) and eco-evolutionary dynamics in the face of environmental change. In a study of montane insect species, Bálint et al. (2011) concluded that species' range contractions may result in significant losses of evolutionary lineages and genetic diversity within those lineages, termed "cryptic biodiversity loss." Such changes in genetic diversity may have cascading effects, altering genetic and/or species diversity at adjacent trophic levels and impacting community and ecosystem processes (Crutsinger et al., 2006). For example, a study manipulating both plant species diversity and genetic diversity within a dominant plant species in a freshwater sand dune found that species diversity and genetic diversity interacted to affect arthropod community structure (Crawford and Rudgers, 2013). Consequently, predicting the effects of biodiversity loss on trophic interactions may require consideration of multiple components of diversity, particularly in dominant and foundation species. This also requires further examination for consumer species (Hanley et al., unpublished data) – especially given the strength of top-down control in many systems – to aid in predictions of how biodiversity loss may affect trophic interactions and the ability of these important species to respond to rapidly changing environmental conditions. Similarly, functional and phylogenetic diversity, which highlight the importance of traits, evolutionary distance, and ecological interactions, may also inform predictions of the effects of biodiversity loss, particularly those involving multiple trophic levels (Srivastava et al., 2012). The reality of biodiversity loss highlights the importance of understanding the interaction of bottom-up and top-down processes, as predicting the impact of a species extinction or a genetic bottleneck on community and ecosystem processes requires a clear understanding of trophic interactions. Across aquatic and terrestrial ecosystems, higher trophic levels may experience disproportionately higher biodiversity losses due to over-exploitation or habitat modification (Dobson et al., 2006), making comparisons across systems particularly relevant and important.

Biological invasion

The global transport and introduction of species beyond historical and native ranges represents one of the most significant components of human-induced environmental change. The impacts of exotic species range from beneficial to detrimental (Simberloff et al., 2013), depending on the time scale being considered (Strayer et al., 2006) and the response variables of interest (e.g., ecological, economic, evolutionary, conservation, or human well-being) (Fig. 14.4; Vellend et al., 2007; Pejchar and Mooney, 2009; Pimentel, 2011; Schlaepfer et al., 2011).

In reality, there are many examples of exotic species that provide economic (e.g., food crops, livestock) and ecological (e.g., by acting as biological control agents or by supplying critical habitat and resources) benefits (Sax et al., 2007). From a conservation perspective, the potential benefits of exotic species include supplying critical resources or habitat for native species, acting as catalysts for recovery and restoration of native species, filling the empty ecological niche of a closely related, extinct species, and providing ecosystem services (Schlaepfer et al., 2011). The beneficial effects of exotic species may manifest at multiple trophic levels via direct and indirect species interactions. For example, in Cape Cod salt marshes, the exotic green crab Carcinus (Fig. 14.4a) has actually facilitated the recovery of cordgrass (a foundation species) by reducing herbivory by the native marsh crab Sesarma through consumptive and non-consumptive effects (Bertness and Coverdale, 2013).

The introduction of exotic species may also stimulate evolutionary diversification as a result of: (1) exotic species diversifying in a new region and/or in response to native species; (2) native species diversifying in response to exotic species; or (3) hybridization between and/or within native and exotic species (Vellend et al., 2007). Interestingly, this diversification may occur in the same trophic level as the exotic species or in adjacent trophic levels: for example, a recent speciation event in a tephritid fruit fly resulted from a host shift onto

Figure 14.4 Invasive species occur at all trophic levels (e.g., predators: a and d; herbivores: b and e; primary producers: c and f) within both aquatic (a–c) and terrestrial (d–f) systems. Invaders can have myriad effects on ecosystems, some positive and some decidedly negative. Shown here are (a) the European green crab, which can facilitate salt marsh restoration efforts by preying on herbivores that consume the foundation species cordgrass; (b) zebra mussels, whose explosive population growth costs billions of dollars in damage, such as the overgrowth of this water current meter; (c) curly-leafed pondweed, which has negative effects on water quality and hinders fish foraging ability due to its dense growth; (d) feral cats, whose presence often results in reduced diversity of native, endemic species, particularly on islands; (e) giant African land snails, whose invasion of Christmas Island has resulted in alteration of nutrient cycling and reduced plant diversity – an example of invasional meltdown; and (f) honeysuckle, whose invasion of North America has led to the hybrid speciation of a fruit fly. See text for more details about each of these invaders. (Photo credits: (a) D. Hazerli, University of Veterinary Medicine, Hannover; (b) M. McCormick, US NOAA; (c) C. Evans, Illinois Department of Natural Resources; (d) K. La Pierre, University of California, Berkeley; (e) S. Kennedy, University of California, Berkeley; (f) S. Kuebbing, University of Tennessee.)

exotic honeysuckle (Fig. 14.4f), and the new fruit fly species is actually a hybrid of two native species (Schwarz et al., 2005). These divergences create a changing evolutionary landscape that includes not only the abiotic environment, which is clearly being altered by human activities, but also the biotic environment (see Chapter 13). Direct and indirect evolutionary responses of native and exotic species may lead to (1) the generation of new taxa via hybridization and introgression (as in the case of the tephritid fruit fly) – which may ultimately increase biodiversity – and (2) consequent changes in behavioral, morphological, and life history traits, niche displacement, and the disruption or formation of mutualistic or parasitic relationships (Mooney and Cleland, 2001).

Despite the benefits of some species introductions, there are numerous examples of invasive species negatively impacting the recipient system, from the level of individual species to community processes to ecosystem functions (Vitousek et al., 1997a). In aquatic and terrestrial systems, exotic weed species often displace native primary producers, change the basic structure of the environment, and alter nutrient cycling. For example, the introduction of European cheatgrass to shrub/steppe habitat in western North America has dramatically altered fire frequencies, changing the disturbance regime and ultimately resulting in the exclusion of native shrub and perennial grass species (D'Antonio and Vitousek, 1992). The introduction of extremely productive, submerged aquatic plant species, like curly-leaf pondweed (Fig. 14.4c) and Brazilian waterweed, via boat traffic has had detrimental effects on water and habitat quality (Santos et al., 2011) and cascading effects across multiple trophic levels by impacting fish behavior and diet (Engel, 1987). And in freshwater habitats throughout the United States, zebra mussels (Fig. 14.4b), and more recently quagga mussels, have had significant ecological and economic costs associated with their invasions, negatively impacting multiple components of the aquatic food web (Caraco et al., 1997), reducing species diversity, altering primary productivity and nutrient cycling, and incurring huge economic costs in damage repairs and lost ecosystem services (Vitousek et al., 1997a; Pimentel, 2011). In total, billions of dollars are spent annually on control measures around the world (Pimentel, 2011).

Given the rate and extent of species introductions, high priority must be given to improving our understanding of the ecological effects of exotic species on community and ecosystem processes at local and regional scales. However, accurate predictions remain surprisingly challenging, despite extensive research and progress in this field (Hulme et al., 2013; Simberloff et al., 2013). This is in part because the emergence of context-dependency as the dominant "hypothesis" regarding the impact of exotic species has prevented the development of a broad framework with which to predict the consequences of invasion (Ricciardi et al., 2013). Nonetheless, Ricciardi et al.'s (2013) comprehensive review of the hypotheses predicting the impact of exotic species identified some important themes, including species traits, community structure, niche differences, spatial and temporal heterogeneity, propagule and colonization pressure, and synergistic effects. Many of these themes highlight the importance of considering the interaction of bottom-up and top-down processes in predicting the effects of exotic species. For example, while trophic position does not consistently predict the impact of exotic species, those species that create new links in the food web or initiate/disrupt a trophic cascade tend to have significant community-level effects in a wide variety of ecosystems (e.g., lakes: van der Zanden et al., 1999; estuaries: Kimbro et al., 2009; stream–forest boundary: Baxter et al., 2004; islands: Fukami et al., 2006). In addition, extensive research on exotic plant species in terrestrial ecosystems demonstrates cascading effects of exotic plants on arthropod abundance and richness (van Hengstum et al., 2014). Similarly, the enemy release hypothesis highlights the importance of trophic interactions in determining invasion success: decreased regulation of exotic primary producer species by novel consumers and natural enemies results in their increased distribution and abundance (Keane and Crawley, 2002). This release from top-down control may also occur at the consumer level (i.e., "uncontrolled consumers"; Ricciardi et al., 2013). Importantly, this escape from control may be particularly detrimental on islands, where uncontrolled consumers like rats and cats (Fig. 14.4d) encounter rare, endemic species that may not have experienced strong competition or predation (Courchamp et al., 2003); for example, 14% of global bird, mammal, and reptile extinctions are attributed to feral cats on islands (Medina et al., 2011).

Given the magnitude of human-facilitated species introductions, ecosystems often experience multiple invasions, which makes it necessary to consider not only native–exotic interactions but also exotic–exotic interactions. Simberloff and Von Holle (1999) posited the invasional meltdown hypothesis, which suggests that mutualistic interactions between exotic species facilitate future species introductions, thereby accelerating the rate of invasion. This idea remains controversial, in part because it is difficult to demonstrate that exotic–exotic interactions facilitate the entry and spread of exotic species. However, a recent study by Green et al. (2011) demonstrated that a mutualism between the exotic yellow crazy ant and honeydew-secreting scale insects resulted in ant supercolonies that killed native land crabs, thereby creating enemy-free space and facilitating a secondary invasion by the giant African land snail (Fig. 14.4e). In contrast, some exotic–exotic interactions may be antagonistic, for example in the case of an exotic herbivore limiting the growth and survival of an exotic primary producer (La Pierre et al., 2010).

In general, the relative importance of resistance and facilitation is a critical aspect of invasion success. In the case of biotic resistance to invasion, interesting differences emerge between producers and consumers, and the associated mechanism of resistance (i.e., competition or consumption) in marine and terrestrial

ecosystems. A meta-analysis of terrestrial plant invasion found that native competitors (i.e., producers) and herbivores (i.e., consumers) inhibited plant invasion
success (Levine et al., 2004). However, a meta-analysis of marine producer invasion found that native competitors (i.e., producers) generally did not inhibit primary producer invasion success – the exception being at high diversity, though
the magnitude of the diversity effect was weaker in marine systems compared to
terrestrial systems (Kimbro et al., 2013). In contrast, native consumers in marine
systems displayed a similar degree of resistance to invasive producers as native
consumers in terrestrial ecosystems (Kimbro et al., 2013). Because a number of
marine experiments tested invasive consumers, Kimbro et al. (2013) also looked
at whether resistance strength depended on invader trophic level; interestingly,
biotic resistance against invasive consumers was greater than that of invasive
producers. A comparison of these results to freshwater and terrestrial systems
would be interesting, especially given intrinsic differences between aquatic and
terrestrial ecosystems (e.g., Shurin et al., 2002).

In addition to biotic resistance, which generally produces a negative native–
exotic richness relationship, facilitation, which generally produces a positive
native–exotic richness relationship, also plays a key role in invasion success (Fridley et al., 2007). Often, patterns of facilitation vs. biotic resistance depend on
the scale at which the invasion is being studied. An example of the importance
of scale in determining facilitation vs. biotic resistance in invasion comes from a
New England intertidal system, where Altieri et al. (2010) identified a facilitation
cascade in which a native foundation species, cordgrass, and its associated facilitator, the ribbed mussel, promoted the invasion success of the Asian shore crab
by reducing thermal stress and substrate instability at a small spatial scale. Interestingly, these same small-scale interactions between the cordgrass and mussels
(i.e., ecosystem engineering) also benefit native diversity at a larger scale (Altieri
et al., 2010). This, and many other examples, highlight the fact that predicting
invasion success depends on interactions across multiple trophic levels and the
spatial scale being considered (see Chapter 11).

The ecological/evolutionary history of a system also likely impacts invasion
success and the resultant effects on trophic interactions. For example, freshwater systems exhibit greater sensitivity to predator introductions than marine
and mainland terrestrial systems (Cox and Lima, 2006). This is potentially due to
prey naïveté in freshwater systems; high heterogeneity in predation regimes at
local and regional scales increases the likelihood of a naïve prey encountering a
novel predator as a result of invasion. This is in direct contrast to prey in marine
and mainland terrestrial systems, which generally have evolutionary experience
with a diverse predator assemblage as a result of historical biotic interchange
(Cox and Lima, 2006). The relative sensitivity of aquatic and terrestrial systems
to exotic primary producers and consumers merits further investigation. Furthermore, linking the ecological/evolutionary history of a community with the

importance of resistance and facilitation in invasion success at different trophic levels across aquatic and terrestrial ecosystems may provide valuable insight into invasion dynamics.

Interaction of multiple drivers of global environmental change

Human-induced environmental change has clearly altered the nature and strength of bottom-up and top-down forces, thus impacting the structure and dynamics of trophic interactions across aquatic and terrestrial ecosystems. Given the extent of global environmental change and the many components of ecosystems impacted by anthropogenic activities, it is important to consider the interactions of multiple stressors, as it is unlikely that any of these processes are acting in isolation (Folt et al., 1999; Christensen et al., 2006; Bulling et al., 2010). The net effect of multiple stressors, whether natural, anthropogenic, or a combination of both, may be additive, synergistic (i.e., greater than expected), or antagonistic (i.e., less than expected) (Folt et al., 1999; Crain et al., 2008). Avoidance of "ecological surprises" (*sensu* Paine et al., 1998) due to non-additive effects of global change requires an understanding of higher order interactions among stressors and species, and the consideration of factors such as the frequency and intensity of stressors, ecological and evolutionary history of communities (Vinebrooke et al., 2004), and context-dependency (Crain et al., 2008). To address the question of how multiple stressors impact the interaction of bottom-up and top-down forces across marine, freshwater, and terrestrial ecosystems, we first examine whether stressor interactions most often result in additive, synergistic, and/or antagonistic effects in each type of system, and then discuss similarities and differences across ecosystems, offering potential explanations for these patterns.

In a meta-analysis of marine and coastal systems, Crain et al. (2008) found that in individual studies of ≥ 2 stressors, cumulative effects of multiple stressors included additive, synergistic, and antagonistic effects (26%, 36%, and 38%, respectively), but the overall interaction effect across studies was synergistic. Furthermore, as the number of stressors increases, interactions become more complex and likely synergistic (Bulling et al., 2010): in the Crain et al. (2008) study, the addition of a third stressor doubled the number of synergistic interactions. Interestingly, the interaction effect depended on the level of biological organization (population: synergistic vs. community: antagonistic) and the trophic position (autotroph: antagonistic vs. heterotroph: synergistic) of the response variable (Crain et al., 2008). For example, inorganic nutrient enrichment (N) and organic matter addition (OM) in the intertidal – both major concerns in coastal ecosystems – interacted to decrease species diversity and alter trophic structure (i.e., increased proportion of basal species and decreased proportion of top species; O'Gorman et al., 2012). A comparison of the combined treatment (N and OM) to the individual treatments (N or OM), which had minimal effects on community and ecosystem properties, highlights the complicated nature of this interaction, emphasizing that the range of community- and ecosystem-level effects may not be detectable when examining stressors independently. In addition, this study demonstrates the importance of considering the interaction of bottom-up and top-down forces, as bottom-up stressors had cascading effects on top species richness and identity, thereby altering food web structure. Similarly, a study looking at the effects of nutrient enrichment and functional diversity loss (including a key predator and primary consumers) in a marine benthic community found that without nutrient enrichment, changes in functional diversity produced a classic trophic cascade, but with nutrient enrichment, cascading effects of functional diversity loss were disrupted, resulting in unexpected shifts in algal abundance and assemblage structure (O'Connor and Donohue, 2012). This study has important implications for predicting the effects of biodiversity loss, especially in the presence of multiple stressors, given that the strength of direct and indirect interactions among consumers and producers in this system depended on environmental context (e.g., nutrient enrichment). In the open ocean, simultaneous modification of top-down (declining cod populations as a result of overfishing) and bottom-up (fluctuating plankton populations as a result of global warming) forces has significant effects on population and community dynamics in the long term as a result of a mismatch between the timing of larval cod occurrence and peak plankton biomass (Beaugrand et al., 2003). The match/mismatch hypothesis (Cushing, 1990), particularly as it applies to fish populations, emphasizes the importance of considering spatial and temporal scale (see Chapter 11) in predicting the cumulative effects of multiple stressors in aquatic systems.

The inherently fragmented nature of freshwater ecosystems combined with extensive human exploitation of this limited resource makes it particularly important to understand the effects of multiple stressors in these systems (Carpenter et al., 1992; Woodward et al., 2010). Because many freshwater systems are exposed to a variety of local anthropogenic stressors (Woodward et al., 2010), it is possible to think about the interaction not only of multiple stressors, but also of local vs. global stressors and how best to mitigate these effects. Notably, a similar approach is also possible for a number of marine and terrestrial ecosystems. Interestingly, in a study of the cumulative effects of a global stressor (i.e., climate change) and local stressors, Brown et al. (2013) found that if an interaction was synergistic, alleviation of local stressors was beneficial, but if an interaction was antagonistic, alleviation of local stressors was ineffective or harmful (Brown et al., 2013). This pattern may be particularly important to consider for lakes, which often experience a suite of local and global stressors simultaneously. For example, warming (a global stressor) may shift bottom-up and top-down control of a pond food web (Shurin et al., 2012) and may interact with local stressors, like nutrient availability and predation, to alter the strength of trophic cascades (Kratina et al., 2012), similar to the patterns observed in a marine benthic community (O'Connor and Donohue, 2012). The resultant shifts in bottom-up and top-down forcing due to warming extend to the microbial community, as a suite of indirect effects in pond food webs resulted in changes in the biomass of bacteria and viruses (Shurin et al., 2012; also, see Chapter 10). This has important implications for host–parasite dynamics in the context of global change. In a study looking at the independent and interactive effects of natural (e.g., parasitism and predation) and anthropogenic (e.g., pesticide exposure) stressors on the model ecological organism *Daphnia*, Coors and De Meester (2008) found a combination of additive and non-additive effects. The nature of the effects depended on the stressor combination and the response trait of interest (life history traits and population growth rates), but of particular note is the synergistic effect of pesticide exposure and parasite challenge on *Daphnia* survival (Coors and De Meester, 2008). Similarly, a study of the cumulative effects of warming, drought, and acidification in boreal lakes identified a number of short- and long-term ecological surprises (*sensu* Paine et al., 1998) due to non-additive effects of these multiple stressors (Christensen et al., 2006). This study also found that the combined effects of multiple stressors depends on the trophic level of interest, as this interaction had antagonistic effects on producers but synergistic effects on consumers (Christensen et al., 2006). Interestingly, similar results were observed in a meta-analysis of marine studies (Crain et al., 2008), indicating some commonality across ecosystems. In contrast, complex interactions among multiple stressors in streams, such as nutrient enrichment and sediment loading (Townsend et al., 2008), make it difficult to accurately assess risk and prioritize management actions.

The combined effects of multiple stressors may be particularly important at the boundary of aquatic and terrestrial ecosystems, with the potential for interactions to impact donor and recipient communities and to alter the carbon balance between adjacent systems (Greig et al., 2012). In a study looking at the effects of warming, eutrophication, and predation, Greig et al. (2012) found that top-down forces (i.e., predatory fish) decoupled aquatic and terrestrial ecosystems, whereas bottom-up forces (i.e., warming and eutrophication) increased cross-system exchange – both via effects on subsidies, such as aquatic insect emergence and terrestrial plant decomposition. This result highlights the importance of considering the cumulative effects of multiple stressors across ecosystem boundaries, particularly given the prevalence of cross-system subsidies and interest in predicting the magnitude and effects of these subsidies (Marczak et al., 2007).

Meta-analyses of the main drivers of global environmental change in terrestrial ecosystems have identified complex interaction effects of multiple stressors that make it difficult to predict future responses and develop management plans (Tylianakis et al., 2008; Lee et al., 2010), similar to the results of the meta-analysis of marine ecosystems (Crain et al., 2008), indicating that this challenge is ubiquitous across systems. Tylianakis et al. (2008) focused on the effects of global environmental change on biotic interactions like herbivory, predation, parasitism, pollination, and mutualism, because these may be particularly susceptible to the effects of anthropogenic stressors. In general, global environmental change disrupts plant mutualisms via phenological shifts due to changing climate, habitat modification and fragmentation reducing pollination, and competition with invasive plants for pollinators. However, the effects on plant–fungal mutualisms vary depending on environmental context and stressor identities (Tylianakis et al., 2008). Similarly, the effects of global change drivers on the soil food web are highly context-dependent; the relative importance of bottom-up and top-down forces depends on the component of interest in the soil food web (see Chapter 10), which results in complex, cascading effects of multiple stressors (Tylianakis et al., 2008). In addition, habitat modification, including loss and fragmentation, is a key driver threatening terrestrial ecosystems. Mantyka-Pringle et al. (2011) looked at the interactive effects of climate change and habitat loss on biodiversity to better inform conservation strategies. Both climate change and current climate determine the effects of habitat loss on species diversity, with current maximum temperature (negative effects of habitat modification were greatest in areas with warmer temperatures) and mean precipitation change (negative effects of habitat modification were least in areas with increasing precipitation) being the most important determinants. Similarly, habitat modification and invasion interact synergistically to produce sometimes unexpected effects on indigenous species and, ultimately, trophic dynamics. For example, in a grassland/shrubland ecosystem subjected to habitat modification, Norbury et al. (2013) introduced a variety of invasive mammals, including top predators, herbivores, and insectivores, and measured their interactive effects on indigenous and non-indigenous fauna. The results identified interesting interactions of bottom-up (habitat modification) and top-down (introduction of invasive species) processes that produced a mix of expected and unexpected patterns (Norbury et al., 2013), indicating that these stressors can operate both additively and non-additively.

The interaction of multiple stressors may have additive, synergistic, or antagonistic effects in aquatic and terrestrial ecosystems. A meta-analysis of marine, freshwater, and terrestrial ecosystems (Darling and Côté, 2008) found results similar to those looking at only one ecosystem, with the majority of multiple stressor interactions being non-additive, indicating the pervasive complexity of these interactions. It is predicted that habitat modification and climate change will have the greatest impact on biodiversity (Sala et al., 2000), which makes it particularly important to understand the interactive effects of these drivers on not just species diversity, but also species interactions (Tylianakis et al., 2008). In an effort to prevent ecological surprises (Lindenmayer et al., 2010), it is necessary to continue short- and long-term studies manipulating multiple stressors (ideally ≥ 3 given the co-occurrence of drivers of global environmental change) and measuring the cumulative effects on biodiversity and species interactions. In addition, there is a pressing need for further cross-system comparison and consideration of ecosystem boundaries in experimental, theoretical, and meta-analytic studies of the combined effects of multiple drivers of environmental change. Lastly, placing these results in a trophic context (e.g., comparing the nature of these effects on producers and consumers) may reveal important patterns in synergistic and antagonistic interactions (e.g., Christensen et al., 2006; Crain et al., 2008).

Current challenges

In a relatively short time, much progress has been made in understanding how global change is impacting bottom-up and top-down factors in a wide variety of systems. While much of the previous work on the effects of global change has focused on direct effects on individuals or populations of species (e.g., climate envelope modeling), current research is incorporating species interactions within and across trophic levels into predictions of the effects of global change factors (e.g., van der Putten et al., 2010; HilleRisLambers et al., 2013; Blois et al., 2013). As outlined in the sections above, as well as throughout this book, global change factors are affecting trophic interactions through changes in individual morphology, behavior, and life history, as well as species ranges and abundance patterns (Fig. 14.1). However, many gaps still remain in our knowledge of how global change factors will affect bottom-up and top-down interactions, and ultimately ecosystem function. Four key concepts of critical importance in furthering our understanding and prediction of the impacts of global change factors are summarized below.

The importance of comparing across systems

As demonstrated in this book, as well as in numerous reviews and meta-analyses, the effects of global change on trophic interactions may depend upon the system within which they occur. In particular, a distinction has often been drawn between the nature of trophic interactions in aquatic and terrestrial systems (e.g., Borer et al., 2005; Shurin et al., 2006). In a similar vein, comparisons between tropical and temperate systems have been extensively theorized and tested (e.g., Menge and Lubchenco, 1981; Dyer and Coley, 2001). These cross-system comparisons are often formulated on the idea that primary producer and consumer metabolic factors, which form the basis of trophic interactions, may differ among systems (Borer et al., 2005). More recently, comparisons between above- and belowground food web responses to global change factors – and importantly the connections between these distinct but fundamentally linked systems – have also begun to receive more attention (Moore et al., 2004; Wardle et al., 2004). Additionally, the species and taxonomic richness of a system, which notably vary among aquatic vs. terrestrial, temperate vs. tropical, green vs. brown, and above- vs. belowground food webs, may have a large impact on trophic responses to global change factors (Borer et al., 2005; Duffy et al., 2007).

The effects of global change factors on community and ecosystem structure and function are often complex and interactive. Historically, much of the work to understand these effects in a variety of ecosystem types can be compared to a reinvention of the wheel, with scientists rediscovering similar patterns in aquatic and terrestrial systems independently, rather than using the knowledge previously gained from one system to inform research in another. Future research examining the commonalities and differences across diverse ecosystem types will facilitate the identification of mechanisms underlying observed patterns of global change effects. Conducting this work collaboratively across a variety of system types and boundaries will improve the efficiency of scientific study and is imperative for predicting ecosystem responses to global change.

The importance of spatial, temporal, and biological scales

To truly understand the effects of global environmental change and to best develop and implement management plans for conservation and restoration, it is necessary to consider multiple scales, specifically spatial scales (local vs. regional vs. global), temporal scales (short- vs. long-term), and biological scales (from genes to individuals to populations to communities to ecosystems; also, see Chapters 11 and 12). Thinking across spatial scales and comparing the importance of local, regional, and global processes is much like considering the interaction of bottom-up and top-down forces: global drivers clearly impact local dynamics (i.e., top-down), but local processes also feedback to influence global processes (i.e., bottom-up; Wilbanks and Kates, 1999). These cross-scale interactions apply to spatial, temporal, and biological scales, with fine-scale processes influencing a broad spatial range, having long-lasting effects, and cascading from the level of the individual to the community to the ecosystem (Peters et al., 2007). Transfer processes, such as fire, water, wind, and plant and animal dispersal (see Chapters 4, 6, and 7 for discussions of these processes in diverse aquatic and terrestrial systems), play a critical role in mediating the link between fine-scale and broad-scale patterns and processes (Peters et al., 2007). In addition, spatial heterogeneity likely determines the relative importance of fine-scale processes in determining local and regional vulnerability to global environmental change (Diffenbaugh et al., 2005; Peters et al., 2007). For example, the predicted increase in frequency and magnitude of extreme temperature and precipitation events as a result of global climate change is likely to have differing effects at the local and regional scales, such as heightened effects of extreme precipitation on the lee side of a rain shadow or in coastal areas driven by convective precipitation (Diffenbaugh et al., 2005). Furthermore, invasion and extinction often initially occur on a local scale, but there is potential for their effects to cascade regionally. Leroux et al. (2013) proposed a mechanistic model to predict species range shifts as a result of climate change and habitat loss – a critical step in linking current and future community dynamics, and local and regional spatial interactions (Harrington et al., 1999). The integration of experimental studies and theoretical models in considering the effects of global change across scales and across systems may be particularly beneficial (see Chapter 1).

The mechanisms that determine individual fitness and response to selection also merit consideration in predictions of global change effects because the responses of individuals to environmental change can scale to the levels of populations, communities, and ecosystems (Gilman et al., 2010). This necessitates the examination not only of genetic diversity, but also of the potential role of rapid or contemporary evolution in determining individual responses to local and global environmental drivers (Parmesan, 2006; Ellner, 2013). While genetic shifts and local adaptation often modulate environmental change at the local scale (Parmesan, 2006), it is unclear whether evolutionary rescue is possible at the regional scale (or beyond) given potentially weak or conflicting selection pressures. The occurrence of evolutionary change on ecological time scales necessitates consideration of the role of evolution, in addition to the role of species interactions, in determining species responses to anthropogenic activities and the resultant effects on food web interactions (Harmon et al., 2009; also, see Chapter 13). This applies not just at the level of the genotype, but also at the level of phenotype, including trait plasticity, and physiological (Hofmann and Todgham, 2010) and life history/phenological (Parmesan, 2006; Ovaskainen et al., 2013) responses. By considering multiple levels of biological organization to better understand the mechanisms behind species' responses to global environmental drivers, it may also be possible to better predict the effects of anthropogenic change across spatial and temporal scales.

Identifying ecological tipping points, applying threshold models, and predicting regime shifts

Given that anthropogenic activities have resulted in ecosystems being subjected to multiple drivers of global environmental change and that it is difficult to predict the often synergistic effects of several stressors, the application of threshold models to identify ecological tipping points (i.e., abrupt shift of a system from one state to a contrasting state in response to a comparatively small environmental change; Scheffer et al., 2001) and predict regime shifts may be particularly useful in conservation and restoration management efforts (Andersen et al., 2009; Suding and Hobbs, 2009). Although it is difficult to identify ecological tipping points, incorporation of observational, experimental, and theoretical data may make it possible to determine how close a given ecosystem is to this key threshold (Scheffer and Carpenter, 2003). For example, in polar ecosystems, the earlier timing of ice melt at high latitudes due to climate warming exponentially increases annual light, eventually exceeding a light threshold that constitutes a tipping point for this system, with effects cascading from producers to consumers, reducing biodiversity and dramatically altering ecosystem function (Clark et al., 2013). The rapid development of this field has resulted in the identification of early warning signals that a system may be approaching a tipping point, including (1) a loss of resilience (Scheffer et al., 2001; 2009); (2) a slowing of recovery rates from small perturbations (Scheffer et al., 2009; Lenton, 2011; Veraart et al., 2012); (3) increased variance (Brock and Carpenter, 2006; Scheffer et al., 2009; Lenton, 2011); and (4) changing skewness (Guttal and Jayaprakash, 2008). A whole-ecosystem experiment that involved adding top predators to a lake to destabilize the food web confirmed that many of these warning signals accurately predicted a regime shift more than 1 year before it happened (Carpenter et al., 2011). There are interesting similarities in these processes across ecosystems (e.g., Scheffer et al., 2001; Suding and Hobbs, 2009). Currently, an understanding of how regime shifts propagate across scales is lacking, making it difficult to predict whether a local regime shift is likely to extend regionally or even globally (Hughes et al., 2013). However, it is clear that the interaction of bottom-up and top-down processes plays a key role in these major state shifts (e.g., Carpenter et al., 2011; Clark et al., 2013). Therefore, it is critically important to consider trophic interactions across aquatic and terrestrial systems in attempts to identify ecological tipping points.

Linking community change with shifts in ecosystem function

Global change factors affect species composition and diversity, which may, in turn, have large impacts on community and ecosystem processes (Chapin, 1997; Duffy et al., 2007). A fundamental goal of BEF research has been to predict how changes in species diversity may affect ecosystem function (see Chapter 12), and identify the subsequent consequences of these effects for human well-being (Díaz et al., 2006; Cardinale et al., 2012). In general, a positive effect of diversity on ecosystem function has been identified in diverse systems, and has been shown to be relatively consistent across trophic levels and ecosystem types (Giller et al., 2004; Balvanera et al., 2006; Cardinale et al., 2006; Worm et al., 2006; Flynn et al., 2011), though exceptions do exist. But, although we have come a long way in determining the effects of biodiversity loss on ecosystem function, open questions still remain. The next challenge in this field is to better relate the wealth of information on how biodiversity loss may impact ecosystem function to realistic global change scenarios, particularly non-random species loss across different scales.

Much of the early BEF research examined the effects of random species loss; however, changes in community composition due to global change occur in a non-random fashion. Consequently, the effects of shifts in community structure on ecosystem function may depend on the extent to which the species that most influence ecosystem function are affected by global change factors (Walker, 1992). Importantly, the types of species that are most likely to be lost due to global change factors vary among trophic levels (Duffy, 2003). Specifically, at the level of primary producers, the majority of species that are lost due to global change factors are rare and their contributions to ecosystem function, as compared to dominant species, are less well understood (Duffy, 2003; Jain et al., 2014). In contrast, at higher trophic levels, the species that are lost due to global change are more likely to be strong interactors, with potentially large consequences for ecosystem function (Duffy, 2003).

The scale at which biodiversity influences ecosystem function also requires more explicit consideration. A recent meta-analysis (Vellend et al., 2013) examining global biodiversity loss found no evidence for diversity declines at local scales – the scale at which most BEF research is conducted – calling into question the relevance of many local-scale BEF studies to conservation at broader scales (Srivastava and Vellend, 2005). However, regional and global biodiversity have decreased (Vellend et al., 2013); thus, BEF studies comparing across multiple scales may greatly increase understanding of how species loss and community shifts impact ecosystem function in the context of global change. Overall, the explicit study of the importance of non-random species loss across multiple scales merits further investigation and may facilitate efforts aimed at the restoration of biodiversity to recover lost ecosystem function, with important implications for the reestablishment of degraded ecosystems (Worm et al., 2006).

Conclusions

Human-induced environmental change has already dramatically affected populations, communities, and ecosystems around the planet, and human impacts are expected to continue to increase at an alarming rate. These global changes are profoundly altering trophic interactions in all ecosystem types at multiple scales (Vitousek et al., 1997b; Rosa et al., 2004). It is no longer realistic to consider the effects of different global change factors (e.g., altered nutrient availability, top predator loss, etc.) independently, as they regularly co-occur within and across ecosystems (Folt et al., 1999; Christensen et al., 2006; Bulling et al., 2010). Therefore, it is imperative to consider how simultaneous modifications of bottom-up and top-down forces may affect community and ecosystem processes. As demonstrated throughout this book, much progress has been made toward this end in a variety of ecosystems and using diverse approaches; however, much remains to be done. Encouragingly, a vast amount of system-specific information about the effects of global change on bottom-up and top-down interactions already exists. Consequently, the current challenge is to synthesize this information across systems and scales. By taking a multi-pronged approach that incorporates theoretical, observational, and experimental studies to explicitly examine the interactions of bottom-up and top-down factors, as well as meta-analyses and reviews aimed at identifying important patterns and processes across systems, we can continue to increase our understanding of the effects of global change on community and ecosystem processes across aquatic and terrestrial systems.

Acknowledgments

We are grateful to C. Chang, A. Jones, and M. Avolio for their thoughtful comments on earlier drafts of this manuscript.

References

Akçay, E. and Simms, E. L. (2011). Negotiation, sanctions, and context dependency in the legume-Rhizobium mutualism. *The American Naturalist*, **178**, 1–14.

Altieri, A. H., van Wesenbeeck, B. K., Bertness, M. D. and Silliman, B. R. (2010). Facilitation cascade drives positive relationship between native biodiversity and invasion success. *Ecology*, **91**, 1269–1275.

Andersen, T., Carstensen, J., Hernández-García, E. and Duarte, C. M. (2009). Ecological thresholds and regime shifts: approaches to identification. *Trends in Ecology and Evolution*, **24**, 49–57.

Avolio, M. L., Beaulieu, J. M. and Smith, M. D. (2013). Genetic diversity of a dominant C4 grass is altered with increased precipitation variability. *Oecologia*, **171**, 571–81.

Avolio, M. L., Koerner, S. E., La Pierre, K. J., et al. (2014). Changes in plant community composition, not diversity, during a decade of nitrogen and phosphorus additions drive above-ground productivity in a tallgrass prairie. *Journal of Ecology*, **102**(6), 1649–1660.

Ayres, M. P. and Lombardero, M. J. (2000). Assessing the consequences of global change for forest disturbance from herbivores and pathogens. *The Science of the Total Environment*, **262**, 263–286.

Baillie, J. E. M., Hilton-Taylor, C. and Stuart, S. N. (eds.) (2004). *2004 IUCN Red List of Threatened Species: A Global Species Assessment*. Cambridge, UK: IUCN Publications Services Unit.

Bálint, M., Domisch, S., Engelhardt, C. H. M., et al. (2011). Cryptic biodiversity loss linked to global climate change. *Nature Climate Change*, **1**, 313–318.

Balmford, A. and Bond, W. (2005). Trends in the state of nature and their implications for human well-being. *Ecology Letters*, **8**, 1218–1234.

Balvanera, P., Pfisterer, A. B., Buchmann, N., et al. (2006). Quantifying the evidence for biodiversity effects on ecosystem functioning and services. *Ecology Letters*, **9**, 1146–1156.

Barton, B. T. (2010). Climate warming and predation risk during herbivore ontogeny. *Ecology*, **91**, 2811–2818.

Baum, J. K. and Worm, B. (2009). Cascading top-down effects of changing oceanic predator abundances. *The Journal of Animal Ecology*, **78**, 699–714.

Baxter, C. V., Fausch, K. D., Murakami, M. and Chapman, P. L. (2004). Fish invasion restructures stream and forest food webs by interrupting reciprocal prey subsidies. *Ecology*, **85**, 2656–2663.

Beaugrand, G., Brander, K. M., Lindley, J. A., Souissi, S. and Reid, P. C. (2003). Plankton effect on cod recruitment in the North Sea. *Nature*, **426**, 661–664.

Bertness, M. D. and Coverdale, T. C. (2013). An invasive species facilitates the recovery of salt marsh ecosystems on Cape Cod. *Ecology*, **94**, 1937–1943.

Blankinship, J. C., Niklaus, P. A. and Hungate, B. A. (2011). A meta-analysis of responses of soil biota to global change. *Oecologia*, **165**, 553–565.

Blenckner, T., Adrian, R., Livingstone, D. M., et al. (2007). Large-scale climatic signatures in lakes across Europe: a meta-analysis. *Global Change Biology*, **13**, 1314–1326.

Blois, J. L., Zarnetske, P. L., Fitzpatrick, M. C. and Finnegan, S. (2013). Climate change and the past, present, and future of biotic interactions. *Science*, 341, 499–504.

Bobbink, R., Hicks, K., Galloway, J., Spranger, T., et al. (2010). Global assessment of nitrogen deposition effects on terrestrial plant diversity: a synthesis. *Ecological Applications*, **20**, 30–59.

Borer, E. T., Seabloom, E. W., Shurin, J. B., et al. (2005). What determines the strength of a trophic cascade? *Ecology*, **86**, 528–537.

Borer, E. T., Seabloom, E. W., Gruner, D. S., et al. (2014). Herbivores and nutrients control grassland plant diversity via light limitation. *Nature*, **508**(7497), 517–520.

Bradshaw, W. E. and Holzapfel, C. M. (2006). Evolutionary response to rapid climate change. *Science*, **312**, 1477–1478.

Bracken, M. E. S. and Low, N. H. N. (2012). Realistic losses of rare species disproportionately impact higher trophic levels. *Ecology Letters*, **15**, 461–467.

Brock, W. A. and Carpenter, S. R. (2006). Variance as a leading indicator of regime shift in ecosystem services. *Ecology and Society*, **11**, 217–231.

Brown, C. J., Saunders, M. I., Possingham, H. P. and Richardson, A. J. (2013). Managing for interactions between local and global stressors of ecosystems. *PLoS One*, **8**, e65765.

Bulling, M. T., Hicks, N., Murray, L., et al. (2010). Marine biodiversity-ecosystem functions under uncertain environmental futures. *Philosophical Transactions of the Royal Society of London B: Biological Sciences*, **365**, 2107–2116.

Butchart, S. H. M., Walpole, M., Collen, B., et al. (2010). Global biodiversity: indicators of recent declines. *Science*, **328**, 1164–1168.

Caraco, N. F., Cole, J. J., Raymond, P. A., et al. (1997). Zebra mussel invasion in a large, turbid river: phytoplankton response to increased grazing. *Ecology*, **78**, 588–602.

Cardinale, B. J., Srivastava, D. S., Duffy, J. E., et al. (2006). Effects of biodiversity on the functioning of trophic groups and ecosystems. *Nature*, **443**, 989–992.

Cardinale, B. J., Duffy, J. E., Gonzalez, A., et al. (2012). Biodiversity loss and its impact on humanity. *Nature*, **486**, 59–67.

Carpenter, S. R., Fisher, S. G., Grimm, N. B. and Kitchell, J. F. (1992). Global change and freshwater ecosystems. *Annual Review of Ecology and Systematics*, **23**, 119–139.

Carpenter, S. R., Cole, J. J., Pace, M. L., et al. (2011). Early warnings of regime shifts: a whole-ecosystem experiment. *Science*, **332**, 1079–1082.

Cebrian, J., Shurin, J. B., Borer, E. T., et al. (2009). Producer nutritional quality controls ecosystem trophic structure. *PLoS One*, **4**.

Chapin, F. S. (1997). Biotic control over the functioning of ecosystems. *Science*, **277**, 500–504.

Chapin, F. S. (2003). Effects of plant traits on ecosystem and regional processes: a conceptual framework for predicting the consequences of global change. *Annals of Botany*, **91**, 455–463.

Chapin, F. S., Zavaleta, E. S., Eviner, V. T., et al. (2000). Consequences of changing biodiversity. *Nature*, **405**, 234–242.

Christensen, M. R., Graham, M. D., Vinebrooke, R. D., et al. (2006). Multiple anthropogenic stressors cause ecological surprises in boreal lakes. *Global Change Biology*, **12**, 2316–2322.

Clark, C. M., Cleland, E. E., Collins, S. L., et al. (2007). Environmental and plant community determinants of species loss following nitrogen enrichment. *Ecology Letters*, **10**, 596–607.

Clark, C. M., Hobbie, S. E., Venterea, R. and Tilman, D. (2009). Long-lasting effects on nitrogen cycling 12 years after treatments cease despite minimal long-term nitrogen retention. *Global Change Biology*, **15**, 1755–1766.

Clark, G. F., Stark, J. S., Johnston, E. L., et al. (2013). Light-driven tipping points in polar ecosystems. *Global Change Biology*, **19**, 3749–3761.

Cleland, E. E. and Harpole, W. S. (2010). Nitrogen enrichment and plant communities. In *Year in Ecology and Conservation Biology 2010, Vol 1195*. Oxford: Blackwell Publishing, pp. 46–61.

Cleland, E. E., Chuine, I., Menzel, A., Mooney, H. A. and Schwartz, M. D. (2007). Shifting plant phenology in response to global change. *Trends in Ecology and Evolution*, **22**, 357–365.

Cleland, E. E., Allen, J. M., Crimmins, T. M., et al. (2012). Phenological tracking enables positive species responses to climate change. *Ecology*, **93**, 1765–1771.

Coors, A. and De Meester, L. (2008). Synergistic, antagonistic and additive effects of multiple stressors: predation threat, parasitism and pesticide exposure in *Daphnia magna*. *Journal of Applied Ecology*, **45**, 1820–1828.

Courchamp, F., Chapuis, J.-L. and Pascal, M. (2003). Mammal invaders on islands: impact, control and control impact, *Biological Reviews*, **78**, 347–383.

Cox, J. G. and Lima, S. L. (2006). Naiveté and an aquatic–terrestrial dichotomy in the effects of introduced predators. *Trends in Ecology and Evolution*, **21**, 674–680.

Crain, C. M., Kroeker, K. and Halpern, B. S. (2008). Interactive and cumulative effects of multiple human stressors in marine systems. *Ecology Letters*, **11**, 1304–1315.

Crawford, K. M. and Rudgers, J. A. (2013). Genetic diversity within a dominant plant outweighs plant species diversity in structuring an arthropod community. *Ecology*, **94**, 1025–1035.

Crutsinger, G. M., Collins, M. D., Fordyce, J. A., et al. (2006). Plant genotypic diversity predicts community structure and

governs an ecosystem process. *Science*, **313**, 966–968.

Cushing, D. H. (1990). Plankton production and year-class strength in fish populations: an update of the match/mismatch hypothesis. In *Advances in Marine Biology*, ed. J. H. S. Blaxter and A. J. Southward. London: Academic Press, pp. 249–293

D'Antonio, C. M. and Vitousek, P. M. (1992). Biological invasions by exotic grasses, the grass/fire cycle, and global change. *Annual Review of Ecology and Systematics*, **23**, 63–87.

Darling, E. S. and Côté, I. M. (2008). Quantifying the evidence for ecological synergies. *Ecology Letters*, **11**, 1278–1286.

Daskalov, G. M., Grishin, A. N., Rodionov, S. and Mihneva, V. (2007). Trophic cascades triggered by overfishing reveal possible mechanisms of ecosystem regime shifts. *Proceedings of the National Academy of Sciences of the USA*, **104**, 10518–10523.

Díaz, S., Fargione, J., Chapin III, F. S. and Tilman, D. (2006). Biodiversity loss threatens human well-being, *PLoS Biology*, **4**, e277.

Diffenbaugh, N. S., Pal, J. S., Trapp, R. J. and Giorgi, F. (2005). Fine-scale processes regulate the response of extreme events to global climate change. *Proceedings of the National Academy of Sciences of the USA*, **102**, 15774–15778.

Dobson, A., Lodge, D., Alder, J., et al. (2006). Habitat loss, trophic collapse, and the decline of ecosystem services. *Ecology*, **87**, 1915–1924.

Douglass, J. G., Duffy, J. E. and Bruno, J. F. (2008). Herbivore and predator diversity interactively affect ecosystem properties in an experimental marine community. *Ecology Letters*, **11**, 598–608.

Downing, J. A., Osenberg, C. W. and Sarnelle, O. (1999). Meta-analysis of marine nutrient-enrichment experiments: variation in the magnitude of nutrient limitation. *Ecology*, **80**, 1157–1167.

Downing, A. L., Brown, B. L. and Leibold, M. A. (2014). Multiple diversity-stability mechanisms enhance population and community stability in aquatic food webs. *Ecology*, **95**, 173–184.

Duffy, J. E. (2003). Biodiversity loss, trophic skew and ecosystem functioning. *Ecology Letters*, **6**, 680–687.

Duffy, J. E., Cardinale, B. J., France, K. E., et al. (2007). The functional role of biodiversity in ecosystems: incorporating trophic complexity. *Ecology Letters*, **10**, 522–538.

Dupont, S., Dorey, N. and Thorndyke, M. (2010). What meta-analysis can tell us about vulnerability of marine biodiversity to ocean acidification? *Estuarine, Coastal and Shelf Science*, **89**, 182–185.

Dyer, L. A. and Coley, P. D. (2001). Latitudinal gradients in tri-trophic interactions. In *Multitrophic Level Interactions*, ed. T. Tscharntke and B. A. Hawkins. Cambridge, UK: Cambridge University Press, pp. 67–88.

Ellis, E. C. (2011). Anthropogenic transformation of the terrestrial biosphere. *Philosophical Transactions A: Mathematical, Physical, and Engineering Sciences*, **369**, 1010–1035.

Ellner, S. P. (2013). Rapid evolution: from genes to communities, and back again? *Functional Ecology*, **27**, 1087–1099.

Elmhagen, B. and Rushton, S. P. (2007). Trophic control of mesopredators in terrestrial ecosystems: top-down or bottom-up? *Ecology Letters*, **10**, 197–206.

Elser, J. J., Bracken, M. E. S., Cleland, E. E., et al. (2007). Global analysis of nitrogen and phosphorus limitation of primary producers in freshwater, marine and terrestrial ecosystems. *Ecology Letters*, **10**, 1135–1142.

Emmerson, M., Martijn Bezemer, T., Hunter, M., et al. (2004). How does global change affect the strength of trophic interactions? *Basic and Applied Ecology*, **5**, 505–514.

Engel, S. (1987). The impact of submerged macrophytes on largemouth bass and bluegills. *Lake and Reservoir Management*, **3**, 227–234.

Estes, J. A. (1998). Killer whale predation on sea otters linking oceanic and nearshore ecosystems. *Science*, **282**, 473–476.

Estes, J. A., Terborgh, J., Brashares, J. S., et al. (2011). Trophic downgrading of planet earth. *Science*, **333**, 301–306.

FAO. (2012). *The State of World Fisheries and Aquaculture (SOFIA)*. Rome: FAO Fisheries and Aquaculture Department.

Field, J. G., Harris, R. P., Hofmann, E. E., Perry, R. I. and Werner, F. E. (eds.) (2010). *Marine Ecosystems and Global Change*. Oxford: Oxford University Press.

Flynn, D. F., Mirotchnick, N., Jain, M., Palmer, M. I. and Naeem, S. (2011). Functional and phylogenetic diversity as predictors of biodiversity-ecosystem-function relationships. *Ecology*, **92**, 1573–1581.

Foley, J. A., Defries, R., Asner, G. P., et al. (2005). Global consequences of land use. *Science*, **309**, 570–574.

Folt, C. L., Chen, C. Y., Moore, M. V. and Burnaford, J. (1999). Synergism and antagonism among multiple stressors. *Limnology and Oceanography*, **44**, 864–877.

Frank, K. T., Petrie, B., Fisher, J. A. D. and Leggett, W. C. (2011). Transient dynamics of an altered large marine ecosystem. *Nature*, **477**, 86–89.

Fridley, J. D., Stachowicz, J. J., Naeem, S., et al. (2007). The invasion paradox: reconciling pattern and process in species invasions. *Ecology*, **88**, 3–17.

Fukami, T., Wardle, D. A., Bellingham, P. J. et al. (2006). Above- and below-ground impacts of introduced predators in seabird-dominated island ecosystems. *Ecology Letters*, **9**, 1299–1307.

Giller, P. S., Hillebrand, H., Berninger, U.-G., et al. (2004). Biodiversity effects on ecosystem functioning: emerging issues and their experimental test in aquatic environments. *Oikos*, **104**, 423–436.

Gilman, S. E., Urban, M. C., Tewksbury, J., Gilchrist, G. W. and Holt, R. D. (2010). A framework for community interactions under climate change. *Trends in Ecology and Evolution*, **25**, 325–331.

Green, P. T., O'Dowd, D. J., Abbott, K. L., et al. (2011). Invasional meltdown:

invader-invader mutualism facilitates a secondary invasion. *Ecology*, **92**, 1758–1768.

Greig, H. S., Kratina, P., Thompson, P. L., et al. (2012). Warming, eutrophication, and predator loss amplify subsidies between aquatic and terrestrial ecosystems. *Global Change Biology*, **18**, 504–514.

Gruber, N. and Galloway, J. N. (2008). An Earth-system perspective of the global nitrogen cycle. *Nature*, **451**, 293–296.

Gruner, D. S., Smith, J. E., Seabloom, E. W., et al. (2008). A cross-system synthesis of consumer and nutrient resource control on producer biomass. *Ecology Letters*, **11**, 740–755.

Guttal, V. and Jayaprakash, C. (2008). Changing skewness: an early warning signal of regime shifts in ecosystems. *Ecology Letters*, **11**, 450–460.

Haddad, N. M., Crutsinger, G. M., Gross, K., et al. (2009). Plant species loss decreases arthropod diversity and shifts trophic structure. *Ecology Letters*, **12**, 1029–1039.

Haddad, N. M., Crutsinger, G. M., Gross, K., Haarstad, J. and Tilman, D. (2011). Plant diversity and the stability of foodwebs. *Ecology Letters*, **14**, 42–46.

Halaj, J. and Wise, D. H. (2001). Terrestrial trophic cascades: How much do they trickle? *The American Naturalist*, **157**, 262–281.

Hall, S. R., Smith, V. H., Lytle, D. A. and Leibold, M. A. (2005). Constraints on primary producer N:P stoichiometry along N:P supply ratio gradients. *Ecology*, **86**, 1894–1904.

Hallock, P. (2001). Coral reefs, carbonate sediments, nutrients, and global change. In *The History and Sedimentology of Ancient Reef Systems*, ed. G. D. Stanley. New York, NY: Kluwer Academic/Plenum Publishers, pp. 388–422.

Harmon, L. J., Matthews, B., Des Roches, S., et al. (2009). Evolutionary diversification in stickleback affects ecosystem functioning. *Nature*, **458**, 1167–1170.

Harrington, R., Woiwod, I. and Sparks, T. (1999). Climate change and trophic interactions. *Trends in Ecology and Evolution*, **14**, 146–150.

Hautier, Y., Seabloom, E. W., Borer, E. T., et al. (2014). Eutrophication weakens stabilizing effects of diversity in natural grasslands. *Nature*, **508**(7497), 521–525.

Hendriks, I. E., Duarte, C. M. and Álvarez, M. (2010). Vulnerability of marine biodiversity to ocean acidification: a meta-analysis. *Estuarine, Coastal and Shelf Science*, **86**, 157–164.

Hillebrand, H., Gruner, D. S., Borer, E. T., et al. (2007). Consumer versus resource control of producer diversity depends on ecosystem type and producer community structure. *Proceedings of the National Academy of Sciences of the USA*, **104**, 10904–10909.

Hillebrand, H., Borer, E. T., Bracken, M. E. S., et al. (2009). Herbivore metabolism and stoichiometry each constrain herbivory at different organizational scales across ecosystems. *Ecology Letters*, **12**, 516–527.

HilleRisLambers, J., Harsch, M. A., Ettinger, A. K., Ford, K. R. and Theobald, E. J. (2013). How will biotic interactions influence climate change-induced range shifts? *Annals of the New York Academy of Sciences*, **1297**, 112–125.

Hoffmann, A. A. and Sgro, C. M. (2011). Climate change and evolutionary adaptation. *Nature*, **470**, 479–485.

Hofmann, G. E. and Todgham, A. E. (2010). Living in the now: physiological mechanisms to tolerate a rapidly changing environment. *Annual Review of Physiology*, **72**, 127–145.

Holt, R. and Loreau, M. (2001). Biodiversity and ecosystem functioning: the role of trophic interactions and the importance of system openness. In *The Functional Consequences of Biodiversity: Empirical Progress and Theoretical Extensions*, ed. A. P. Kinzig, S. W. Pacala and D. Tilman. Princeton, NJ: Princeton University Press, pp. 246–262.

Hooper, D. U., Adair, E. C., Cardinale, B. J., et al. (2012). A global synthesis reveals biodiversity loss as a major driver of ecosystem change. *Nature*, **486**, 105–108.

Hughes, A. R. and Stachowicz, J. J. (2004). Genetic diversity enhances the resistance of a seagrass ecosystem to disturbance.

Proceedings of the National Academy of Sciences of the USA, **101**, 8998–9002.

Hughes, A. R., Inouye, B. D., Johnson, M. T. J., Underwood, N. and Vellend, M. (2008). Ecological consequences of genetic diversity. *Ecology Letters*, **11**, 609–623.

Hughes, T. P., Carpenter, S., Rockström, J., Scheffer, M. and Walker, B. (2013). Multiscale regime shifts and planetary boundaries. *Trends in Ecology and Evolution*, **28**, 389–395.

Hulme, P. E., Pyšek, P., Jarošík, V., Pergl, J., Schaffner, U. and Vilà, M. (2013). Bias and error in understanding plant invasion impacts. *Trends in Ecology and Evolution*, **28**, 212–218.

IPCC. (2013). In *Climate Change 2013: The Physical Basis*, ed. T. F. Stocker, D. Qin, G.-K. Plattner, et al. Cambridge, UK: Cambridge University Press.

Isbell, F., Tilman, D., Polasky, S., Binder, S. and Hawthorne, P. (2013). Low biodiversity state persists two decades after cessation of nutrient enrichment. *Ecology Letters*, **16**, 454–460.

Ives, A. R. and Carpenter, S. R. (2007). Stability and diversity of ecosystems. *Science*, **317**, 58–62.

Jain, M., Flynn, D. F. B., Prager, C. M., et al. (2014). The importance of rare species: a trait-based assessment of rare species contributions to functional diversity and possible ecosystem function in tall-grass prairies. *Ecology and Evolution*, **4**, 104–112.

Jamieson, M. A., Trowbridge, A. M., Raffa, K. F. and Lindroth, R. L. (2012). Consequences of climate warming and altered precipitation patterns for plant-insect and multitrophic interactions. *Plant Physiology*, **160**, 1719–1727.

Jeppesen, E., Merrhoff, M., Holmgren, K., et al. (2010). Impacts of climate warming on lake fish community structure and potential effects on ecosystem function. *Hydrobiologia*, **646**, 73–90.

Johnson, N. C., Angelard, C., Sanders, I. R. and Kiers, E. T. (2013). Predicting community and ecosystem outcomes of mycorrhizal responses to global change. *Ecology Letters*, **16**, 140–153.

Johnston, E. L. and Roberts, D. A. (2009). Contaminants reduce the richness and evenness of marine communities: a review and meta-analysis. *Environmental Pollution*, **157**, 1745–1752.

Keane, R. M. and Crawley, M. J. (2002). Exotic plant invasions and the enemy release hypothesis. *Trends in Ecology and Evolution*, **17**, 164–170.

Kimbro, D. L., Grosholz, E. D., Baukus, A. J., et al. (2009). Invasive species cause large-scale loss of native California oyster habitat by disrupting trophic cascades. *Oecologia*, **160**, 563–575.

Kimbro, D. L., Cheng, B. S. and Grosholz, E. D. (2013). Biotic resistance in marine environments. *Ecology Letters*, **16**, 821–833.

Knapp, A. K., Smith, M. D., Hobbie, S. E., et al. (2012). Past, present, and future roles of long-term experiments in the lter network. *BioScience*, **62**, 377–389.

Knight, T. M., McCoy, M. W., Chase, J. M., McCoy, K. A. and Holt, R. D. (2005). Trophic cascades across ecosystems. *Nature*, **437**, 880–883.

Knops, J. M. H., Naeem, S. and Reich, P. B. (2007). The impact of elevated CO_2, increased nitrogen availability and biodiversity on plant tissue quality and decomposition. *Global Change Biology*, **13**, 1960–1971.

Kratina, P., Greig, H. S., Thompson, P. L., Carvalho-Pereira, T. S. A. and Shurin, J. B. (2012). Warming modifies trophic cascades and eutrophication in experimental freshwater communities. *Ecology*, **93**, 1421–1430.

Krauss, J., Bommarco, R., Guardiola, M., et al. (2010). Habitat fragmentation causes immediate and time-delayed biodiversity loss at different trophic levels. *Ecology Letters*, **13**, 597–605.

Kroeker, K. J., Kordas, R. L., Crim, R. N. and Singh, G. G. (2010). Meta-analysis reveals negative yet variable effects of ocean acidification on marine organisms. *Ecology Letters*, **13**, 1419–1434.

La Pierre, K. J. and Smith, M. D. (2014). Functional trait expression of grassland species shift with short- and long-term nutrient additions. *Plant Ecology*, DOI: 10.1007/511258-014-0438-4.

La Pierre, K. J., Harpole, W. S. and Suding, K. N. (2010). Strong feeding preference of an exotic generalist herbivore for an exotic forb: a case of invasional antagonism. *Biological Invasions*, **12**, 3025–3031.

Larson, A. J. and Paine, R. T. (2007). Ungulate herbivory: indirect effects cascade into the treetops. *Proceedings of the National Academy of Sciences of the USA*, **104**, 5–6.

Layman, C. A., Quattrochi, J. P., Payer, C. M. and Allgeier, J. E. (2007). Niche width collapse in a resilient top predator following ecosystem fragmentation. *Ecology Letters*, **10**, 937–944.

Lee, M., Manning, P., Rist, J., Power, S. A. and Marsh, C. (2010). A global comparison of grassland biomass responses to CO_2 and nitrogen enrichment. *Philosophical Transactions of the Royal Society of London B: Biological Sciences*, **365**, 2047–2056.

Lenton, T. M. (2011). Early warning of climate tipping points. *Nature Climate Change*, **1**, 201–209.

Leroux, S. J. and Loreau, M. (2008). Subsidy hypothesis and strength of trophic cascades across ecosystems. *Ecology Letters*, **11**, 1147–1156.

Leroux, S. J., Larrivée, M., Boucher-Lalonde, V., et al. (2013). Mechanistic models for the spatial spread of species under climate change. *Ecological Applications*, **23**, 815–828.

Levine, J. M., Adler, P. B. and Yelenik, S. G. (2004). A meta-analysis of biotic resistance to exotic plant invasions. *Ecology Letters*, **7**, 975–989.

Lindenmayer, D. B., Likens, G. E., Krebs, C. J. and Hobbs, R. J. (2010). Improved probability of detection of ecological "surprises." *Proceedings of the National Academy of Sciences of the USA*, **107**, 21957–21962.

Lindroth, R. L. (2010). Impacts of elevated atmospheric CO_2 and O_3 on forests: phytochemistry, trophic interactions, and ecosystem dynamics. *Journal of Chemical Ecology*, **36**, 2–21.

Loreau, M., Naeem, S., Inchausti, P., et al. (2001). Biodiversity and ecosystem functioning: current knowledge and future challenges. *Science*, **294**, 804–808.

Lyons, S., Smith, F. and Brown, J. (2004). Of mice, mastodons and men: human-mediated extinctions on four continents. *Evolutionary Ecology Research*, **6**, 339–358.

Mantyka-Pringle, C. S., Martin, T. G. and Rhodes, J. R. (2011). Interactions between climate and habitat loss effects on biodiversity: a systematic review and meta-analysis. *Global Change Biology*, **18**, 1239–1252.

Marczak, L. B., Thompson, R. M. and Richardson, J. S. (2007). Meta-analysis: trophic level, habitat, and productivity shape the food web effects of resource subsidies. *Ecology*, **88**, 140–148.

Matson, P., Lohse, K. A. and Hall, S. J. (2002). The globalization of nitrogen deposition: consequences for terrestrial ecosystems. *Ambio*, **31**, 113–119.

Medina, F. M., Bonnaud, E., Vidal, E., et al. (2011). A global review of the impacts of invasive cats on island endangered vertebrates. *Global Change Biology*, **17**, 3503–3510.

Menge, B. A. and Lubchenco, J. (1981). Community organization in temperate and tropical rocky intertidal habitats: prey refuges in relation to consumer pressure gradients. *Ecological Monographs*, **51**, 429–450.

Milchunas, D. G. and Lauenroth, W. (1995). Inertia in plant community structure: state changes after cessation of nutrient-enrichment stress. *Ecological Applications*, **5**, 452–458.

Milner-Gulland, E. J. and Bennett, E. L. (2003). Wild meat: the bigger picture. *Trends in Ecology and Evolution*, **18**, 351–357.

Mooney, H. A. and Cleland, E. E. (2001). The evolutionary impact of invasive species. *Proceedings of the National Academy of Sciences of the USA*, **98**, 5446–5451.

Moore, J. C., Berlow, E. L., Coleman, D. C., et al. (2004). Detritus, trophic dynamics and biodiversity. *Ecology Letters*, **7**, 584–600.

Mouillot, D., Bellwood, D. R., Baraloto, C., et al. (2013). Rare species support vulnerable functions in high-diversity ecosystems. *PLoS Biology*, **11**, e1001569.

Murphy, G. E. P. and Romanuk, T. N. (2012). A meta-analysis of community response predictability to anthropogenic disturbances. *The American Naturalist*, **180**, 316–327.

Musolin, D. L. (2007). Insects in a warmer world: ecological, physiological and life-history responses of true bugs (Heteroptera) to climate change. *Global Change Biology*, **13**, 1565–1585.

Myers, R. A. and Worm, B. (2003). Rapid worldwide depletion of predatory fish communities. *Nature*, **423**, 280–283.

Norbury, G., Byrom, A., Pech, R., et al. (2013). Invasive mammals and habitat modification interact to generate unforeseen outcomes for indigenous fauna. *Ecological Applications*, **23**, 1707–1721.

O'Connor, N. E. and Donohue, I. (2012). Environmental context determines multi-trophic effects of consumer species loss. *Global Change Biology*, **19**, 431–440.

O'Gorman, E. J., Yearsley, J. M., Crowe, T. P., et al. (2011). Loss of functionally unique species may gradually undermine ecosystems. *Proceedings of the Royal Society B: Biological Sciences*, **278**, 1886–1893.

O'Gorman, E. J., Fitch, J. E. and Crowe, T. P. (2012). Multiple anthropogenic stressors and the structural properties of food webs. *Ecology*, **93**, 441–448.

Ostfeld, R. S. and Holt, R. D. (2004). Are predators good for your health? Evaluating evidence for top-down regulation of zoonotic disease reservoirs. *Frontiers in Ecology and the Environment*, **2**, 13.

Ovaskainen, O., Skorokhodova, S., Yakovleva, M., et al. (2013). Community-level phenological response to climate change. *Proceedings of the National Academy of Sciences of the USA*, **110**, 13434–13439.

Pace, M. L., Cole, J., Carpenter, S. R. and Kitchell, J. F. (1999). Trophic cascades revealed in diverse ecosystems. *Trends in Ecology and Evolution*, **14**, 483–488.

Packer, C., Brink, H., Kissui, B. M., et al. (2011). Effects of trophy hunting on lion and leopard populations in Tanzania. *Conservation Biology*, **25**, 142–153.

Paine, R. T., Tegner, M. J. and Johnson, E. A. (1998). Compounded perturbations yield ecological surprises. *Ecosystems*, **1**, 535–545.

Parmesan, C. (2006). Ecological and evolutionary responses to recent climate change. *Annual Review of Ecology, Evolution, and Systematics*, **37**, 637–669.

Pauls, S. U., Nowak, C., Bálint, M. and Pfenninger, M. (2013). The impact of global climate change on genetic diversity within populations and species. *Molecular Ecology*, **22**, 925–946.

Pejchar, L. and Mooney, H. A. (2009). Invasive species, ecosystem services and human well-being. *Trends in Ecology and Evolution*, **24**, 497–504.

Peters, D. P. C., Bestelmeyer, B. T. and Turner, M. G. (2007). Cross-scale interactions and changing pattern–process relationships: consequences for system dynamics. *Ecosystems*, **10**, 790–796.

Petrin, Z., Englund, G. and Malmqvist, B. (2008). Contrasting effects of anthropogenic and natural acidity in streams: a meta-analysis. *Proceedings of the Royal Society B: Biological Sciences*, **275**, 1143–1148.

Pimentel, D. (ed.) (2011). *Biological Invasions: Economic and Environmental Costs of Alien Plant, Animal, and Microbe Species*, 2nd edn. Boca Raton, FL: Taylor and Francis Group.

Pinsky, M. L., Worm, B., Fogarty, M. J., Sarmiento, J. L. and Levin, S. A. (2013). Marine taxa track local climate velocities. *Science*, **341**, 1239–1242.

Power, M. E. (1992). Top-down and bottom-up forces in food webs: do plants have primacy? *Ecology*, **73**, 733–746.

Ricciardi, A., Hoopes, M. F., Marchetti, M. P. and Lockwood, J. L. (2013). Progress toward understanding the ecological impacts of nonnative species. *Ecological Monographs*, **83**, 263–282.

Ripple, W. J. and Beschta, R. L. (2012). Trophic cascades in Yellowstone: the first 15 years after wolf reintroduction. *Biological Conservation*, **145**, 205–213.

Ripple, W. J., Estes, J. A., Beschta, R. L., et al. (2014). Status and ecological effects of the world's largest carnivores. *Science*, **343**, 1241484.

Root, T. L., Price, J. T., Hall, K. R., et al. (2003). Fingerprints of global warming on wild animals and plants. *Nature*, **421**, 57–60.

Rosa, E. A., York, R. and Dietz, T. (2004). Tracking the anthropogenic drivers of ecological impacts. *Ambio*, **33**, 509–512.

Rosenzweig, M. (1971). Paradox of enrichment: destabilization of exploitation ecosystems in ecological time. *Science*, **171**, 385–387.

Sala, O. E., Chapin, F. S., Armesto, J. J., et al. (2000). Global biodiversity scenarios for the year 2100. *Science*, **287**, 1770–1774.

Santos, M. J., Anderson, L. W. and Ustin, S. L. (2011). Effects of invasive species on plant communities: an example using submersed aquatic plants at the regional scale. *Biological Invasions*, **13**, 443–457.

Sardans, J., Rivas-Ubach, A. and Peñuelas, J. (2011). The C:N:P stoichiometry of organisms and ecosystems in a changing world: a review and perspectives. *Perspectives in Plant Ecology, Evolution and Systematics*, **14**, 33–47.

Sax, D. F., Stachowicz, J. J., Brown, J. H., et al. (2007). Ecological and evolutionary insights from species invasions. *Trends in Ecology and Evolution*, **22**, 465–471.

Scheffer, M. and Carpenter, S. R. (2003). Catastrophic regime shifts in ecosystems: linking theory to observation. *Trends in Ecology and Evolution*, **18**, 648–656.

Scheffer, M., Carpenter, S., Foley, J. A., Folke, C. and Walker, B. (2001). Catastrophic shifts in ecosystems. *Nature*, **413**, 591–596.

Scheffer, M., Bascompte, J., Brock, W. A., et al. (2009). Early-warning signals for critical transitions. *Nature*, **461**, 53–59.

Schiffers, K., Bourne, E. C., Lavergne, S., Thuiller, W. and Travis, J. M. J. (2012). Limited evolutionary rescue of locally adapted populations facing climate change. *Philosophical Transactions of the Royal Society of London B: Biological Sciences*, **368**, 20120083.

Schlaepfer, M. A., Sax, D. F. and Olden, J. D. (2011). The potential conservation value of non-native species. *Conservation Biology*, **25**, 428–437.

Schmitz, O. J. (2013). Global climate change and the evolutionary ecology of ecosystem functioning. *Annals of the New York Academy of Sciences*, **1297**, 61–72.

Schmitz, O. J., Hamback, P. A. and Beckerman, A. P. (2000). Trophic cascades in terrestrial systems: a review of the effects of carnivore removals on plants. *The American Naturalist*, **155**, 141–153.

Schwarz, D., Matta, B. M., Shakir-Botteri, N. L. and McPheron, B. A. (2005). Host shift to an invasive plant triggers rapid animal hybrid speciation. *Nature*, **436**, 546–549.

Shurin, J. B., Borer, E. T., Seabloom, E. W., et al. (2002). A cross-ecosystem comparison of the strength of trophic cascades. *Ecology Letters*, **5**, 785–791.

Shurin, J. B., Gruner, D. S. and Hillebrand, H. (2006). All wet or dried up? Real differences between aquatic and terrestrial food webs. *Proceedings of the Royal Society B: Biological Sciences*, **273**, 1–9.

Shurin, J. B., Clasen, J. L., Greig, H. S., Kratina, P. and Thompson, P. L. (2012). Warming shifts top-down and bottom-up control of pond food web structure and function. *Philosophical Transactions of the Royal Society of London B: Biological Sciences*, **367**, 3008–3017.

Simberloff, D. and Von Holle, B. (1999). Positive interactions of nonindigenous species: invasional meltdown? *Biological Invasions*, **1**, 21–32.

Simberloff, D., Martin, J.-L., Genovesi, P., et al. (2013). Impacts of biological invasions: what's what and the way forward. *Trends in Ecology and Evolution*, **28**, 58–66.

Sistla, S. A. and Schimel, J. P. (2012). Stoichiometric flexibility as a regulator of carbon and nutrient cycling in terrestrial ecosystems under change. *New Phytologist*, **196**, 68–78.

Smith, M. D., Knapp, A. K. and Collins, S. L. (2009). A framework for assessing ecosystem dynamics in response to chronic resource alterations induced by global change. *Ecology*, **90**, 3279–3289.

Srivastava, D. S. and Bell, T. (2009). Reducing horizontal and vertical diversity in a foodweb triggers extinctions and impacts functions. *Ecology Letters*, **12**, 1016–1028.

Srivastava, D. S. and Vellend, M. (2005). Biodiversity-ecosystem function research: is it relevant to conservation? *Annual Review of Ecology, Evolution, and Systematics*, **36**, 267–294.

Srivastava, D. S., Cadotte, M. W., MacDonald, A. A. M., Marushia, R. G. and Mirotchnick, N. (2012). Phylogenetic diversity and the functioning of ecosystems. *Ecology Letters*, **15**, 637–648.

Stachowicz, J. J. (2001). Mutualism, facilitation, and the structure of ecological communities. *BioScience*, **51**, 235–246.

Sterner, R. W. and Elser, J. J. (2002). *Ecological Stoichiometry: The Biology of Elements from Molecules to the Biosphere*. Princeton, NJ: Princeton University Press.

Strayer, D. L., Eviner, V. T., Jeschke, J. M. and Pace, M. L. (2006). Understanding the long-term effects of species invasions. *Trends in Ecology and Evolution*, **21**, 645–651.

Strong, D. R. and Frank, K. T. (2010). Human involvement in food webs. *Annual Review of Environment and Resources*, **35**, 1–23.

Suding, K. N. and Hobbs, R. J. (2009). Threshold models in restoration and conservation: a developing framework. *Trends in Ecology and Evolution*, **24**, 271–279.

Suding, K. N., Collins, S. L., Gough, L., et al. (2005). Functional- and abundance-based

mechanisms explain diversity loss due to N fertilization. *Proceedings of the National Academy of Sciences of the USA*, **102**, 4387–4392.

Thackeray, S. J., Sparks, T. H., Frediksen, M., et al. (2010). Trophic level asynchrony in rates of phenological change for marine, freshwater and terrestrial environments. *Global Change Biology*, **16**, 3304–3313.

Thébault, E. and Loreau, M. (2005). The relationship between biodiversity and ecosystem functioning in food webs. *Ecological Research*, **21**, 17–25.

Theurillat, J.-P. and Guisan, A. (2001). Potential impact of climate change on vegetation in the European Alps: a review. *Climatic Change*, **50**, 77–109.

Throop, H. L. and Lerdau, M. (2004). Effects of nitrogen deposition on insect herbivory: implications for community and ecosystem processes. *Ecosystems*, **7**, 109–133.

Tilman, D. (1996). Biodiversity: population versus ecosystem stability. *Ecology*, **77**, 350–363.

Tilman, D., May, R. M., Lehman, C. L. and Nowak, M. A. (1994). Habitat destruction and the extinction debt. *Nature*, **371**, 65–66.

Townsend, C. R., Uhlmann, S. S. and Matthaei, C. D. (2008). Individual and combined responses of stream ecosystems to multiple stressors. *Journal of Applied Ecology*, **45**, 1810–1819.

Travis, J., Coleman, F. C., Auster, P. J., et al. (2013). Integrating the invisible fabric of nature into fisheries management. *Proceedings of the National Academy of Sciences of the USA*, **111**, 581–584.

Treseder, K. K. (2004). A meta-analysis of mycorrhizal responses to nitrogen, phosphorus, and atmospheric CO_2 in field studies. *New Phytologist*, **164**, 347–355.

Tylianakis, J. M., Didham, R. K., Bascompte, J. and Wardle, D. A. (2008). Global change and species interactions in terrestrial ecosystems. *Ecology Letters*, **11**, 1351–1363.

van Asch, M. and Visser, M. E. (2007). Phenology of forest caterpillars and their host trees: the importance of synchrony. *Annual Review of Entomology*, **52**, 37–55.

van der Putten, W. H., Ruiter, P. C. de, Martijn Bezemer, T., et al. (2004). Trophic interactions in a changing world. *Basic and Applied Ecology*, **5**, 487–494.

van der Putten, W. H., Macel, M. and Visser, M. E. (2010). Predicting species distribution and abundance responses to climate change: why it is essential to include biotic interactions across trophic levels. *Philosophical Transactions of the Royal Society of London B: Biological Sciences*, **365**, 2025–2034.

van der Zanden, M. J., Casselman, J. M. and Rasmussen, J. B. (1999). Stable isotope evidence for the food web consequences of species invasions in lakes. *Nature*, **401**, 464–467.

van Hengstum, T., Hooftman, D. A. P., Oostermeijer, J. G. B. and van Tienderen, P. H. (2014). Impact of plant invasions on local arthropod communities: a meta-analysis. *Journal of Ecology*, **102**, 4–11.

Vellend, M., Harmon, L. J., Lockwood, J. L., et al. (2007). Effects of exotic species on evolutionary diversification. *Trends in Ecology and Evolution*, **22**, 481–488.

Vellend, M., Baeten, L., Myers-Smith, I. H., et al. (2013). Global meta-analysis reveals no net change in local-scale plant biodiversity over time. *Proceedings of the National Academy of Sciences of the USA*, **110**, 19456–19459.

Veraart, A. J., Faassen, E. J., Dakos, V., et al. (2012). Recovery rates reflect distance to a tipping point in a living system. *Nature*, **481**, 357–359.

Vinebrooke, D. R., Cottingham, K. L., Norberg, M. S., et al. (2004). Impacts of multiple stressors on biodiversity and ecosystem functioning: the role of species co-tolerance. *Oikos*, **104**, 451–457.

Vitousek, P. M., D'Antonio, C. M., Loope, L. L., Rejmanek, M. and Westbrooks, R. (1997a). Introduced species: a significant component of human-caused global change. *New Zealand Journal of Ecology*, **21**, 1–16.

Vitousek, P. M., Mooney, H. A., Lubchenco, J. and Melillo, J. M. (1997b). Human domination of Earth's ecosystems. *Science*, **277**, 494–499.

Walker, B. (1992). Biodiversity and ecological redundancy. *Conservation Biology*, **6**, 18–23.

Walker, B., Kinzig, A. and Langridge, J. (1999). Plant attribute diversity, resilience, and ecosystem function: the nature and significance of dominant and minor species. *Ecosystems*, **2**, 95–113.

Walther, G., Post, E., Convey, P. and Menzel, A. (2002). Ecological responses to recent climate change. *Nature*, **416**, 389–395.

Wardle, D. A., Bardgett, R. D., Klironomos, J. N., et al. (2004). Ecological linkages between aboveground and belowground biota. *Science*, **304**, 1629–1633.

Wilbanks, T. J. and Kates, R. W. (1999). Global change in local places: how scale matters. *Climatic Change*, **43**, 601–628.

Winder, M. and Schindler, D. E. (2004). Climate change uncouples trophic interactions in an aquatic ecosystem. *Ecology*, **85**, 2100–2106.

Winkler, D. W., Dunn, P. O. and McCulloch, C. E. (2002). Predicting the effects of climate change on avian life-history traits. *Proceedings of the National Academy of Sciences of the USA*, **99**, 13595–13599.

Woodward, G., Perkins, D. M. and Brown, L. E. (2010). Climate change and freshwater ecosystems: impacts across multiple levels of organization. *Philosophical Transactions of the Royal Society of London B: Biological Sciences*, **365**, 2093–2106.

Worm, B. and Duffy, J. E. (2003). Biodiversity, productivity and stability in real food webs. *Trends in Ecology and Evolution*, **18**, 628–632.

Worm, B., Lotze, H. K., Hillebrand, H. and Sommer, U. (2002). Consumer versus resource control of species diversity and ecosystem functioning. *Nature*, **417**, 848–851.

Worm, B., Barbier, E. B., Beaumont, N., et al. (2006). Impacts of biodiversity loss on ocean ecosystem services. *Science*, **314**, 787–790.

Xia, J. Y. and Wan, S. (2008). Global response patterns of terrestrial plant species to nitrogen addition. *New Phytologist*, **179**, 428–439.

Index

adjacent ecosystems, 7, 68, 151, 163, 169–170, 188–189, 245, 247, 260, 389
Africa, 88, 94–95, 114–115, 117, 170, 171, 239
algae, 44, 60, 73, 74, 148, 161, 163–164, 172, 223, 237, 248, 271, 327, 371
  abundance, 20, 171, 388
  community composition, 20, 45, 60, 62, 65, 158, 169–170, 302, 388
  defense, 210, 211–212, 214
  diversity, 60–61, 170, 323–324, 326, 329, 332–333
  productivity, 19, 58, 61, 150, 171, 173, 188, 223, 301, 304, 310, 326, 332, 346
  stoichiometry, 59, 60–62
allochthonous, 16–18, 68–69, 164, 170, 244–245, 248–250, 261, 270, 272–274, 376
alternate states, 38–39, 43, 47–49, 74, 93, 125, 393, 394
anthropogenic effects, 31, 34, 46, 75, 94–95, 115, 125–126, 163–164, 174, 176, 177, 180, 225, 250, 265, 308, 365
ant-plant, 289, 291, 293, 296, 298
ants, 123, 240, 267, 291, 293, 294, 296, 297, 310, 352, 385
apex predator. *See* top predator
apparent competition, 17, 139, 142, 148–151
apparent mutualism, 148, 150
Atlantic, 9, 10, 35, 37, 39, 46, 48, 174–176, 186, 189, 247, 293
autochthonous, 135, 270, 271, 274

bacteria, 237, 239, 242, 244, 248, 263, 371
  abundance, 274, 278, 324
  community composition, 264, 266, 269, 272, 275–276
  diversity, 266, 269, 278, 324
  productivity, 57, 265, 269, 275–276, 389
  stoichiometry, 236, 267
Baltic Sea, 43, 324
barnacles, 160, 161, 164–167, 173
beetles, 111, 291, 293, 294, 296, 297, 345
behavior, 74, 277, 301, 304, 325, 353, 373, 384, 391
  herbivore, 93, 95, 115, 184, 208, 344
  prey, 115, 121, 150, 163, 172, 243, 344, 348, 353

Białowieża Primeval Forest (BPF), 115–121
biodiversity loss, 72–73, 376, 377–382, 388, 394
Biodiversity-Ecosystem Function (BEF), 56, 72, 318, 377, 394–395
biotic resistance, 334, 343, 385, 387
birds, 44, 149, 163, 171, 188, 249, 345, 352
browse traps, 94, 113–114, 116, 118
browsers, 93, 94, 95, 111, 113–114, 126, 241, 376
browsing pressure, 94, 95, 118, 121

carnivores, 61–62, 67, 115, 123, 125, 136, 184, 204
carrying capacity, 34, 48, 93, 120
caterpillars, 293, 294, 296, 297, 298, 345, 346, 352, 353
climate, 35, 92, 120, 240, 296
climate change, 44, 98, 225, 372–375, 377, 390, 392, 394
competition–defense tradeoff, 330–332, 333
competitive exclusion, 92, 122, 269, 371
complementarity effect, 321–323
conservation, 56, 134, 161, 163, 382, 392, 395
crab, 41, 172, 177, 179, 328, 382, 385–386
cross-ecosystem effects, 16–17, 18, 21, 69, 134, 139, 149, 150, 151, 167, 170, 173, 182, 189, 243, 246, 248, 249, 250, 389
cyanobacteria, 44, 59, 61, 67, 71, 170, 237, 327

Daphnia, 75, 277, 278, 309, 346, 389
decomposition rate, 57, 151, 221, 235, 261, 263
defense, 136, 297, 342, 344, 345, 351, 368
  allocation, 214, 370
  chemical, 205, 210–213, 214, 267, 269, 293, 294, 298
  constitutive, 214
  induced, 210–211, 213, 219, 241
  resistance, 210–212, 213–215
  resistance traits, 118, 124, 205–206, 208–209, 215, 219, 221, 224–225, 343
  structural, 95, 205, 210–212, 214, 269
detritivore, 12, 70, 73, 174, 182, 233, 242, 244, 248, 272, 376
  biomass, 185

community composition, 263
  diversity, 260, 327, 379
  food web, 182, 186, 188, 265, 327
disease, 95, 122, 275, 340, 374, 376
disturbance, 71, 95, 97, 111, 113, 125, 160,
  180, 262, 278, 324, 384
diversification, 298, 382
diversity
  beta, 288, 296
  functional, 97, 242, 266, 334, 382, 388
  horizontal, 7, 379
  intra-specific, 318, 374, 381, 393
  phylogenetic, 318, 382
  vertical, 7, 329, 379
dominant species, 35, 43, 45, 46, 170, 172,
  176, 185, 221, 328, 334, 378, 381
donor ecosystem, 15, 139, 141, 150, 389

eastern Scotian Shelf, 38, 41–43, 44, 47–49
eco-evolutionary feedback, 358, 381
ecosystem engineer, 14, 56, 63, 70–72, 74, 260
ecosystem recovery, 38, 47–49, 75, 94, 111,
  180, 304, 381, 382, 394
ecosystem resilience, 38, 46, 152, 378, 394
ecosystem resistance, 38, 378
energy flow, 3, 4, 11, 15, 55, 60, 149, 160, 264,
  274, 318, 329, 333
environmental gradients, 4, 59, 65, 88, 89,
  96–97, 109, 124, 136, 158–159, 167, 175,
  178–180, 214, 262, 268, 296, 304, 343
Europe, 116, 126, 135, 174, 175, 181, 186, 189,
  377
evolution, 35, 92, 95, 213, 224, 298–299,
  308–310, 340, 371, 382, 386, 393
excretion, 13, 15, 64–67, 74, 164, 238–239,
  242, 244
exotic species, 45, 73–75, 301, 334, 377,
  382–387, 390
Exploitation Ecosystems Hypothesis (EEH), 4,
  109, 136, 184, 203, 219, 224
extinction debt, 377

facilitation, 170, 178, 186, 221, 264, 322, 385
feeding guild, 41, 223
fire, 86–88, 91–93, 94–98, 111–115, 126, 376,
  384
fire trap, 113
fish, 9, 17, 20, 36, 38, 41, 46, 58, 61, 69, 70–75,
  135, 148, 150, 163, 242, 243, 244, 248,
  249, 301, 309, 328, 373, 384
  abundance, 44
  community composition, 41, 44, 304
  diversity, 35, 70
  evolution, 345, 358
  productivity, 31, 34, 35, 41, 48, 65, 69, 75
fisheries, 34–35, 41, 48, 70, 73, 375
Flood Pulse Concept (FPC), 68, 135
floodplain, 68, 70, 135
forest gaps, 111, 116, 121, 124
foundation species, 175, 381, 386
functional groups, 32, 46, 87, 89, 90–91, 94,
  211, 223, 371, 372

functional redundancy, 73, 379
functional response, 8, 11, 13, 138, 150, 351,
  353–357
functional trait, 38, 209, 213, 224, 225, 250,
  334
fungus, 73, 237, 240, 263, 272, 277, 371, 390
  community composition, 264, 267–269,
  274, 278
  diversity, 265, 267–269, 274
  productivity, 57, 264, 267–269

geese, 177, 178, 181, 183–185, 186–187
generalist, 7, 45, 140, 151, 223, 268, 277, 293,
  294, 327, 346, 348
genetic diversity. See diversity, intraspecific
global environmental change, 98, 250,
  308–311, 365
grazing lawn, 93–95, 113, 114
grazing pressure, 41, 45, 75, 94, 187, 278
Green World Hypothesis. See Hairston Smith
  Slobodkin (HSS)
growth rate, 46, 55, 59, 61, 90, 96, 113, 119,
  170, 173, 205, 210, 222, 327
Growth Rate Hypothesis (GRH), 219, 224
Growth/Defense Tradeoff Hypothesis, 213,
  214, 343, 344, 348

habitat fragmentation, 378
habitat modification, 31, 126, 365, 375, 377,
  382, 390, 393
Hairston Smith Slobodkin (HSS), 3, 32, 135,
  268, 341
hare, 184–187
herbivore
  abundance, 97, 109, 112, 135, 148, 167, 171,
    178, 180, 183, 187, 207, 291, 368, 370
  community composition, 62, 266
  diversity, 96, 97, 295, 329
  populations, 33, 94, 109, 115, 165, 177, 184,
    203, 225, 296, 333, 344
  productivity, 61, 62, 136, 137, 185, 329,
    333
herbivory rate, 10, 13, 208, 354
host plant, 17, 296, 342, 345, 346, 352, 382

inorganic nutrients, 3, 7–9, 18, 163, 221, 275,
  318–320, 324, 330, 387
invasion. See exotic species
invertebrate, 111, 122, 135, 148, 161, 170, 173,
  189, 239, 248, 323, 326, 381
  abundance, 149, 291, 332, 370
  community composition, 158, 266, 310
  diversity, 295
  productivity, 61, 68, 71, 183, 301

Janzen–Connell Hypothesis (JC), 122

keystone species, 15, 71, 277, 295

life history, 121, 169, 210, 223, 309, 342, 373
light, 60, 70, 75, 89, 92, 93, 116, 118, 173, 186,
  212, 293, 394

litter, 12, 68, 91, 111, 209, 221–223, 235, 238, 268, 274, 277, 376
livestock, 174, 175, 176, 186, 239

MacArthur–Rosenzweig Consumer Resource model (MRCR), 138–139
management, 20, 47–48, 56, 92, 377, 389
marine protected areas (MPAs), 161, 303–308
Mechanism Switching Hypothesis (MSH), 206–209, 213–214, 223, 224
meta-analysis
    bottom-up control, 9, 323
    bottom-up/top-down interactions, 36, 46, 204, 289, 304, 305, 341, 343
    defense, 214, 224
    global change, 378, 386, 387, 389–390, 395
    nutrient cycling, 12
    scale, 309
    top-down control, 20, 39, 212, 224
metabolic processes, 64, 243–244, 264, 279, 370, 376, 391
    secondary metabolites, 111, 124, 210, 267, 269, 293, 299
meta-communities, 245, 302, 324, 378
meta-ecosystems, 7, 18, 19, 21, 162, 169, 172–173, 245, 250
meta-populations, 160
micronutrients, 57, 236
migratory species, 69–70, 71–73, 183, 248–250
monoculture, 267, 322–324, 326–328
moose, 111, 112, 116, 126, 241
multiple stressors, 367, 387–391
mussels, 65, 72, 74, 161, 163–167, 172, 188, 301, 384, 386
mutualism, 240, 293, 358, 371, 385, 390

nitrogen, 56, 88, 110, 176, 178, 180, 182, 215, 221, 236, 237, 239, 242, 244, 266, 302, 310, 323, 347, 367, 371
nitrogen fixation, 59, 67, 71, 237, 241
non-additive effects, 341, 348–353, 387, 389
non-consumptive effects (NCEs), 4, 14, 21, 163, 242, 382
North America, 71, 88, 110, 111, 125, 165, 167, 171, 173, 174, 248, 310, 384
North Sea, 46, 47
nutrient cycling, 4, 12–15, 17, 21, 56, 60, 63–67, 71, 72, 74, 111, 124, 164, 173, 219–223, 225, 233, 374
nutrient enrichment, 10, 44–45, 56, 57, 58, 67, 75, 88, 148, 149, 164, 171, 180, 300, 302, 305, 310, 367–372, 387–389
nutrient immobilization, 235–237, 241, 242, 263, 267
nutrient limitation, 57, 61
nutrient mineralization, 4, 12, 56, 70, 74, 111, 182, 185, 186, 239–240, 242–245, 260, 262, 264, 267, 269, 272, 275
nutrient ratios, 56, 59–61, 64–67, 188, 210, 219, 224, 235, 237, 238, 244, 265, 267, 310, 327, 346, 347

organic matter (OM), 12, 43, 68, 71, 134–135, 139, 150, 170, 221, 235, 236, 247, 260, 387
    dissolved, 247, 271
    particulate, 73, 247, 271
organic nutrients, 4, 68, 71, 148, 238, 264, 269, 277, 279

Pacific, 15–16, 34, 39, 170, 247, 248
palatability, 111, 118, 124, 185, 188, 204, 205, 211, 212, 220–221, 241–242, 263, 268, 277
parasitism, 205, 275, 278, 293, 296–299, 310, 341, 353–357, 389
phosphorus, 56, 73, 74, 88, 219, 236, 240, 242, 244, 247, 266, 310, 324, 347, 368, 371
Piper, 291–299, 304, 306
plant
    community composition, 109, 122, 124, 135, 175, 180, 186, 188, 221, 241, 243, 249, 264
    diversity, 88, 91, 94, 97, 122, 124, 178, 180–181, 288, 295, 299, 322, 323, 330, 332, 371, 378
    productivity, 8, 18, 34, 44, 55, 56, 70, 73, 74, 90, 93, 109, 110, 123, 135, 137, 148, 149, 176, 183, 188, 203, 208, 239, 241, 243, 244, 247, 249, 264, 293, 294, 299, 318, 322, 327, 329, 332, 341, 344, 368, 371, 373, 378
    stoichiometry, 61
plant volatiles, 206, 211, 351, 353
polar ecosystems, 393
pollinators, 150, 353, 358
precipitation, 88, 247, 372, 390, 392
predation pressure, 136, 149, 300, 303, 304, 308, 352
predation rate, 10, 120, 165, 328, 351
predation risk, 14, 94, 121, 172, 243, 249, 344, 352
predator
    abundance, 135, 142, 149, 165, 167, 172, 250, 300, 349, 352, 375
    diversity, 115, 125, 250, 295, 321, 323, 329, 379, 386
    populations, 115, 151, 165, 296, 331, 377
    productivity, 58, 68, 75, 142, 333
predator avoidance, 10, 39, 121, 344, 353
productivity–diversity relationship, 88, 302, 322, 325, 331, 332

rainfall, 15, 88, 90, 93, 96, 97, 107, 113, 114, 123, 125, 372
rare species, 122, 378, 385
recipient ecosystem, 15, 17, 18, 73, 139, 141, 150, 151, 245, 248, 389
recruitment, 9, 34, 38, 48, 92, 110–111, 113–114, 115, 116, 118, 119, 121–124, 164–167, 173, 349–353, 354–357
Resource Availability Hypothesis, 343, 344, 357
resource limitation, 3, 7, 33, 55, 56, 67, 74, 88, 96, 110, 173, 203, 215, 236, 250, 261, 268, 276, 318, 344, 368

absolute, 206–208, 225, 261, 368
relative, 206–208, 261, 368
restoration, 134, 152, 304, 382, 392, 393, 395
Rhizobia, 237, 371
River Continuum Concept (RCC), 68, 135

salinity, 175, 177–179, 181, 372
salmon, 15–17, 69, 71, 73, 248
scale
    spatial, 15, 34–35, 37, 59, 64, 114, 151, 160,
        162, 165, 169, 171, 173, 175, 245, 266,
        279, 288, 367, 386, 392–393, 395
    temporal, 59, 62, 90, 248, 288, 392–393,
        395
seagrass, 19, 237, 240, 248, 249, 328
sediment, 43, 62, 71–72, 158, 175–176, 178,
    179, 180–182, 185–187, 243, 245, 250
seed predation, 122, 170, 341, 353–356
selection, 118, 212, 269, 309–310, 341, 347,
    348–358, 371, 374, 393
selection effect, 322, 323, 324, 325, 331
Slow-Growth/High-Mortality Hypothesis, 345,
    346, 351, 352
source–sink dynamics, 18, 74, 240, 244–245
South America, 72, 88, 167, 171, 174, 189, 247
specialist, 122, 276–277, 291, 293, 297, 326,
    346, 348, 371
spiders, 10, 148–149, 301, 310
stability, 4, 31–32, 35, 48, 123, 137, 152, 250,
    303, 318, 371, 373, 378–379
stoichiometry, 20, 59–64, 74, 235, 236, 244,
    368
storage effect, 90–91, 92, 96
subsidies, 10, 16–19, 21, 56, 58, 68–70, 72–74,
    142, 148, 164–170, 172, 188, 211,
    247–250, 278, 376, 389

succession, 111, 158, 164, 165, 169, 175,
    180–187, 211, 263, 291, 342

temperature, 38, 46, 107, 303, 372, 374, 390,
    392
termites, 236, 239–241, 267
tolerance, 124, 158, 205, 209, 210–212, 214,
    215–219, 279, 343, 348, 374
tolerance traits, 118, 205, 208, 213, 221
top predator, 4, 32, 35, 37–38, 46, 48, 62–63,
    67, 76, 119, 120, 125, 137, 184, 248, 293,
    295, 304, 309, 375, 376, 394
top predator loss, 37, 41, 43, 45, 177, 299, 367,
    375–377
trophic cascade, 9–10, 35, 38, 45, 58, 63, 67,
    68, 94, 135, 150, 172, 204, 208, 214, 224,
    243, 262, 268, 294, 334, 344, 375, 385
trophic complexity, 136, 329, 379
trophic diversity. *See* diversity, vertical
trophic skew, 379

ungulates, 112, 115, 116–121, 125, 239
urchins, 171, 303–304, 348, 376

vertical accretion, 158, 174–176

water column, 74, 210, 236, 249, 271, 272, 275
wolves, 9, 20, 112, 116, 119, 375

zonation, 158–159, 173, 175–176
zooplankton, 34, 41, 43, 47, 61, 65, 210, 214,
    239, 245, 275, 276
    abundance, 45, 327
    community composition, 41, 300
    diversity, 378
    productivity, 61